T0206164

A First Course in Engineering Drawing

K. Rathnam

A First Course in Engineering Drawing

K. Rathnam
Annamalai University
Cuddalore, Tamil Nadu
India

ISBN 978-981-13-5376-5 ISBN 978-981-10-5358-0 (eBook)
DOI 10.1007/978-981-10-5358-0

Softcover re-print of the Hardcover 1st edition 2018

Printed on acid-free paper

This Springer imprint is published by Springer Nature
The registered company is Springer Nature Singapore Pte Ltd.
The registered company address is: 152 Beach Road, #21-01/04 Gateway East, Singapore 189721, Singapore

Dedicated to The Divine Mother

Preface

The primary object of the book is to provide an easy approach to the basic principles of Engineering Drawing which is one of the core subjects for undergraduate students of all branches in engineering. Years of experience in teaching this subject convinced the author to prepare an exercise book for this course. This booklet underwent several revisions with the help of his colleagues. This book is an attempt to present methods of solutions for the exercise book from the students' point of view and with the sole object of keeping a book of reasonable size.

Emphasis is placed on the precise and logical presentation of the concepts and principles which are essential for understanding the subject. The methods presented help students to grasp the fundamentals more easily. The book offers comprehensive coverage of topics required for a first course in this subject. Illustrations are original drawings reduced in size for conceptual clarity. The problem-solving strategies detailed in certain chapters will induce independent learning skills in individuals.

The Introduction chapter begins with a mention of various instruments required for drawing works, sizes of drawing sheets, methods of dimensioning and lettering exercise. Geometrical constructions which are needed in the succeeding chapters are explained in Chap. 2. The constructions of plain, diagonal and vernier scales are presented in Chap. 3. Chapter 4 deals with the construction of curves used in engineering practice. Mastering the methods of construction of curves in general and conic curves in particular develops analytical skill which will be useful later in the professional career of the learners. Orthographic projection which lies at the heart of engineering drawing is presented in Chap. 5. Chapter 6 discusses the projections of points located in any of the four quadrants. The importance of profile plane is also stressed to complete the projections of a point. Projections of lines are covered under three categories in Chap. 7. In the first type, the projection of one view is easily obtained. In the second type, projections of a line are given and its true length, inclination to co-ordinate planes and traces are obtained following the trapezium method. In the third category, the method of obtaining the projections of line inclined to both the co-ordinate planes is explained. Students who spend more time concentrating on the solution techniques detailed in this chapter will find it easy to understand other topics in the succeeding chapters. Chapter 8 presents the methods of projections of plane figures inclined to both the co-ordinate planes. Projections of solids are detailed in Chap. 9, and grouping has been based upon the type of solids. Chapter 10 discusses auxiliary projections of solids using auxiliary planes. Chapter 11 addresses the importance of sectioning of solids and explains the procedure of obtaining the true shape of section formed. The methods of obtaining intersecting curves or lines based on the type of intersecting surfaces are discussed in Chap. 12. Chapter 13 presents the methods of development of solid surfaces. The principle of isometric projection is explained in Chap. 14. The advantage of showing the forms of cubical objects in their isometric projections more clearly than the corresponding orthographic projections is brought to light in a few problems in this chapter. Chapter 15 is devoted to the perspective projection. Chapter 16 provides 100 objective type questions with keys.

Cuddalore, India K. Rathnam

Acknowledgment

I express my sincere thanks to my teacher Professor K.S. Venkatasubban for his inspiring lectures with elegant blackboard drawings in general and the drawing of helical spring of square in cross section in particular. I wish to record my sincere thanks to my former colleague (Late) Professor V. Saravanaperumal from whom I generously received help and guidance during the formative years of my teaching career. I thank my children Mrs. Tharani Sabarigirisan and Mr. R. Vigneshwaran for their help. I also thank my wife Mrs R. Savithiri for her patience during the course of this undertaking. Last but not the least, I thank the editorial team at Springer Nature for their timely help approving the manuscript for publication.

Contents

Chapter 1
Introduction

The successful teaching of any engineering subject rests 49% with the students and 51% with the teacher. This is true in subjects like 'Engineering Drawing' wherein the student is learning the skill to communicate his ideas through drawings since the beginning of the class. It is our universal language even today when some our drawings are prepared by computers and plotters. Hence, we need not overemphasize the fact that engineering student must be proficient in 'Engineering Drawing' for a full and complete exchange of ideas with colleagues.

We begin the course with the introduction and uses of the instruments needed for drawing works. Lettering the alphabets and numerals shall form the first exercise together with printing of a few sentences/anecdotes like 'Small things make perfection, but perfection is no small thing'. Drawing of geometrical figure is introduced with the object of making the student to think and draw and also learn the art of using pencils. It is worthwhile to recollect that child begins to draw curves rather than straight lines.

The principle of orthographic projections is introduced and a lot of time should be spent on mastering orthographic drawings. The value of pictorial drawing should also be emphasized to supplement multi-view drawing, for, even today, the chief executive conveys his ideas only through pictorial sketches. To speed up the work we can introduce the graph papers/sheets for solving problems on projections of points and straight lines.

For the preparation of a neat and correct drawings, good drawing instruments are to be used. A list of drawing instruments and materials required is given below:

1. Drawing board
2. Mini-drafter
3. Instrument box
4. Engineer's scale set/Scale
5. Protractor
6. Drawing sheets
7. Eraser (good quality)

K. Rathnam, *A First Course in Engineering Drawing*,
DOI 10.1007/978-981-10-5358-0_1

8. Sand paper
9. Drawing pencils (HB, H and 2H)
10. Pocket knife/Blade
11. Drawing clips
12. Padding sheet

Drawing Sheets

Drawing sheets of different sizes are available in the market. A good quality drawing is always made on a tough, strong and glossy sheet with perfect white colour. The most common size of the sheet is imperial one (760 mm × 560 mm). However, sheets are also available in standard sizes ranging from A0 through A3; see Fig. 1.1.

The drawing sheets should be gently rolled while handling them. During sketching, pencils should be used with light handed pressure so that the pencil tip impression is not made on the drawing sheet (the use of padding sheet is recommended in this regard). The use of eraser should, as far as possible, be avoided.

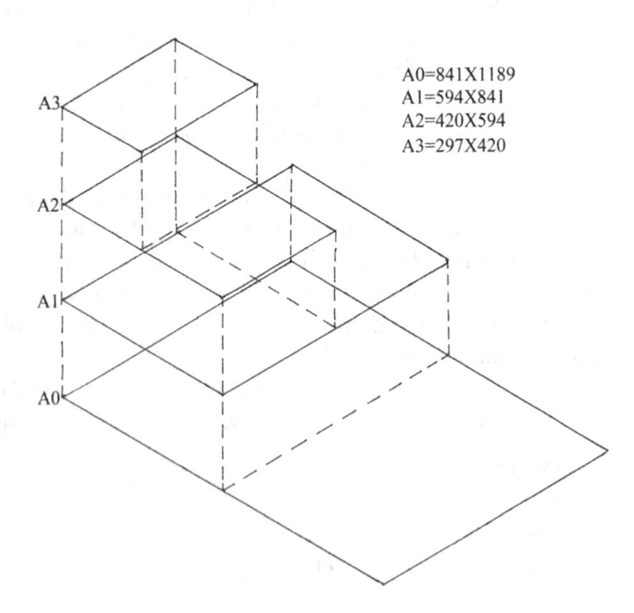

Fig. 1.1 Different sizes of drawing sheets

Pencils

Drawings make sense when they are made with correct types of lines drawn using the correct grades of pencils. Although an equal grade of pencils manufactured by any brand possesses same softness, it is advisable to purchase pencils of different grades of the same brand. The pencils must be peeled off with a knife/blade, so that a length of 10 mm lead is left projected out of wood. The projecting lead should be grounded with a sand paper to get proper tips. The line thickness is controlled by the conical tip which is recommended for pencils. However, the chisel-type tip is recommended for the lead in the compass.

Lines

For general engineering drawings, various types of lines are recommended. Each line has a specific meaning and function. The types of lines required are shown in Fig. 1.2.

Type A—used to represent visible outlines, visible edges
Type B—used to represent projectors, dimension lines, hatching lines, locus lines
Type C—used to represent hidden lines, hidden edges
Type D—used to represent imaginary lines
Type E—used to represent centre line, axis of solids
Type F—used to represent cutting planes

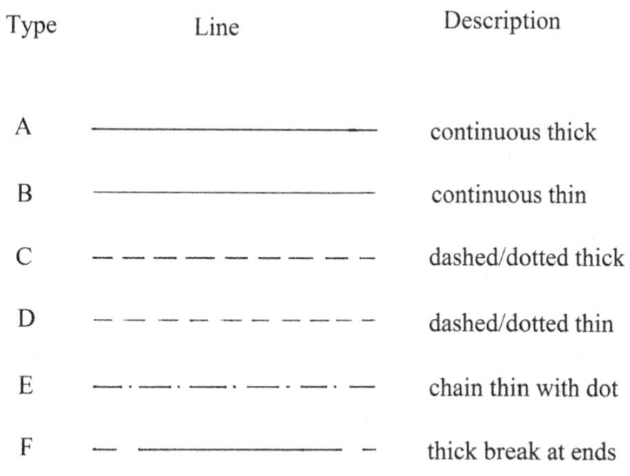

Fig. 1.2 Types of lines

Type	Line	Description
A	————————————	continuous thick
B	————————————	continuous thin
C	— — — — — — — — —	dashed/dotted thick
D	— — — — — — — — — —	dashed/dotted thin
E	—·—·—·—·—·—	chain thin with dot
F	— ——————— –	thick break at ends

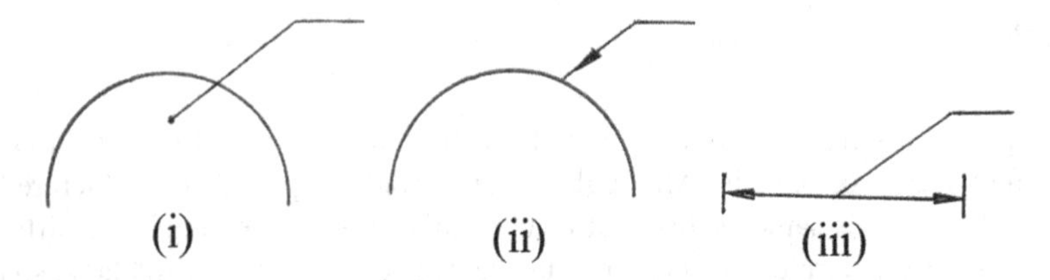

Fig. 1.3 Leader lines

Leader Lines

These are used to connect a note representing a special feature (Object, outline, dimensions, etc.) on the drawings. These are drawn at an angle not less than 30° with the horizontal. A leader line is terminated in a horizontal bar with a note at one end. The other end is terminated with a dot if it ends within the outline of the object; see Fig. 1.3(i). An arrow head is used to terminate its end if the end is on the outline of the object; see Fig. 1.3(ii). The end is terminated without a dot or an arrow head if it ends on a dimension line; see Fig. 1.3(iii).

Dimensioning

Drawing describes the shape of an object. For manufacturing of the object, size description is required. Dimensions are shown on the drawings. The same unit of measurement (i.e. millimetres) is adopted for all dimensioning but without showing the unit symbol. Unit symbol on a drawing is shown in a note [ALL DIMENSIONS ARE IN mm].

Elements of dimensioning

1. Projection line
2. Dimension line
3. Leader line
4. Termination of dimension line
5. Dimensions

Projection lines and dimension lines are shown in Fig. 1.4.

Leader lines are drawn as continuous thin lines with termination of leader by an arrow head. An arrow head leader line applied to an arc should be in line with arc centre; see Fig. 1.5.

Termination of dimension lines are indicated by arrow heads (closed and filled or open, >) or oblique strokes 45°; see Fig. 1.6.

Dimension is a numerical value expressed in appropriate unit of measurement.

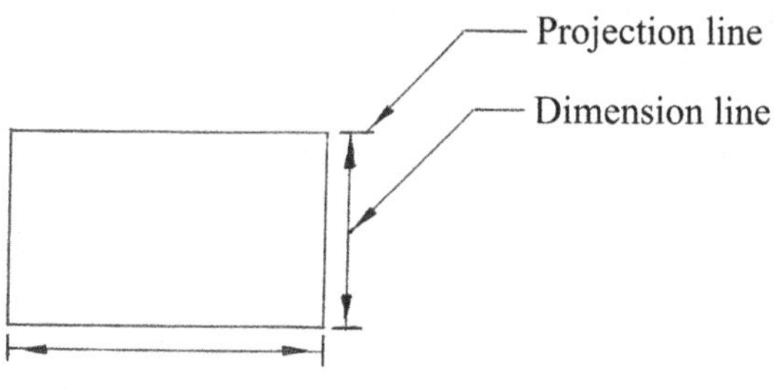

Fig. 1.4 Projection and dimension lines

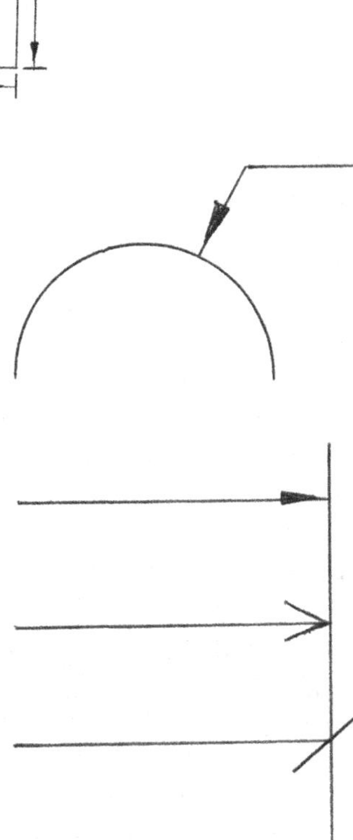

Fig. 1.5 Arrow head leader

Fig. 1.6 Termination of dimension lines

Methods of Dimensioning

Method I Aligned system: Dimensional values are placed parallel to the dimension lines. Dimensional values are placed above the dimension lines as shown in Fig. 1.7. Dimensional values should not touch dimension line.

Method II Unidirectional system: Dimensions are placed in such a way that they can be read from the bottom. Dimensional values are placed at the middle of horizontal dimension line. For non-horizontal (vertical and oblique) dimension line, the dimensional value is placed at the middle by interrupting the dimension line as shown in Fig. 1.8.

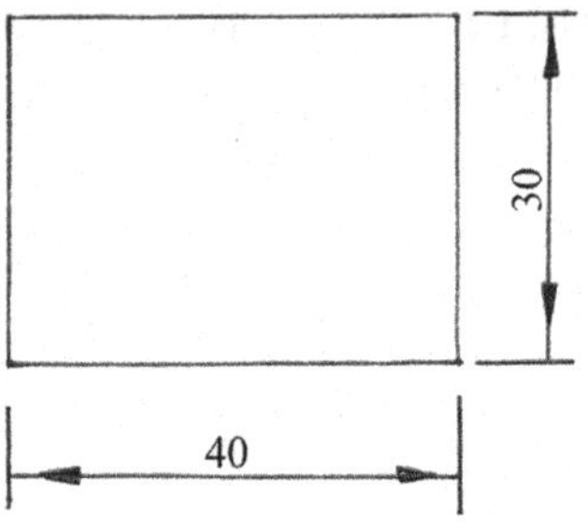

Fig. 1.7 Aligned system

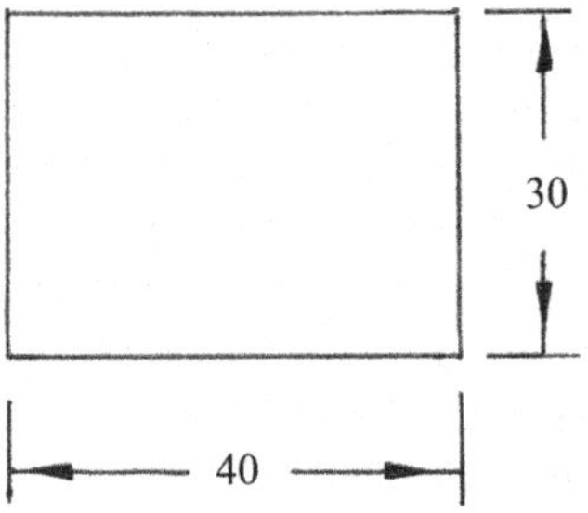

Fig. 1.8 Unidirectional system

Lettering

The description of an object or machine component requires the use of graphic language to show the shape and of the written language to explain sizes and other information. The written language used on drawings is in the form of lettering. Freehand lettering, perfectly legible and quickly made, is an important part of engineering drawings.

The ability to letter well and rapidly can be gained by constant and careful practice. The forms and preparations of each of the letters must be mastered by study and practice. Particular attention should be given to numerals as they form a very important part of every working drawing.

Single-Stroke Vertical Capitals

An alphabet of vertical capitals and the vertical numerals are shown in Fig. 1.9. Each letter is shown in a square so that the ratios of its width to height may be easily learned. Letters A, O, Q, T, X and Z fill the square. Letters M and W are wider than their height. Other letters E, H, L, N, etc. are narrower.

The first step is to study the shapes and proportions of the individual letters and the order in which the strokes are made. Guidelines drawn lightly with a sharp pencil should always be drawn for both the top and bottom of each line of letters. The semi-arrows and numbers give the order and direction of the strokes. Vertical

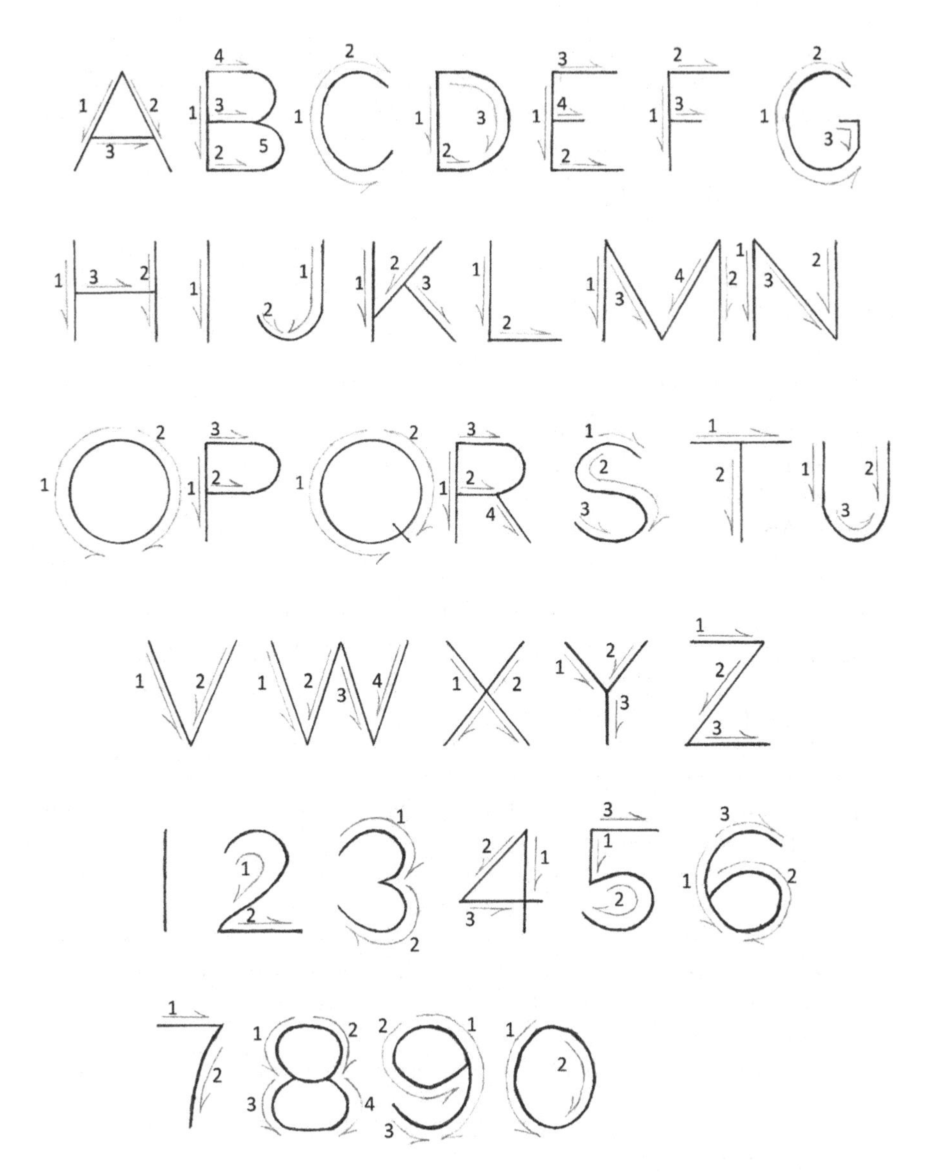

Fig. 1.9 Single-stroke vertical capitals and numerals

strokes are always made from top to bottom. Horizontal strokes are made from left to right. Numerals require special attention and practice. Uniform height is obtained by having each letter meet the top and bottom guidelines. Uniform weight is obtained by making all strokes of same thickness. Single stroke does not imply that the letter should be made in one stroke without lifting the pencil.

All letters should be written/printed in capitals. In spacing words, the clear distance between them should not be more than the height of the letters. Lower case letters are used for abbreviations. For example, we can use mm for millimetre.

The sizes of letters and numerals are designated by their heights. The height of the capital letter is taken as the base dimension. The height of numerals in dimensioning is equal to the height of the capital letters. The range of standard height h for lettering is given below:

2.5, 3.5, 5, 7, 10, 14 and 20 mm.

Lettering Exercise

1. Print the following using single-stroke vertical capital letters of height 10 mm:

   (a) ENGINEERING GRAPHICS
   (b) LETTERS AND NUMERALS
   (c) CAPITAL LETTERS
   (d) PLAN
   (e) ELEVATION

2. Print the following sentences using 7 mm height guidelines.

   (a) ENGINEERING DRAWING IS THE LANGUAGE OF ENGINEERS.
   (b) LETTERING SHOULD BE DONE PROPERLY IN FREEHAND CLEAR, LEGIBLE AND UNIFORM STYLE.
   (c) THE LETTERS SHOULD BE SO SPACED THAT THEY DO NOT APPEAR TOO CLOSE TOGETHER OR TOO MUCH APART.
   (d) UNIFORMITY IN HEIGHT, INCLINATION AND STRENGTH OF LINE IS ESSENTIAL FOR GOOD LETTERING.
   (e) WHEN THE TOP AND BOTTOM OF EACH LETTER TOUCHES THE GUIDE LINE, LETTERING LOOKS FINE.
   (f) NO DECORATIONS ARE USED ON ENGINEERING LETTERS WHERE PRECISION AND EASE OF READING ARE IMPORTANT.
   (g) LETTERING BECOMES A REAL PLEASURE WHEN THE LETTERS BEGIN TO LOOK WELL FORMED.
   (h) LETTERING IS A SKILL THAT YOU LEARN SLOWLY AND ONLY BY LONG AND STEADY PRACTICE.
   (i) SMALL THINGS MAKE PERFECTION, BUT PERFECTION IS NO SMALL THING.
   (j) ONLY A DISCIPLINED PERSON WITH LOFTY IDEALS AND REGULAR HABITS CAN BECOME A SUCCESSFUL ENGINEER.

Drawing Exercises

1. In a square of 50 mm sides, draw thick horizontal lines with 10 mm spacing as shown in Fig. 1.10.
2. In a square of 50 mm sides, draw thick vertical lines with 10 mm spacing as shown in Fig. 1.11.

3. Draw Fig. 1.12
4. Draw Fig. 1.13
5. Copy Fig. 1.14
6. Copy Fig. 1.15

Fig. 1.10 ■

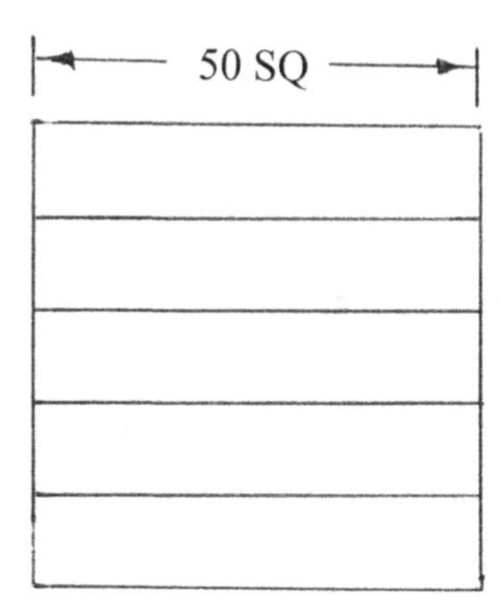

Fig. 1.11 ■

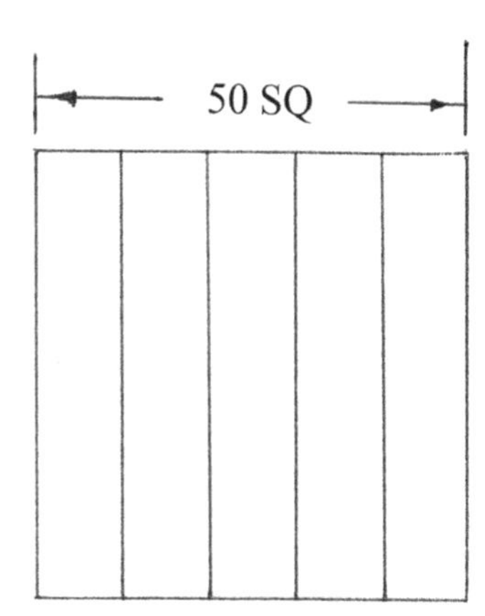

Fig. 1.12 ■

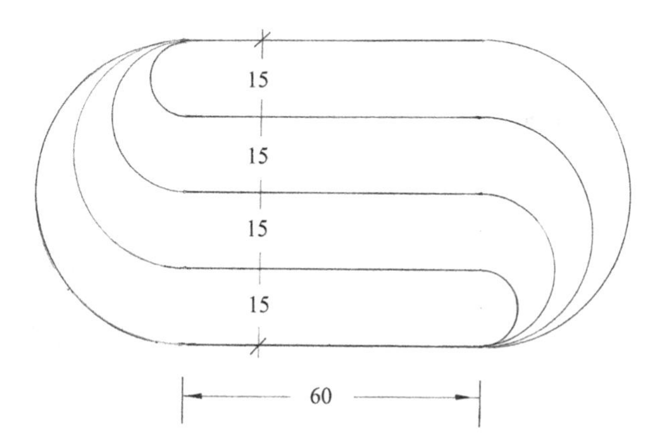

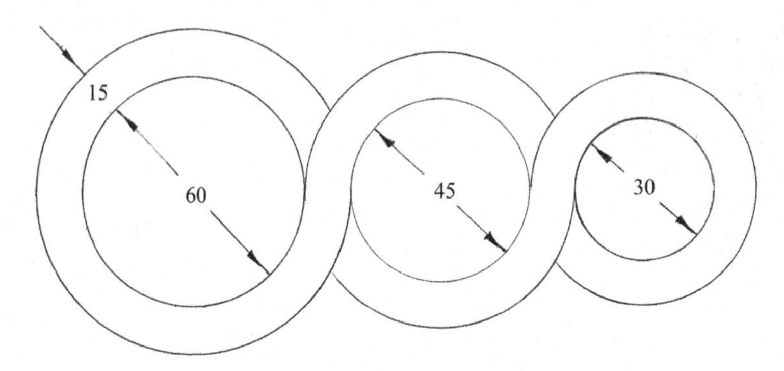

Fig. 1.13 ■

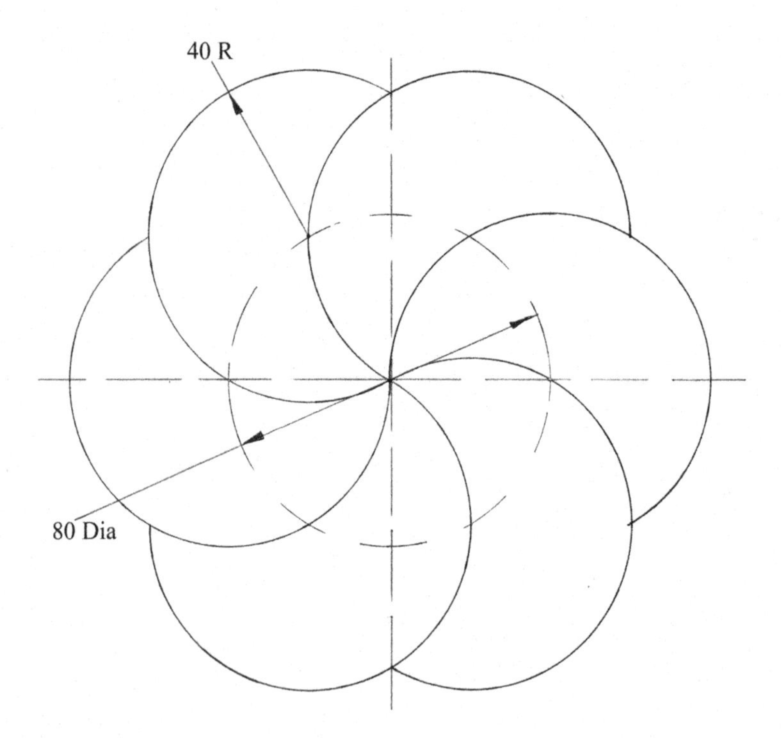

Fig. 1.14 ■

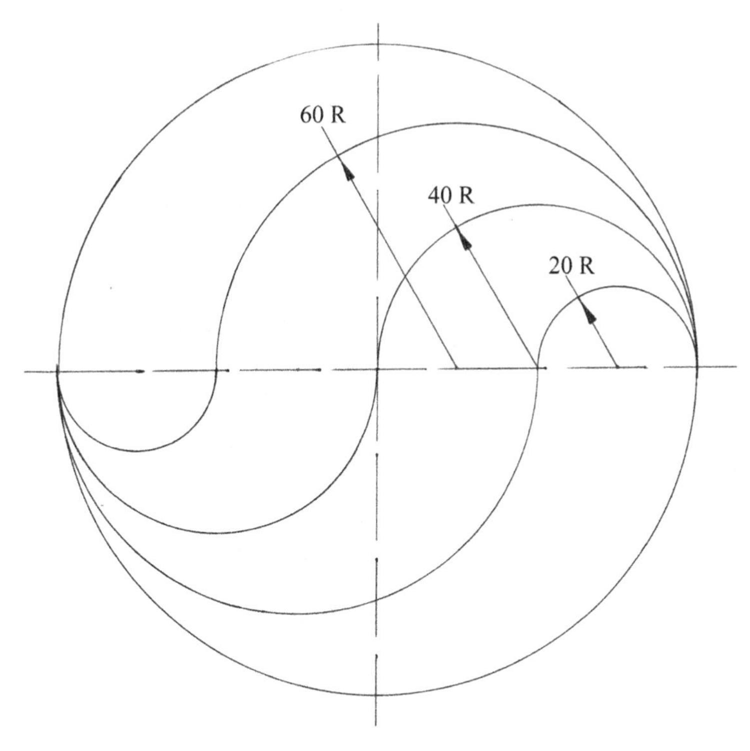

Fig. 1.15 ■

Chapter 2
Geometrical Construction

1. Divide a line AB 100 mm long into nine equal parts.
   Draw a horizontal line AB of 100 mm long as shown in Fig. 2.1. From A draw AP at any convenient angle. Along AP mark off with divider or scale nine equal distances and mark the ninth division point as C. Join CB and from the points division on AC draw straight lines parallel to CB to meet AB. The line AB is now divided into nine equal parts.
2. A point P is 15 mm from a line AB of 60 mm length. Draw an arc of radius 25 mm passing through this point and tangential to the line AB.
   Draw a horizontal line AB of 60 mm in length and locate a point P 15 mm above it as shown in Fig. 2.2. Draw a line CD parallel to AB at a distance of 25 mm above the line. The centre of the arc lies on this line. Since the arc passes through the point P, an arc of radius 25 mm is drawn from P to cut the line CD at O. An arc ST of radius 25 mm is drawn with centre O. This arc ST passes through P and touches the line AB.
3. A line OA, 30 mm long, is a radius of a circle whose centre is O. AB is a line 60 mm long making an angle of 120° with OA. Draw a circle which touches the given circle at A and passes through B.
   Draw a horizontal line OX and draw the circle of radius 30 mm with centre O as shown in Fig. 2.3. Mark point A and draw the line AB, 60 mm long, making an angle of 120° with OA as shown in the figure. Since the circle to be drawn passes through B and touches the circle already drawn at A, AB will be the chord of the circle to be drawn and circle centre will be on the extension of the line OA. The centre P is located by drawing a perpendicular bisector of the chord AB. The centre can also be located by drawing a line from B making an angle of $\angle$ABP $= \angle$ XAB. The circle of radius PA or PB drawn with centre P touches the circle of radius 30 mm and passes through the point B.
4. In a circle of 80 mm in diameter inscribe a triangle whose angles are to one another as

K. Rathnam, *A First Course in Engineering Drawing*,
DOI 10.1007/978-981-10-5358-0_2

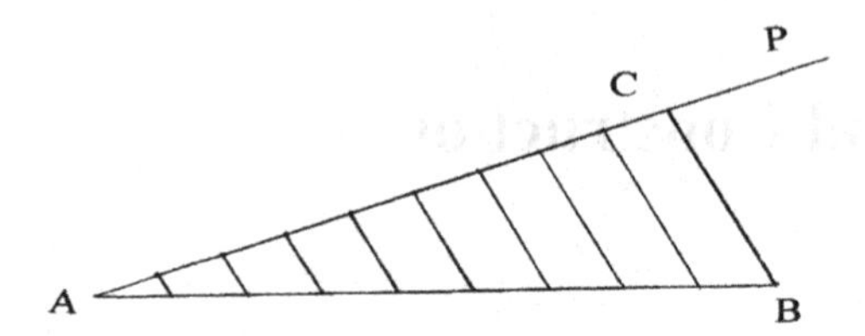

Fig. 2.1 Dividing line AB in equal parts

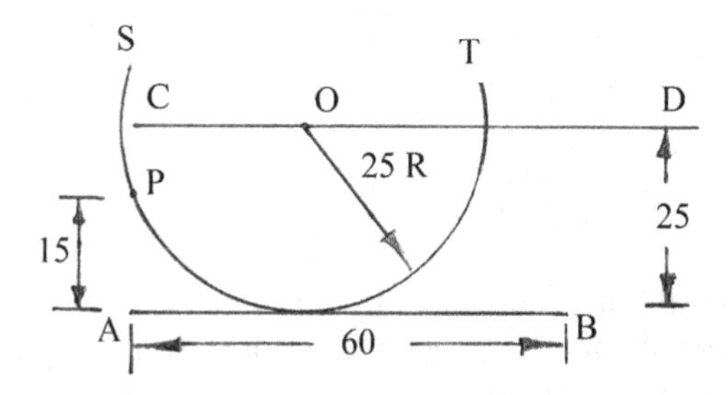

Fig. 2.2 Construction of an arc touching a line and passing through a point

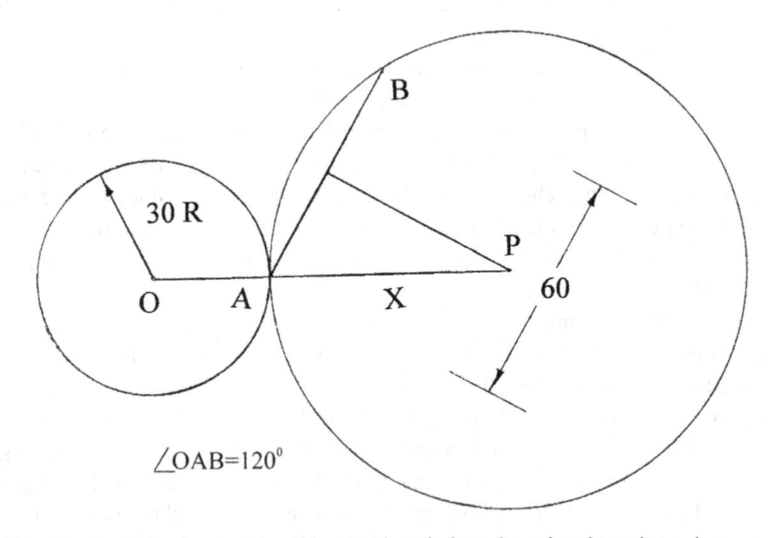

Fig. 2.3 Construction of a circle touching another circle and passing through a point

2: 3: 4.

Draw a circle of diameter 80 mm with centre O as shown in Fig. 2.4. Since the angles are 2: 3: 4, the individual angles are respectively 40°, 60° and 80°. Draw three radial lines from O keeping the angles of 80°, 120° and 160° (twice the individual angles of triangle) between them to meet the circle at A, B and C as shown in the figure. Join the points A, B and C. Then ABC is the required triangle with angles of 40°, 60° and 80°, respectively, at A, B and C.

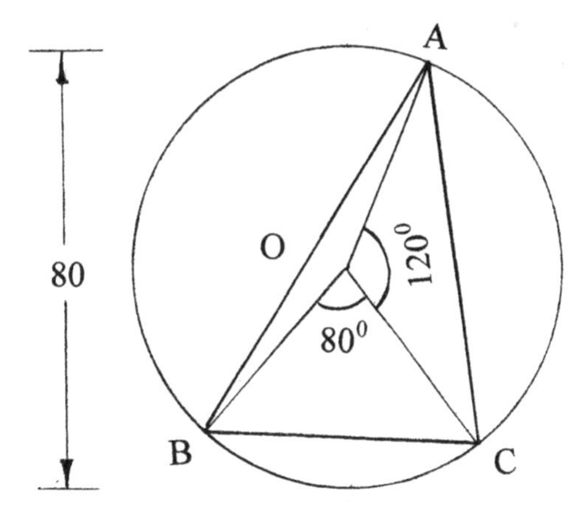

Fig. 2.4 Construction of triangle in a circle

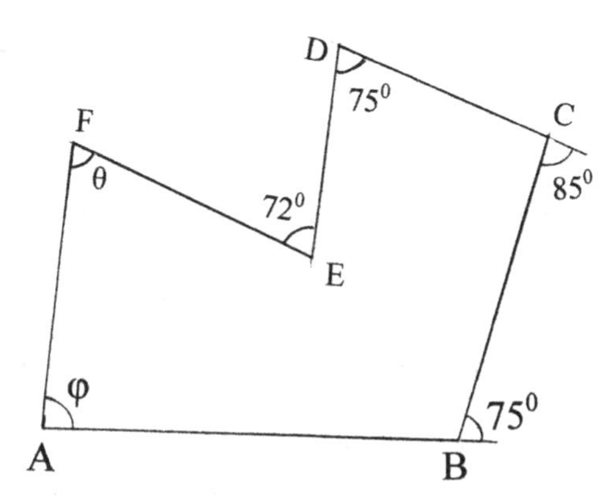

Fig. 2.5 Construction of polygon for the given data

5. On the base AB construct the polygon ABCDEF to the dimensions and angles given. AB = 100 mm; BC = 75 mm; CD = 55 mm; DE = 50 mm and EF = 63 mm. Then measure the side AF and the angle θ and φ.

   Draw a horizontal line and locate points A and B keeping the distance of 100 mm between them as shown in Fig. 2.5. At B set an angle of 75° and draw a line to locate point C at a distance of 75 mm from B. Set an angle of 85° at C and draw a line to the left of C as shown in the figure. Mark point D at a distance of 55 mm from C. At point D set an angle of 75° and draw a line from D. The point E is located on this line at a distance of 50 mm from D. From the point E draw a line inclined to ED at an angle of 72° to the left of E as shown in the figure. On this line point, F is located at a distance of 63 mm from E. Join FA and its length is measured. The angles θ and φ are also measured.

   Answers: AF = 67 mm; θ = 71°; and φ = 86°.

6. Draw an equilateral triangle of 80 mm side and in it place three equal circles, each one touching the other two circles and one side of the triangle.

   Draw an equilateral triangle of side 80 mm as shown in Fig. 2.6. Draw the angular bisector from A, B and C to meet the opposite sides at L, M and N. The

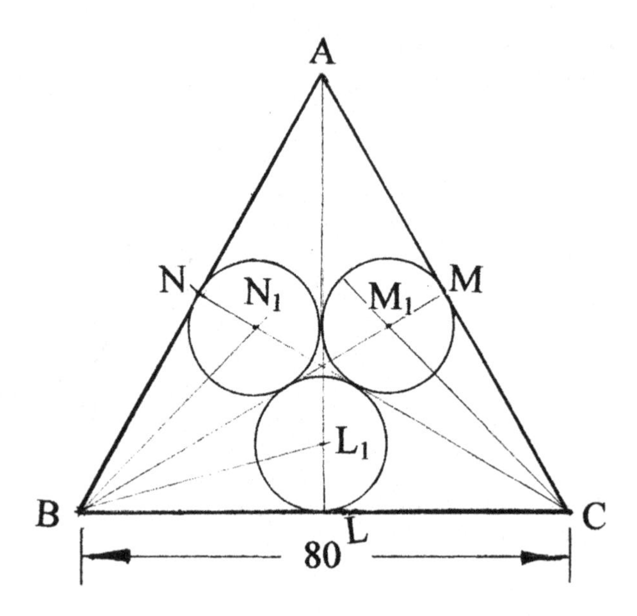

Fig. 2.6 Construction of three equal circles

angular bisectors also intersect at O, the centre of the triangle. Since the circles to be constructed touch one side of the triangle and the other circles to be drawn, their centres should lie on OL, OM and ON. Draw the bisector ∠LBO to intersect the line AL at L_1. Similarly, points M_1 and N_1 are located as shown in the figure. A circle is drawn with centre L_1 and radius L_1L. This circle will touch side BC at L and lines BM and CN will be the tangents to it. Two circles are drawn with centres M_1 and N_1 and radius L_1L. Each circle touches one side of the triangle and the other two circles.

7. Draw a triangle ABC. AB = 75 mm. BC = 63 mm. CA = 50 mm. Draw a circle of radius 25 mm to touch the side AB and pass through the point C.

   The triangle is drawn taking the side AB along a horizontal line as shown in Fig. 2.7. Draw a line parallel to AB at a distance 25 mm from and above it. The centre of the circle should lie on this line drawn parallel to AB. The centre O is located on this line by an arc of radius 25 mm drawn from C. A circle of radius 25 mm is now drawn with centre O.

   This circle touches the line AB and passes through the point C.

8. Draw two circles having their centres 80 mm apart. The diameter of one circle (A) is to be 50 mm and the diameter of the other (B) to be 80 mm. Draw a circle (C), 100 mm in diameter, to touch the circles A and B so that A is inside and B outside C.

   Draw a horizontal line and locate the centres O1 and O2 of the circles A and B keeping a distance of 80 mm between them as shown in Fig. 2.8. Draw the two circles A and B. The centre of circle C, which touches both the circles A and B, is located from O1 and O2. Since the circle A lies inside the circle C, the centre O3 of circle C is 25 mm (i.e. $r_c - r_a$) from O1. As the circle B lies outside the circle C, the centre O3 is 90 mm (i.e. $r_c + r_b$) from O2. Since the distance of O3

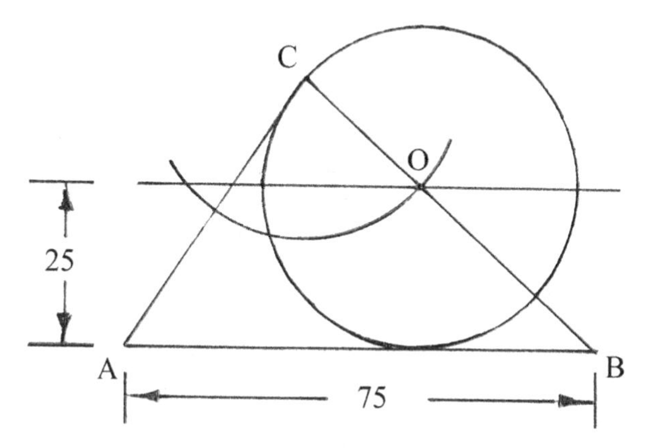

Fig. 2.7 Circle touching a side of triangle and passing through opposite point

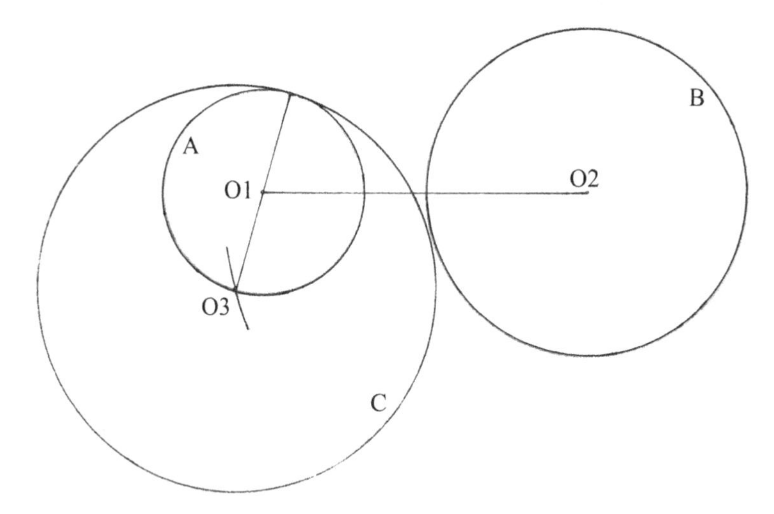

Fig. 2.8 Circle C touching circle A internally and circle B externally

from O1 is 25 mm which is the radius of circle A in this case, the centre O3 lies on the circumference of the circle A. An arc of radius 90 mm is drawn from O2, the centre of the circle B, to cut the circle A giving the centre O3 of the circle C. The circle C is drawn with a radius of 50 mm from centre O3. This circle C touches both the circles A and B as shown in the figure.

9. Construct a regular pentagon of 30 mm side.

   A regular polygon is one in which all the sides and all the interior angles are equal. Consider a pentagon shown in Fig. 2.9. The five isosceles triangles with the sides of the pentagon as bases and vertices at O are all equal and the angles at the centre are equal. The sum of the angles at the centre is 360°. This is true for all regular polygons.

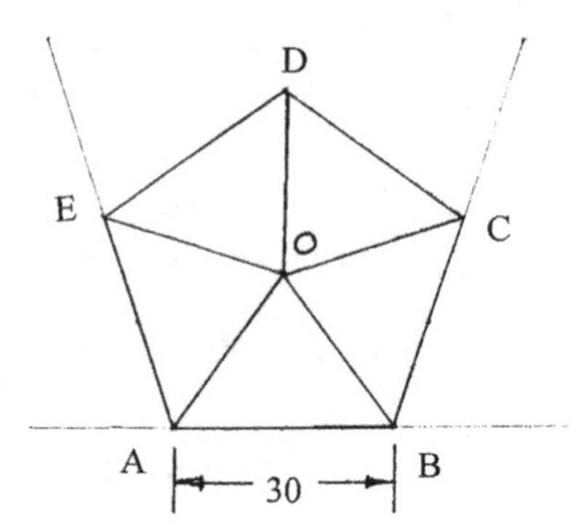

Fig. 2.9 Construction of pentagon

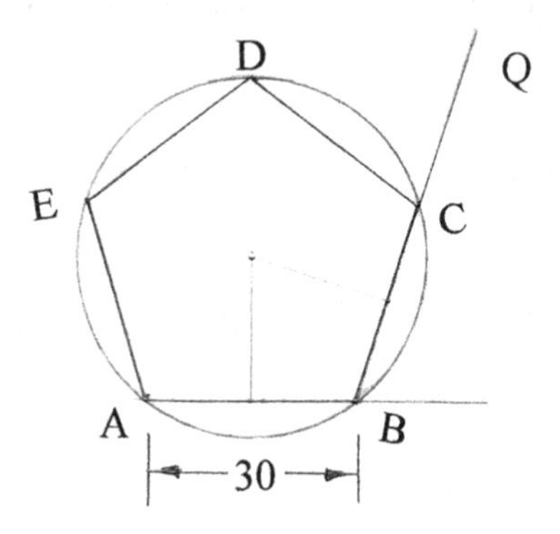

Fig. 2.10 Construction of pentagon

For the pentagon, $\angle AOB = \frac{1}{5}(360^\circ) = 72^\circ$, $\angle OAB = \angle OBA$

$$= \left(\frac{180^\circ - 72^\circ}{2}\right) = 54^\circ, \qquad \text{Hence,} \ \angle ABC = 54^\circ + 54^\circ = 108^\circ$$

Draw a horizontal line and locate point A and B 30 mm apart. At B construct a line BQ at an interior angle 108° as shown in Fig. 2.10. Along BQ mark off BC = AB. BC is the side of the pentagon. The same procedure is repeated at C to get side CD. Similarly, point E is obtained to complete the pentagon ABCDE. Alternately, the centre of the circumscribing circle can be located by drawing the perpendicular bisectors of the two sides AB and BC. The circumscribing circle is drawn and the remaining angular points of the pentagon are located on the circumference of the circle.

10. Construct a regular hexagon of 30 mm side.

    A regular hexagon has six sides and each side subtends an angle of 60° at the vertex of the equilateral triangle (for a regular polygon an isosceles triangle).

    A circle of 30 mm radius is drawn and two points A and D are located along the horizontal diameter as shown in Fig. 2.11. Four more points B, C, E and F

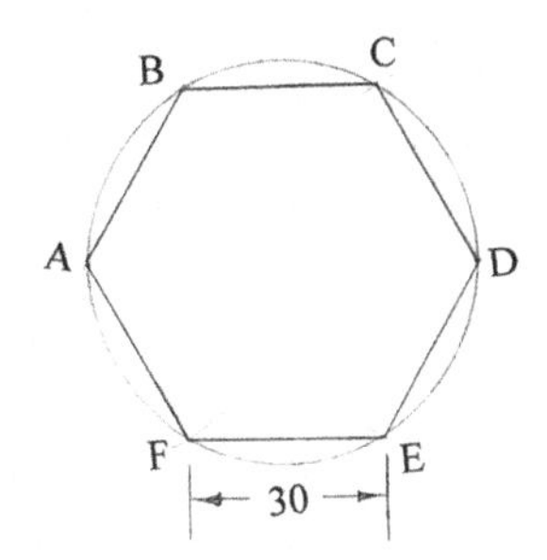

Fig. 2.11 Construction of hexagon

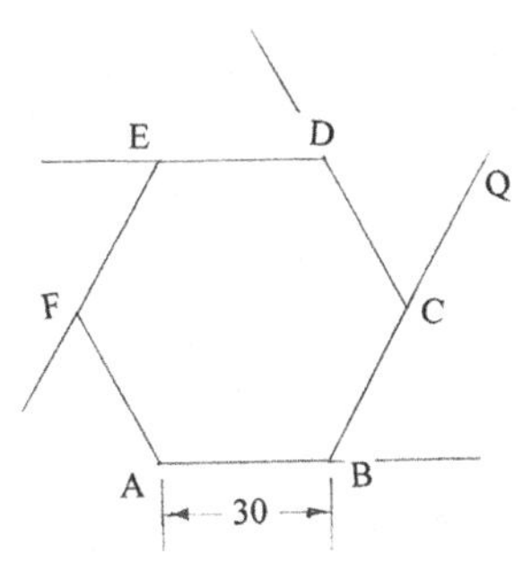

Fig. 2.12 Construction of hexagon

are located on the circumference of the circle using the radius of the circle. These points are joined to show the hexagon ABCDEF.

Draw a horizontal line and locate points A and B 30 mm apart representing one side of the hexagon as shown in Fig. 2.12. At B construct an interior angle $\angle ABQ = 120^{\circ}$ (This can be done easily by a 30° set square). Along BQ mark off BC = AB. BC is another side of the hexagon. The same procedure is extended to obtain subsequent points to complete the construction of the hexagon.

11. ABC is a triangle. AB = 40 mm, BC = 35 mm and CA = 25 mm. C is the centre of the circle of 50 mm radius. Draw two circles to touch this circle and pass through the points A and B.

    Draw the triangle ABC and draw the circle of radius 50 mm with centre C as shown in Fig. 2.13. Draw a line from C perpendicular to AC to meet the circle at D. Draw a perpendicular bisector of the chord AD to meet DC at E. A circle of radius ED or EA with centre E is drawn which touches the larger circle at D and passes through the point A. Draw a line from C perpendicular to BC to meet the larger circle at F. Draw a perpendicular bisector of the chord BF to meet CF at G. A circle of radius GF or GB with centre G is drawn which touches the larger circle at F and passes through the point B.
12. AB (Fig. 2.14) is an arc of circle of 50 mm radius. BC is an arc of 40 mm radius. The centres of these circles are 30 mm apart. ADC is an arc of circle of 15 mm radius which touches the arcs AB and BC. Draw the figure ABCD.

    Let the centres of the arc of the circles AB, BC and ADC be designated as a, c and d. The distance between c and a is 30 mm. The distance between d and a

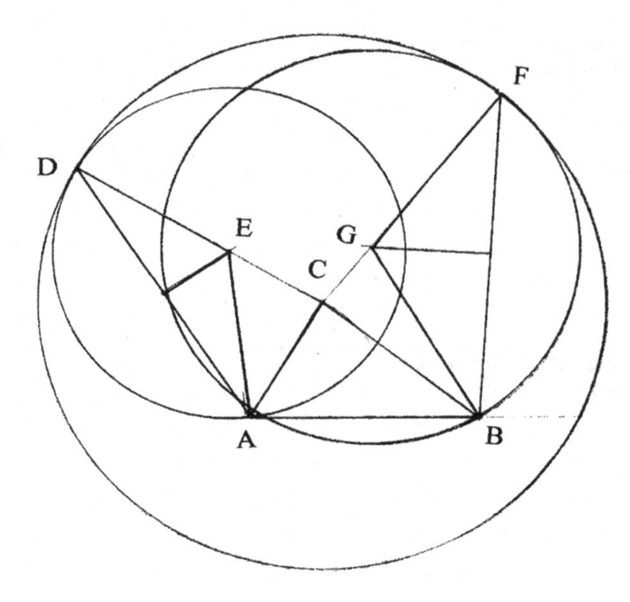

Fig. 2.13 Construction of two circles for the given data

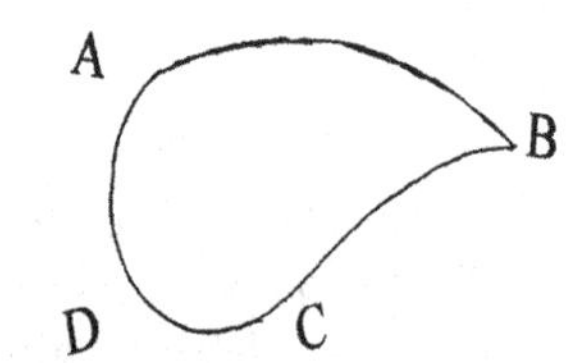

Fig. 2.14 Aerofoil section

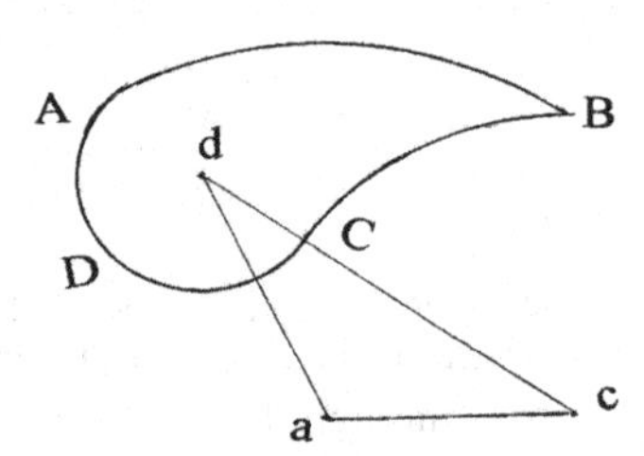

Fig. 2.15 Construction of circular arcs

will be 35 mm ($r_a - r_d$) and the distance between d and c will be 55 mm ($r_c + r_d$). The triangle acd is drawn knowing the sides ac, da and dc, respectively, 30 mm, 35 mm and 55 mm as shown in Fig. 2.15. The line ad is extended. With centre a, draw an arc of radius 50 mm starting from the point A on the extension of the line ad. With centre c, draw an arc of radius 40 mm to intersect the previous arc at B and the line cd at C. With centre d draw an arc of radius 15 mm to meet the points A and C. The figure ABCD is obtained as shown in Fig. 2.15.

13. Two pulleys of diameter 70 mm and 40 mm, respectively, with their centres 80 mm apart are connected by a flat belt in (i) open and (ii) crossed systems. Draw the line diagram of the drive in both cases, assuming pulleys and belt are of line thickness.

The two pulleys are drawn for both the cases on horizontal lines as shown in Fig. 2.16. Two semi-circles are drawn passing through the centres of the pulleys as shown in the figure. In the case of open belt, an arc of radius equal to the difference in the radii of the pulleys is drawn with its centre at the larger pulley to cut the semi-circle already drawn. A radial line is drawn from the centre of the larger pulley through the intersecting point to cut the larger circle. Another line is drawn parallel to this radial line from the centre of smaller pulley to cut the smaller circle. The meeting points of these radial lines with circles are joined to show the common external tangent. The other common external tangent is drawn as shown in the figure. This completes the line diagram of the open system. In the case of crossed system, an arc of radius equal to the sum of the radii of the pulleys is drawn from the centre of the larger pulley to cut the semi-circle already drawn. The intersecting point is connected to the centre of the larger pulley. The radial line cuts the larger circle to locate the meeting point of the common internal tangent. A line parallel to this radial line is drawn from the centre of smaller pulley to cut the smaller circle in the opposite direction as shown in the figure. The two common internal tangents are drawn to show the line diagram of the crossed system.

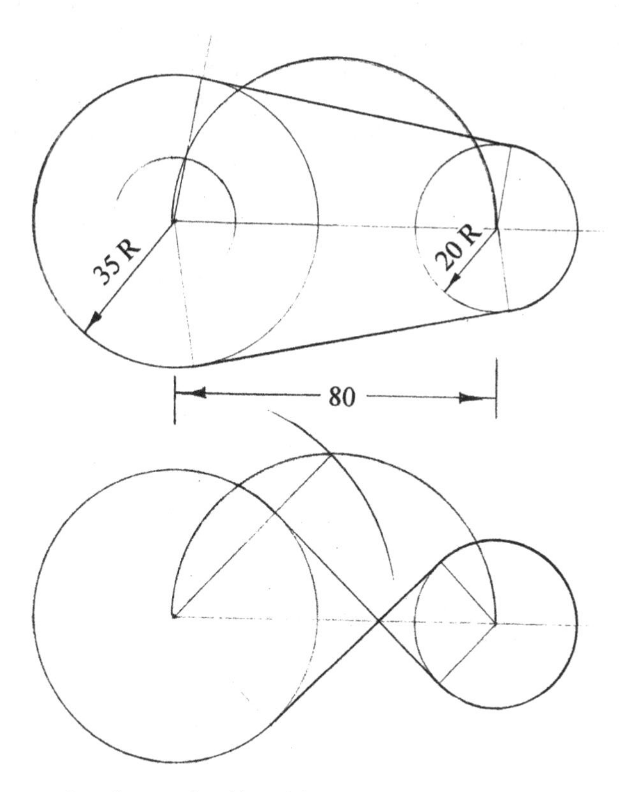

Fig. 2.16 Construction of external and internal tangents to two circles

14. Two straight lines include an angle of 60°. Draw a circle 80 mm in diameter, cutting one of the lines at points 44 mm apart and then the other at points 58 mm apart.

    On a horizontal line, an isosceles triangle COD, base 58 mm and equal sides 40 mm, is drawn as shown in Fig. 2.17. Another isosceles triangle AOB, base 44 mm and equal sides 40 mm, is drawn and its altitude OS is measured. Draw the circle with centre O and radius 40 mm passing through the points C and D. Draw the radial OY inclined at 60° to the vertical diameter of the circle. Locate a point S on this radial line at a distance equal to the altitude of the isosceles triangle AOB from the centre of the circle. Draw a line perpendicular to OS, and this line cuts the circle at points A and B which are 44 mm apart. The line AB is extended to cut the horizontal line already drawn. The angle between these lines is 60°, and the horizontal line cuts the circle at points 58 mm apart.
15. In a circle of 120 mm in diameter draw six equal circles, each one to touch the original circle and two of the others.

    Draw a circle of radius 60 mm and six radial lines OA, OB, OC, OD, OE and OF as shown in Fig. 2.18. The centres of the six equal circles lie on these radial lines. Draw angular bisectors OH_1 and OB_1 on both sides of the radial line OA as shown in the figure. From A draw a line perpendicular to OH_1 to meet it at H_2. Draw the bisector of the angle OAH_2 to meet OH_1 at H_3. Draw a line perpendicular to OH_1 from H_3 to meet the radial line OA at A_2. With A_2 as centre and radius AA_2, draw a circle which touches the original circle and the two radial lines OH_1 and OB_1. On the radial line OB, locate a point B_2 such that $AA_2 = BB_2$. A circle is drawn using the same radius with centre B_2. This procedure is repeated for other radial lines and the remaining four circles are drawn. Each one of the six circles touches the original circle and two adjacent circles.

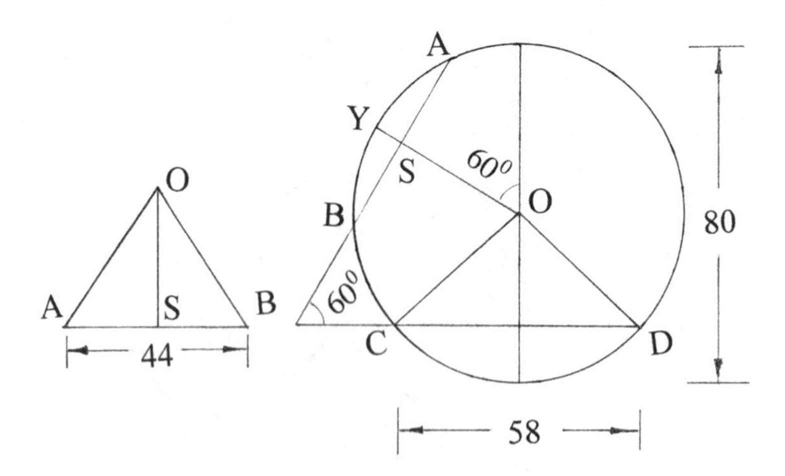

Fig. 2.17 Construction of a circle passing through four points on two lines

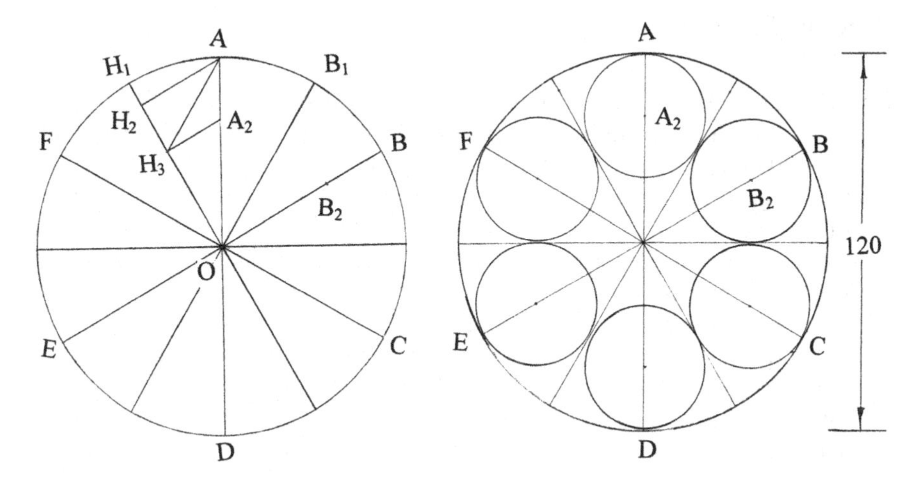

Fig. 2.18 Construction of six equal circles inside a given circle

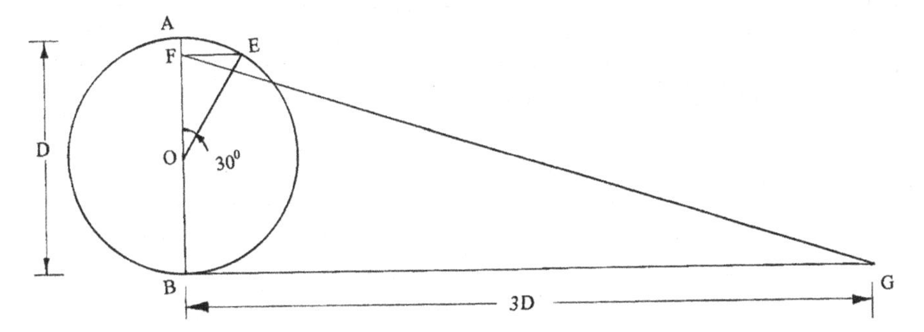

Fig. 2.19 Approximate length of the circumference of a circle of diameter D

The radius of the six equal circles can also be obtained from the following equation:

$r/(R - r) = \text{Sin}\ (\theta/2)$

where R is the radius of the given circle, r is the radius of the circle to be drawn and θ is angle subtended by each circle at the centre of the original circle. The centres of the circles can be located on the radial lines and circles are drawn using the radius r.

16. Determine graphically the circumference of a circle of diameter D.

Draw a circle of diameter D with centre at O and also the diameter AB as shown in Fig. 2.19. A tangent is drawn at B. Mark point G on the tangent such that the distance of G from B is equal to three times the diameter (BG = 3 D). Draw a radial line OE at 30° to AB. Draw EF perpendicular to AB from E intersecting AB at F. Join FG. The length FG gives a close approximation to the circumference of the circle, π D.

Practice Problems

1. Divide a line AB 120 mm long into seven equal parts.
2. A line AB is inclined to another line AC, angle BAC = 60°. Draw an arc of radius 30 mm tangential to the lines AB and AC.
3. Draw tangents to a circle of diameter 40 mm from a point 70 mm from the centre of the circle.
4. Draw a common internal tangent to two circles of equal radii 40 mm and having their centres 100 mm apart.
5. Construct a regular octagon of 35 mm side.
6. Draw a sector of a circle of radius 60 mm and angle of sector 60°. In this sector inscribe a circle.
7. ABC is a triangle. AC = 50 mm. BC = 25 mm. Angle C = 90°. Draw a circle which touches AC and passes through B.
8. Draw an arc of radius 50 mm tangential externally to two circles having their centres 80 mm apart and radii 30 mm and 15 mm.
9. Draw an arc of radius 100 mm tangential internally to two circles having their centres 80 mm apart and radii 30 mm and 15 mm.
10. In a circle of 140 mm in diameter draw eight equal circles, each one to touch the original circle and two of the other.

Chapter 3
Scales

Drawings of buildings/machines are prepared adopting a scale, and the scale adopted is to be mentioned below the drawing. If a drawing is made of the same size as the object, the view obtained will have the same size as the object. The drawing thus obtained is called a full size drawing and the scale used is called full size scale. If the object is larger in size, its drawing cannot be accommodated in the standard size drawing sheets. Hence, the drawing is made smaller in size. The scale used for this type is called a reduced scale. Objects of smaller sizes require drawings of larger in size. The scale used for drawing such components is called an enlarged scale. A scale is used to prepare a reduced or an enlarged size drawing.

Standard draftsman scales are available in sets of 8 scales or 12 scales. The scale used for the drawing should be mentioned in the appropriate place of the title block. If more than one scale is used in a drawing sheet, the scale is to be printed under each drawing, and the actual dimensions of the object should be marked on the drawings.

Representative Fraction

The ratio of size of a component of an object in the drawing to its actual size is called representative fraction (or RF). A scale is designated by its RF. The RF of a full size drawing is 1:1, RF of a reducing scale is less than one and that of an enlarging scale is greater than one.

RF = (Length of the object in the drawing)/(Length of the object)

For example, if a 5 m distance is represented on a drawing by a line of 50 mm, then the RF of the scale on which the drawing is made will be (50 mm) ÷ (5000 mm) = 1 ÷ 100 (a reducing scale), i.e. RF = 1:100. If a 5 mm distance is represented on a drawing by a line of 50 mm, then the RF of the scale will be (50 mm) ÷ (5 mm) = 10 ÷ 1 (an enlarging scale), i.e. RF = 10:1.

K. Rathnam, *A First Course in Engineering Drawing*,
DOI 10.1007/978-981-10-5358-0_3

Table 3.1 Measuring units

Unit	Equivalent value in destination unit
1 centimetre	10 millimetre
1 decimetre	10 centimetre
1 metre	10 decimetre
1 decametre	10 metre
1 hectometre	10 decametre
1 kilometre	10 hectometre
1 inch	2.54 centimetre
1 foot	12 inches
1 yard	3 feet
1 chain	22 yards
1 furlong	10 chains
1 mile	8 furlongs
1 mile	1.609 kilometre

The standard recommended scales are given below:
Full size
1:1
Reducing scales (drawings smaller than full size)
1 : 2   1 : 5   1 : 10   1 : 20   1 : 50   1 : 100   1 : 200   1 : 500   1 : 1000
Enlarging scales (drawings larger than full size)
2 : 1   5 : 1   10 : 1   20 : 1   50 : 1

The distances and dimensions are always measured in standard units. The relationship between some of the measurement units is given in Table 3.1.

When a scale other than the standard one is required for the preparation of a drawing, such scale is to be constructed. The information required for the construction of a scale are

(i) The RF of the scale
(ii) The maximum length the scale is supposed to measure
(iii) A close examination of the units (two or three units) required for measurement

The construction of Plain, Diagonal and Vernier scales are explained with examples.

Plain Scales

A plain scale is the simplest scale in which a line is used to represent dimensions. The line is divided into a suitable number of equal parts or units, and the first division is subdivided into equal number of smaller parts or units. A plain scale represents either two units or a unit and its subdivisions.

Solved Problems

1. Construct a plain scale of RF = 1:60 to show metres and decimetres and long enough to measure up to 6 m. Mark on the scale a dimension representing 5.7 m.

Length of the scale = (1/60) × (6000) = 100 mm

Draw a line PQ of length 100 mm and divide it into six equal parts as shown in Fig. 3.1. Each main division represents 1 m. Subdivide the first left-most division into 10 equal number of parts. Each subdivision represents 1 decimetre. In this scale, 0 (zero) is marked at the end of the first left-most main division. The subdivisions are numbered to the left starting from 0 as shown in the figure. The main units are numbered to the right starting from 0. The scale is shown in the form of a rectangle PQRS of convenient width (say 10 mm). To distinguish the divisions, draw thick horizontal lines in the alternate divisions representing both metres and decimetres. The RF is indicated at the centre of the scale preferably at the bottom. The units of the main divisions and subdivisions are indicated in capitals at the right and left ends respectively. The length measuring 5.7 m is marked by dimension line as shown in the figure.

2. The distance between the centres of two drilled holes which are 0.8 m apart is shown by a line of 2 cm on the drawing. Construct a scale to read up to 5 m. Show on the scale a length of 2.3 m.

   RF of the scale = (20)/(0.8 × 1000) = 1/40; Length of the scale = (1/40) × (5000) = 125 mm

   Draw a line PQ of length 125 mm and divide it into five equal parts as shown in Fig. 3.2. Each main division represents 1 m. Subdivide the first left-most main division into 10 equal parts. Each subdivision represents 1 decimetre. In this scale 0 (zero) is marked at the end of first left-most main division. The subdivisions are numbered to the left starting from 0. The main divisions are numbered to the right starting from 0. The scale is shown in the form of a

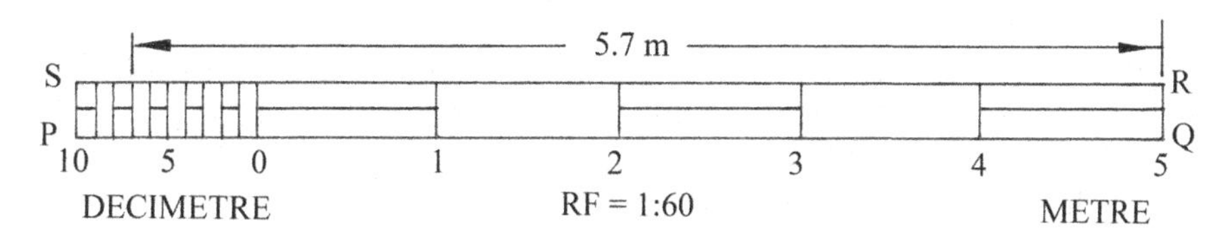

Fig. 3.1 Plain scale

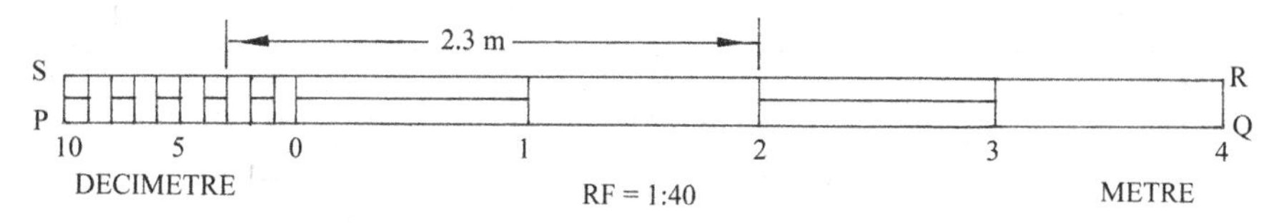

Fig. 3.2 Plain scale

rectangle PQRS of convenient width (say 10 mm). To distinguish the divisions, draw thick horizontal lines in the alternate divisions representing both metres and decimetres. The RF is indicated at the centre of the scale preferably at the bottom. The units of the main divisions and subdivisions are indicated in capitals at the right and left ends respectively. The length measuring 2.3 m is marked by dimension line as shown in the figure.

3. An actual distance of 6 miles is represented on a map by 120 mm long line. Construct a plain scale to read miles and furlongs and long enough to measure 8 miles. Mark on the scale a distance of 5 miles and 4 furlongs.

   RF of the scale $= (120)/(6 \times 1.6 \times 1000 \times 1000) = 1/80{,}000$ (1 mile = 1.6 km)
   Length of the scale $= (1/80{,}000) \times (8 \times 1.6 \times 1000 \times 1000) = 160$ mm

   Draw a line PQ of length 160 mm and divide it into eight equal number of parts as shown in Fig. 3.3. Each main division represents 1 mile. Subdivide the first left-most main division into eight equal number of parts. Each subdivision represents 1 furlong. In this scale 0 (zero) is marked at the end of first left-most main division. The subdivisions are numbered to the left starting from 0. The main divisions are numbered to the right starting from 0. The scale is shown in the form of rectangle PQRS of convenient width (say 10 mm). To distinguish the divisions, draw thick horizontal lines in the alternate divisions representing both miles and furlongs. The RF is indicated at the centre of the scale preferably at the bottom. The units of the main divisions and subdivisions are indicated in capitals at the right and left ends. A distance of 5 miles and 4 furlongs is furnished by dimension line as shown in the figure.

4. A field of 125 ha is shown on a map by a quadrilateral. One of the diagonals of the quadrilateral is 8 cm long and the perpendiculars on the diagonal from the other two corners are 2 cm and 3 cm in length. Construct a plain scale to show kilometre and hectometre. Show on the scale a length of 3 km and 6 hm. What is the RF of the scale?

   1 hectare $= 10{,}000\ \text{m}^2$; Area of the quadrilateral $= (1/2) \times (2 + 3) \times (8) = 20\ \text{cm}^2$
   RF of the scale = Square root of (area of the quadrilateral/area of the field)
   $= \text{Sq.rt}\{(20 \times 100)/(125 \times 10{,}000 \times 1000 \times 1000)\}$
   $= 1/25{,}000$ or $1:25{,}000$
   Length of the scale $= (1/25{,}000) \times (4 \times 1000 \times 1000) = 160$ mm

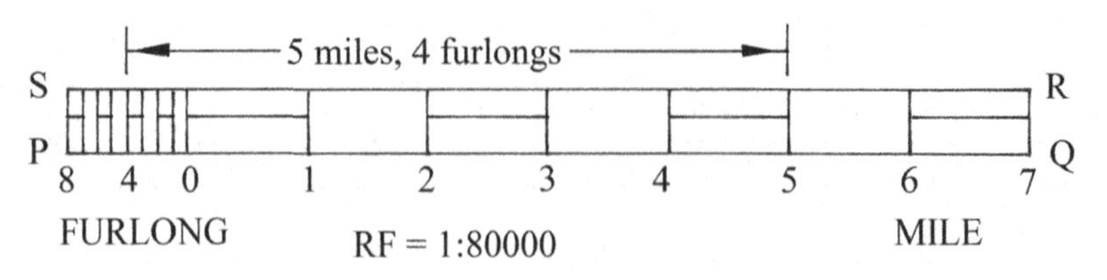

Fig. 3.3 Plain scale

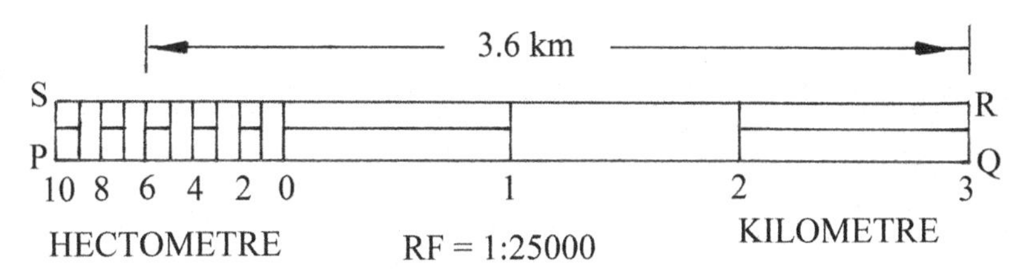

Fig. 3.4 Plain scale

Draw a line PQ = 160 mm long and divide it into four equal parts as shown in Fig. 3.4. Each main division represents 1 km. Subdivide the first left-most main division into 10 equal parts. Each subdivision represents 1 hm. In this scale, 0 (zero) is marked at the end of the first left-most main division. The subdivisions are numbered to the left starting from 0 as shown in the figure. The main divisions are numbered to the right starting from 0. The scale is shown in the form of a rectangle PQRS. A length measuring 3 km and 6 hm is marked by the dimension line as shown in the figure.

Diagonal Scales

Diagonal scales are used to represent three successive units such as metre, decimetre and centimetre or to the accuracy correct to two decimal places i.e. to measure a length of 5.83 m. In a diagonal scale, a line is divided into suitable number of equal parts or units and the first division is subdivided into smaller parts or units diagonally.

Solved Problems

5. Construct a diagonal scale to read up to (1/100)th of a metre and long enough to measure up to 6 m. Take RF = 1:50 and mark on the scale a distance of 4.58 m.

Length of the scale = (1/50) × (6000) = 120 mm

Draw a line PQ of length 120 mm and divide it into six equal parts as shown in Fig. 3.5. Each part or main division represents 1 m. Subdivide the first left-most part into 10 equal parts. Each subdivision represents (1/10)th of a metre. In this scale, 0 (zero) is marked at E, the end of the first left-most main division. Draw perpendiculars PS and QR of convenient length (say 50 mm) at points P and Q. Complete the rectangle PQRS. Divide PS into 10 equal parts and number the division points from P. Draw horizontal lines through these points to meet the line QR. Divide PE into 10 equal parts and transfer these points from PE to SF. Join ninth subdivision on PE with S and draw lines parallel to the line S-9 through subdivision points

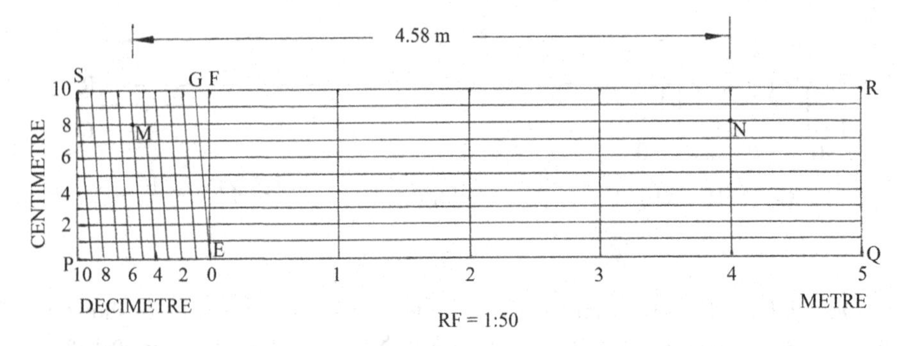

Fig. 3.5 Diagonal scale

eighth, seventh, sixth, fifth, etc. on line PE as shown in the figure. These parallel lines are called diagonal lines. GF represents 1 dm and the diagonal GE of the right angled triangle GEF is divided into 10 equal parts by horizontal lines already drawn. The length of the horizontal intercepts within the triangle progressively increases from 0 at E to 1 dm at GF. The increase of length of successive horizontal lines is by 0.1 dm = 1 cm. To mark a distance of 4.58 m move to the fifth subdivision point to the left of 0 and then move up on the diagonal through the fifth subdivision point to intersect the eighth horizontal line to locate point M on the scale. Move along horizontal line through this point towards right to the fourth main division and mark this point N. The distance between the points M and N is 4.58 m as shown by the dimension line in the figure.

6. Distance between Delhi and Chennai is 1800 km. On a railway map, it is represented by 36 cm length. Calculate the RF. Construct a diagonal scale to read up to a single kilometre. Mark on it the following distances: (1) 76 km; (2) 593 km.

$$\text{RF of the scale} = (360)/(1800 \times 1000 \times 1000) = 1:5{,}000{,}000$$
$$\text{Length of the scale} = (1/5{,}000{,}000) \times (600 \times 1000 \times 1000) = 120\ \text{mm}$$

Draw a line PQ of length 120 mm and divide it into six equal parts as shown in Fig. 3.6. Each part or main division represents 100 km. Subdivide the first left-most part into 10 equal parts. Each subdivision represents (1/10)th of 100 km or 10 km. In this scale, 0 (zero) is marked at E, the end of the first left-most main division. Draw perpendiculars PS and QR of convenient length (say 50 mm) at points P and Q. Complete the rectangle PQRS. Divide PS into 10 equal parts and number the division points from P. Draw horizontal lines through these points to meet the line QR. Transfer the subdivision points from PE to the line SF. Join ninth subdivision on PE with S and draw lines parallel to the line S-90 through subdivision points eighth, seventh, sixth, etc. on the line PE. These parallel lines are called diagonal lines. GF represents 10 km and the diagonal GE of the right angled triangle GEF is divided into 10 equal parts by horizontal lines already drawn. The length of the horizontal intercepts within the triangle increases progressively from 0 at E to 10 km at GF. The increase in length of each horizontal intercept is given by

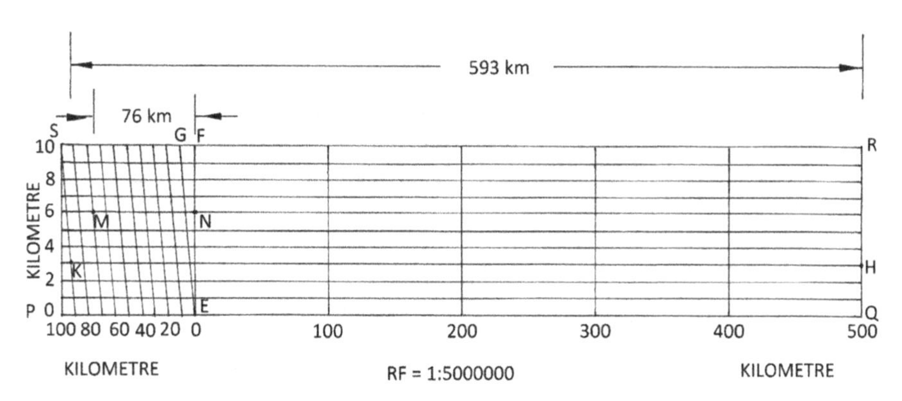

Fig. 3.6 Diagonal scale

0.1 × (10) = 1 km. Points M, N, K and H are marked following the method detailed in the previous problem. The distance between the marked points M and N is 76 km and that of points K and H is 593 km.

7. Distance between IIT Chennai and Guindy is 2.5 km. On a road map, this is represented by a distance of 10 cm. Draw a diagonal scale to read a smallest distance of 5 m and long enough to measure 4 km. Show on this scale a distance of 3.375 km.

$$\text{RF of the scale} = (10 \times 10)/(2.5 \times 1000 \times 1000) = 1/25,000 \text{ or } 1 : 25,000$$
$$\text{Length of the scale} = (1/25,000) \times (4 \times 1000 \times 1000) = 160\,\text{mm}$$

Draw a line PQ = 160 mm long and divide it into four equal parts as shown in Fig. 3.7. Each main division represents 1 km. Subdivide the first left-most main division into 10 equal parts. Each subdivision represents 1 hm. In this scale, 0 (zero) is marked at E, the end of the first left-most main division. Draw perpendiculars PS and QR of convenient length (say 100 mm) at points P and Q. Complete the rectangle PQRS. Divide PS into 20 equal parts and number the alternate division points as 1, 2, 3, etc. so that each division represents 0.5 dm or 5 m. Draw horizontal lines through these points to meet the line QR. Transfer subdivisions from PE to SF. Join ninth subdivision on PE with S and draw lines parallel to the line S-9 through subdivision points eighth, seventh, sixth, etc. on PE. These parallel lines are called diagonal lines. GF represents 1 hm, and the diagonal GE of the right angled triangle GEF is divided into 20 equal parts by horizontal lines already drawn. The length of the horizontal intercept within the triangle increases progressively from 0 at E to 1 hm at GF. The increase of length of the successive horizontal intercept is given by 0.05 hm (0.5 dm or 5 m). The dimension line MN shown in the figure furnishes the required distance of 3.375 km (3 km, 3 hm, 7.5 dm).

8. Construct a diagonal scale of RF = 1/24 capable of reading 4 yards and showing yards, feet and inches. Show on the scale a length of 3 yards, 2 ft and 10 in.

Length of the scale = (1/24) × (4 × 3 × 12) = 6 in.

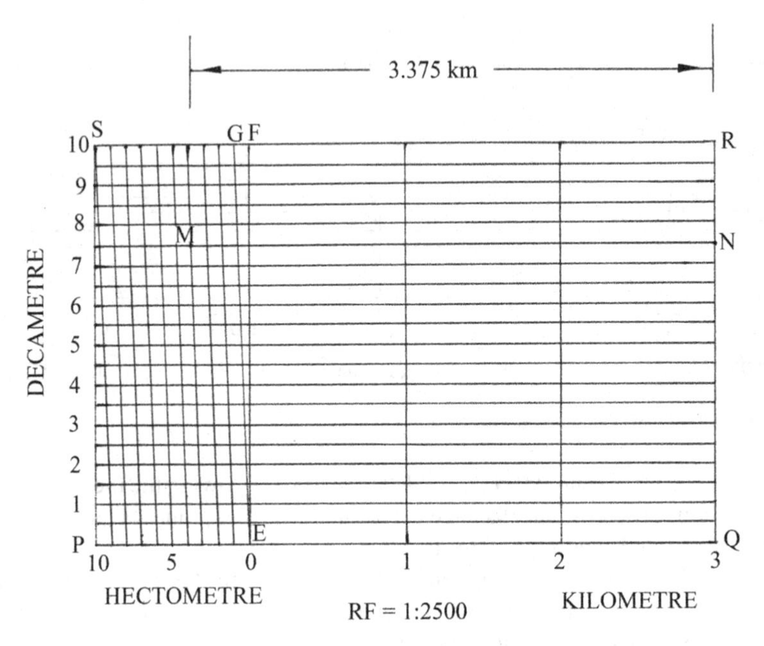

Fig. 3.7 Diagonal scale

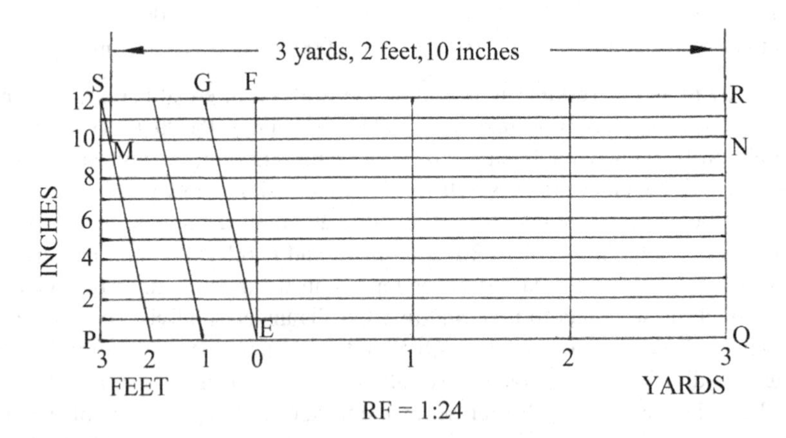

Fig. 3.8 Diagonal scale

Draw a line PQ = 6 in. long and divide it into four equal parts as shown in Fig. 3.8. Each part of main division represents 1 yard. Subdivide the first left-most main part into three equal parts. Each subdivision represents 1 ft. In the scale, 0 (zero) is marked at E, the end of the first left-most main division. Draw perpendiculars PS and QR of convenient length (say 2″) at points P and Q. Complete the rectangle PQRS. Divide PS into 12 equal parts and number the alternate division points as 2, 4, 6, etc. from P. Draw horizontal lines through these division points on PS to meet the line QR. Transfer the subdivision points from PE to SF. Join second subdivision on PE with S and draw lines parallel to the line S-2 through points 1 and 0 on PE as shown in the figure. These parallel lines are called as diagonal lines. GF represents 1 ft, and the diagonal GE of the right angled triangle GEF is divided into 12 equal parts by horizontal lines already drawn. The increase of length of successive horizontal intercept within the triangle GEF is given by (1/12) × (1 ft × 12) = 1 in. The required length of 3 yards, 2 ft and 10 in. is marked by the dimension line MN as shown in the figure.

Vernier Scale

Vernier scale is a modified form of diagonal scale. It is used to measure very small units with great accuracy over a small area. A vernier scale consists of a primary scale similar to a plain scale and a secondary scale called vernier. A line is divided into suitable number of equal parts or units, and the first division is subdivided into smaller n equal parts or units. For constructing a vernier scale, a length of $(n + 1)$ smaller subdivisions on the main scale is taken as the length of vernier scale and this length is divided into n equal parts.

9. Construct a vernier scale to read up to (1/100)th of a metre and long enough to measure up to 3 m. Take RF = 1:20 and mark on it a distance of 2.58 m.

   Length of the scale = (1/20) × (3 × 1000) = 150 mm

   Draw a line PQ = 150 mm long and divide it into three equal parts as shown in Fig. 3.9. Each main division represents 1 m. Subdivide the first left-most part and each main part into 10 equal parts. Each subdivision represents (1/10)th of a metre or 1 dm. In this scale, 0 (zero) is marked at the end of the first left-most main division. The subdivisions are numbered to the left starting from 0 and the main divisions are numbered to the right starting from 0. The scale is shown in the form of a rectangle PQRS. Draw a line AB equal to a length of 11 subdivisions and divide it into 10 equal parts. This scale is the vernier scale and is shown in the form of a rectangle ABCD. Each of this subdivision on the vernier scale represents (11/10) of

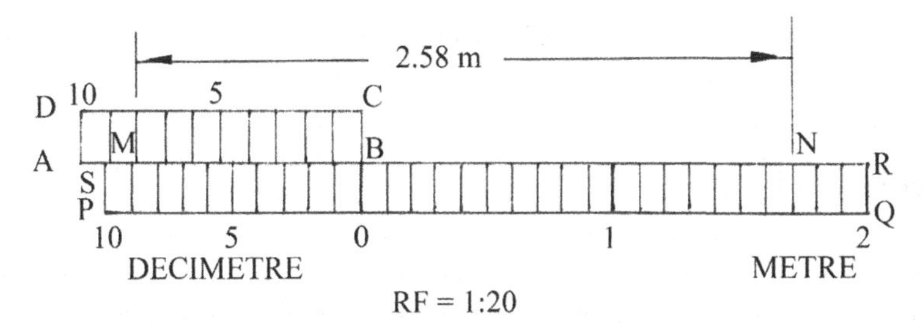

Fig. 3.9 Vernier scale

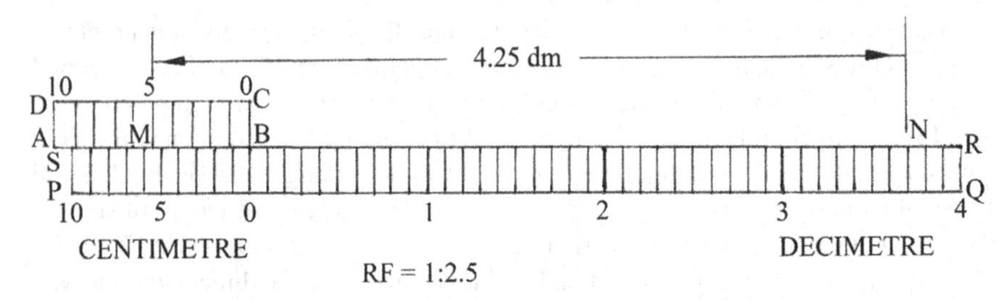

Fig. 3.10 Vernier scale

each subdivision on the main scale, i.e. 1.1 dm or 11 cm.To mark the required distance MN = 2.58 m, split it as (0.88 + 0.7 + 1.0) m. Take one main division representing 1 m, seven subdivisions beyond one main division on the main scale representing 7 dm and eight divisions on the vernier scale representing 8.8 dm or 88 cm.

Distance between MN = 1 m + 7 dm + 8.8 dm = 2.58 m

10. Construct a vernier scale of RF = (1/2.5) to show decimetre, centimetre and millimetre. The scale should be long enough to measure 5 dm. Mark on it a distance of 4.25 dm.

Length of the scale = (1/2.5) × (5 × 100) = 200 mm

Draw a line PQ = 200 mm long and divide it into five equal parts as shown in Fig. 3.10. Each part or main division represents 1 dm. Subdivide the first left-lost part and each main part into 10 equal parts. In this scale, 0 (zero) is marked at the end of first left-most main division. The subdivisions are numbered to the left starting from 0 in the first main part, and the main divisions are numbered to the right starting from 0. The scale is shown in the form a rectangle PQRS. Draw a line AB equal to a length of 11 subdivisions and divide it into 10 equal parts. This is the Vernier scale and is shown in the form of a rectangle ABCD. Each of the subdivision on the Vernier scale represents (11/10) of each subdivision on the main scale, i.e. 1.1 cm.To mark the required distance MN = 4.25 dm, split it as (0.55 dm + 0.7 dm + 3.0 dm). Take three main division on the main scale representing 3 dm, seven subdivisions beyond the three main division representing 0.7 dm and five divisions on the Vernier representing 0.55 dm.

Distance between MN = 3.0 + 0.7 + 0.55 = 4.25 dm

Practice Problems

1. Draw a scale of RF 1:400 to show metre. Show 57 m on the scale.
2. On a survey map, the distance between two places 1 km apart is 2.5 cm. Construct a scale to read 1.8 km. What is the RF of the scale?

3. Draw a diagonal scale to read kilometre, hectometre and decametre given that 1 km is represented by 5 cm on the drawing. Mark on the scale a distance of 1.54 km. What is the RF of the scale?
4. Draw a diagonal scale of RF = 1/20,000 to show kilometre and decimals of kilometre. Mark on the scale the distances 1.64 km and 2.37 km.
5. An area of 144 cm^2 on a map represents an area of 36 km^2 on the field. Find the RF of the scale for the map. Draw a diagonal scale to show kilometre, hectometre and decametre and to measure up to 10 km. Indicate on the scale a distance of 7 km, 5 hm and 6 dm.

Chapter 4
Curves Used in Engineering Practice

Conic Curves

Conic curves are obtained cutting a right circular double cone by different section planes as shown in Fig. 4.1. Straight lines connecting the base points of the cone with the apex are called generators or elements of cone. Conic curves are plane curves.

When the cutting plane 1-1 is perpendicular to the axis of the cone, the curve obtained is a circle as shown in Fig. 4.2(i).

If the cutting plane 2-2 makes a greater angle with the axis than the generators or elements of the cone, the curve obtained is an ellipse as shown in Fig. 4.2(ii).

If the cutting plane 3-3 makes the same angle with the axis as generators or elements of the cone, the resulting curve is a parabola as shown in Fig. 4.2(iii).

Finally, if the cutting plane makes a smaller angle with the axis than do the generators or elements of the cone or is parallel to the axis of the cone, the plane cuts both the parts of cone. The curves obtained are hyperbolae as shown in Fig. 4.2 (iv).

The conic curves may also be defined with reference to their properties as plane curves. Let us assume F a fixed point and AB a fixed straight line as shown in Fig. 4.3. A point P moves in a plane containing F and AB in such a manner that the distance FP always bears the same ratio to the perpendicular PM to the fixed line AB. The curve traced out by the point P is called a conic curve or conic. The fixed point F is called the focus and the fixed straight line AB is called the directrix of the conic. A straight line through the focus perpendicular to the directrix is called the axis. The point V where the axis cuts the curve is called the vertex of the conic. The constant ratio of FP to PM is called the eccentricity, e, of the conic.

When FP is less than PM, the conic is an ellipse.
When FP is equal to PM, the conic is a parabola.
When FP is greater than PM, the conic is a hyperbola.

K. Rathnam, *A First Course in Engineering Drawing*,
DOI 10.1007/978-981-10-5358-0_4

Fig. 4.1 Right circular double cone

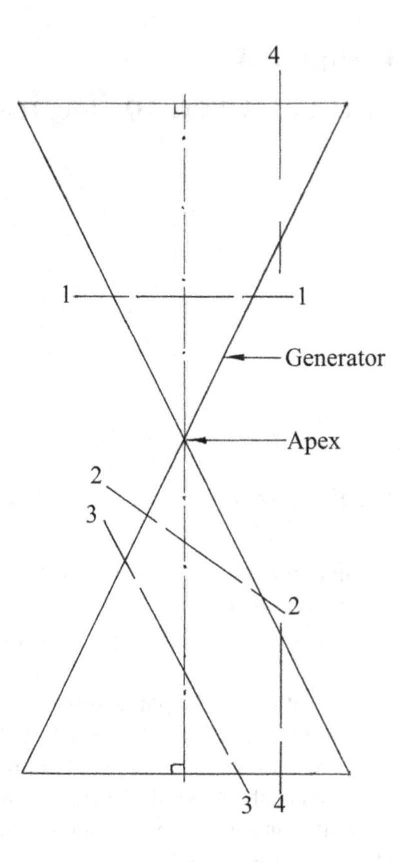

A straight line joining two points on a conic is called a chord. If the chord passes through the focus, it is called a focal chord. The mid-points of parallel chords which lie in a straight line is called a diameter. A perpendicular from a point on the conic to the axis is called an ordinate and if produced to meet the conic again it is called a double ordinate. The double ordinate through the focus is called the latus rectum.

To draw tangent and normal at a point on the conic curve

The conic curve is drawn with the directrix AB, axis OX, focus F and eccentricity e as shown in Fig. 4.4. Let P be the point on the curve at which the tangent and normal are to be constructed. Join PF and draw a line perpendicular to PF at F. This line intersects the directrix at T. Join PT and produce it. The line PT is the tangent to the conic curve at P. The normal PN is drawn perpendicular to PT as shown in Fig. 4.4.

Fig. 4.2 Types of conic sections

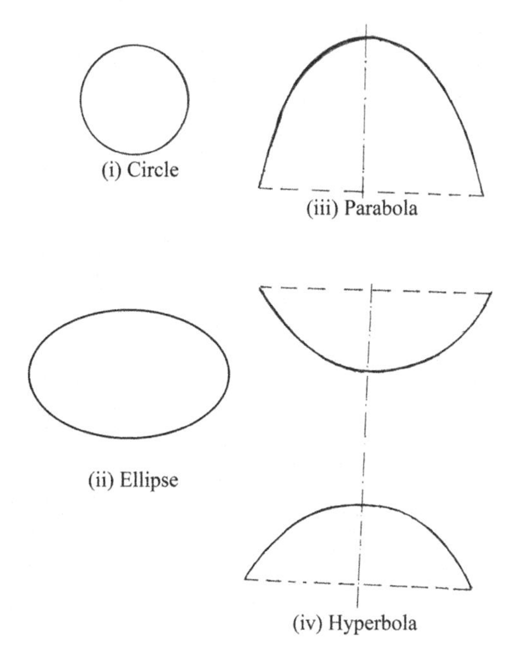

Fig. 4.3 Conic curve

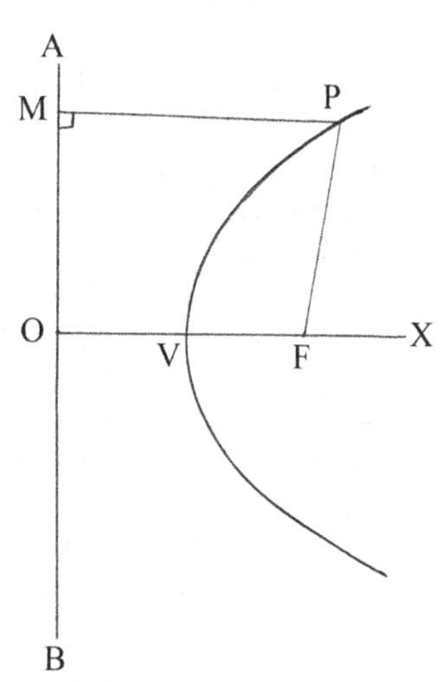

Solved Problems

1. Taking a focus 50 mm from the directrix construct (a) an ellipse, eccentricity, e = 2/3, (b) a parabola, eccentricity, e = 1 and (c) one branch of hyperbola, eccentricity, e = 3/2.

   (a) Draw the directrix AB and the axis OX perpendicular to AB as shown in Fig. 4.5. The focus is located on the axis 50 mm from O. Since the eccentricity

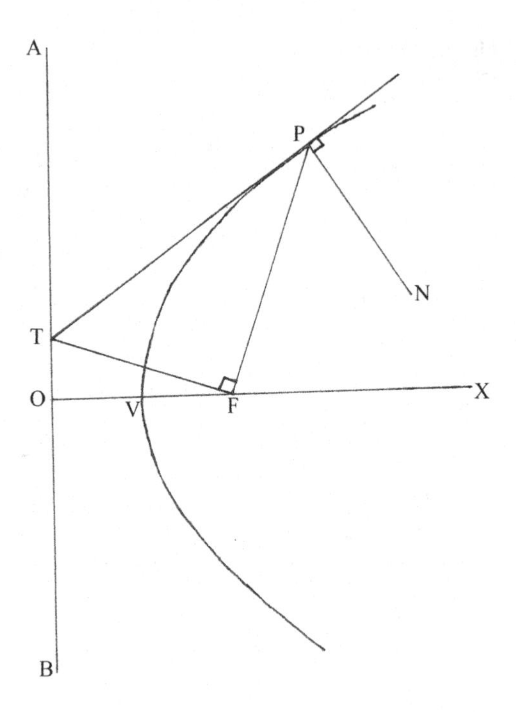

Fig. 4.4 Tangent and normal to conic curve

is 2/3, the distance OF is divided into five equal parts. The vertex V is located on OF such that (VF)/(VO) = 2/3. Draw VE perpendicular to the axis such that VE = VF. Join OE and extend it. Draw a line inclined at 45° to the axis at F to cut the extension of OE at G as shown in the figure. Draw a line perpendicular to the axis from G to locate the other vertex V1 of the ellipse. Mark points 1, 2, 3, 4, 5 and 6 between V and V1 along the axis. Draw perpendiculars through these points above and below the axis. The perpendiculars intersect the line OEG at 1′, 2′, 3′, 4′, 5′ and 6′. With centre F and radius 11′, draw arcs to cut the perpendicular drawn through 1 at P1 above and P1′ below the axis. Similarly, points P2, P2′, P3, P3′, P4, P4′, P5, P5′, P6 and P6′ are obtained following the same procedure. Draw a smooth curve passing through the points V, P1, P2, P3, P4, P5, P6, V1, P6′, P5′, P4′, P3′, P2′, P1′ and V to obtain the curve ellipse.

(b) Draw the directrix AB and the axis OX perpendicular to AB as shown in Fig. 4.6. The focus is located on the axis 50 mm from O. Locate the vertex, 25 mm from O (mid-point of OF). Mark six points 1, 2, 3, 4, 5 and 6 on VX and draw perpendiculars through these points. With centre F and O1 as radius, draw arcs to cut the perpendicular through point 1 at P1 above and P1′ below the axis as shown in the figure. Similarly, obtain points P2, P2′, P3, P3′, P4, P4′, P5, P5′, P6 and P6′ following the same procedure. Draw a smooth curve passing through points P6, P5, P4, P3, P2, P1, V, P1′, P2′, P3′, P4′, P5′ and P6′ to obtain the curve parabola.

(c) Draw the directrix AB and the axis OX perpendicular to AB as shown in Fig. 4.7. The focus is located on the axis 50 mm from O. Since the eccentricity

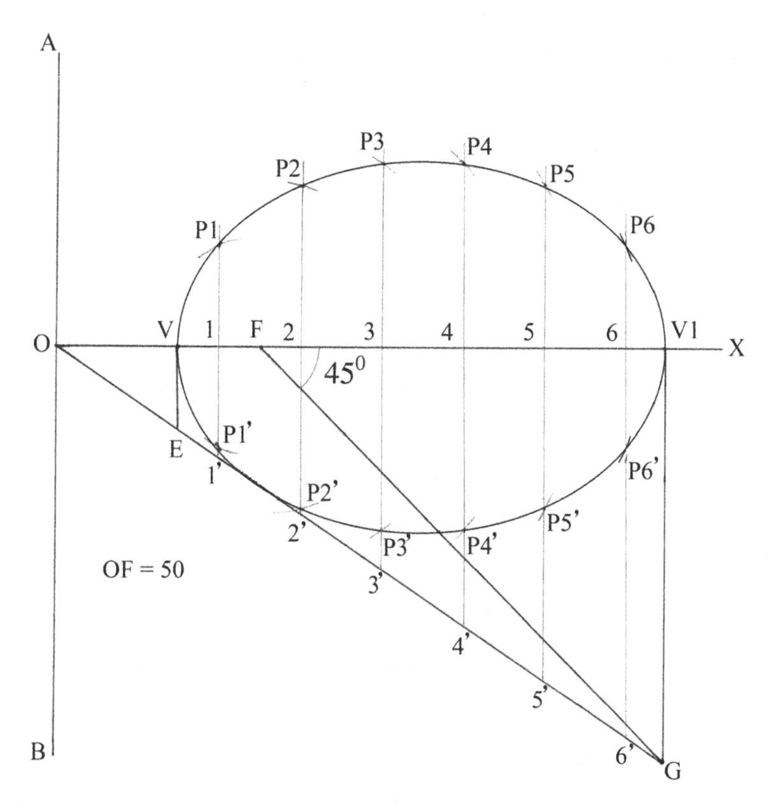

Fig. 4.5 Ellipse (eccentricity method)

is 3/2, the distance OF is divided into five equal parts. The vertex V is located on OF such that (VF)/(VO) = 3/2. Draw VE perpendicular to the axis such that VE = VF. Join OE and extend it up to G. Mark four points 1, 2, 3 and 4 along the axis between V and X and draw perpendiculars through these points. The perpendiculars intersect the line OEG at $1'$, $2'$, $3'$ and $4'$. With centre F and radius $11'$, draw arcs to cut the perpendicular drawn through point 1 at P1 above and $P1'$ below the axis. Similarly, points P2, $P2'$, P3, $P3'$, P4 and $P4'$ are obtained following the same procedure. Draw a smooth curve passing through P4, P3, P2, P1, V, $P1'$, $P2'$, $P3'$ and $P4'$ to obtain the curve hyperbola.

2. A point moves such that the sum of its distances from two fixed points is always equal to 100 mm. The distance between the fixed points is 70 mm. Draw the curve using the intersecting arcs method. Draw tangent and normal to the curve at any point on it.

   The ellipse is a plane curve generated by a point moving in space such that the sum of its distances from two fixed points is a constant (equal to the major axis). Draw the major axis AB of length 100 mm and locate the foci F and G 15 mm from the ends of the major axis as shown in Fig. 4.8. Draw a line

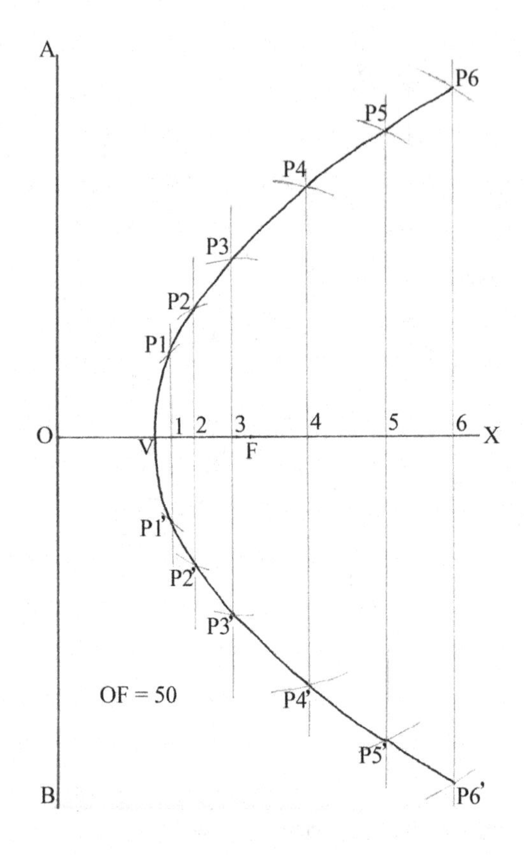

Fig. 4.6 Parabola (eccentricity method)

perpendicular to the major axis through O, the centre of the ellipse. With centre F and radius OA (one half of the major axis), draw arcs to cut the perpendicular drawn through O at C and D. The minor axis is CD. Four points are chosen between F and O as shown in the figure. With F and G as centres and the distances A1 and B1 as radii, draw intersecting arcs to locate points J on the curve ellipse. Radii A2 and B2 are used to locate points K. Points L and M are also located following the same procedure. A smooth curve is drawn passing through these intersecting points. Point Q is chosen on the curve. Join QF and QG and draw the bisector of the angle FQG. The bisector QN is the required normal, and the line QT drawn perpendicular to QN is the tangent to the curve ellipse at point Q.

3. The major axis of an ellipse is 100 mm long and the minor axis is 60 mm long. Draw the ellipse by concentric circles or auxiliary circles method. Draw a tangent at any point on the ellipse using auxiliary circles.

   Draw two concentric circles having the major and minor axes as diameters using the centre of ellipse O as shown in Fig. 4.9. The major and minor axes are, respectively, AB and CD. Draw a radial line through O cutting the auxiliary circles at P and Q. Draw a line parallel to CD through Q and another line parallel to AB through P. These lines which are parallel, respectively, to the minor and major axes intersect at point R. The point R lies on the ellipse.

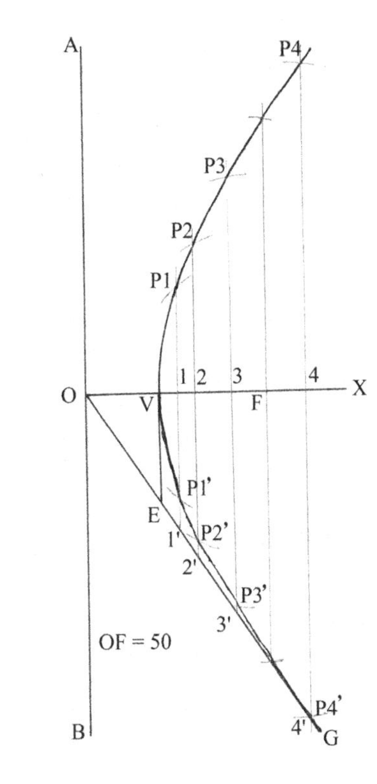

Fig. 4.7 Hyperbola (eccentricity method)

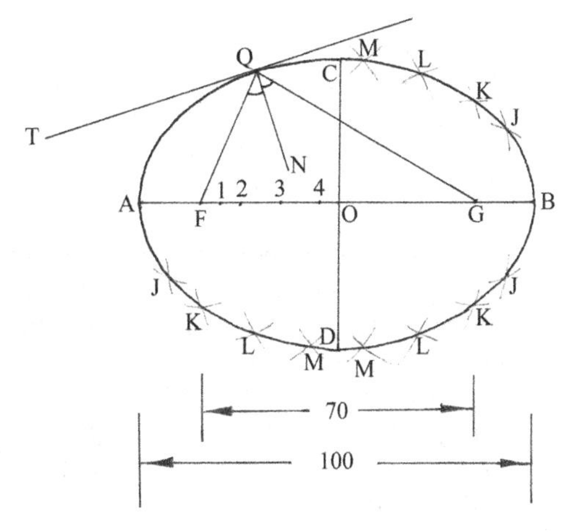

Fig. 4.8 Construction of ellipse (intersecting arcs method)

The same procedure is repeated to locate a few more points in each quadrant. A smooth curve is drawn passing through these points.

A point S is chosen on the ellipse as shown in Fig. 4.9. Points E and F are located on the concentric circles by drawing lines parallel to CD and AB from S. Draw the radial line OFE. Draw tangent ET at E to meet the major axis

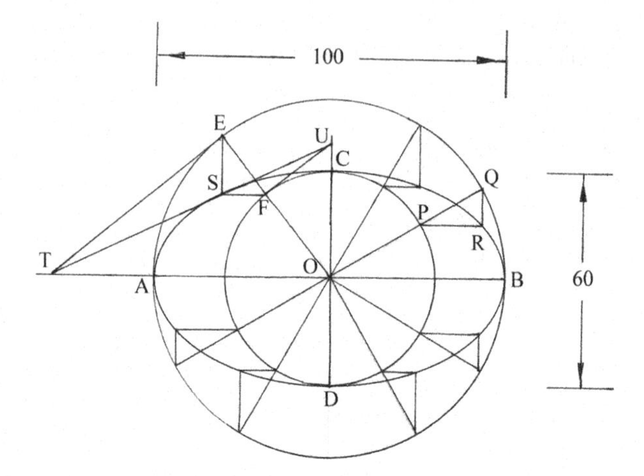

Fig. 4.9 Construction of ellipse (auxiliary circles method)

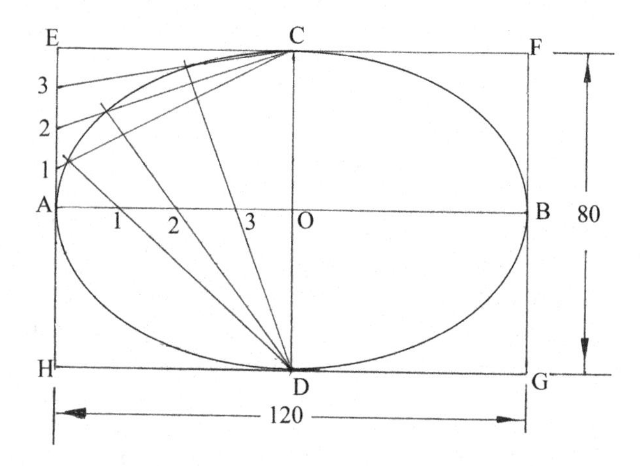

Fig. 4.10 Construction of ellipse (oblong/rectangle method)

extension at T. Another tangent FU at F is drawn to meet the minor axis extension at U. Join TSU which is the required tangent to the ellipse.

4. Draw an ellipse whose major axis is 120 mm and minor axis is 80 mm using oblong/rectangle method.

   Construct a rectangle EFGH using the given major axis and minor axis as its sides as shown in Fig. 4.10. Join the mid-points on the opposite sides of the rectangle to show the major axis AB and the minor axis CD of the ellipse. The centre of the ellipse is designated as O. Divide AO and AE into the same number of equal parts (four in this case) and number the division points from A. Draw a line from C through point 1 on the line AE and draw another line from D through point 1 on the line AO. The point of intersection of these lines is a point on the ellipse. Similarly, the intersections of lines from C and D through points 2 and 3 will also be on the ellipse. Points are located in the other three quadrants following the same procedure. A smooth curve is drawn passing through these points to get the required ellipse.

5. The lengths of the conjugate axes or diameters of an ellipse are 100 mm and 80 mm and the included angle is 60°. Draw the ellipse. Determine the principal axes and the angle which the major axis makes with the longer conjugate axis or diameter.

   A parallelogram EFGH is drawn of sides 100 mm and 80 mm with the included angle of 60° as shown in Fig. 4.11. The mid-points of the opposite sides are joined to show the conjugate axes PQ and RS and also the centre of the ellipse O. Divide PO and PE into the same number of equal parts (four in this case) and number the division points from P. Join R to point 1 along PE and draw a line from S through point 1 along PO. The point of intersection lies on the ellipse. Similarly, points are located at the intersections of lines from R and S through points 2 and 3. These points lie on the ellipse. Points are also located on the other three parts of the parallelogram following the same procedure. A smooth curve is drawn passing through these points besides the extremities of the conjugate axes.

   A semi-circle is drawn on RS as diameter intersecting the ellipse at U. A line is drawn parallel to UR through O, centre of the ellipse. This line intersects the ellipse at C and D. The line CD is the minor axis of the ellipse. Another line is drawn parallel to SU through O. This line intersects the ellipse at A and B. The line AB is the major axis of the ellipse. The inclination Θ of the major axis with the longer conjugate axis, angle ∠QOB, is 18°.

6. The major axis of an ellipse is 120 mm long and the minor axis is 70 mm long. Draw the ellipse by the 'Four entre method'.

   The major axis AB and the minor axis CD with centre of ellipse O are drawn as shown in Fig. 4.12. Draw line AC. With O as centre and radius OA, draw an arc AE as shown in the figure. With C as centre and radius EC, draw an arc EF. Draw the perpendicular bisector of AF to cut the major axis at G and also CD (extended) at H. Points K and L are located on AB and CD (extended) such that KB = AG and OL = OH. With H and L as centres and radius CH, draw two

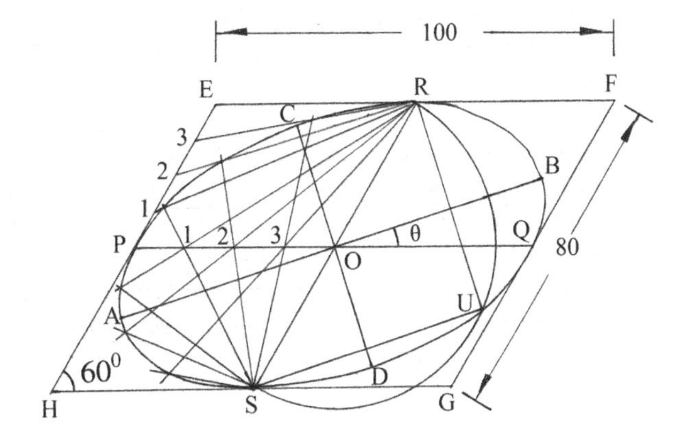

Fig. 4.11 Construction of ellipse (parallelogram method)

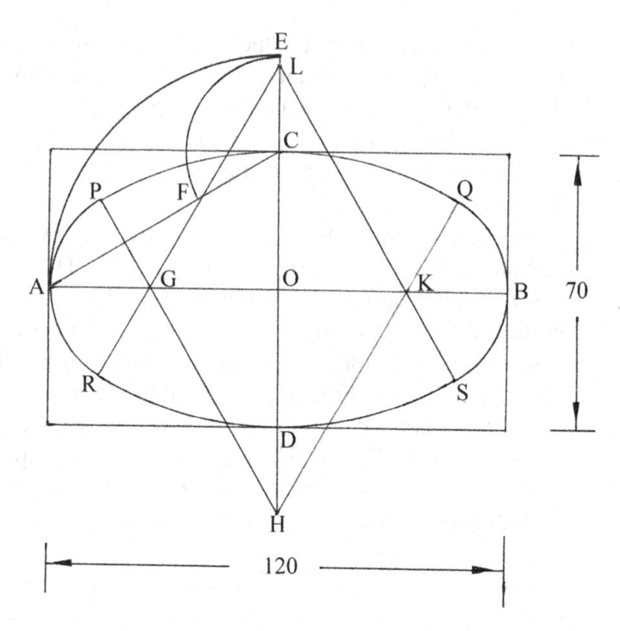

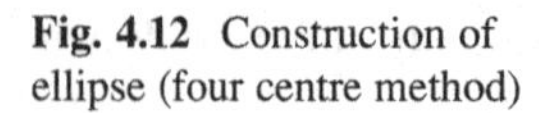
Fig. 4.12 Construction of ellipse (four centre method)

arcs PCQ and RDS as shown in the figure. With G and K as centres and radius GA, draw two arcs PAR and QBS. The points G, H, K and L are four centres that are used to draw the arcs to complete the construction of the ellipse.

7. Draw an ellipse whose minor axis is 70 mm and the major axis is 120 mm. Draw a pair of tangents from a point P 40 mm above the major axis and 70 mm to the left of the minor axis.

   Draw the major axis AB and the minor axis CD as shown in Fig. 4.13. The foci F and G are located on the major axis. The curve ellipse is drawn following the intersecting arc method. The point P is located 70 mm from the minor axis and 40 mm from the major axis. With centre P and radius PF (distance of nearest focus from P), draw an arc. With centre G and radius AB, draw an arc to intersect the previously drawn arc at H and K. Join HG and KG intersecting the ellipse at Q and T. Join PQ and extend it. Join PT and extend it. The lines PQ and PT are the tangents to the ellipse from the point P.

8. A line AB represents the major axis and measures 120 mm. A point P is 90 mm from A and 50 mm from B. Draw the elliptical curve through the points P, A and B.

   Draw a circle taking the major axis as diameter and locate the centre of the ellipse O as shown in Fig. 4.14. AB is the major axis of the ellipse. With centre A and radius 90 mm, draw an arc. With centre B and radius 50 mm, draw an arc to intersect the previously drawn arc to locate the point P. A line is drawn perpendicular to AB through the point P to cut the major circle at E. Join the radial line OE. Draw a line parallel to AB through the point P to intersect the radial line at F. With centre O and radius OF, draw the concentric circle. This

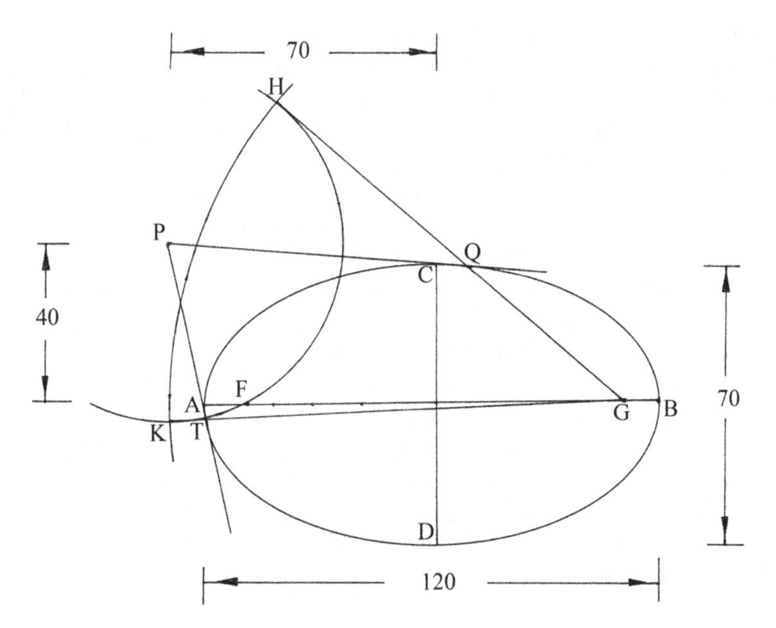

Fig. 4.13 Construction of a pair of tangents to the ellipse

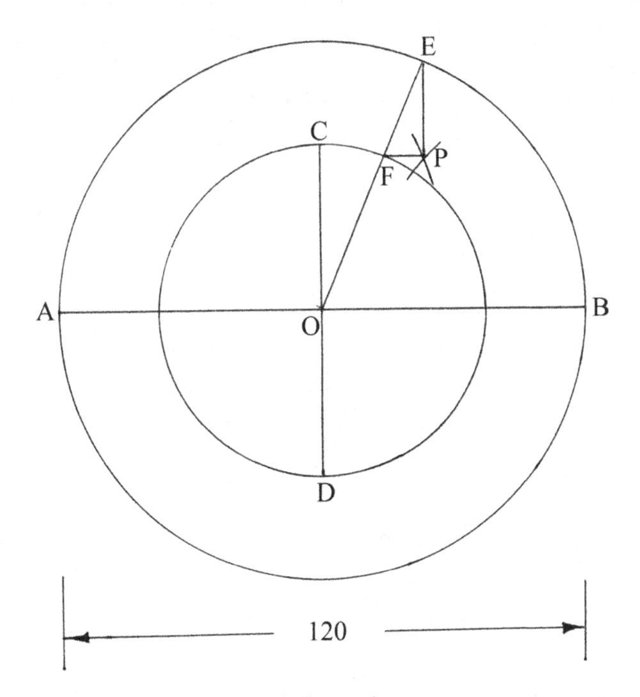

Fig. 4.14 Construction of an ellipse through three points

circle is the minor circle. Other points on the ellipse can be located following the auxiliary circles method detailed in problem 3.

9. The double ordinate through the focus F of an ellipse is 60 mm. If the eccentricity of the curve is 2/3, find the location of the directrix and the lengths of major and minor axes.

   Draw the axis of the ellipse and locate the focus F on it as shown in Fig. 4.15. Draw a line perpendicular to the axis at F and represent the double ordinate PQ such that PF = FQ. A right angled triangle RST is constructed on the axis such that (ST)/(RS) is 2/3. Draw a line parallel to RT through Q to intersect the axis at U. Draw a line perpendicular to the axis through U. This line is the directrix of the ellipse. Draw two lines from F inclined at 45° to the axis of the ellipse to intersect the line UQ and its extension at E and G. Draw lines perpendicular to the axis through E and G to locate the vertices A and B of the ellipse. The line AB is the major axis of the ellipse. The centre of the ellipse O is located at the mid-point of AB. A line perpendicular to the axis of the ellipse is drawn through O. With centre F and radius AO, an arc is drawn to cut the perpendicular line drawn through O at C and D. The line CD is the minor axis of the ellipse. The lengths of the major axis AB and the minor axis CD are, respectively, 102 mm and 72 mm.

10. Determine the length of the minor axis of an ellipse whose distance between the directrices is 120 mm and that of the vertices is 90 mm.

    Draw the axis of the ellipse and locate the directrices EGH and JKL 120 mm apart as shown in Fig. 4.16. The centre of the ellipse O is located mid-way between the directrices along the axis. The vertices are located on the axis such

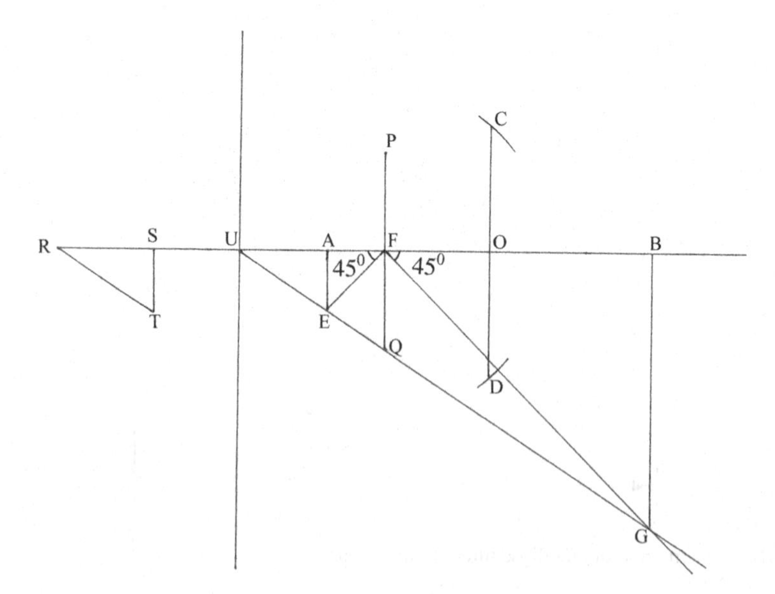

Fig. 4.15 Determination of principal axes of an ellipse

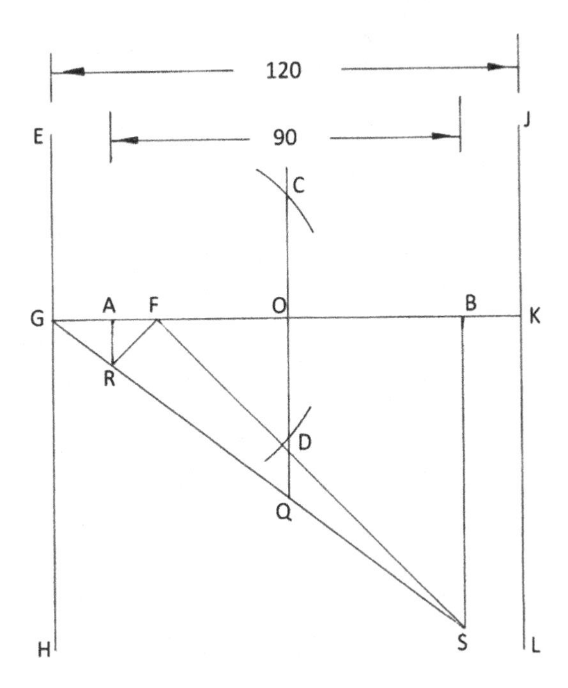

Fig. 4.16 Determination of length of minor axis

that AO = BO = 45 mm. A line is drawn perpendicular to the axis through O, and point Q is located on it such that OQ = AO. Join GQ and extend it. Two lines are drawn perpendicular to the axis through points A and B. The intersections of the lines with the line through GQ locate points R and S. The focus F is located on the axis from A such that AF = AR. With centre F and radius AO, draw an arc to cut the perpendicular through O at C and D. The line CD is the minor axis of the ellipse and its length is 60 mm.

11. A stone thrown from the ground level reaches a maximum height of 20 m and falls on the ground at a distance of 35 m from the point of projection. Trace the path of the stone. Determine the angle of projection.

    The trajectory of the stone is a parabola. Adopt a scale 1:250. With this scale, a length of 80 mm in the drawing represents the maximum height 20 m reached by stone, and a length of 140 mm in the drawing represents the horizontal range 35 m covered by the stone. A rectangle ABCD is drawn as shown in Fig. 4.17. The mid-point of BC is vertex V of the parabola, and a vertical line through V represents axis of the parabola, VO. Divide AO and AB into the same number of equal parts (four in the present case) and number the division points from A. Draw a line from V through point 1 on line AB and draw a line parallel to VO through point 1 on line AO. The point of intersection lies on the parabola. Draw lines from V through points 2 and 3 on line AB. Draw lines parallel to VO through points 2 and 3 on line AO. The intersections of lines locate two more points on the curve. The same procedure is repeated to locate points on the other part of the rectangle. A smooth curve is

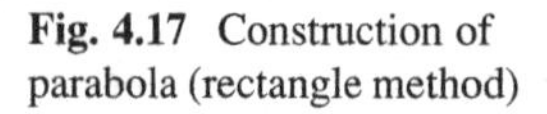

Fig. 4.17 Construction of parabola (rectangle method)

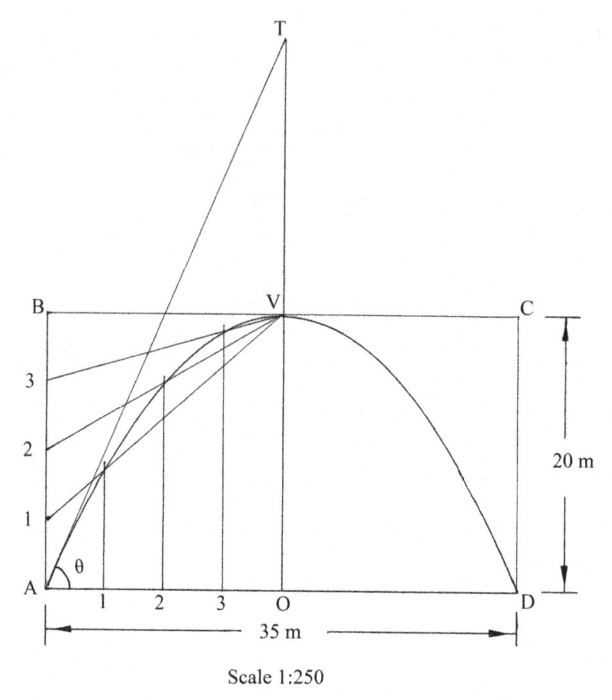

drawn passing through these points. The angle of projection is obtained by drawing tangent to the parabola at the initial point A of the projection. Extend the line OV and locate point T on it such that TV = VO. Join AT. AT is the required tangent. Measure the angle ∠TAO, and the angle of projection Θ is found to be 67°.

12. Draw a parabola with the axis vertical, for a rise of 80 mm and a span of 80 mm. Locate the focus and the directrix.

    Draw a rectangle ABCD as shown in Fig. 4.18. The mid-point of BC is the vertex of the parabola, and a vertical line through V represents the axis of the parabola. Divide AO and AB into the same number of parts (four in the present case) and number the division points from A. Draw a line from V through point 1 on line AB and draw a line parallel to VO through point 1 on line AO. The point of intersection lies on the parabola. Draw lines from V through points 2 and 3 on line AB. Draw lines parallel to VO through points 2 and 3 on line AO. The intersections of lines locate two more points on the curve. The same procedure is repeated to locate points on the other part of the rectangle. A smooth curve is drawn passing through these points.

    Extend OV upwards and locate point T on it such that TV = VO. Join AT and measure the angle ∠BAT. Draw a line AG such that the angle ∠TAG = the angle ∠BAT. The line intersects the axis VO and locates the focus F. Locate point H on VT such that VH = VF. A line perpendicular to the axis of the parabola through H represents the directrix XY of the parabola.

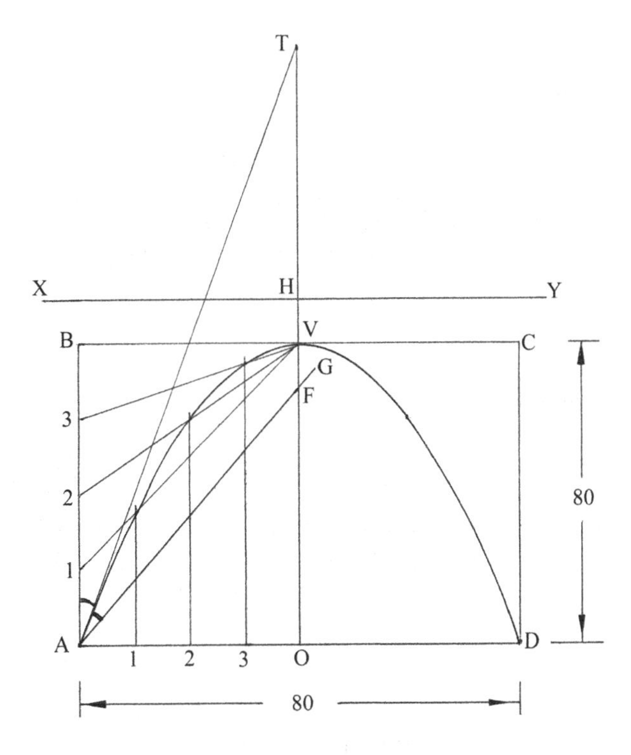

Fig. 4.18 Construction of parabola (rectangle method)

13. Inscribe a parabola in a parallelogram whose sides are 120 mm and 60 mm with an included angle of 60°. Locate the focus and draw a tangent to the curve at any point on it.

A parallelogram PQRS is drawn with sides 120 mm and 60 mm and included angle 60° as shown in Fig. 4.19. Mark the point U, the mid-point of QR, and also point O, the mid-point of PS. Join UO. Divide PO and PQ into the same number of equal parts (four in the present case) and number the division points from P. Draw a line from U through point 1 on line PQ and draw a line parallel to UO through point 1 on line PO. The point of intersection is on the curve parabola. Draw lines from U through points 2 and 3 on line PQ. Draw lines parallel to UO through points 2 and 3 on line PO. The intersections of lines locate two more points on the curve. The same procedure is repeated to locate points on the other part of the parallelogram. A smooth curve is drawn passing through these points on the parabola.

The parabola is redrawn in Fig. 4.20 to locate the focus.

Draw two parallel chords AB and CD to the parabola. Mark the mid-points G and H of the chords AB and CD. Draw a line through G and H. At any point K on this line GH, draw MN perpendicular to the line GH. Draw the perpendicular bisector of MN through O, and this bisector is the axis of the parabola. The intersection of the axis with the curve locates the vertex V of the parabola. Extend OV and point T is located such that TV = OV. Join MT and this is the

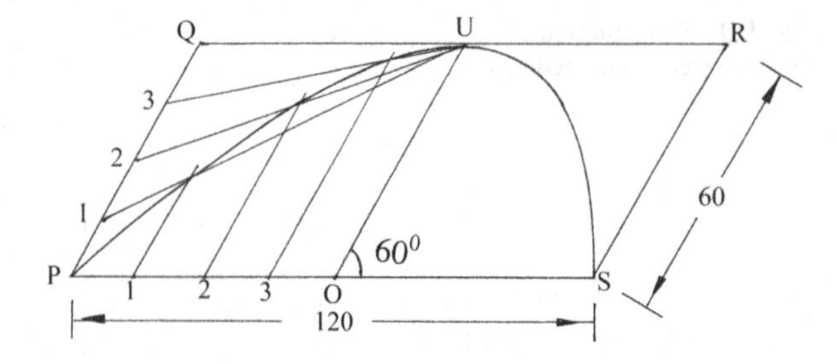

Fig. 4.19 Construction of parabola (parallelogram method)

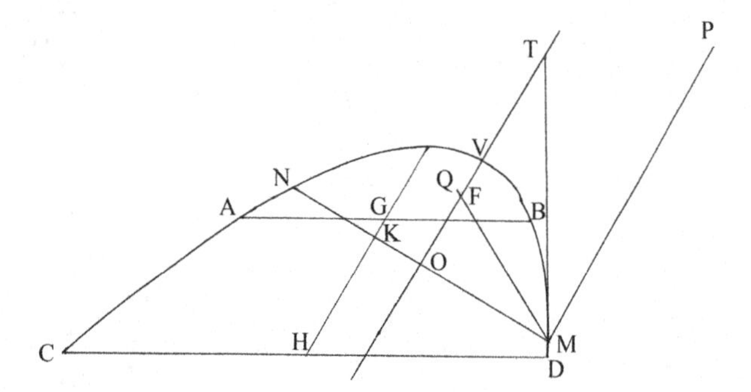

Fig. 4.20 Drawing of tangent to parabola and location of its focus

tangent to the parabola at point M. Draw a line MP parallel to the axis OV. Measure the angle ∠TMP. A line MQ is drawn such that the angle ∠TMQ = angle ∠TMP. This line MQ intersects the axis and locates the focus F.

14. A stone is thrown up from a building of 10 m high. The initial direction of velocity makes an angle of 50° with the horizontal. The maximum height reached by the stone is 25 m above the ground. Trace the path of the stone. Find the distance from the building at which the stone will reach the ground.

    Adopt a scale of 1:250. Then, a length of 40 mm in the drawing represents the height of the building, and a length of 60 mm in the drawing represents the maximum height reached by the stone above the building. Draw a vertical line HA representing the height of the building as shown in Fig. 4.21. The line HA is extended upwards and a point S is located on it such that AS = 2 × 15 m = 30 m (120 mm in the drawing). The tangent to the parabola is drawn at A making an angle of 50° to the horizontal. A horizontal line is drawn through point S to intersect the tangent at T. A vertical line is drawn through T to intersect the horizontal line through A at O. The mid-point of OT locates the vertex V of the parabola. A rectangle ABCD is constructed as shown in the figure. Divide AO and AB into the same number of equal parts (four in the present case) and number the division points from A. Draw a line from V through point 1 on line AB and draw a line parallel to VO through 4 on line AO. The point of intersection is on the parabola. Draw lines from V through 2 and 3 on line AB. Draw lines parallel to VO through 5 and 6 on line AO. The

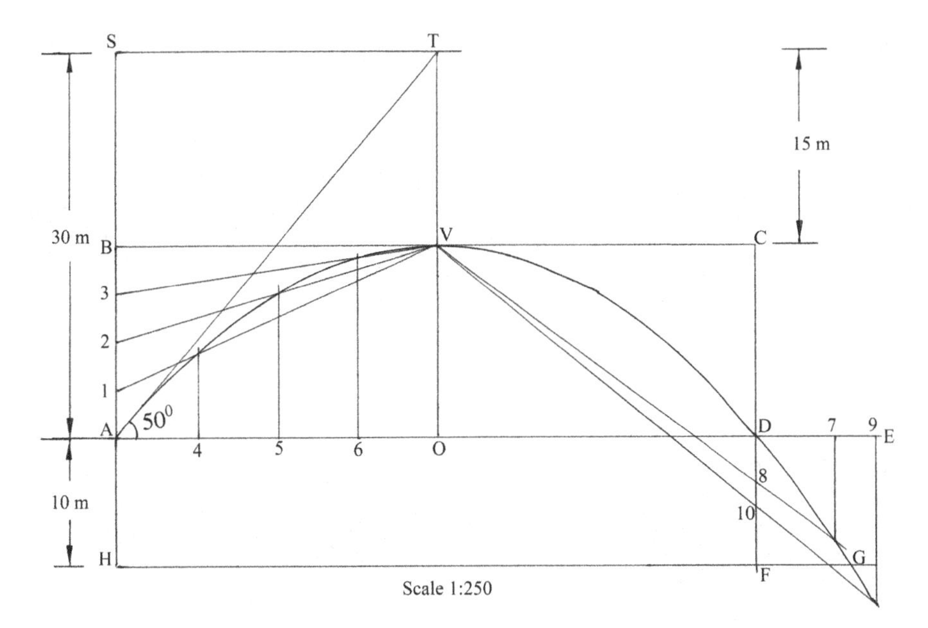

Fig. 4.21 Trajectory of the stone

intersections of lines locate two more points on the parabola. The same procedure is repeated to locate points in the other part of the rectangle. The parabola is to be extended beyond the point D. Extend the line AD up to E and also the line CD up to F. Point 7 is marked on DE such that D7 = A4 on line AO. Point 8 is marked on DF such that D8 = A1 on line AB. A line is drawn from V through point 8. A line is drawn parallel to VO through 7. The point of intersection lies on the parabola. Since the point of intersection is above the ground, the parabola is to be extended further. Point 9 is marked on DE such that D9 = 1.5 times A4. Point 10 is marked on DF such that D10 = 1.5 times A1. A line is drawn from V through point 10. A line is drawn parallel to VO through point 9. The point of intersection lies on the parabola. The parabola is extended through these two points. The horizontal line through H intersects the parabola at G. The horizontal line HG represents range of the stone on the ground, and the distance of G from the building is 58.5 m.

15. The tangents PT and QT of a parabola are, respectively, 120 mm and 60 mm long. P and Q are points of tangency. The distance between P and Q is 100 mm. Draw the parabola.

Draw a horizontal line PQ 100 mm long as shown in Fig. 4.22. With centre P and radius 120 mm, draw an arc. With centre Q and radius 60 mm, draw an arc to intersect the former arc at T. Join PT and QT. PT and QT are tangents to the parabola passing through points P and Q. Divide PT and TQ into the same number of equal parts (six in the present case) and number the division points, respectively, from P and T as shown in the figure. Join the points 1 and 1, 2 and

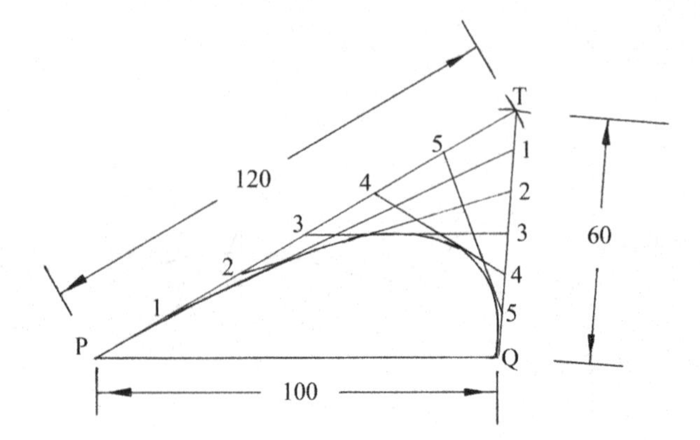

Fig. 4.22 Construction of parabola (tangent method)

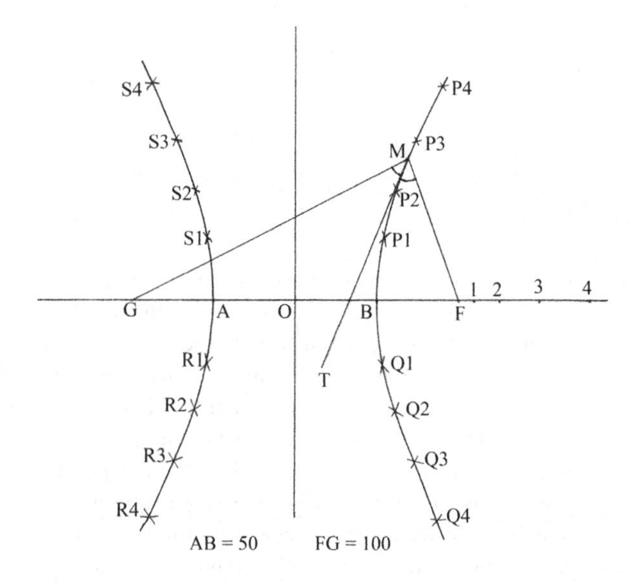

Fig. 4.23 Construction of hyperbola (intersecting arcs method)

2, 3 and 3, etc. Develop a curve tangential to these lines. This curve is the required parabola.

16. A point moves such that the difference of its distances from two fixed points 100 mm apart is always a constant and equal to 50 mm. Draw the locus of the point and name the curve. Draw a tangent to the curve at any point on it.

    The locus of the point describes the curve hyperbola. The difference of the distance is its transverse axis and the fixed points are the foci of the curve.

    Draw AB equal to the transverse axis and extend it on both sides as shown in Fig. 4.23. Locate the centre of the curve at O. Mark F and G equal to the distance between the foci such that FO = GO. On BF produced, mark points 1, 2, 3, 4, etc. arbitrarily. With centres F and G and radius equal to B1, describe arcs of circles. With centres F and G and radius A1, describe arcs of circles to cut the former arcs at four points P1, Q1, R1 and S1. These four points are on

Table 4.1 Co-ordinates of hyperbolae

X	±50.00	±60.00	±70.00	±80.00	±90.00	±100.00
Y	0	±33.17	±48.99	±62.45	±74.83	±86.60

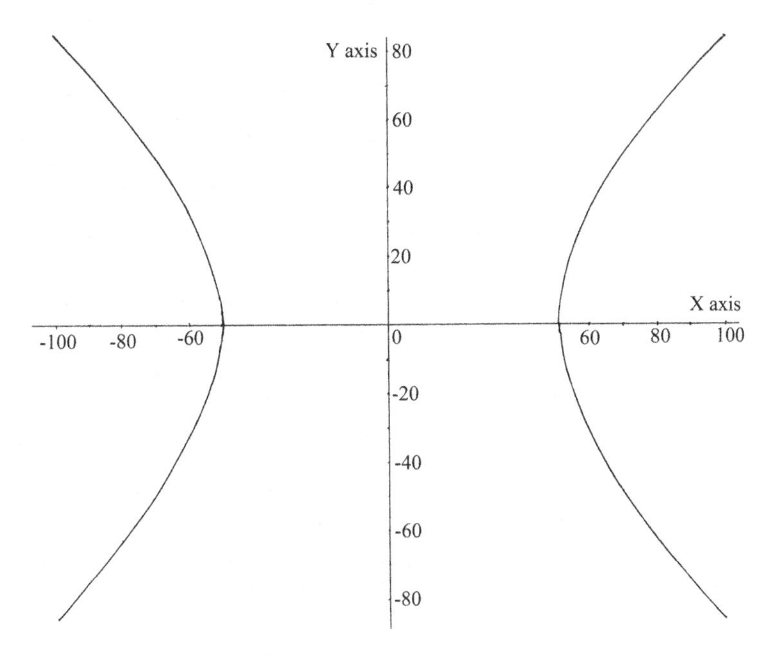

Fig. 4.24 Construction of hyperbolae from co-ordinate points

the curve. The same procedure is repeated for other points 2, 3 and 4. There will be 16 points of intersections. These points are joined to obtain both the branches of hyperbola. Point M is chosen on the curve as shown in the figure. Join MF and MG. The bisector MT of the angle ∠FMG between the focal radii MF and MG is drawn. MT is the required tangent.

17. Trace the curve represented by the equation

$$\left(X^2\right)/(2500) - \left(Y^2\right)/(2500) = 1$$

The above equation represents the curve hyperbola. Table 4.1 furnishes co-ordinates for drawing the curve, and the hyperbolae are drawn as shown in Fig. 4.24.

18. A hyperbola has a transverse diameter of 100 mm, a rise of 50 mm and a span of 180 mm. Draw one branch of the hyperbola. Locate its asymptotes and the focus.

Draw AB equal to the transverse diameter, BC equal to the rise and RQ equal to the span with the given data as shown in Fig. 4.25. Complete the rectangle PQRS as shown in the figure. Divide QC and QP into the same

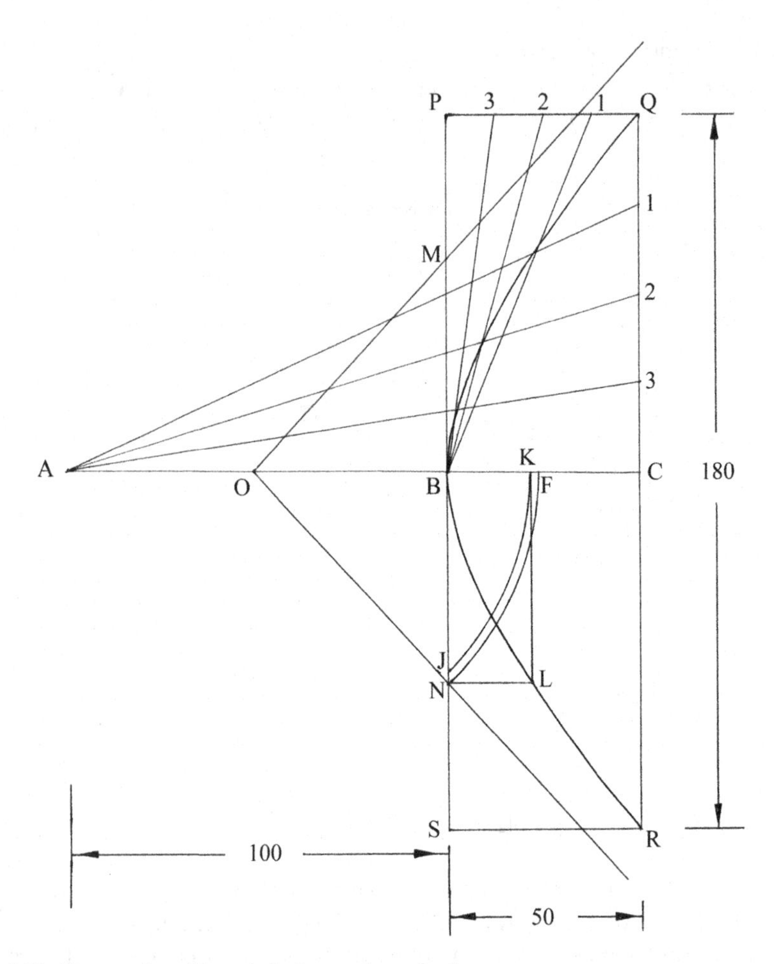

Fig. 4.25 Construction of hyperbola (rectangle method)

number of equal parts (four in the present case) and number the division points from Q. Draw a line from A through point 1 on line QC. Draw a line from B through point 1 on line QP. The point of intersection lies on the curve hyperbola. Draw lines from A through points 2 and 3 on line QC. Draw lines from B through points 2 and 3 on line QP. The intersections of lines locate two more points on the curve. Similarly, points are located in the other part of rectangle BCRS. The centre of the conic O is located at the mid-point of AB. Mark point J on BS such that BJ = BO. With centre O and radius OJ, draw an arc to cut the axis at K. Draw a line perpendicular to the axis through point K to cut the hyperbola at L. Draw a line parallel to the axis through L to cut BS at N. Join ON and extend it. The line ON represents one asymptote, and draw OM the other asymptote such that ∠COM = ∠CON. With centre O and radius ON, draw an arc to cut the axis at F. Point F is the focus of the hyperbola.

19. The angle between the asymptotes of a hyperbola is 60°, and a point P on it is 10 mm from one asymptote and 20 mm from the other. Draw the hyperbola and find its focus.

    Draw a horizontal line ON and a line OM inclined at 60° to ON to represent the asymptotes of the hyperbola as shown in Fig. 4.26. The point P is located such that its perpendicular distances from ON and OM are, respectively, 10 mm and 20 mm. Draw lines AB and CD, respectively, parallel to the asymptotes ON and OM through point P. Draw a radial line through O to cut PA at point 1 and PD at point 1′. Draw a line parallel to OM through point 1 and another line parallel to ON through point 1′. The point of intersection lies on the hyperbola. Similarly, obtain points of intersections by drawing radial lines through O and lines parallel to the asymptotes. A smooth curve is drawn through these points and also the given point P as shown in the figure. Draw the bisector of the angle ∠MON to locate the axis OX. The intersection of the axis with the hyperbola locates the vertex V. Draw a line perpendicular to the axis at V to cut the asymptotes at K and L. With centre O and radius OK, draw an arc to cut the axis at F. Point F is the focus of the hyperbola.
20. A gas is compressed from a volume of 3 m^3 and a pressure of 1 bar to a volume of 0.2 m^3 according to the Boyle's law, pV = constant. Draw the curve of compression and find the final pressure.

    The Boyle's law, pV = constant, represents a hyperbolic curve or more correctly a rectangular hyperbola. The asymptotes of a rectangular hyperbola are mutually perpendicular. The pressure, p, is represented along the Y axis with a scale 10 mm = 2 bar and the volume, V, is represented along the X axis with a scale 10 mm = 0.25 m^3. Let OX and OY be the asymptotes with the included angle 90°.

    Draw the asymptotes OX and OY representing, respectively, the volume and pressure of the gas as shown in Fig. 4.27. The initial point P is located for a

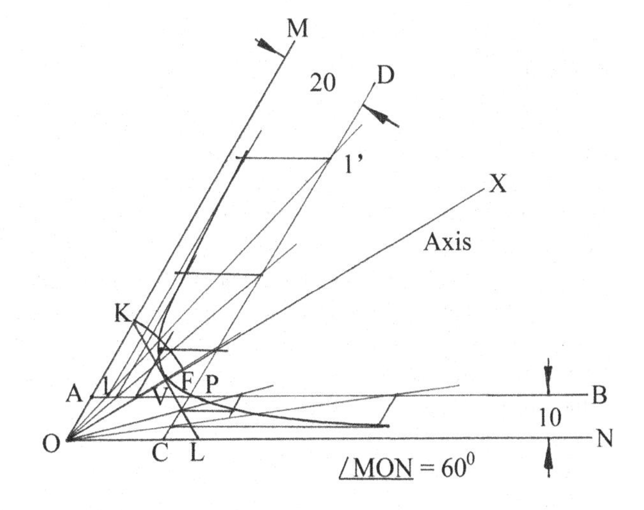

Fig. 4.26 Construction of hyperbola from its asymptotes

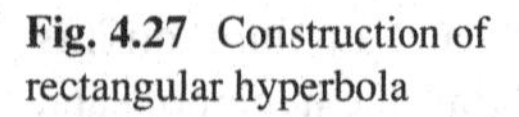
Fig. 4.27 Construction of rectangular hyperbola

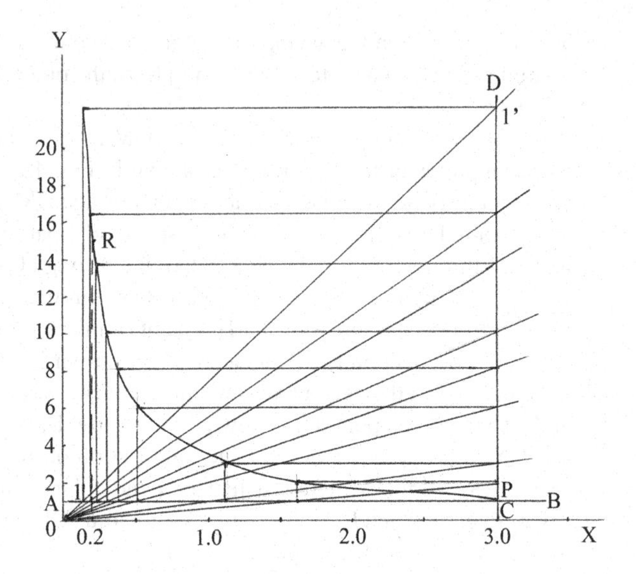

pressure of 1 bar and a volume of 3 m^3. Draw lines AB and CD, respectively, parallel to the asymptotes OX and OY. Draw a radial line through O to cut line PA at point 1 and line PD at point 1′. Draw a line parallel to OY through point 1 and another line parallel to OX through point 1′. The point of intersection lies on the curve hyperbola. Similarly, obtain points of intersections by drawing radial lines through O and lines parallel to the asymptotes. A smooth curve is drawn passing through the points of intersections. The rectangular hyperbola represents the curve of compression. The final pressure is obtained by drawing a line parallel to the asymptote OY through the volume line 0.2 m^3 to cut the hyperbola at R. The final pressure is 15 bar.

Cycloidal Curves

Cycloidal curves are generated by points on the circumference of a rolling circle when the circle rolls along a straight line or on another circle. The rolling circle is called a generating circle. The straight line on which the circle rolls is called the directing or base line. The circle on which the generating circle rolls is called the directing or base circle. When the circle rolls on a straight line the curve obtained is a cycloid. An epicycloid is the curve generated by a point on the circumference of a circle when the circle rolls in a plane on the outside of another circle. A hypocycloid is the curve generated by a point on the circumference of a circle that rolls in a plane on the inside of another circle.

Solved Problems

21. A circle of 50 mm diameter rolls along a straight line without slipping. Draw the curve generated by a point P on the circumference of a circle for one complete revolution. Draw also a tangent to the curve at any point on it.

    Draw a circle of diameter 50 mm as shown in Fig. 4.28. Draw the base line AB tangential to the circle and equal in length to its circumference. Let P be the generating point on the rolling circle and O centre of the circle in the initial position. Draw OQ parallel to AB and equal to it in length. Divide OQ and the circle into eight equal parts and mark the division points as shown in the figure. Draw horizontal lines through the points 1, 2, 3, etc. With C1 as centre and 25 mm as radius, describe an arc to intersect the horizontal line through point 1 on the circle. The point of intersection lies on the curve cycloid. The same procedure is repeated with centres C2, C3, C4, etc. and 25 mm as radius to get other points on the curve. A smooth curve is drawn passing through these points of intersections to get the required curve cycloid. Point R is chosen on the curve to draw the tangent to the curve. With centre R and 25 mm as radius, draw an arc to intersect the line OQ at C. Draw a line perpendicular to AB through C. This line intersects AB at N. Join RN and draw a line RT perpendicular to RN. RT is the tangent and RN is the normal to the cycloid at point R.

22. A circle of 40 mm diameter rolls over a straight line without slipping. In the initial position, the diameter RS of the circle is parallel to the line on which the circle rolls. Draw the locus of points R and S for one complete revolution of the circle.

    Draw a circle of diameter 40 mm as shown in Fig. 4.29. Draw the base line AB tangential to the circle and equal in length to its circumference. Let O be the centre of the circle in the initial position. Draw OQ parallel to AB and equal to its length. Divide OQ and the circle into eight equal parts and mark the division points as shown in the figure. RS represents the initial position of the diameter parallel to AB. Point R coincides with point 2 and point S coincides with point 6. Draw horizontal lines through the points 1, 2, 3, 4, etc. With centre C1 and radius 20 mm, draw an arc to intersect the horizontal line through point 3. The intersecting point locates the position of R when the circle centre occupies the

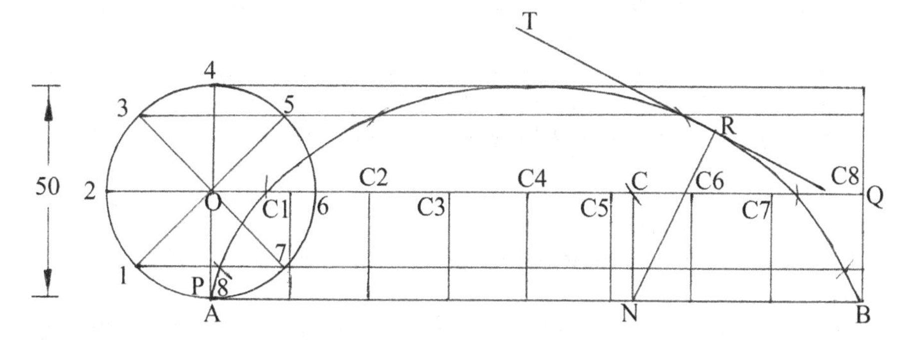

Fig. 4.28 Construction of cycloid

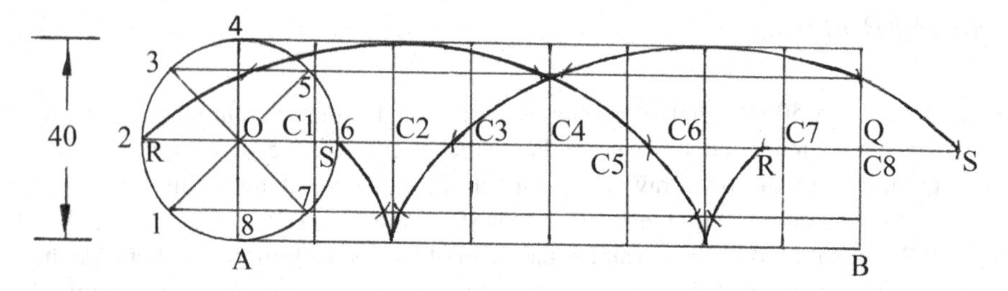

Fig. 4.29 Construction of cycloids

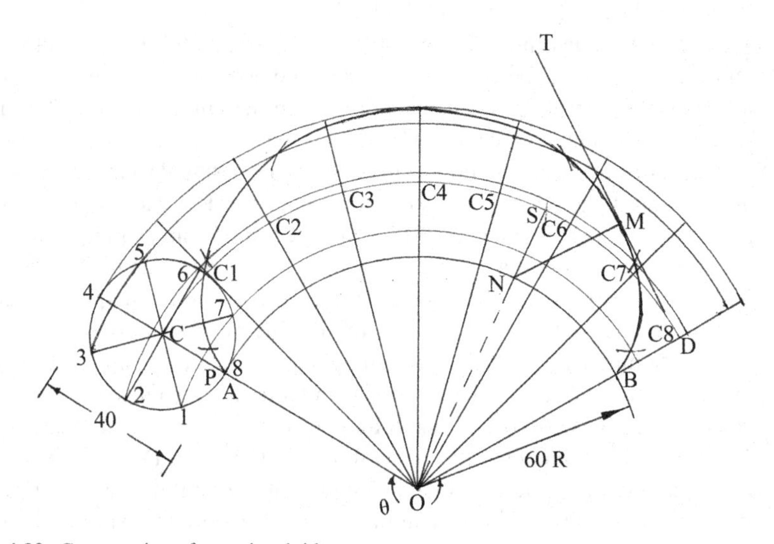

Fig. 4.30 Construction of an epicycloid

position C1. With the same centre C1 and radius 20 mm, draw an arc to intersect the horizontal line through point 7 locating the position of S. The same procedure is repeated with centres C2, C3, C4, etc. to locate the positions of points R and S. Two smooth curves are drawn passing through the two sets of the points of intersections. The curves obtained are cycloids.

23. A circle of 40 mm diameter rolls on another circle of 120 mm diameter. Draw an epicycloid and also a tangent and normal at any point on the curve.

    For one complete revolution of generating circle of radius r, the arc length travelled by the generating circle on the base circle of radius R would be equal to the circumference of the generating circle. The angle subtended by an arc of base circle at its centre is given by

$$\theta = (2\Pi \mathrm{r}) \times (360)/(2\Pi \mathrm{R}) = (360 \times \mathrm{r})/\mathrm{R}$$
$$\text{For } \mathrm{r} = 20 \text{ mm and } \mathrm{R} = 60 \text{ mm}, \theta = (360 \times 20)/(60) = 120^{\circ}$$

Draw an arc AB of base circle of radius (R) 60 mm such that the angle ∠AOB is 120° (Θ) as shown in Fig. 4.30. Produce OA to C such that AC is equal to the radius (r) of the rolling circle, 20 mm. Draw the rolling circle with centre C. With centre O and radius OC, draw an arc CD specifying the locus of centre of the rolling circle. Point P lies initially at A. Divide the rolling circle and arc CD into eight equal parts. Mark the division points as shown in the figure. With centre O describe concentric arcs through points 1, 2, 3, etc. With centre C1 and radius 20 mm (r), draw an arc to cut the concentric arc through point 1. The point of intersection lies on the curve epicycloid described by the locus of point P. The same procedure is followed to get the remaining points of intersections. A smooth curve is drawn passing through these points. Point M is chosen on the curve. With centre M and radius 20 mm (r), draw an arc to intersect the arc CD at S. Join S to O. This cuts the base line AB at N. Join M to N and draw line MT perpendicular to MN at M. MT is the tangent and MN is the normal to the epicycloid at point M.

24. Draw a hypocycloid whose diameter of rolling circle is 60 mm and the diameter of the base circle is 180 mm. Draw a tangent and normal at any point on the curve.

    For one complete revolution of the rolling circle of radius r, the arc length travelled by the rolling circle on the base circle of radius R would be equal to the circumference of the rolling circle. The angle subtended by the arc of the base circle at its centre is given by

$$\theta = (2\Pi \mathrm{r}) \times (360)/(2\Pi \mathrm{R}) = (360 \times \mathrm{r})/\mathrm{R}$$
$$\text{For } \mathrm{r} = 30 \text{ mm and } \mathrm{R} = 90 \text{ mm}, \theta = (360 \times 30)/(90) = 120^\circ$$

Draw an arc AB of base circle of radius 90 mm such that the angle ∠AOB is equal to 120° as shown in Fig. 4.31. Locate point C on OA such that AC is equal to the radius of rolling circle, 30 mm. Draw the rolling circle with centre C. With centre O and radius OC, draw an arc CD specifying the locus of centre of the

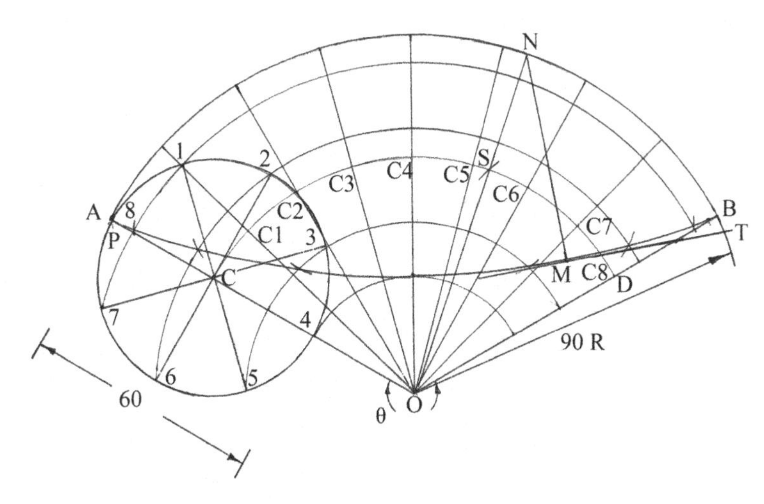

Fig. 4.31 Construction of a hypocycloid

rolling circle. Point P lies initially at A. The rolling circle rolls in counter-clockwise direction. Divide the rolling circle and the arc CD into eight equal parts and mark the division points as shown in the figure. With centre O, describe concentric arcs through points 1, 2, 3, etc. With centre C1 and radius 30 mm (r), describe an arc cutting the concentric arc through point 1. The point of intersection lies on the curve hypocycloid described by the locus of the point P. The same procedure is repeated to get the remaining points of intersections. A smooth curve is drawn passing through these points. With centre M and radius 30 mm (r), draw an arc to intersect the arc CD at S. Join O to S and produce it to cut the base line AB at N. Join M to N and draw line MT perpendicular to MN at M. MT is the tangent and MN is the normal to the hypocycloid at point M.

25. Draw the hypocycloid when the diameter of the directing circle is twice the diameter of the rolling circle. Take the diameter of the rolling circle as 70 mm.

    The angle subtended by the arc of the directing circle at its centre for one revolution of the rolling circle is given by

$$\theta = (r) \times (360/R)$$

When $r = 35$ mm and $R = 2 \times 35 = 70$ mm, $\theta = (35) \times (360/70) = 180^{\circ}$

Draw a semi-circle AB of radius 70 mm as shown in Fig. 4.32. Locate point C on OA such that AC is equal to 35 mm. Draw the rolling circle with centre C. With centre O and radius OC, draw an arc CD specifying the locus of the centre of the rolling circle. Point P lies initially at A. The rolling circle rolls in counter-clockwise direction. Divide the rolling circle and the arc CD into eight equal parts and mark the division points as shown in the figure. With centre O, describe concentric arcs through points 1, 2, 3, etc. With centre C1 and radius 35 mm (r), describe an arc cutting the concentric arc through point 1. The point of intersection lies on the curve hypocycloid described by the locus of the point P. The

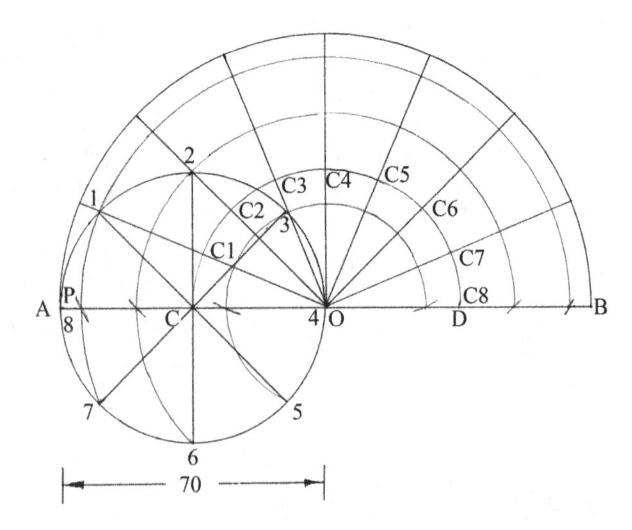

Fig. 4.32 Construction of a hypocycloid

same procedure is repeated to get the remaining points of intersections. All points lie on the diameter of the directing circle. The hypocycloid is a straight line when the diameter of the directing circle is twice the diameter of the rolling circle.

Involute

The spiral curve traced by a point on a thread as it unwinds from around a polygon or a circle keeping the thread always taut is an involute curve. The involute of a polygon is obtained by extending the sides and drawing arcs using the corners as centres. The circle may be considered to be a polygon having an infinite number of sides.

Solved Problems

26. Draw one turn of an involute of an equilateral triangle of side 20 mm.

    Draw the equilateral triangle ABC of side 20 mm and extend its sides as shown in Fig. 4.33. With centre A and radius AB, draw an arc to intersect CA produced at 1. With centre C and radius C1, draw an arc to intersect BC produced at 2. With centre B and radius B2, draw an arc to intersect AB produced at 3. The curve B-1-2-3 consisting of arcs is the required involute of the equilateral triangle.
27. Draw an involute of a square of side 20 mm.

    Draw a square ABCD of side 20 mm and extend its sides as shown in Fig. 4.34. With centre A and radius AB, draw an arc to intersect DA produced at point 1. With centre D and radius D1, draw an arc to intersect CD produced at point 2. With centre C and radius C2, draw an arc to intersect BC produced at

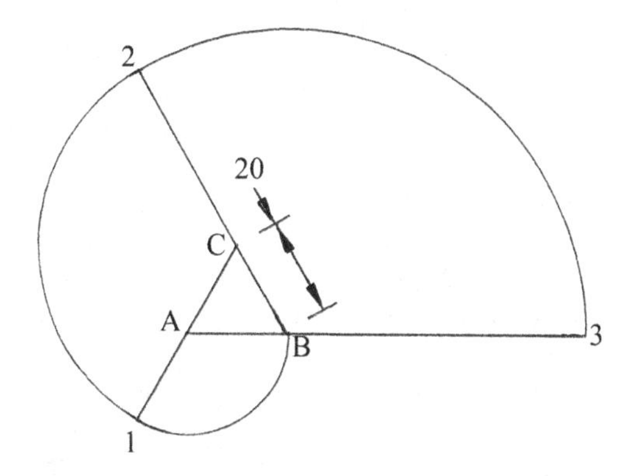

Fig. 4.33 Involute of an equilateral triangle

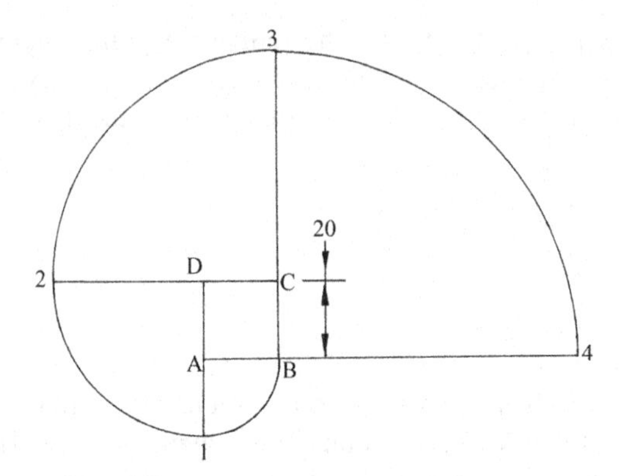

Fig. 4.34 Involute of a square

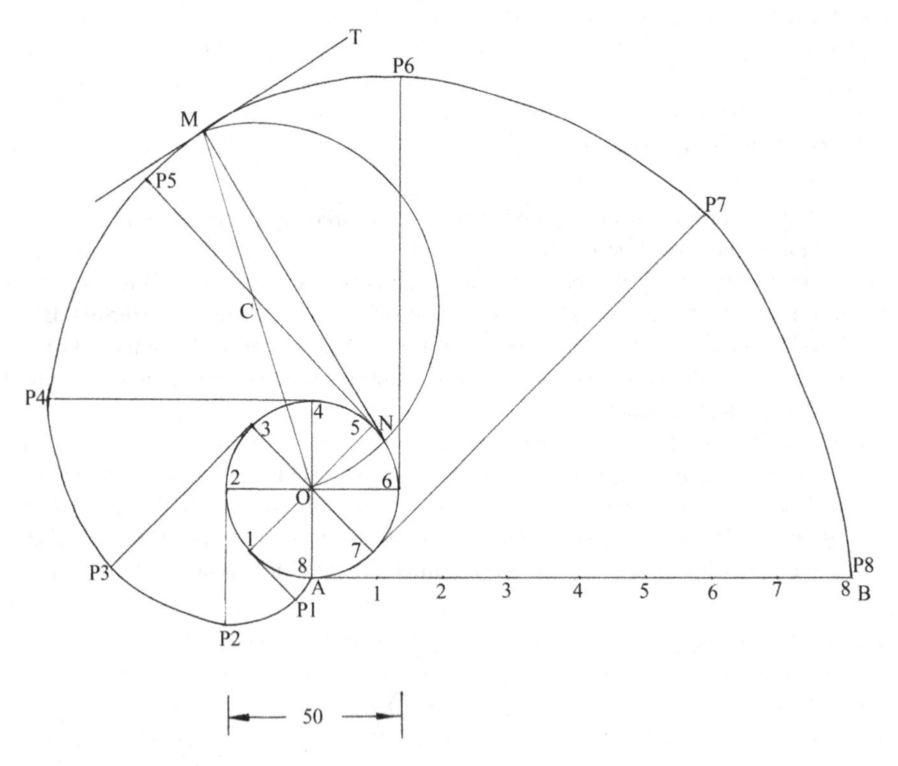

Fig. 4.35 Involute of a circle

point 3. With centre B and radius B3, draw an arc to intersect AB produced at point 4. The curve B-1-2-3-4 is the involute of a square.

28. Draw an involute of a circle 50 mm diameter and also tangent and normal to this involute at a point on the curve 100 mm from the centre of the circle.

    The length of the circumference of the circle of diameter 50 mm is found graphically as given in Fig. 2.19. Draw the circle, centre O and the base line AB tangential to the circle and equal in length to its circumference as shown in Fig. 4.35. Divide the line AB into eight equal parts. Divide the circumference of the circle into eight equal parts and mark the points 1, 2, 3, etc. and at these points draw tangents to the circle. Mark along the tangent drawn at point 1 a length equal to A1 (one-eighth of the length AB) giving point P1. Mark along the tangent drawn at point 2 a length A2 (two-eighth of AB) giving point P2. The same procedure is followed to get the remaining points P3, P4, P5, P6, P7 and P8. A smooth curve is drawn passing through these points. This curve is the involute of the circle. Mark point M on the curve 100 mm from O, the centre of the circle. Join MO and mark the mid-point of MO at C. With centre C and radius MC, describe a semi-circle which cuts the circle at N as shown in the figure. Join MN and draw a line MT perpendicular to MN. MT is the tangent and MN is the normal to the involute at point M.

29. AB the diameter of a circle is 80 mm. A piece of string is tied tightly round the circumference of the semi-circle, starting from A and finishing at B. The string is unwound from B until it lies along the tangent at A. Trace the path of the moving end of the string.

 The length of half the circumference of the circle of diameter 80 mm is found graphically as given in Fig. 2.19. Draw the semicircle, centre O and the vertical tangential line AB at A equal to half the circumference of the semi-circle as shown in Fig. 4.36. Divide the line AB into four equal parts. Divide the circumference of the semi-circle into four equal parts and mark the points 1, 2 and 3. Draw tangents at these points. Mark along the tangent drawn at point 1 a length equal to B1 giving the point P1. Mark along the tangent drawn at point

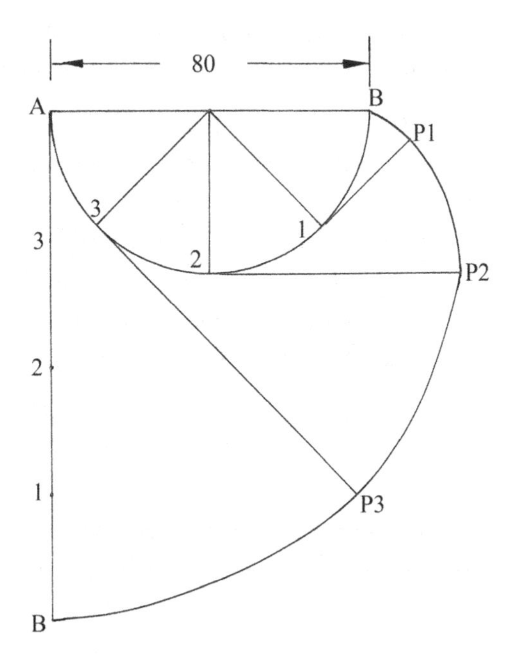

Fig. 4.36 Involute of a semi-circle

2 a length equal to B2 giving the point P2. Mark along the tangent drawn at point 3 a length equal to B3 giving the point P3. A smooth curve is drawn passing through the points B-P1-P2-P3-B giving the path of the moving end B. The curve obtained is the involute of the semi-circle.

Spiral Curves

Assume a line rotates about one of its ends and a point moves continuously in one direction along the line. The locus of the moving point is a spiral. The point about which the line rotates is called the pole or axis. The part of the line between the moving point and the pole is called the radius vector, and the angle between the line and its initial position is called the vectorial angle. The moving point traces one convolution for each complete revolution of the line.

Archimedean Spiral

It is the locus of a moving point along a line while the line rotates at uniform angular velocity in such a way that equal increases in the angular displacement accompany equal movements of the point towards or away from the fixed point called pole.

30. Draw one convolution of an Archimedean spiral, least radius 20 mm and the greatest radius 80 mm. Draw the tangent and normal at a point on the curve 60 mm from the pole.

    The initial radius OP (least radius, 20 mm) and the final radius OQ (greatest radius, 80 mm) are drawn as shown in Fig. 4.37. The line PQ is divided into 12 equal parts. Draw 12 equidistant radial lines OA, OB, OC, etc. through O starting from OP. With centre O and radius O1, draw an arc to cut the radial line OA at point P1. With centre O and radius O2, draw an arc to cut the radial line OB at point P2. The same procedure is followed to obtain other points P3, P4, P5, P6, - - - P12. A smooth curve is drawn passing through the points P, P1, P2, P3, - - - P12 (Q). The curve represents one convolution of an Archimedean spiral.

    The polar equation of the curve is given by

$$r = a + c\,\theta \tag{4.1}$$

where r is the radius vector, a is the initial radius vector or the least radius and c is a constant. When $\theta = 2\Pi$, $r = 80$, $a = 20$, the constant c is obtained from equation (4.1)

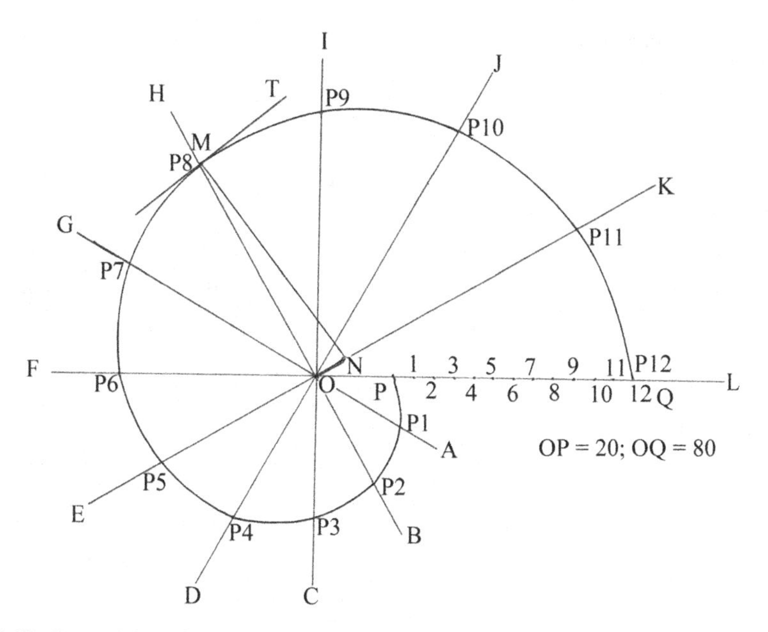

Fig. 4.37 Construction of an Archimedean spiral

$$c = (80 - 20)/(2\Pi) = 9.5\,\text{mm}$$

Point M is chosen 60 mm from O. Points M and P8 coincide in this case. Draw a perpendicular to the radial line OH at O and locate point N, 9.5 mm from O as shown in the figure. Join MN and draw MT perpendicular to MN. MT is the tangent and MN is the normal to the curve.

31. A circular disc of diameter AB = 120 mm rotates about its centre with uniform angular velocity. During one complete revolution, a point P moves along the diameter AB from A to B. Draw the path traced by the point P.

    Draw the circular disc of diameter 120 mm and locate its diameter AB as shown in Fig. 4.38. The point P is initially at A and moves towards B along the diameter AB and reaches B during one revolution of the disc. Divide the diameter AB into 12 equal parts and number the division points from A. Draw 12 equidistant radial lines A1, A2, A3, etc. as shown in the figure. Locate point P1 along the radial line A1 such that the distance of P1 from A1 is equal to the distance of point 1 from A. The same procedure is followed to locate points P2, P3, P4, etc. A smooth curve is drawn passing through the points P, P1, P2, - - - P12 to show the locus of the point P which is the required curve Archimedean spiral. It is to be noted that the point P moves towards the pole (centre of the circular disc) during the first half of the revolution and moves away from the pole during the second half of the revolution.

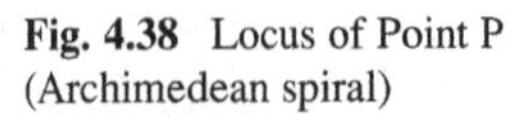

Fig. 4.38 Locus of Point P (Archimedean spiral)

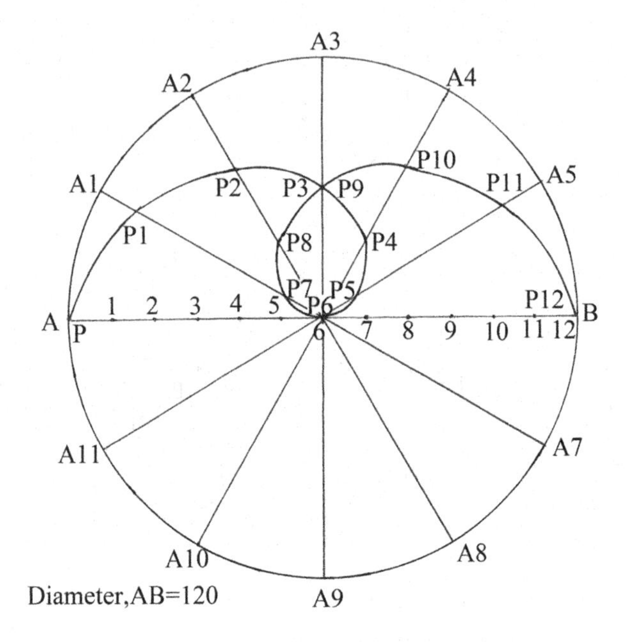

Logarithmic Spiral

Logarithmic spiral is the curve traced by a point moving along a line such that for equal angular displacements of the line the ratio of the lengths of consecutive radius vectors is constant.

32. The shortest radius vector in a logarithmic spiral is 15 mm. The lengths of adjacent radius vectors enclosing an angle of 40° are in the ratio 5:4. Construct one convolution of the curve and determine α, the angle tangent makes with the radius vector at any point on the curve.

    A logarithmic scale is constructed to determine the lengths of successive radius vectors. Draw two lines AB and AC enclosing an angle of 40° as shown in Fig. 4.39. Along AB locate point 0 at radius r0 (15 mm) and along AC locate point 1′ at radius r1 ((15 × 5)/4 = 19 mm). Join point 0 to 1′. Transfer point 1′ to point 1 on AB by circular arc with centre A. At point 1 draw a line parallel to 01′ and locate point 2′ on AC. Transfer point 2′ to point 2 on AB by circular arc with centre A. Points 3, 4, 5, 6, 7, 8 and 9 are located on AB following the same procedure. Points 0, 1, 2, 3, 4, 5, etc. will be at distances of r0 (shortest radius), r1, r2, r3, r4, etc. from A.

    Draw OA equal to the shortest radius vector (r0) as shown in Fig. 4.40 for the construction of one convolution of the logarithmic spiral. Draw nine equidistant radial lines OC, OD, OE, OF, OG, OH, OI, OJ and OK through O. With centre O and radius r1, draw an arc to cut the radial line OC at point P1. With centre O and radius r2, draw an arc to cut the radial line OD at P2. The same procedure is followed to locate other points P3, P4, P5, etc.

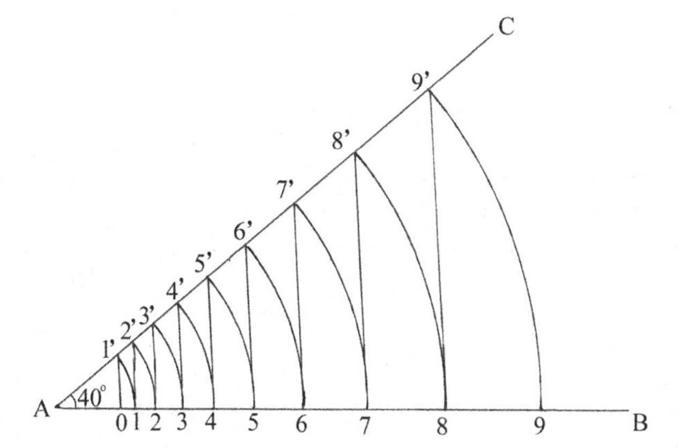

Fig. 4.39 Logarithmic scale

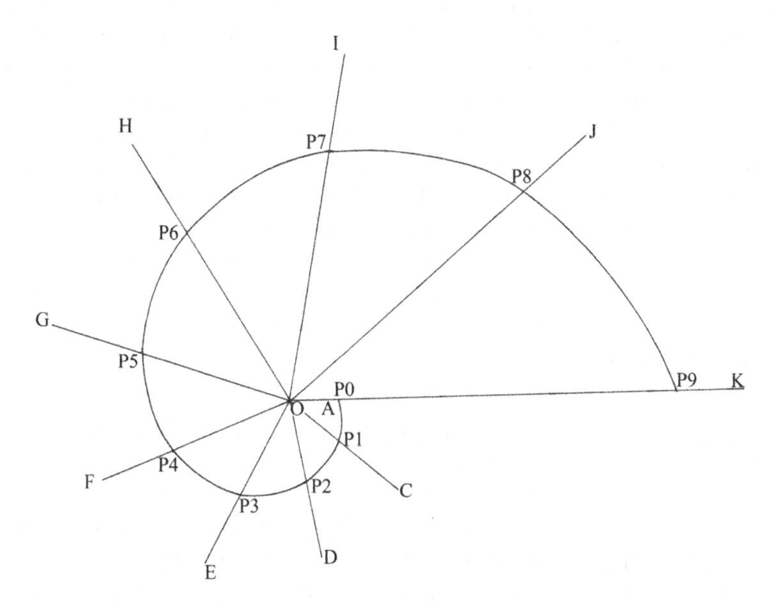

Fig. 4.40 Logarithmic spiral

A smooth curve is drawn passing through these points. This curve represents one convolution of logarithmic spiral. The angle α which the tangent makes with the radius vector at any point is constant. Hence, the logarithmic spiral is also called equiangular spiral. The value of tan (α) is given by the ratio $\log_{10}(e) \div \log_{10}(a)$, where $e = 2.718$ and $\log_{10}(a) = \{(180)/(\prod \times 40)\} \log_{10}(5/4)$.

Hence, the angle $\alpha = 72^{\circ}$

Note: The construction of helical curves and helical springs are explained in Chap. 13.

Applications

The curve ellipse finds its application in the construction of arches of stone and concrete bridges in the form of semi-ellipses. The manholes of boilers and flanges of pipes are generally designed to be elliptical in shape. Elliptic gears are used in certain kinds of machinery to obtain varying speed rate. The orbits in which planets revolve about the sun are ellipses. Reflector of head lamps of motor car is made in the shape of parabola (paraboloid of revolution). Parabolic reflectors are used in solar concentrators. Any material point moving with an acceleration constant in value and direction (e.g. in the gravity field) will describe a parabola and, hence, this curve finds its application in ballistics. The liquid level contained in a vessel rotating round a vertical axis with constant velocity takes the shape of a parabola. The curve hyperbola finds its application in the construction of natural draft cooling towers in thermal power plants. Another application of the hyperboloid of revolution is in the design of skew gearing to transmit power between two non-parallel, non-intersecting shafts. Cycloidal curves are used for tooth profiles of small gears. Involute is employed in the design of gear tooth profiles as it is easy to manufacture. Involute also finds its application in the design of the casings of blowers and centrifugal pumps. The heart-shaped disc used on sewing machines in connection with the coiling of the spool consists of two spirals of Archimedean. Spiral springs are used in clocks for storing the mechanical energy and supply it when required. Insects approach a burning candle along a logarithmic spiral.

Practice Problems

1. Draw an ellipse when the distance of the focus from the directrix is equal to 45 mm and the eccentricity is 4/5. Draw the tangent and normal at any point on the curve.
2. Two points P and Q are 120 mm apart. The third point R is 80 mm from P and 90 mm from Q. Draw an ellipse passing through P, Q and R.
3. The major axis of an ellipse is 100 mm long and the minor axis is 80 mm long. Draw one-half of the ellipse by rectangle method and the other half by concentric circles method.
4. The major and minor axes of an ellipse are, respectively, 120 mm and 100 mm. Draw the ellipse and locate its foci.
5. The sides of a parallelogram measure 120 mm and 100 mm and the included angle is 60°. Inscribe an ellipse and determine the lengths of the major and minor axes.
6. A cricket ball is thrown and reaches a maximum height of 20 m and falls on the ground at a distance of 40 m from the point of projection. Draw the path of the cricket ball and name the curve. Determine the angle of projection.

7. The sides of a rectangle are 100 mm and 80 mm. Inscribe two parabolas with axes perpendicular to each other.
8. A suspension bridge has a total span of 20 m. The dip at the centre is 2 m. Draw the profile of the bridge and also tangents to the curve at the two ends. Adopt a scale 1:100
9. Draw the curve given by the equation $Y^2 = 16X$. Draw the tangent to this curve at any point on it and also locate its focus.
10. Draw the curve representing the shape of the principal section of the parabolic motor car head lamp reflector having an aperture of 140 mm and a depth of 70 mm.
11. The focus and the corresponding vertex of a hyperbola are 30 mm and 10 mm distant from the directrix. Draw the curve and also the tangent and normal at any point on the curve.
12. Draw both branches of a hyperbola, given eccentricity = 3/2 and focus 40 mm from the directrix. Mark the second directrix and focus. Draw the asymptotes and measure the angle between them.
13. Draw a rectangular hyperbola, given that a point on the curve is 20 mm from one asymptote and 30 mm from the other.
14. The asymptotes of a hyperbola intersect at 70° and at a distance of 40 mm from the vertices of both branches of hyperbola. Draw one branch of the hyperbola and locate its focus.
15. Draw a cycloid, given rolling circle diameter 60 mm. Draw the tangent and normal for a point 40 mm from the base line and on the descending side of the curve.
16. A circle of 40 mm diameter rolls for the first half of a revolution on a horizontal line and the second half of the revolution on a vertical line. Draw the curve traced out by the point P which is initially at the point of contact of the horizontal line and the circle.
17. A circle of 50 mm diameter rolls on another circle of 50 mm diameter. Draw the epicycloid.
18. A circus man rides a motor-cycle inside a globe of 4 m diameter. The motor-cycle wheel diameter is 0.8 m. Draw the locus of a point spot on the circumference of the motor-cycle wheel for its one complete turn. Adopt a suitable scale.
19. Draw an involute of a hexagon of side 20 mm.
20. Trace the paths of the ends of a straight line AP 100 mm long when it rolls without slipping on a semi-circle of 80 mm diameter. Assume the line AP to be the tangent to the semi-circle in the starting position.
21. Draw two convolutions of an Archimedean spiral, least radius 10 mm and greatest radius 58 mm. Also draw the tangent and normal at a point on the curve 40 mm from the pole.
22. The shortest radius vector in a logarithmic spiral is 10 mm. The lengths of adjacent radius vectors enclosing 30° are in the ratio 10:9. Draw one convolution of the curve.

23. Draw a series of triangles of base 100 mm and perimeter 240 mm and plot the curve passing through the vertices. [Hint: The curve passing through the vertices of the triangle is an ellipse. The base points of the triangle are the foci of the ellipse and the length of the major axis of the ellipse is 140 mm].
24. AP, a chord of a parabola, is 50 mm long and is inclined at 50° to the axis. A is the vertex of the parabola. Draw the curve parabola.
25. PNP, a double ordinate of a parabola, is 90 mm long. A is the vertex of the parabola. The area bounded by the curve PAP and the double ordinate PNP is 12,000 mm^2. Draw the curve. [Hint: Area bounded by the parabola PAP and the double ordinate PNP is given by the formula, Area = (2/3) X (axis or abscissa) X (double ordinate)].
26. A toy rocket thrown up in the air reaches a maximum horizontal range of 50 m. Trace the path of the rocket choosing a suitable scale. [Hint: The path of the rocket in the air is a parabola. The angle of projection for maximum horizontal range is 45° to the ground. The tangent to the curve at the point of projection makes an angle of 45° to the ground.]
27. The transverse and conjugate axes of a hyperbola are 80 mm and 60 mm, respectively. Draw the hyperbola and locate its focus.
28. For a perfect gas, the relation between the pressure p and volume V in an isothermal expansion is given by pV = constant. Draw the curve of isothermal expansion of an enclosed volume of gas if the pressure of 0.2 m^3 of gas is 40 bar. Determine the final volume of the gas for a pressure of 2 bar.
29. ABC is an equilateral triangle of side 60 mm. Trace the loci of the vertices A, B and C when the circle circumscribing circle rolls without slipping along a fixed straight line for one complete revolution.
30. Trace the simultaneous locus of the two extremities of initially vertical diameter of a circle of diameter 50 mm, which rolls on a straight line without slipping.
31. A circle of 60 mm diameter rolls without slipping on the outside of another circle of diameter 240 mm. Show the path of a point on the periphery of the rolling circle diametrically opposite to the initial point of contact between the circles.
32. Draw a hypocycloid developed by a rolling circle of diameter 40 mm on a base circle of diameter 240 mm Draw also a tangent and normal at any point on the curve.
33. A rod 240 mm long rolls without slipping on the periphery of a disc of diameter 70 mm. Trace the locus of the two ends of the rod.
34. A spider is sitting at the centre of second's hand of a wall clock. Trace and name the curve followed by the spider if it moves radially out such that it travels 60 mm during 1 min time.

Chapter 5
Orthographic Projections

An engineer must be able to visualize in his mind how an object looks like without actually having the object. He must also be able to describe the object so that others could build it from the information provided on his drawing. The description of an object with lines requires a thorough knowledge of the principles of orthographic projections.

The problem of representing a three-dimensional solid object on a sheet of paper to show the exact shape is done by drawing views of the object as seen from different positions and arranging these views in a systematic manner.

A rectangular prism is viewed in three directions as shown in Fig. 5.1. Views in each of these directions are shown and named as in Fig. 5.2.

A cylinder is viewed in three directions as shown in Fig. 5.3. Views of the cylinder are shown and named as in Fig. 5.4. The plan shows the circular shape of the cylinder. The elevation and side elevation appear as rectangles and are of the same size.

The cylinder is placed on the prism, and these parts are viewed as shown in Fig. 5.5. Complete views composed of views of separate parts are shown in Fig. 5.6.

Orthographic Projection

The use of different views to describe an object, as seen previously, is based upon the principles of orthographic projection. Ortho- means 'straight or right angles' and -graphic means 'written or drawn'. Projection comes from two Latin words, 'pro' meaning 'forward' and 'jacere' meaning 'to through'. Thus, orthographic projection means 'through forward, drawn at right angles'. The following definition has been given: Orthographic projection is the method of representing the exact form of an object in two or more views on planes generally at right angles to each other, by dropping perpendiculars from the object to the planes.

K. Rathnam, *A First Course in Engineering Drawing*,
DOI 10.1007/978-981-10-5358-0_5

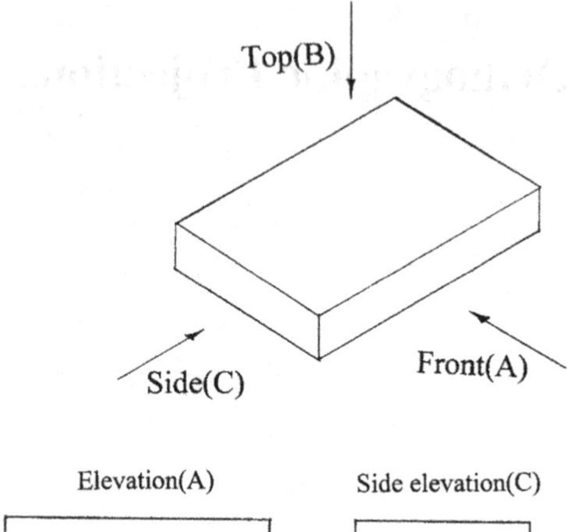

Fig. 5.1 Rectangular prism

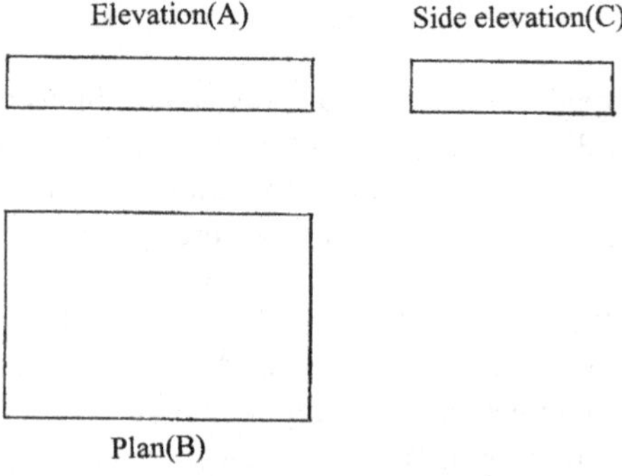

Fig. 5.2 Views of the prism

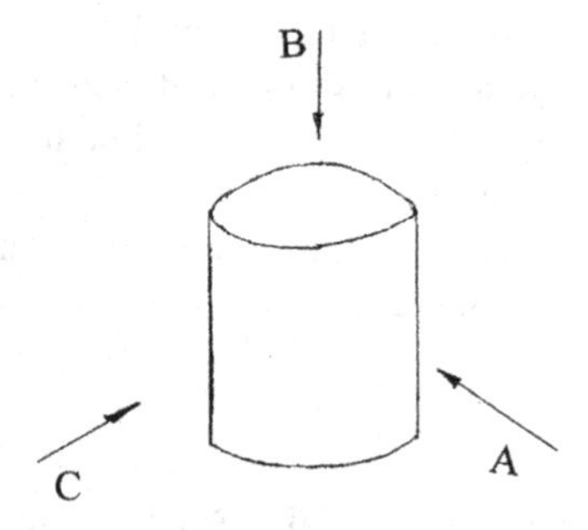

Fig. 5.3 Cylinder

Planes of Reference

Two principal planes are used in orthographic projection, one horizontal and the other vertical, intersecting and dividing space into four quadrants numbered as in Fig. 5.7. These planes are denoted as the horizontal plane (HP) and the vertical plane (VP). The line of intersection of the planes of projection is called as the reference line xy. The object is supposed to be situated in one of the quadrants. It is represented by its orthographic projection on the HP and VP, the views giving, respectively, its plan and elevation. To show these views on a plane surface, the HP

Fig. 5.4 Views of the cylinder

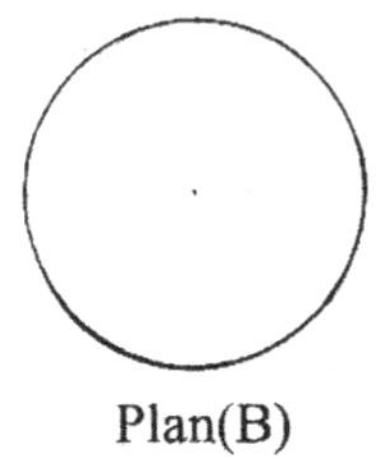

Fig. 5.5 Combination of solids

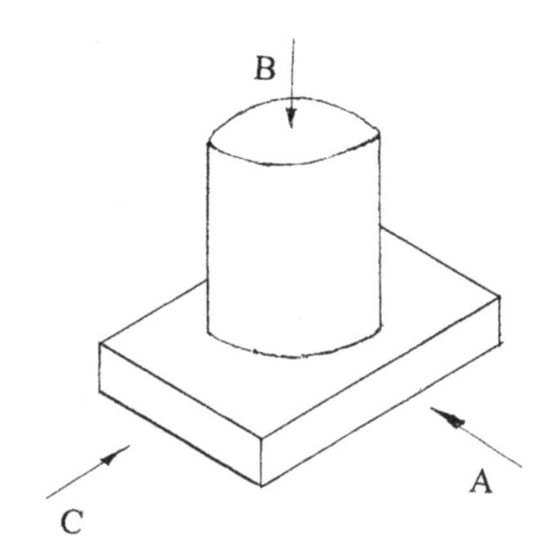

is rotated about xy line until it coincides with the VP. The convention is that the first quadrant is always opened out. For projection in the second, third and fourth quadrants, the planes are assumed to be transparent and the projections are drawn looking from the first quadrant. For practical purposes, the first and third quadrants only are used for the projection; in the second and fourth quadrants, the views overlap and cause confusion. An auxiliary vertical plane perpendicular to both the horizontal plane and vertical plane is also used to get a third view of the object. The third view can also be obtained from the two views by applying the principles of auxiliary projections. This topic will be covered later in Chap. 9.

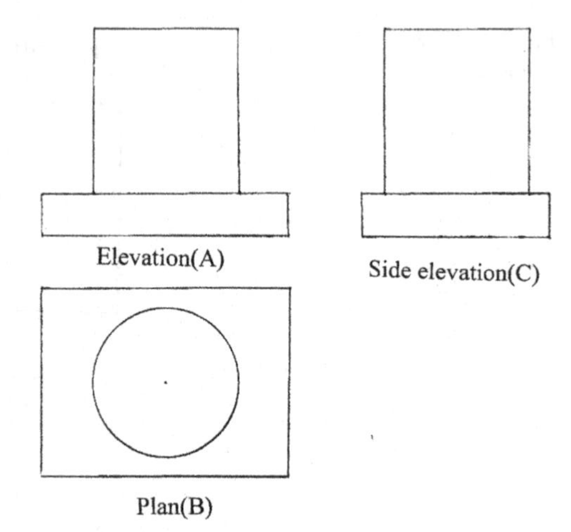

Fig. 5.6 Views of the solids

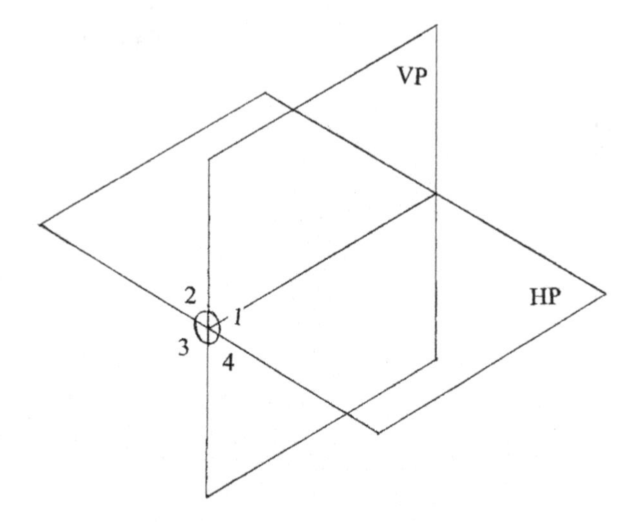

Fig. 5.7 Pictorial views of the principal planes of projection

First Quadrant Projection (Figs. 5.8 and 5.9)

The projection on the HP is obtained by projectors drawn perpendicular to the HP from the corners of the object as shown in Fig. 5.8. This will give the horizontal projection or the plan or the top view of the object. This view will also show the depth (D) and width (W) of the object. The projection on the VP is obtained by drawing projectors perpendicular to the VP from the corners of the object as shown in Fig. 5.8. This will give the vertical projection or the elevation or the front view of the object with its height (H) and width (W). The projection on the auxiliary vertical plane (AVP) is obtained by projectors drawn perpendicular to the AVP from the

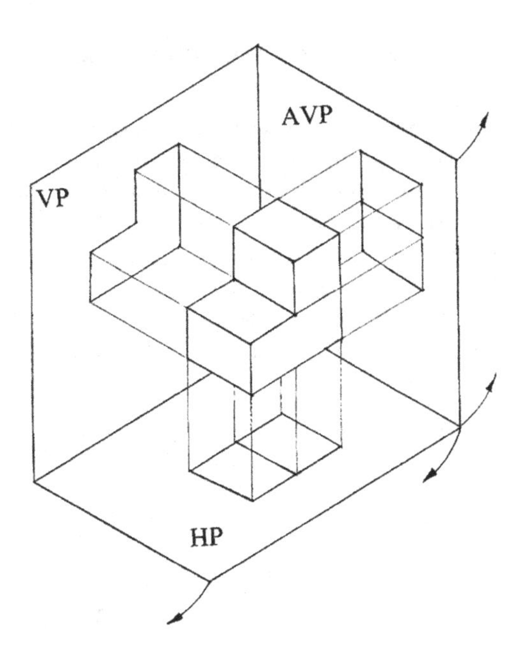

Fig. 5.8 Pictorial view

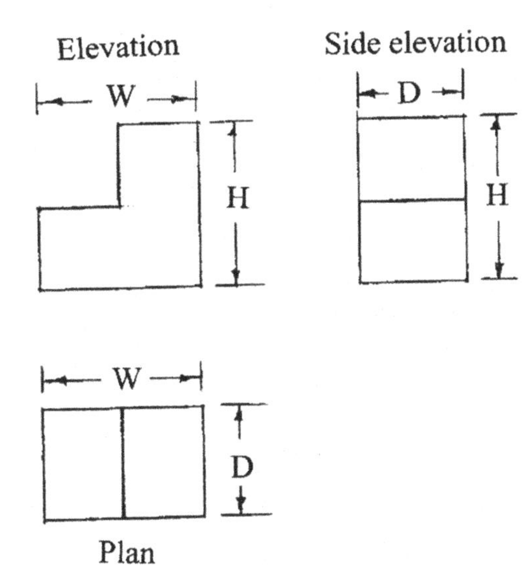

Fig. 5.9 Orthographic projection

corners of the object. This will give the side elevation or the side view of the object with its height (H) and depth (D).

When the first quadrant is opened out until the HP and AVP coincide with the VP, the projections take the position as shown in Fig. 5.9. The plan is placed beneath the elevation and the side elevation looking from the left is placed on the right. The elevation shows the width (W) and the height (H) of the object. The plan shows the width (W) and depth (D) of the object. The side elevation shows the depth (D) and height (H) of the object.

Third Quadrant Projection (Figs. 5.10 and 5.11)

The principal planes which are transparent lie between the observer and the object as shown in Fig. 5.10. The projections are viewed through the planes and are obtained on them as shown in Fig. 5.10. The projection on the VP is named as the front view, the projection on the HP is named as the top view and the projection on the auxiliary vertical plane (AVP) is named as the end view of the object. The HP and AVP are opened out until they coincide with the VP. The projections or views of the object take the positions as shown in Fig. 5.11. The front view is beneath the top view and the end view is adjacent to the end of the object that it

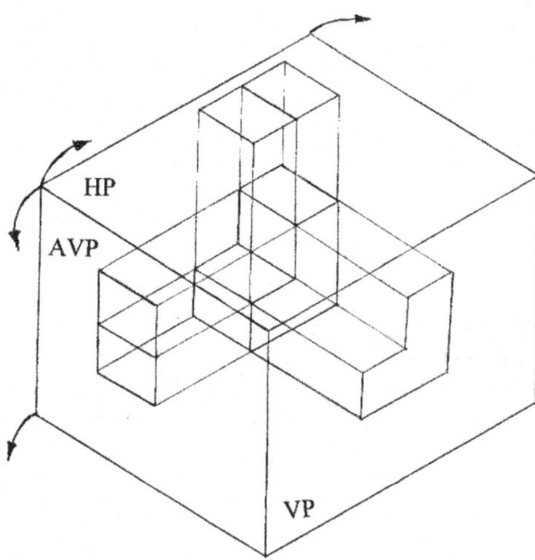

Fig. 5.10 Pictorial view

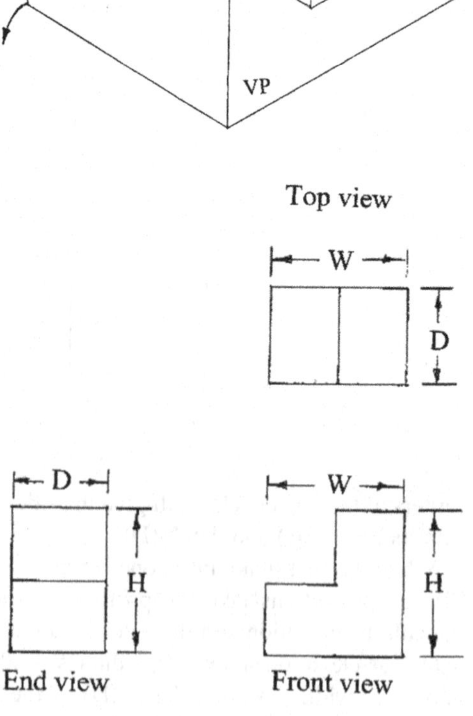

Fig. 5.11 Orthographic views

describes.The front view shows the height (H) and width (W), the top view shows the width (W) and depth (D) and the end view shows the height (H) and depth (D) of the object.

The views in Figs. 5.9 and 5.11 describe the object completely. The first angle projection is used in this book. The first and third angle projections are identifiable by their relative arrangement of views. The system of projection used in a drawing is indicated by standard symbol. Figure 5.12 illustrates standard symbol for the first angle projection. Figure 5.13 illustrates standard symbol for the third angle projection. The method of projection used should be indicated in the space provided for this purpose in the title block of the drawing sheet.

The pictorial sketches and projections of certain objects are furnished in the following figures to understand the theory of projection.

The pictorial sketch of an object is shown in Fig. 5.14. The orthographic projections of the object are shown in Fig. 5.15. Each projection represents a single surface. A normal surface is a plane surface which is parallel to a plane of projection. A normal surface will appear true size in one view and as a vertical or horizontal line in each of the other two views. All surfaces are normal in this case.

Figure 5.16 shows the pictorial sketch of an object with surfaces at different levels. The orthographic projections of the object are given in Fig. 5.17. The plan

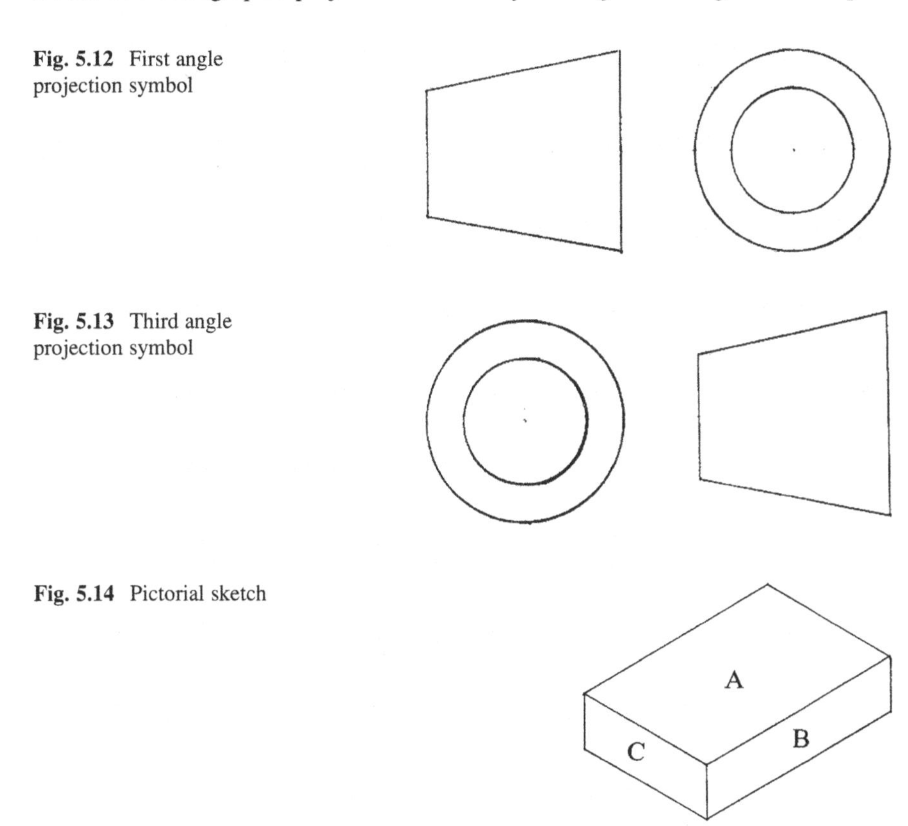

Fig. 5.12 First angle projection symbol

Fig. 5.13 Third angle projection symbol

Fig. 5.14 Pictorial sketch

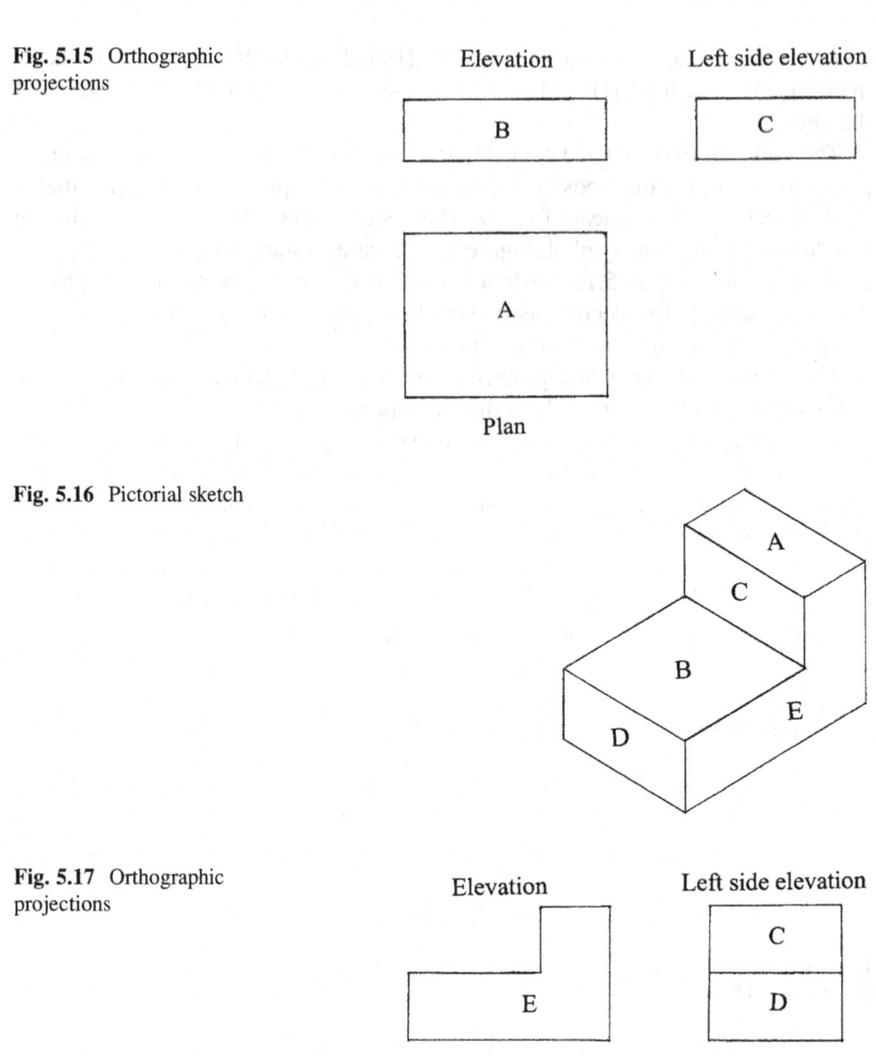

Fig. 5.15 Orthographic projections

Fig. 5.16 Pictorial sketch

Fig. 5.17 Orthographic projections

shows two surfaces A and B at different levels and surface A is above surface B as shown by the elevation. In side elevation surfaces C and D are shown, but it is certain from the elevation the surface C is closer to the AVP, which is behind the object.

Figure 5.18 shows the pictorial sketch of an object with an inclined surface. The orthographic projections of the object are given in Fig. 5.19.

The surface B is shown slightly shortened in the plan and very much shortened in the side elevation. To obtain the true size of surface B, the distance de must be taken from the elevation and the distance ef must be taken from the side elevation. The inclined surface B appears as a line in the elevation (surface is perpendicular to the VP).

Figure 5.20 shows the pictorial sketch of an object with a surface inclined to all three planes (oblique surface). The orthographic projections of the object are as in Fig. 5.21.

Fig. 5.18 Pictorial sketch

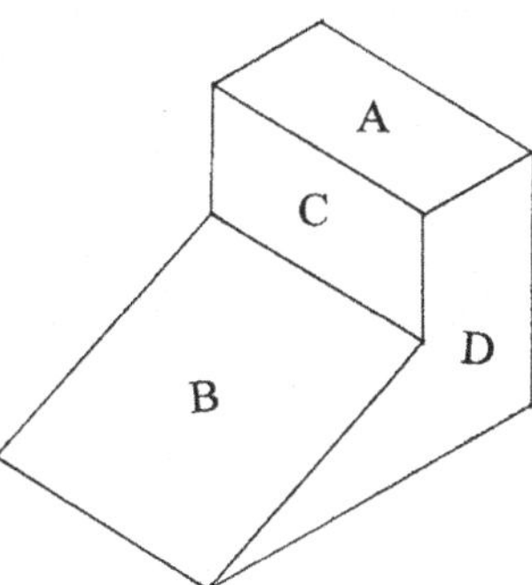

Fig. 5.19 Orthographic projections

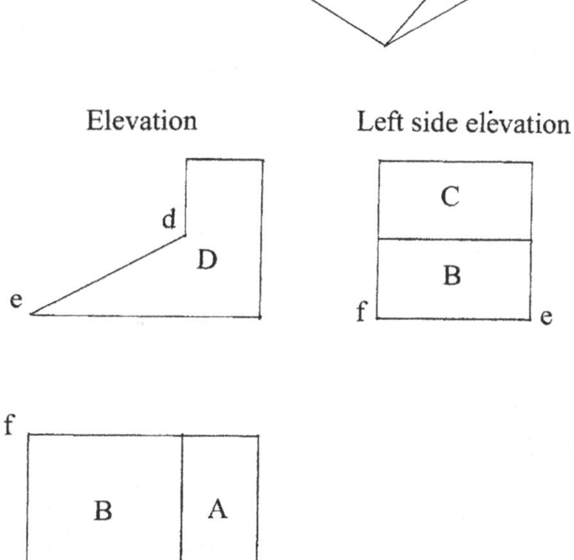

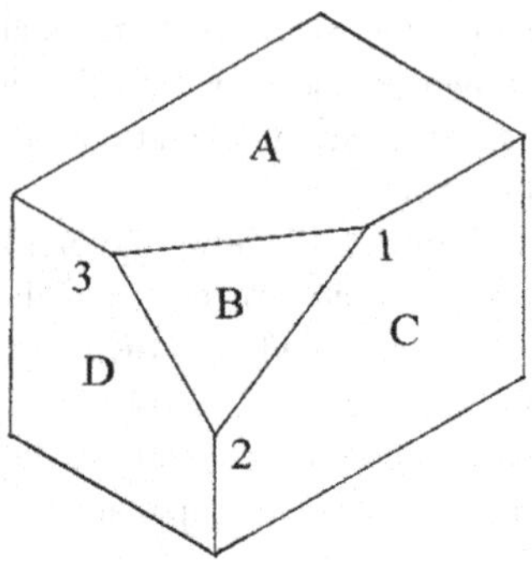

Fig. 5.20 Pictorial sketch

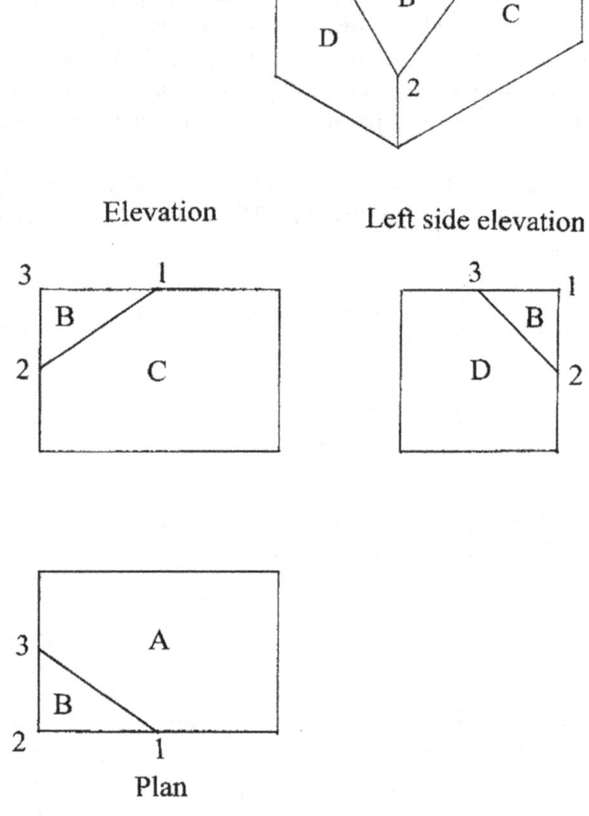

Fig. 5.21 Orthographic projections

The surface B is inclined to all three planes and appears in all three projections. The true size of surface can be obtained knowing distances 12 from elevation, 23 from the side elevation and 31 from the plan.

Hidden Lines

Parts which cannot be seen in the views are represented by hidden lines composed of short dashes or dotted. Figure 5.22 shows the pictorial sketch of an object. The orthographic projections of the object are shown in Fig. 5.23. The surface A is shown as an edge view in the elevation. Since the edge view is invisible in the

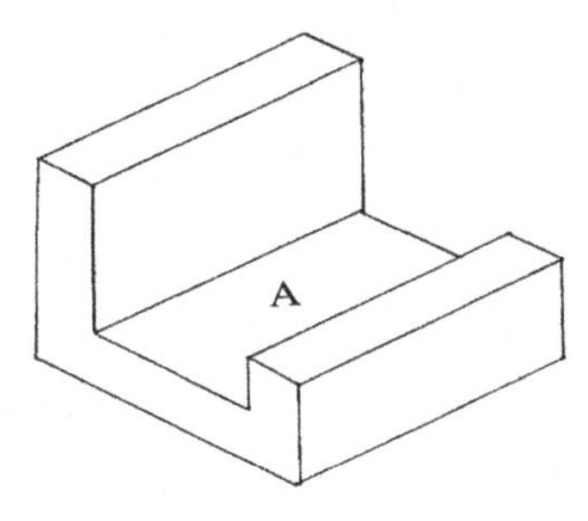

Fig. 5.22 Pictorial sketch

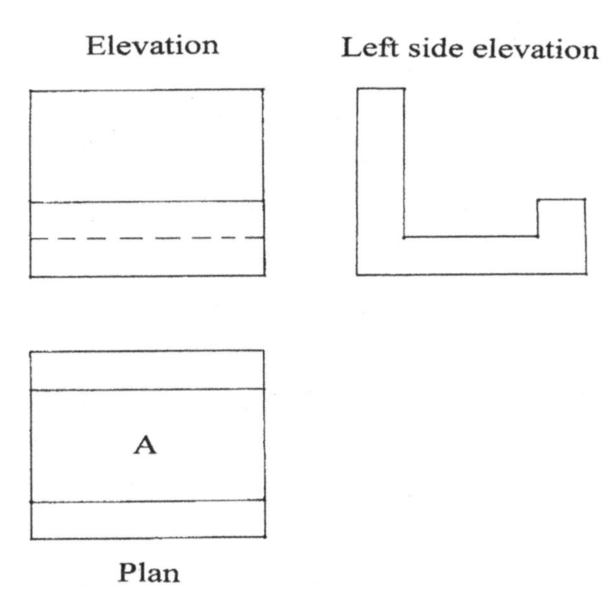

Fig. 5.23 Orthographic projections

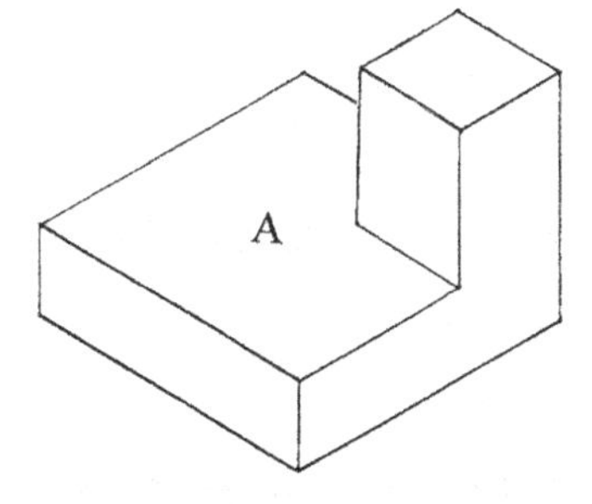

Fig. 5.24 Pictorial sketch

elevation, it is shown as hidden line. The first dash of a hidden line touches the line at which it starts.

Figure 5.24 shows the pictorial sketch of an object. Its orthographic projections are shown in Fig. 5.25. The surface A is shown as an edge view in the elevation. A portion of this edge view is invisible in the elevation. In this case, the invisible

Fig. 5.25 Orthographic projections

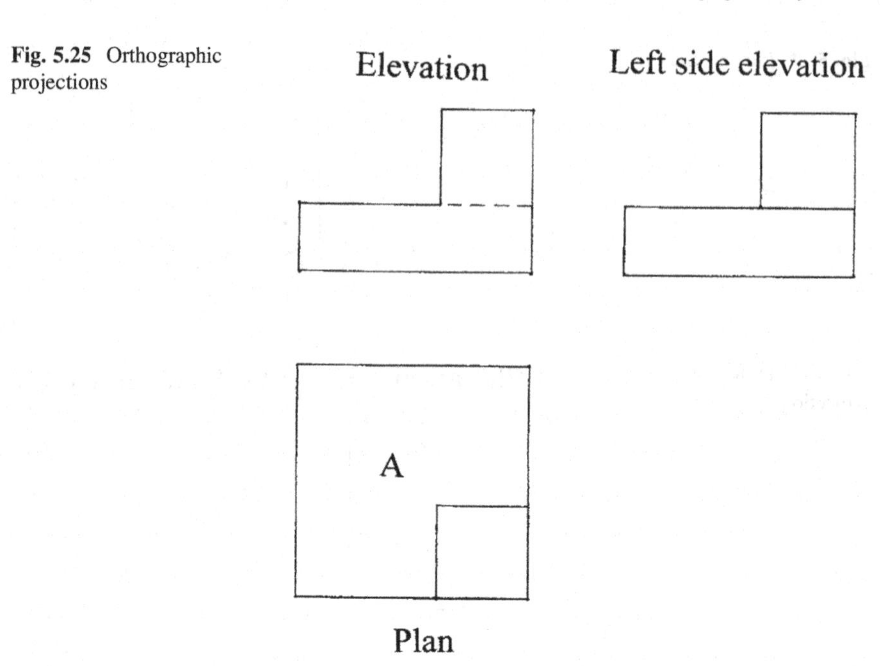

Fig. 5.26 Object

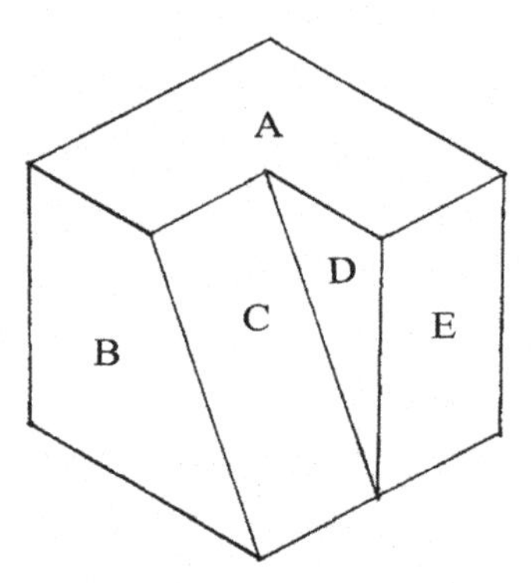

portion of the hidden line is continuation of the visible full line. Hence, a space is left between the full line and the first dash of the hidden line.

Solved Problems

1. Sketch freehand the orthographic projections of the object shown in Fig. 5.26 following the first angle projection. Identify the surfaces on the projections by marking the corresponding alphabets.

   The elevation, plan and left side elevation of the object are shown in Fig. 5.27.

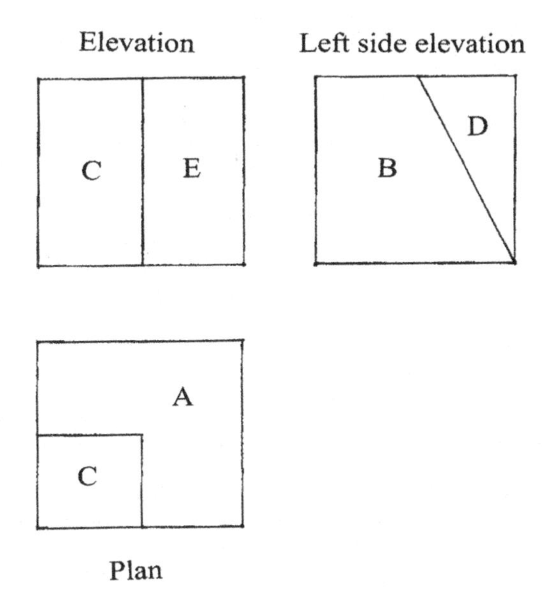

Fig. 5.27 Projections of the object

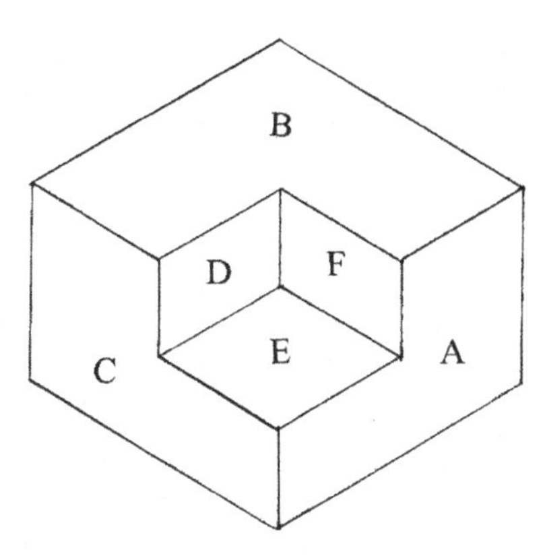

Fig. 5.28 Object

2. Sketch freehand the orthographic projections of the object shown in Fig. 5.28 following the first angle projection. Identify the surfaces on the projections by marking the corresponding alphabets.

   The elevation, plan and left side elevation of the object are shown in Fig. 5.29.
3. Sketch freehand the orthographic projections of the object shown in Fig. 5.30 following the first angle projection. Identify the surfaces on the projections by marking the corresponding alphabets.

   The elevation, plan and left side elevation of the object are shown in Fig. 5.31.

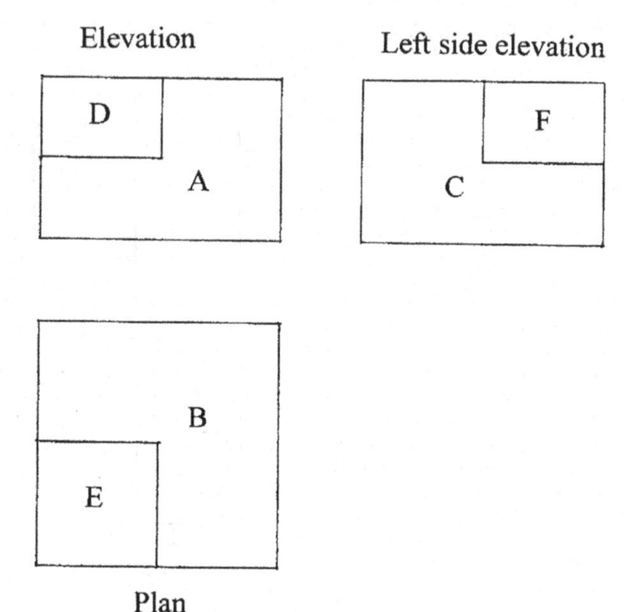

Fig. 5.29 Projections of the object

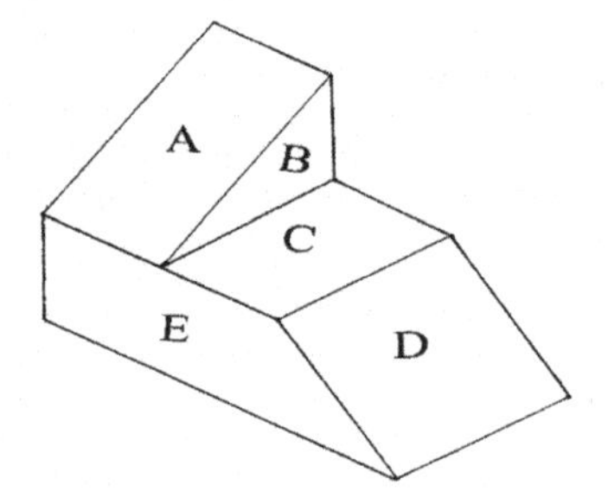

Fig. 5.30 Object

4. Sketch freehand the orthographic views of the object shown pictorially in Fig. 5.32 following the third angle projection. Identify the surfaces on the projection by marking the corresponding alphabets.

   The front view, top view and left end view of the object are shown in Fig. 5.33.
5. The orthographic projections of an object are shown in Fig. 5.34. Visualize the shape of the object and sketch its pictorial view.

   Solution: Fig. 5.35 shows the pictorial view of the object.
6. The orthographic projections of an object are shown in Fig. 5.36. Sketch its pictorial view.

   Solution: Fig. 5.37 shows the pictorial view of the object
7. The orthographic projections of an object are shown in Fig. 5.38. Sketch its pictorial view.

   Solution: Fig. 5.39 shows the pictorial view of the object

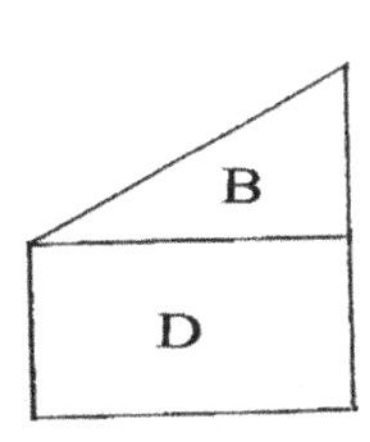

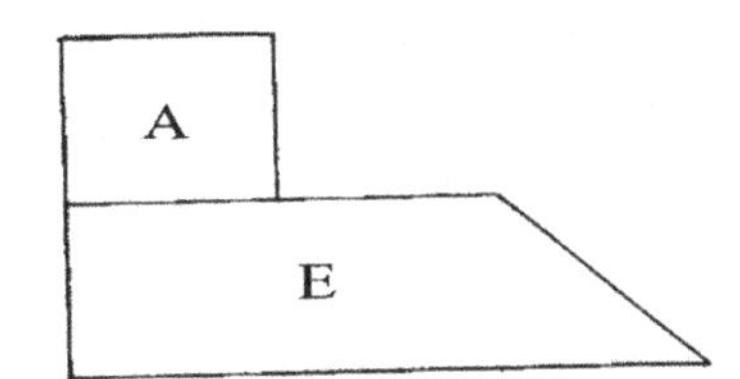

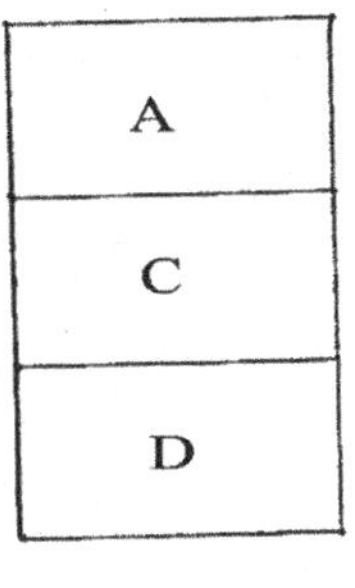

Fig. 5.31 Projections of the object

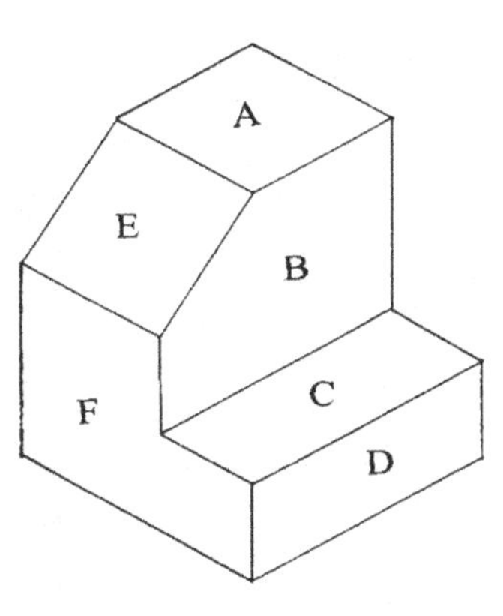

Fig. 5.32 Object

8. The orthographic projections of an object are shown in Fig. 5.40. Sketch its pictorial view.

   Solution: Fig. 5.41 shows the pictorial view of the object.
9. The orthographic projections of an object are shown in Fig. 5.42. Sketch its pictorial view.

   Solution: Fig. 5.43 shows the pictorial view of the object.
10. The orthographic views of an object in third angle projection are shown in Fig. 5.44. Sketch its pictorial view.

    Solution: Fig. 5.45 shows the pictorial view of the object.

Fig. 5.33 Orthographic views

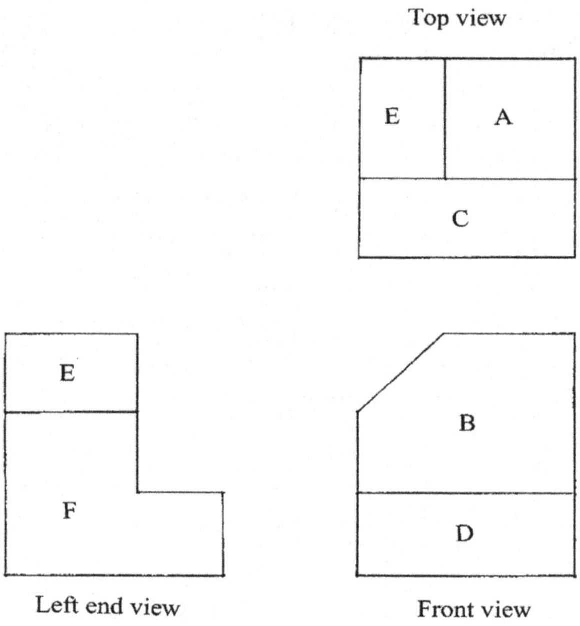

Top view

E A

C

E

F

B

D

Left end view

Front view

Fig. 5.34 Orthographic projections of an object

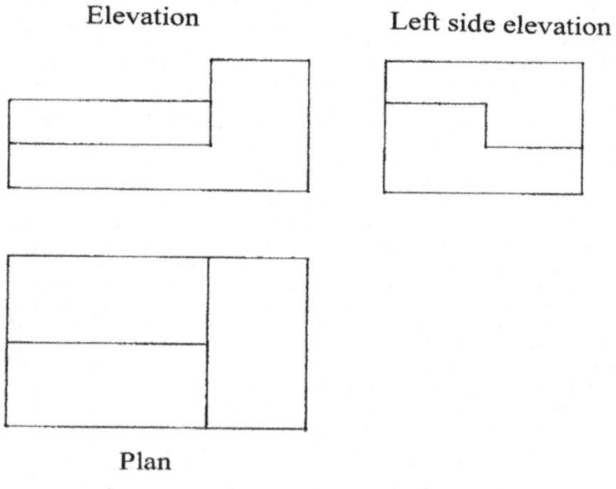

Elevation

Left side elevation

Plan

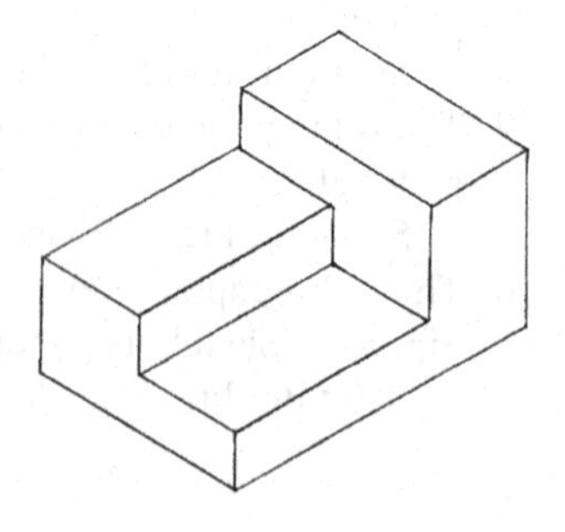

Fig. 5.35 Pictorial view of the object

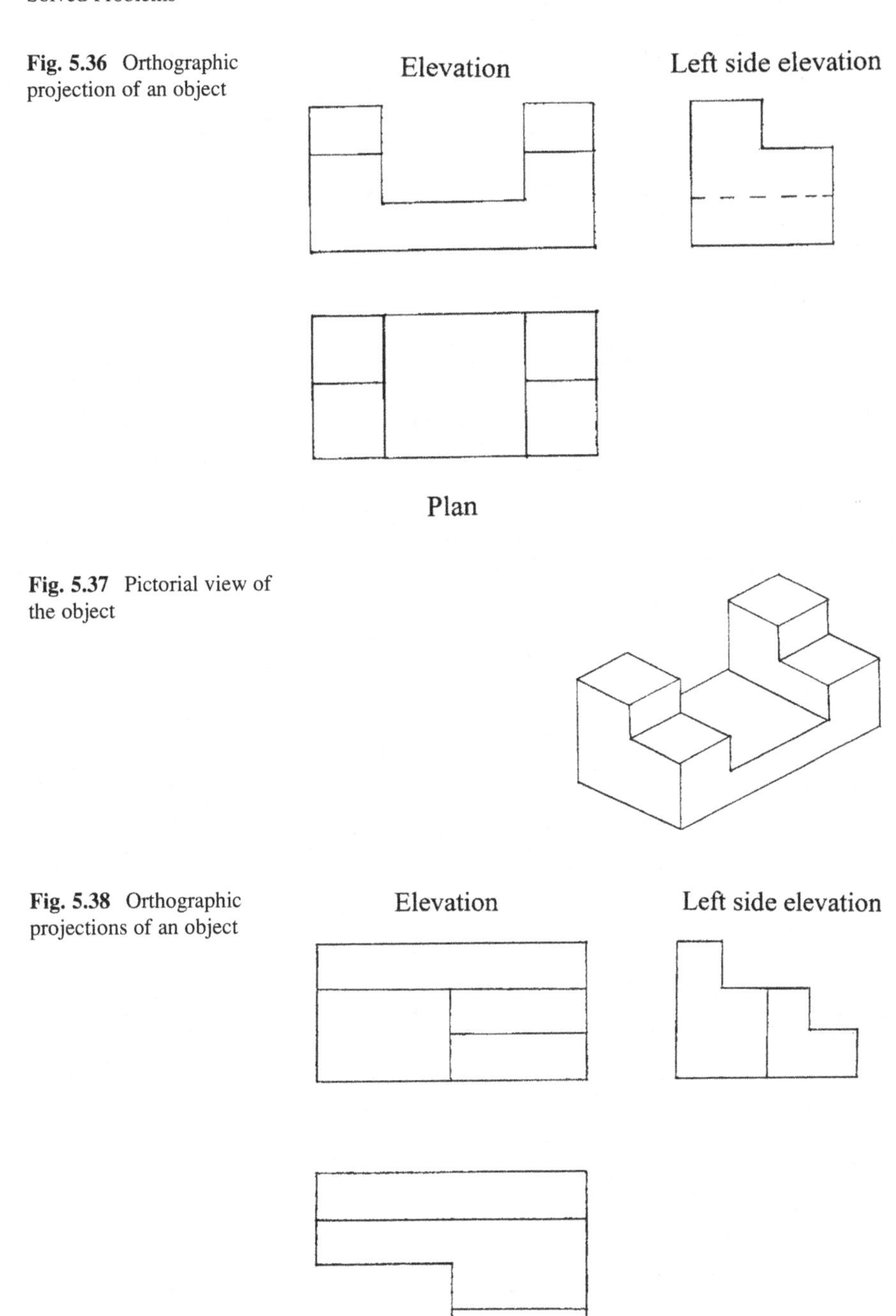

Fig. 5.36 Orthographic projection of an object

Fig. 5.37 Pictorial view of the object

Fig. 5.38 Orthographic projections of an object

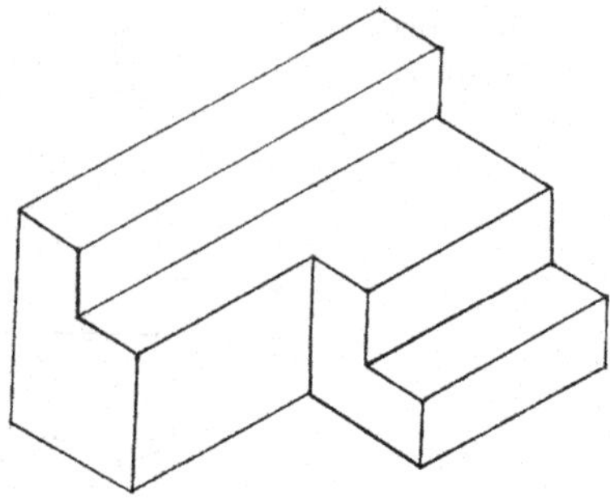

Fig. 5.39 Pictorial view of the object

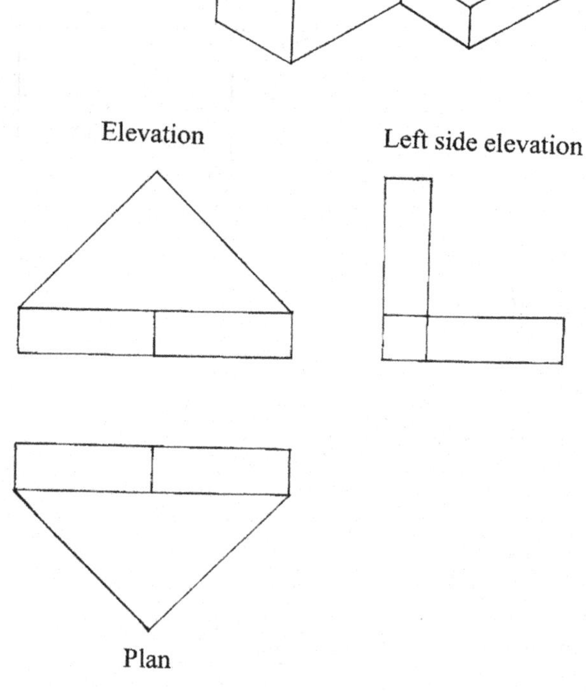

Fig. 5.40 Orthographic projections of an object

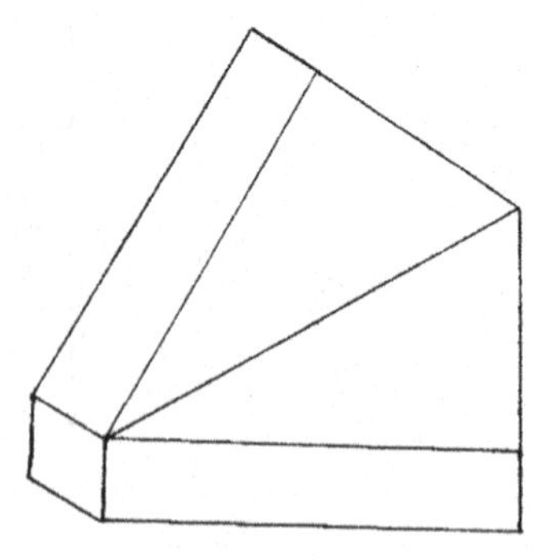

Fig. 5.41 Pictorial view of the object

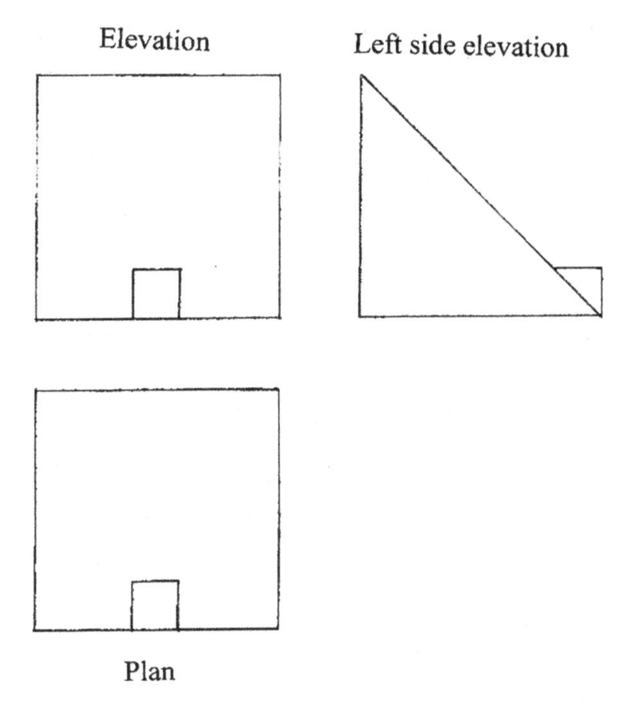

Fig. 5.42 Orthographic projections of an object

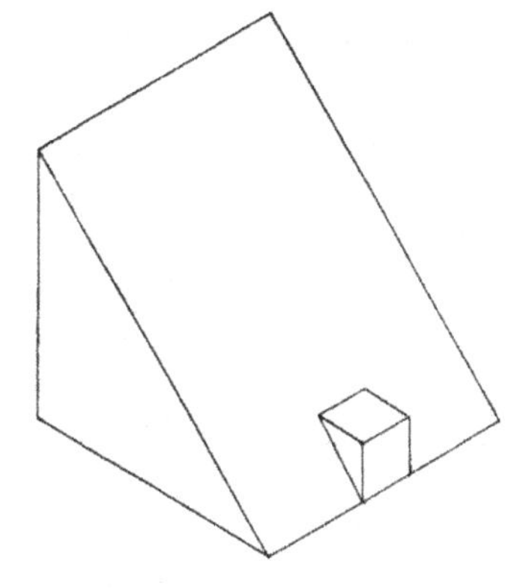

Fig. 5.43 Pictorial view of the object

11. A line is missing in one of the projections of the object whose projections are shown in Fig. 5.46. Identify the projection and sketch it with the missing line.

    Solution: Fig. 5.47 shows the plan of the object with the missing line inserted in the appropriate place.

12. A line is missing in one of the projections of the object whose projections are shown in Fig. 5.48. Identify the projection and sketch it with the missing line.

    Solution: Fig. 5.49 shows the elevation of the object with the missing line inserted in the appropriate place.

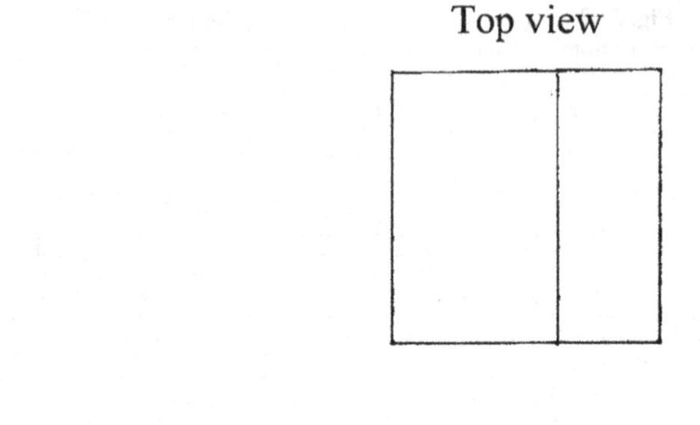

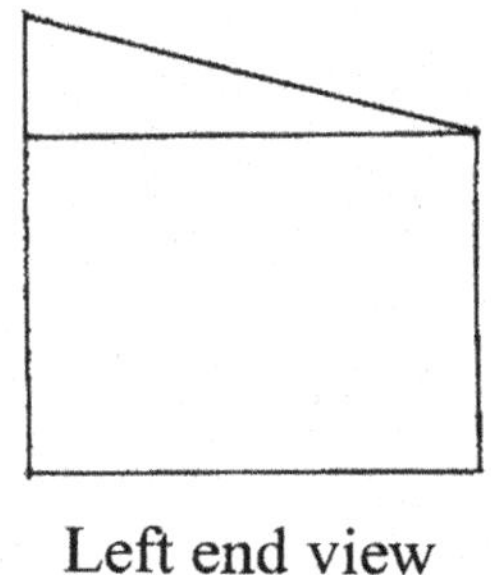

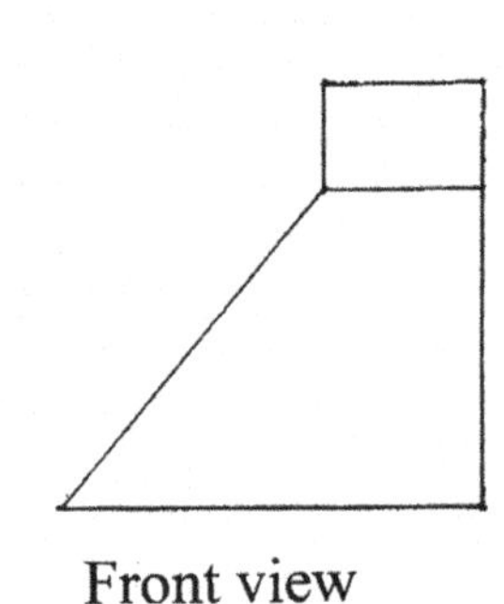

Fig. 5.44 Orthographic views of an object

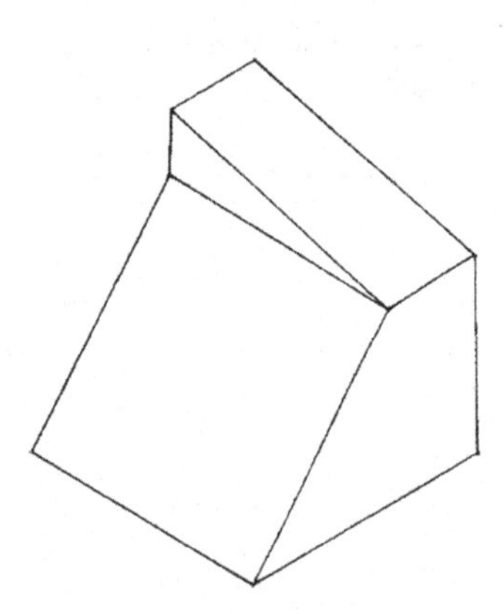

Fig. 5.45 Pictorial view of the object

13. A line is missing in one of the projections of the object whose projections are shown in Fig. 5.50. Identify the projection and sketch it with the missing line.
    Solution: Fig. 5.51 shows the left side elevation of the object with the missing line inserted in the appropriate place.
14. A line is missing in one of the projections of the object whose projections are shown in Fig. 5.52. Identify the projection and sketch it with the missing line.
    Solution: Fig. 5.53 shows the left side elevation of the object with the missing line inserted in the appropriate place.

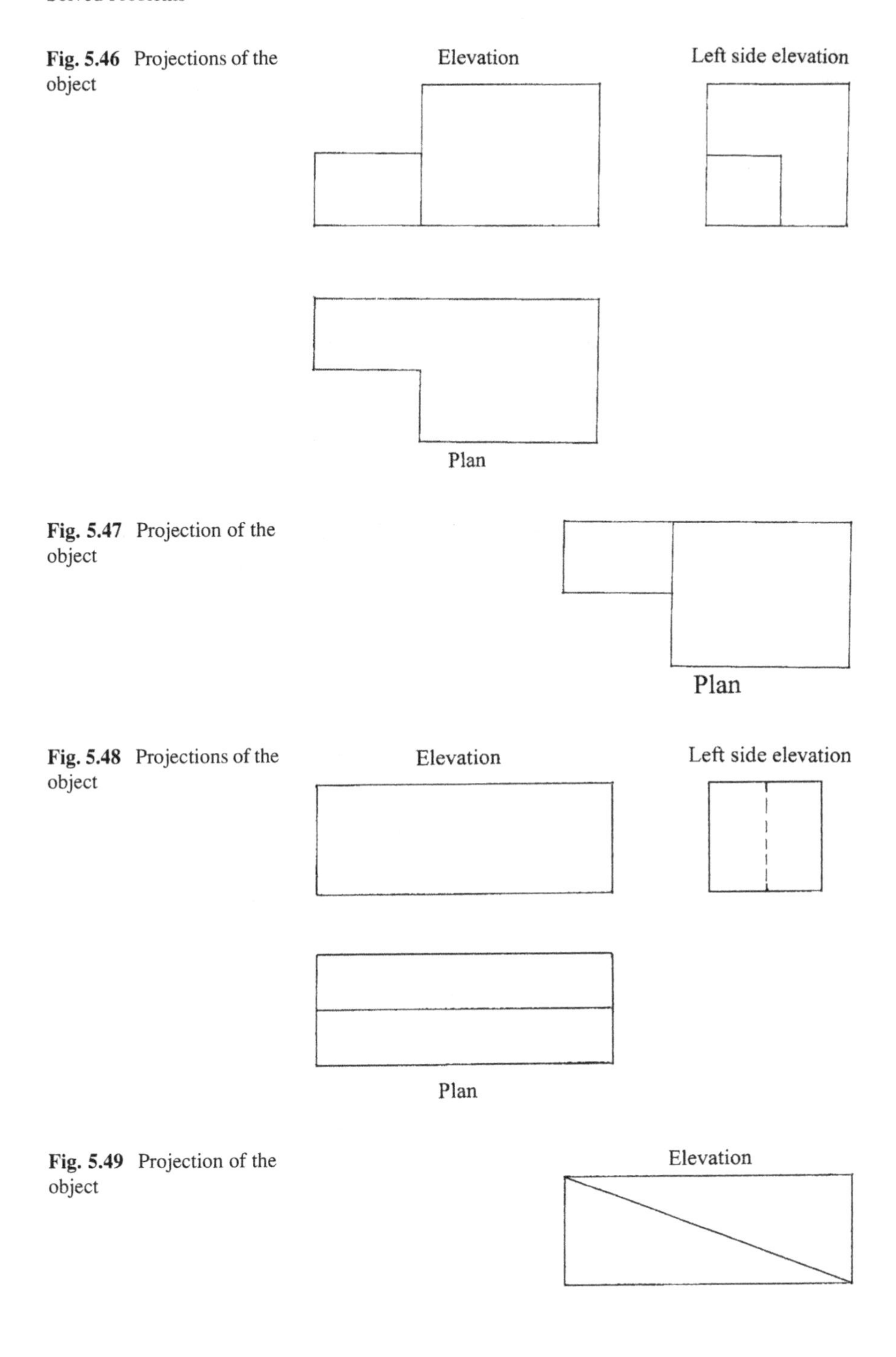

Fig. 5.46 Projections of the object

Fig. 5.47 Projection of the object

Fig. 5.48 Projections of the object

Fig. 5.49 Projection of the object

Fig. 5.50 Projections of the object

Elevation

Left side elevation

Plan

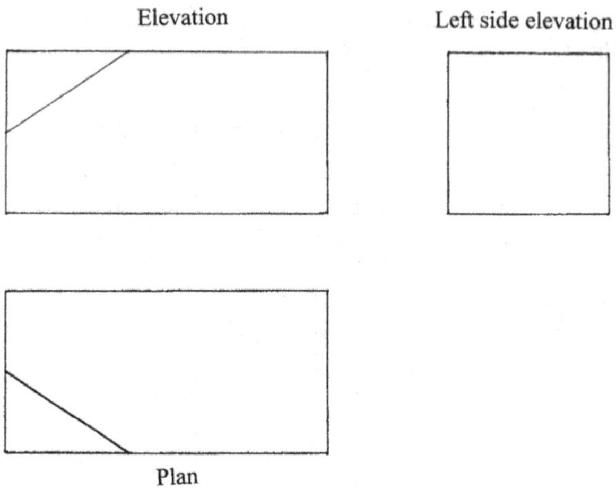

Fig. 5.51 Projection of the object

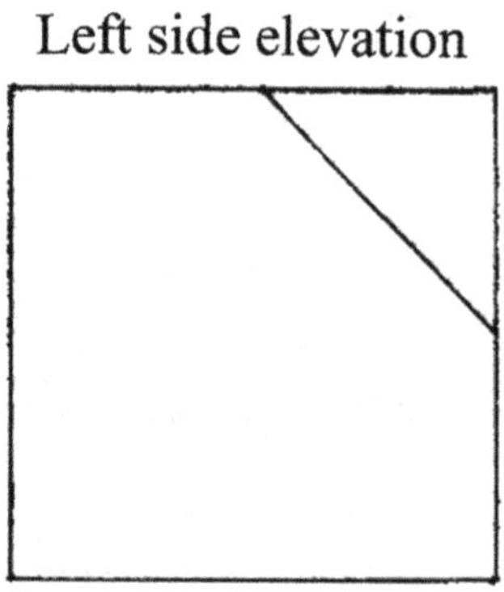

Left side elevation

Fig. 5.52 Projections of the object

Elevation

Left side elevation

Plan

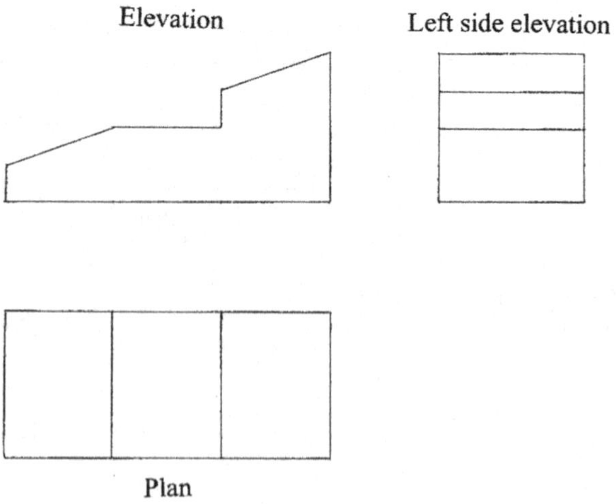

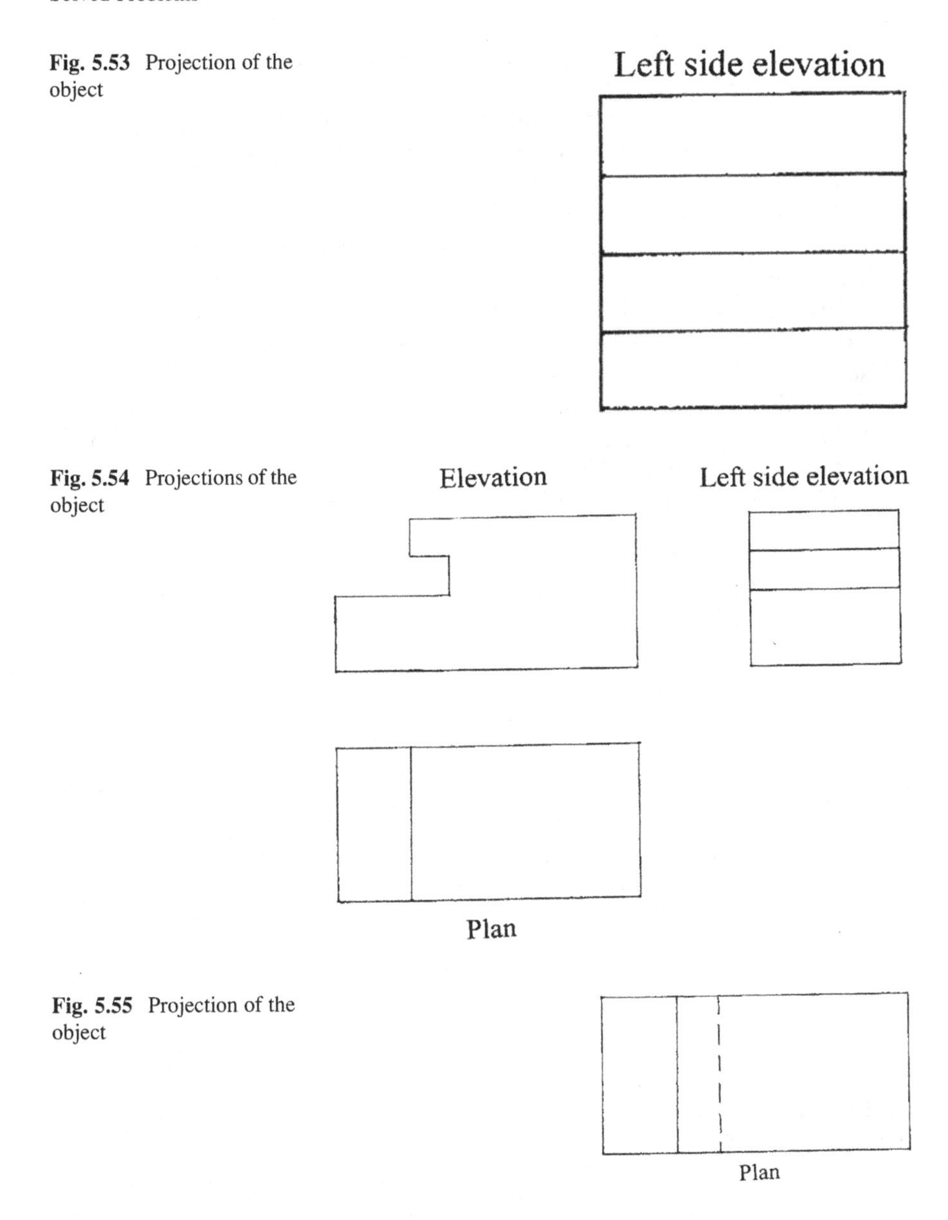

Fig. 5.53 Projection of the object

Fig. 5.54 Projections of the object

Fig. 5.55 Projection of the object

15. A line is missing in one of the projections of the object whose projections are shown in Fig. 5.54. Identify the projection and sketch it with the missing line.

    Solution: Fig. 5.55 shows the plan of the object with the missing line inserted in the appropriate place.

Note: The following five problems are intended primarily for supplementary practice in visualizing projections.

16. Complete the orthographic projections of the object as assigned in Fig. 5.56 supplying the missing projection.
    Solution: The missing left side elevation is shown in Fig. 5.57.
17. Complete the orthographic projections of the object as assigned in Fig. 5.58 supplying the missing projection.
    Solution: The missing elevation is shown in Fig. 5.59.
18. Complete the orthographic projections of the object as assigned in Fig. 5.60 supplying the missing projection.
    Solution: The missing left side elevation is shown in Fig. 5.61.
19. Complete the orthographic projections of the object as assigned in Fig. 5.62 supplying the missing projection.
    Solution: The missing plan is shown in Fig. 5.63
20. Complete the orthographic projections of the object as assigned in Fig. 5.64 supplying the missing projection.
    Solution: The missing plan is shown in Fig. 5.65.

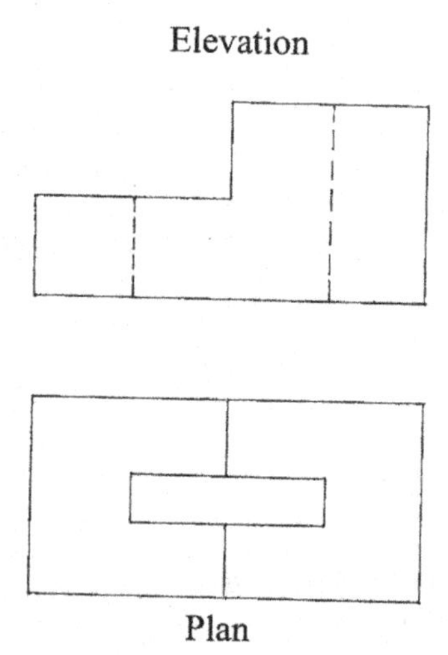

Fig. 5.56 Orthographic projections of the object

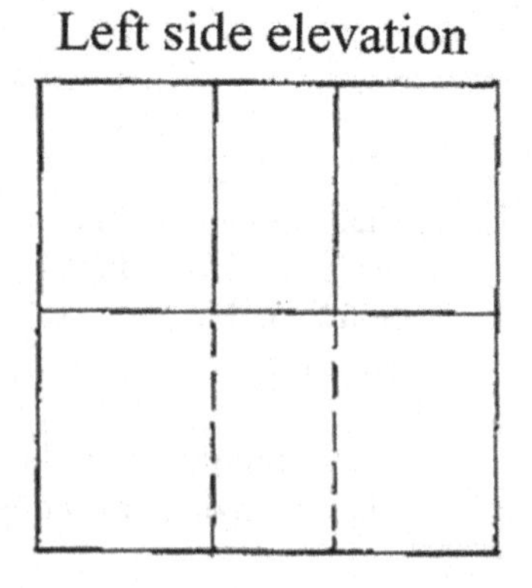

Fig. 5.57 Orthographic projection of the object

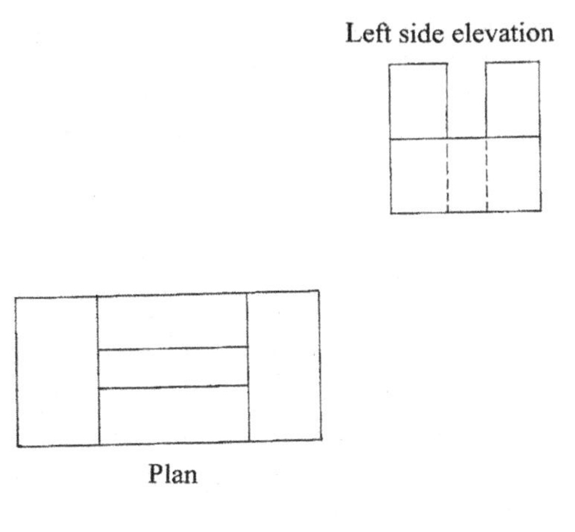

Fig. 5.58 Orthographic projections of the object

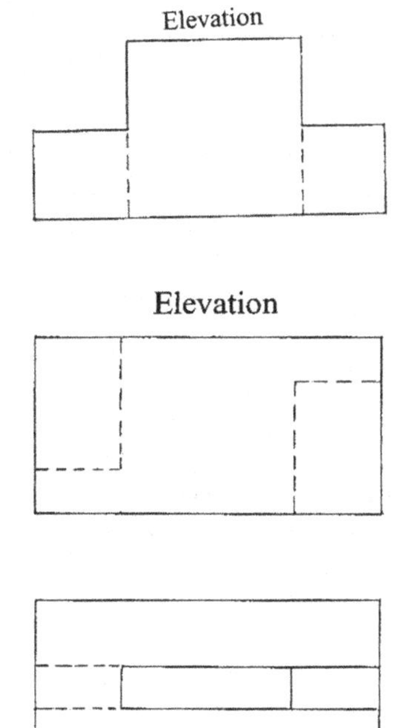

Fig. 5.59 Orthographic projection of the object

Fig. 5.60 Orthographic projections of the object

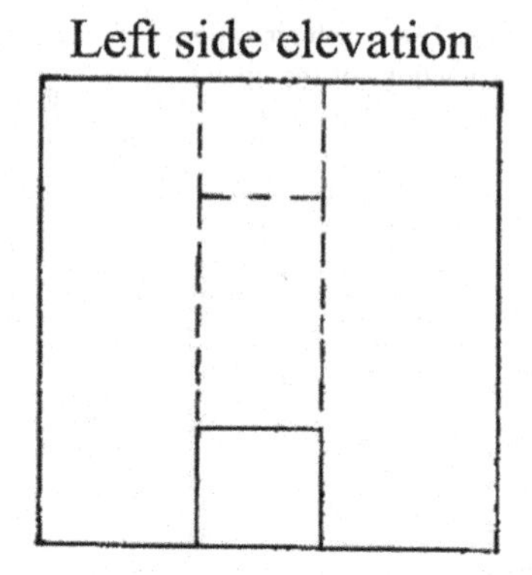

Fig. 5.61 Orthographic projection of the object

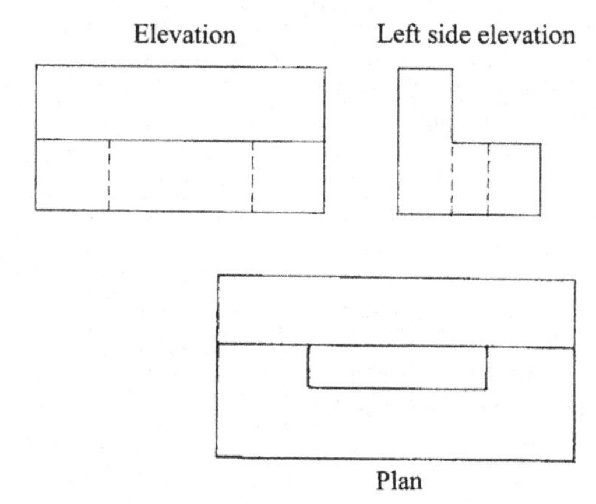

Fig. 5.62 Orthographic projections of the object

Fig. 5.63 Orthographic projection of the object

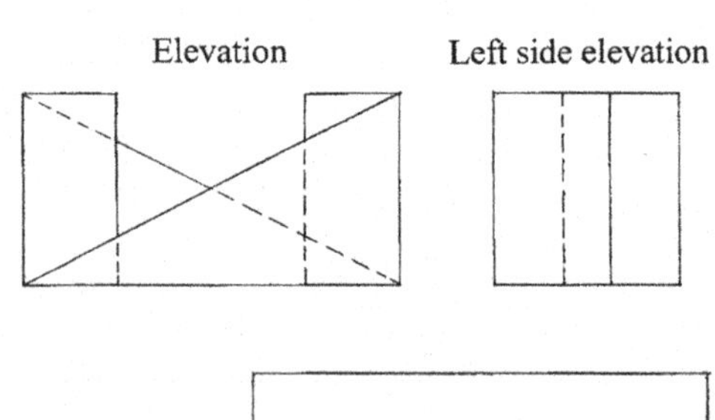

Fig. 5.64 Orthographic projections of the object

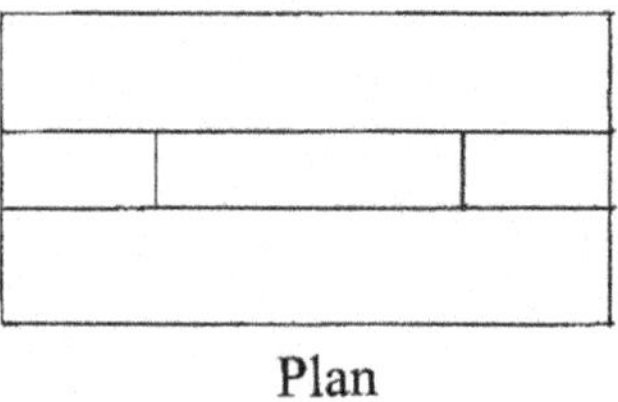

Fig. 5.65 Orthographic projection

Practice Problems

1. Sketch freehand the orthographic projections of the objects shown in Figs. 5.66, 5.67, 5.68 and 5.69 following the first angle projections.
2. Complete the orthographic projections of the objects as assigned in Figs. 5.70, 5.71, 5.72, 5.73, 5.74 and 5.75 supplying the missing projection.

Fig. 5.66 ■

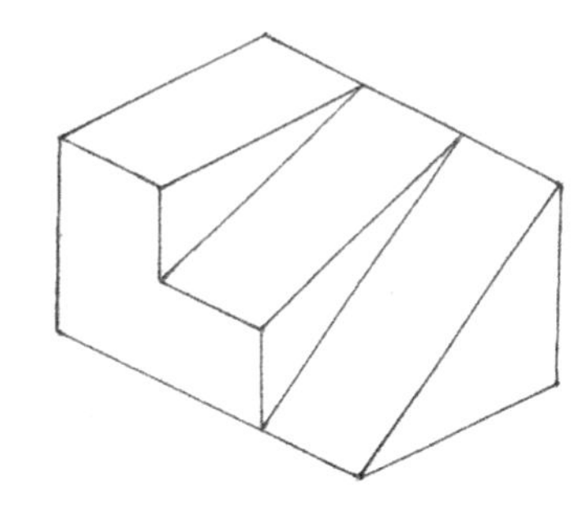

Fig. 5.67 ■

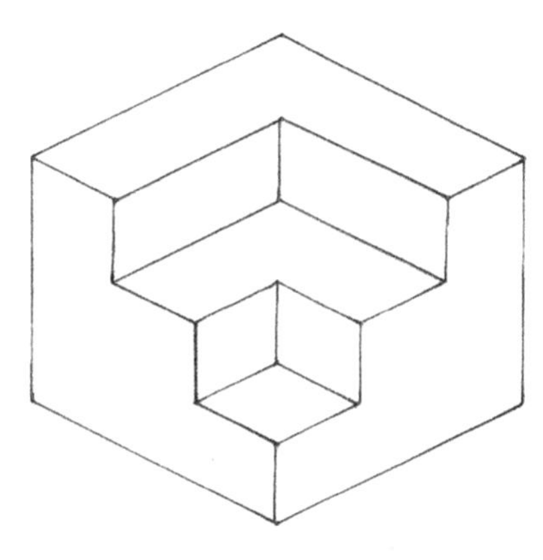

Fig. 5.68 ■

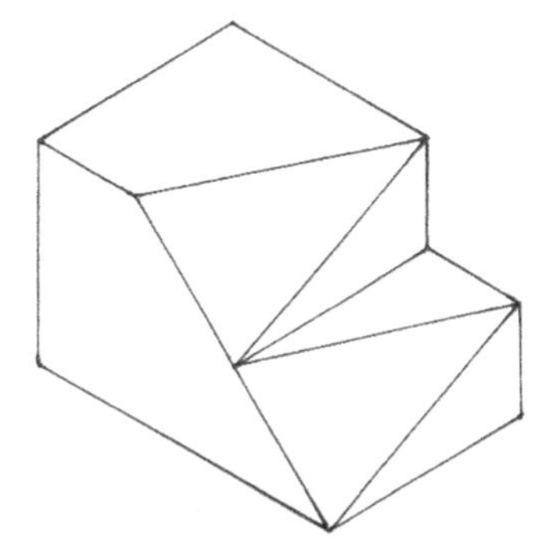

Fig. 5.69 ■

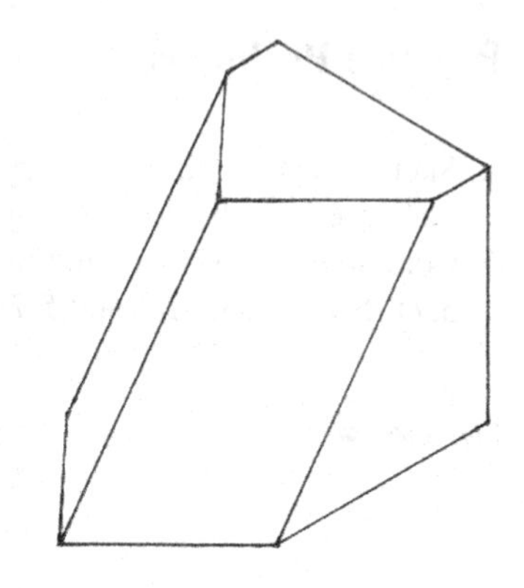

Fig. 5.70 ■

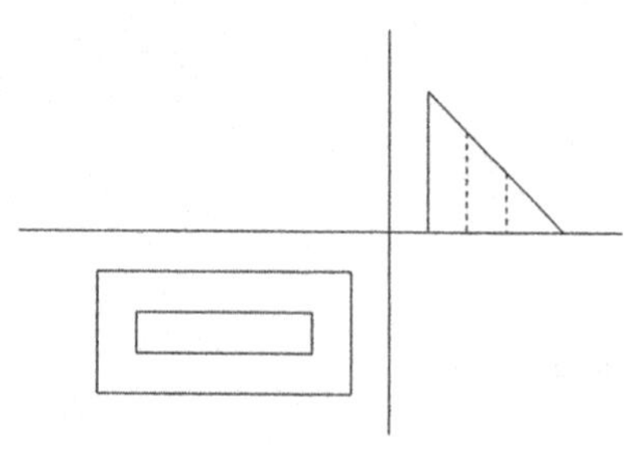

Fig. 5.71 ■

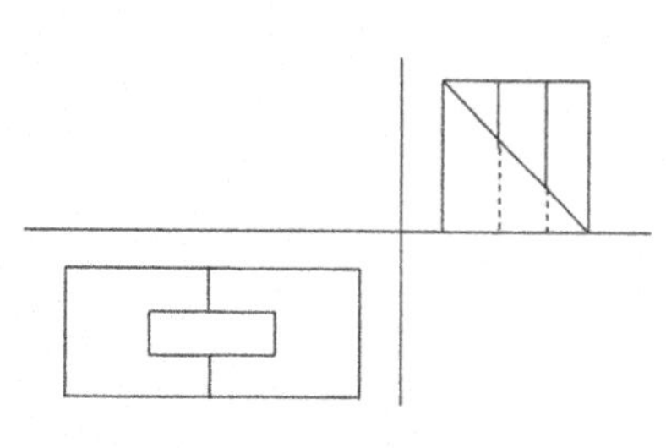

Fig. 5.72 ■

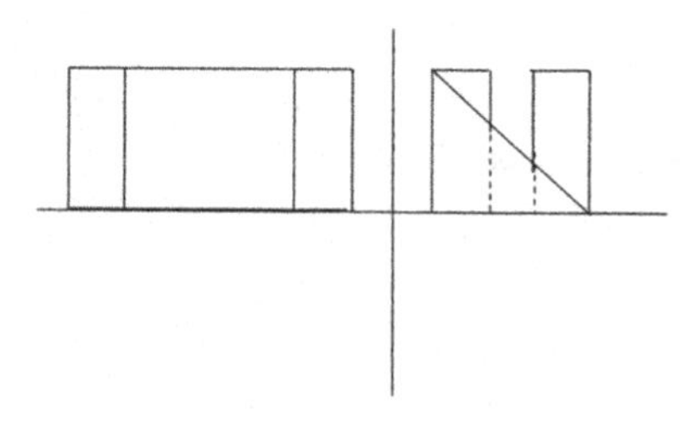

Fig. 5.73 ■

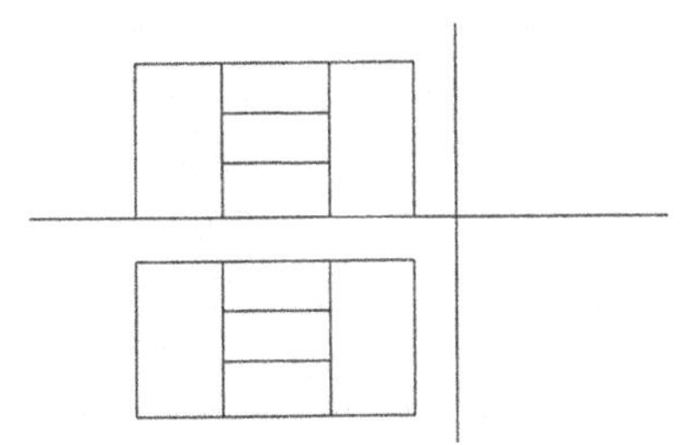

Fig. 5.74 ■

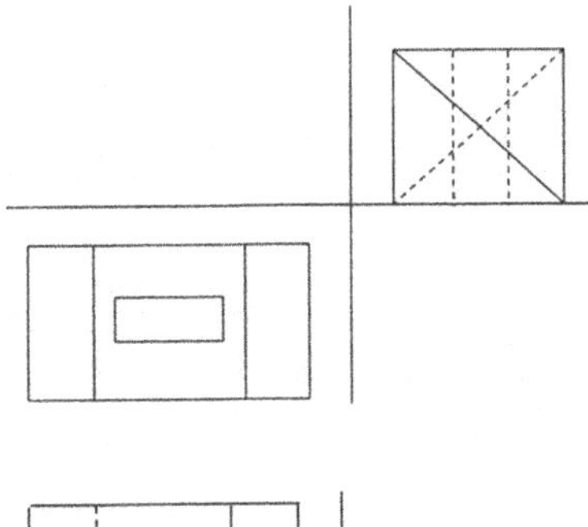

Fig. 5.75 ■

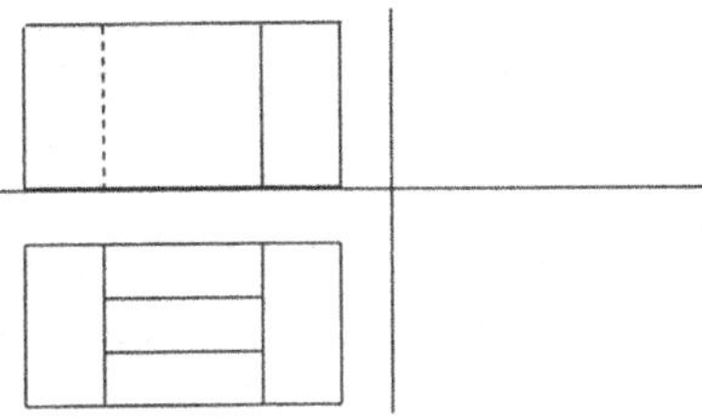

Chapter 6
Projections of Points

In orthographic projection, an exact view of an object is obtained on a plane by projectors drawn from the end points of the object to meet the plane of projection at right angles. The view obtained in general consists of line representing edge of the object and lines representing surfaces of the object. A line on the view is obtained by the projections of two end points of the edge. Hence, the projection of points shall be the starting point to understand the theory of projection of lines and surfaces of the object. As the point has no dimension, its location is specified with reference to the co-ordinate planes. The point can be located in any of the four quadrants.

Projections of a Point

Refer to Fig. 6.1 showing the pictorial representation of a point A and the co-ordinate planes VP and HP. The point A is above the HP and in front of the VP and is situated in the First quadrant. A horizontal projector is drawn from A to meet the VP giving the elevation a′ of the point. The height of the point A above the HP shall be the distance of a′ from XY, the reference line/ground line. A vertical projector is drawn from A to meet the HP giving the plan a of the point. The distance of the point A from the VP shall be the distance of a from XY. To show the projections of the point on a plane surface, the HP is rotated as indicated about XY line until it coincides with the VP. The convention followed is that the first quadrant is always opened out. The projections a′ and a lie in one plane and appear as shown in Fig. 6.2.

K. Rathnam, *A First Course in Engineering Drawing*,
DOI 10.1007/978-981-10-5358-0_6

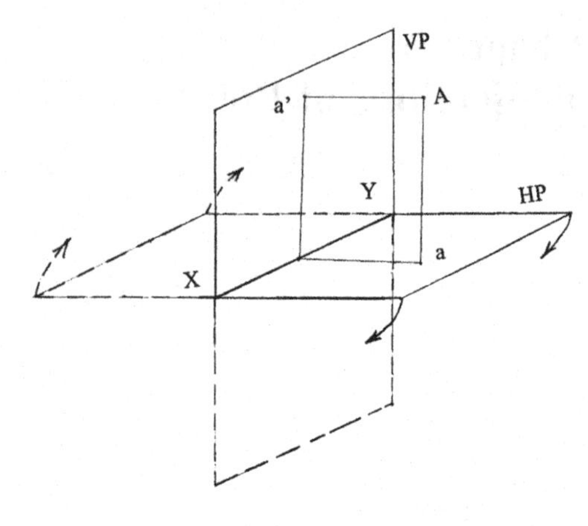

Fig. 6.1 Pictorial representation

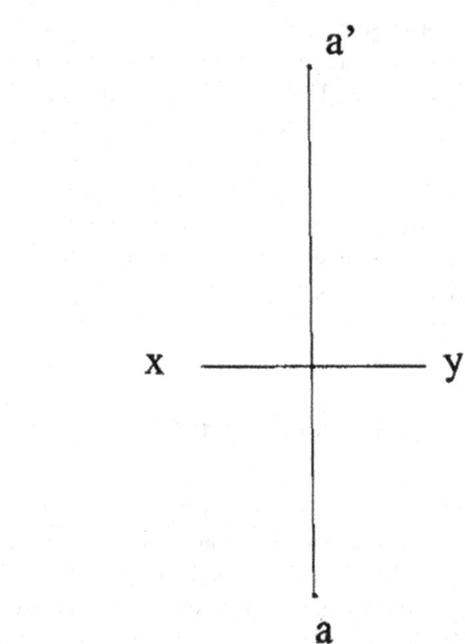

Fig. 6.2 Projections of point A

Notation in Projections

In the text space point / end point of a line / a point in solid is denoted by an upper case alphabet, its plan by its lower case, its elevation by its lower case with a dash over it and its side elevation by its lower case with double dash over it. In the figures space point / end point of a line / a point in solid is denoted by an upper case alphabet, its plan by its lower case, its elevation by its lower case with an apostrophe over it and its side elevation by its lower case with an end quotation mark over it.

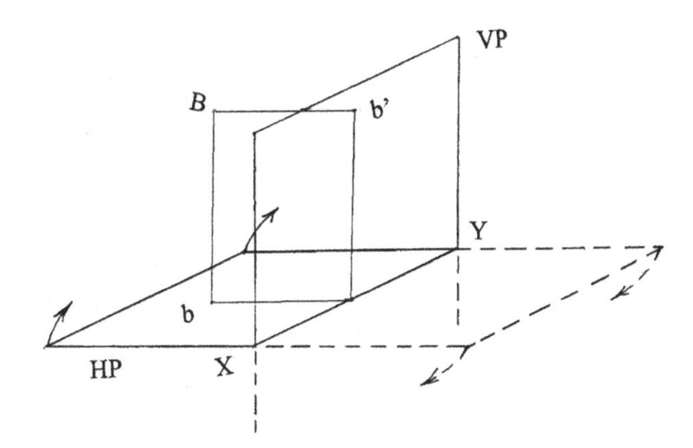

Fig. 6.3 Pictorial representation

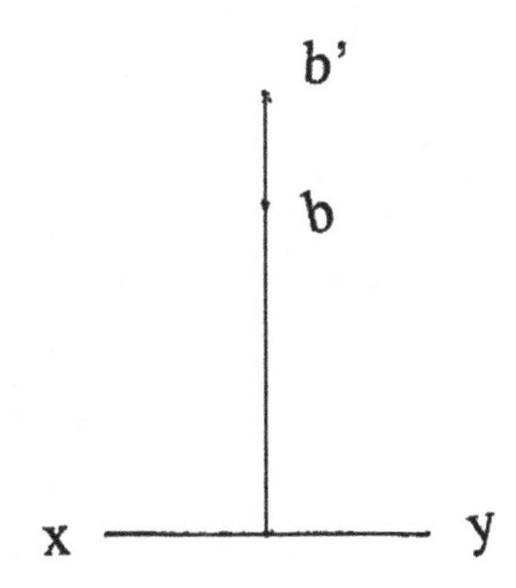

Fig. 6.4 Projections of point B

Figure 6.3 shows the pictorial representation of a point B which is above the HP and behind the VP and is situated in the Second quadrant. The corresponding projections b′ and b are shown in Fig. 6.4.

Figure 6.5 shows the pictorial representation of a point C which is below the HP and behind the VP and is situated in the Third quadrant. The corresponding projections c′ and c are shown in Fig. 6.6.

Figure 6.7 shows the pictorial representation of a point D which is below the HP and in front of the VP and is situated in the Fourth quadrant. The corresponding projections d′ and d are shown in Fig. 6.8.

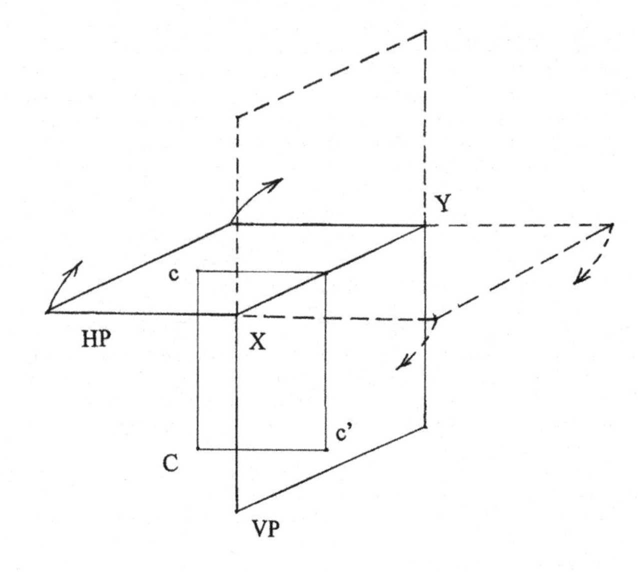

Fig. 6.5 Pictorial representation

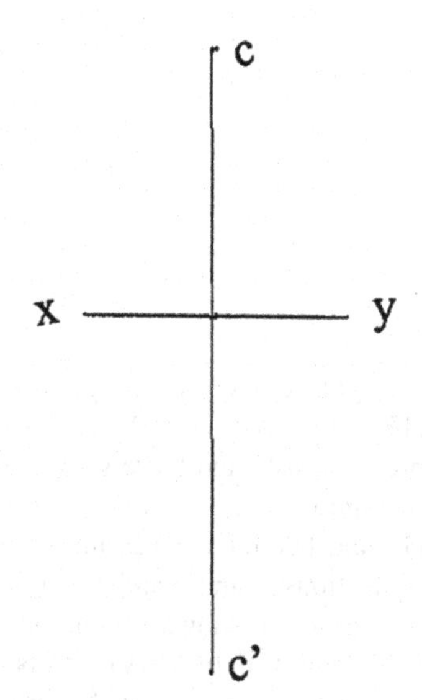

Fig. 6.6 Projections of point C

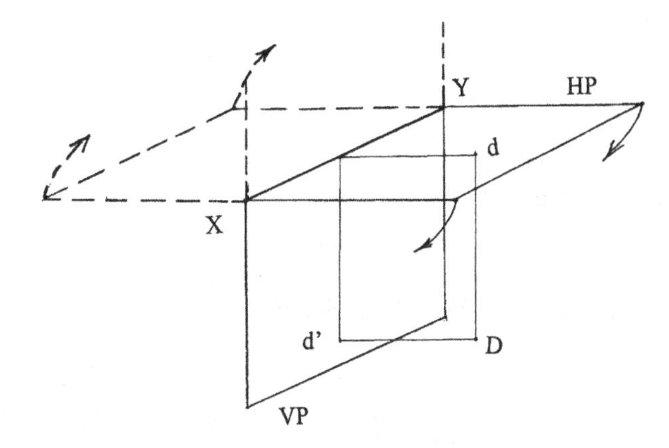

Fig. 6.7 Pictorial representation

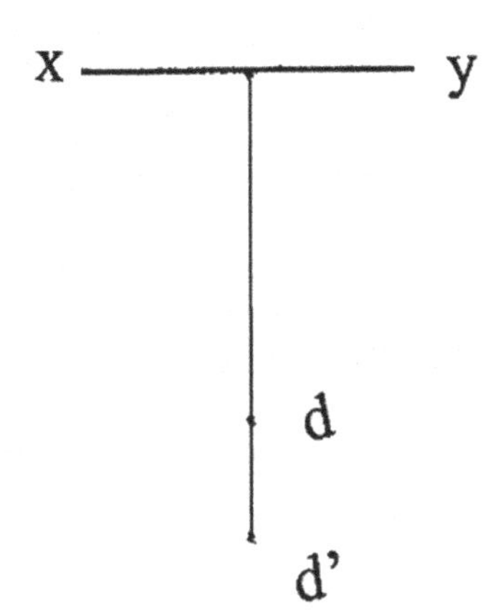

Fig. 6.8 Projections of point D

The following points are to be noted from Figs. 6.2, 6.4, 6.6 and 6.8:

1. The line joining the elevation and plan of a point or its extension in Figs. 6.4 and 6.8 is called a projector and is perpendicular to the reference line xy.
2. The elevation of a point is above or below xy according as the point is above or below the horizontal plane (HP) of projection.
3. The plan of a point is below or above xy according as the point is in front of or behind the vertical plane (VP) of projection.
4. The distance of the elevation of a point from xy is equal to the distance of the point from the horizontal plane (HP) of projection.
5. The distance of the plan of a point from xy is equal to the distance of the point from the vertical plane (VP) of projection.
6. The elevation and plan of a point lie on the same projector. Hence, the plan can be projected from its elevation or the elevation can be projected from its plan.

Solved Problems

(Hint: Graph sheets can be used for showing the projections of the points)

1. Draw the projections of the following points on the same reference line keeping the projectors 20 mm apart.

   A: 40 mm above the HP and 25 mm in front of the VP.
   B: 15 mm above the HP and 50 mm behind the VP.
   C: 25 mm below the HP and 25 mm behind the VP.
   D: 40 mm below the HP and 25 mm in front of the VP.
   E: In the HP and 20 mm behind the VP.
   F: 40 mm above the HP and in the VP.
   G: In both the HP and the VP.

Draw the reference line xy long enough to accommodate the projections of all points keeping a distance of 20 mm between the projectors as shown in Fig. 6.9. Capital letters are marked at the top of the projectors to show the corresponding projections. The projections of the points are as follows:

A: The elevation a′ is 40 mm above xy and the plan a is 25 mm below xy.
B: The elevation b′ is 15 mm above xy and the plan b is 50 mm above xy.
C: The elevation c′ is 25 mm below xy and the plan c is 25 mm above xy.
D: The elevation d′ is 40 mm below xy and the plan d is 25 mm below xy.
E: The elevation e′ is on xy and the plan e is 20 mm above xy.
F: The elevation f′ is 40 mm above xy and the plan f is on xy.
G: The elevation g′ and the plan g are both on xy.

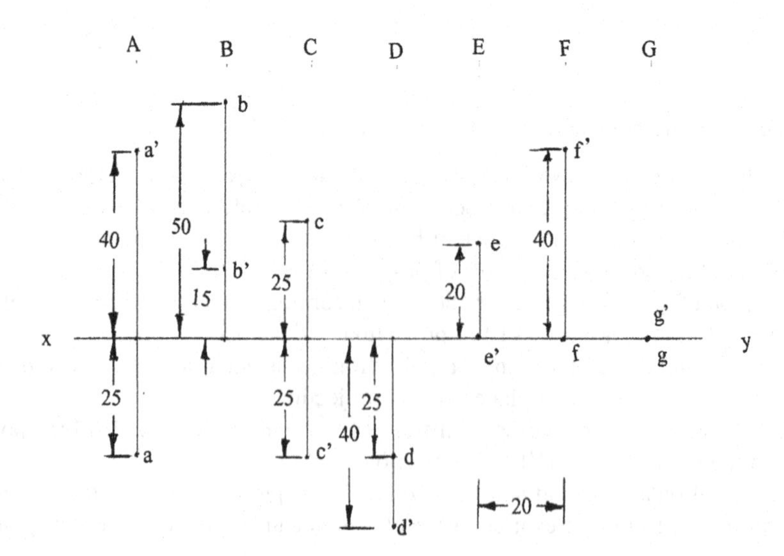

Fig. 6.9 Projections of points A, B, C, D, E, F and G

The projections of the points are shown in the figure.

2. A point P is 50 mm from the HP and 30 mm from the VP. Draw its projections in all possible positions.

   Let P1, P2, P3 and P4 be the four points in the four quadrants with the given distances from the co-ordinate planes. The point P1 is 50 mm above the HP and 30 mm in front of the VP. Its elevation p1′ is 50 mm above xy and plan p1 is 30 mm below xy. Similarly, the projections of other three points P2, P3 and P4 are obtained with their positions from xy. The projections of the four points are shown in Fig. 6.10 keeping a distance of 20 mm between their projectors.
3. State the quadrants in which the following points are situated:

   (i) A point R; its plan is 40 mm above xy line; the elevation is 20 mm below the plan.
   (ii) A point S; its projections coincide and lie 40 mm below xy line.

The projections of the points are shown in Fig. 6.11 on the same reference line xy.

Point R: The elevation r′ is 20 mm above xy and the point R is 20 mm above the HP. The plan r is 40 mm above xy and the point R is 40 mm behind the VP. So, the point R is situated in the second quadrant.

Point S: The elevation s′ is 40 mm below xy and the point is 40 mm below the HP. The plan s is 40 mm below xy and the point is 40 mm in front of the VP. So, the point S is situated in the fourth quadrant.

4. State the exact positions of the points whose projections are given in Fig. 6.12 with reference to the planes of projections giving the distances in mm.

   The following are the positions of the points whose projections are shown in Fig. 6.12:

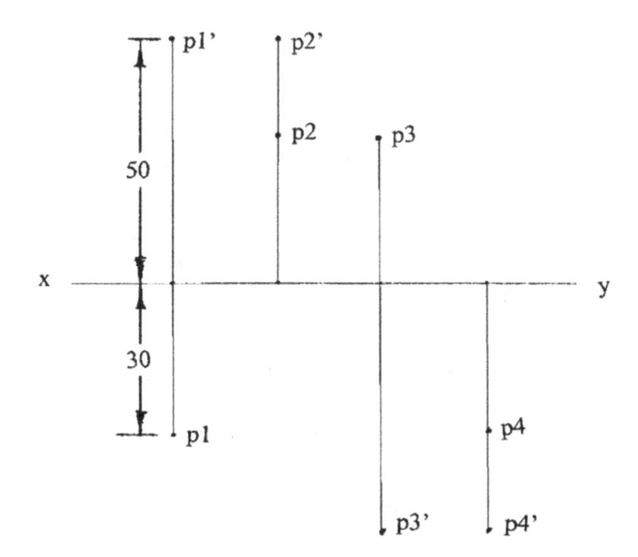

Fig. 6.10 Projections of points P1, P2, P3 and P4

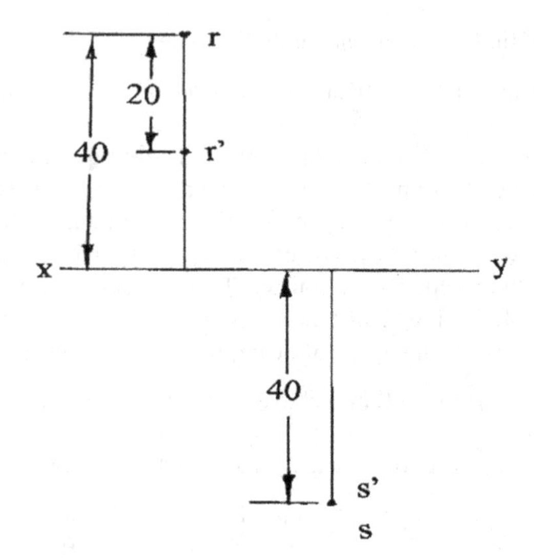

Fig. 6.11 Projections of points R and S

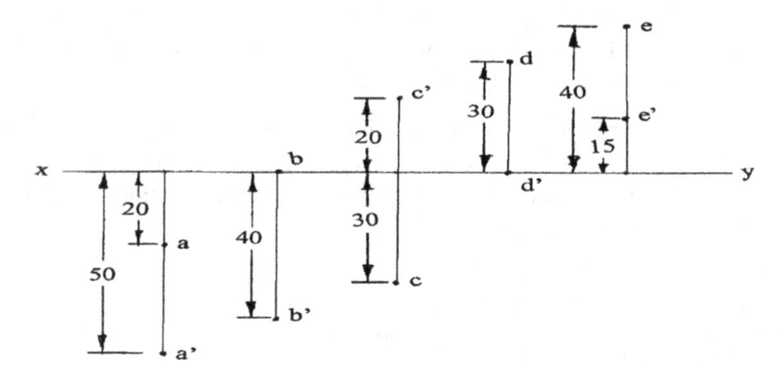

Fig. 6.12 Projections of points A, B, C, D and E

A: 50 mm below the HP and 20 mm in front of the VP.
B: 40 mm below the HP and in the VP.
C: 20 mm above the HP and 30 mm in front of the VP.
D: In the HP and 30 mm behind the VP.
E: 15 mm above the HP and 40 mm behind the VP.

5. A point 40 mm above xy line is the plan of three points A, B and C. The point A is 20 mm above, the point B is 30 mm below and the point C is on the HP. Determine the projections of A, B and C.

   The plans of the given points are located 40 mm above the reference line xy as shown in Fig. 6.13. The elevations of these points lie on the projector drawn from the plan of the points. The elevation a′ is 20 mm above xy. The elevation b′ is 30 mm below xy. The elevation c′ lies on xy. The projections of the points are shown in the same figure.

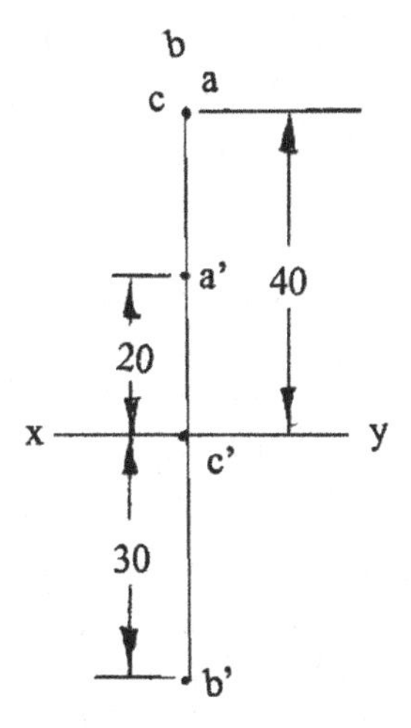

Fig. 6.13 Projections of points A, B and C

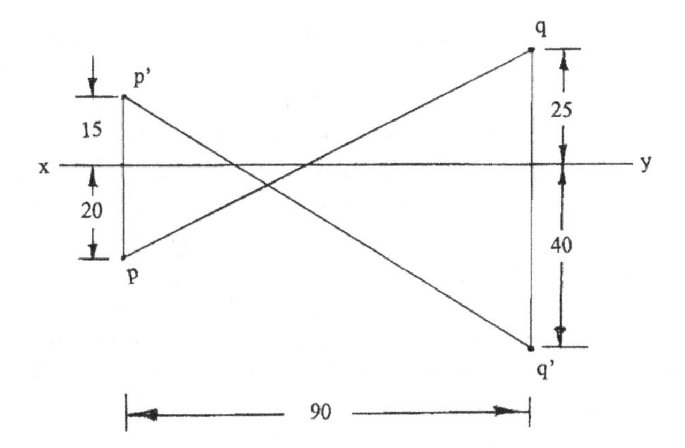

Fig. 6.14 Projections of points P and Q

6. A point P is 15 mm above the HP and 20 mm in front of the VP. Another point Q is 40 mm below the HP and 25 mm behind the VP. Draw the projections of P and Q keeping the distance between their projectors equal to 90 mm. Draw straight lines joining (i) their plans and (ii) their elevations.

   Draw the reference line xy as shown in Fig. 6.14. The elevation p′ is located on the projector drawn perpendicular to xy and 15 mm above xy. The plan p is located on the same projector 20 mm below xy. The elevation q′ is 40 mm below xy and the plan q is 25 mm above xy. Another projector at a distance of 90 mm from the projector P is drawn, and the points q′ and q are located on this projector. A line is drawn connecting p and q. Another line is drawn connecting p′ and q′. The elevation p′q′ and the plan pq of the line PQ are shown in the same figure.

7. Show the projections of the points A, B, C, D and E using the same reference line xy for all projections and making the projectors 20 mm apart.
   (i) The plan of A is 30 mm below xy and the point A is 40 mm above the HP.
   (ii) The plan of B is 35 mm above the xy and the point B is 35 mm above the HP.
   (iii) The elevation of C is on xy and the point C is 45 mm behind the VP.
   (iv) The plan d is on xy and the point D is on the HP.
   (v) The point E is 30 mm below the HP and e is 20 mm above xy.

The reference line xy is drawn, and the projectors are fixed on xy at intervals of 20 mm as shown in Fig. 6.15. The projections of the points are as follows:

A: The plan a is 30 mm below xy and the elevation a′ is 40 mm above xy.
B: The plan b is 35 mm above xy and the elevation b′ is 35 mm above xy.
C: The plan c is 45 mm above xy and the elevation c′ is on xy.
D: The plan d is on xy and the elevation d′ is also on xy.
E: The plan e is 20 mm above xy and the elevation e′ is 30 mm below xy.

The projections of the points are shown in Fig. 6.15.

8. A point touches all the three principal planes of projections. Draw its projections.

   The vertical plane, the horizontal plane and the profile plane are shown pictorially in Fig. 6.16; the intersection of the VP and the HP is shown by the reference line XY and the intersection of the VP, and the profile plane is shown by the reference line X_1Y_1. The HP is rotated about the line XY until it coincides with the VP. The profile plane is also rotated about X_1Y_1 until it coincides with the VP. The three planes are shown in the figure after rotation with the reference lines. The intersection of xy and x_1y_1 gives a point common to the three planes and hence this point P represents its projection.
9. Two points A and B are in the HP. The point A is 30 mm in front of the VP while B is behind the VP. The distance between their projectors is 75 mm, and

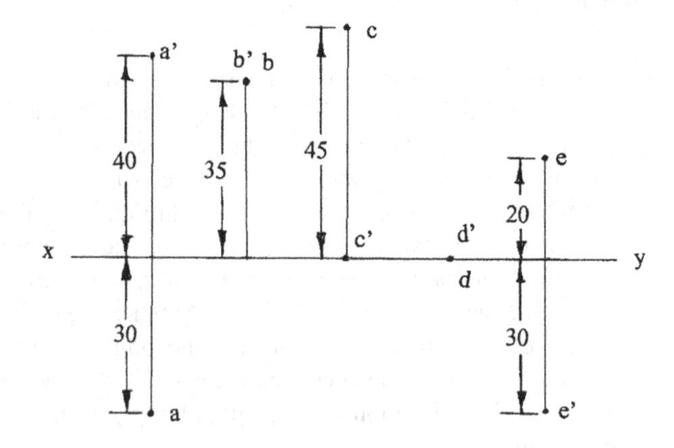

Fig. 6.15 Projections of points A, B, C, D and E

Fig. 6.16 Projections of P touching VP, HP and PP

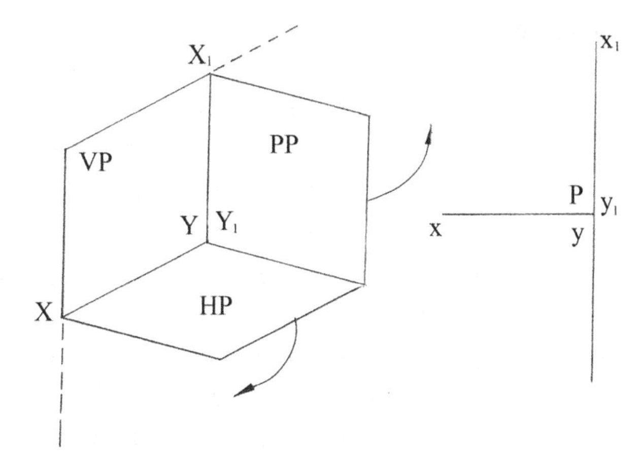

Fig. 6.17 Projections of points A and B

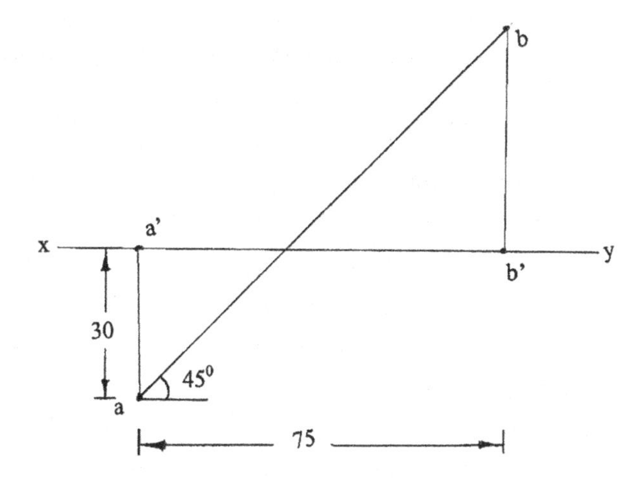

the line joining their plans makes an angle of 45° with xy. Find the distance of the point B from the VP.

Draw the reference line xy as shown in Fig. 6.17. Two projectors are drawn keeping a distance of 75 mm between them and the elevations a′ and b′ are located on xy as shown in the figure. The plan a is located 30 mm below xy from a′. Since the point B is behind the VP, its plan will be above xy on the projector drawn from b′. As the plan makes an angle of 45° with xy, a line is drawn from a making an angle of 45° as shown in the figure. The intersection of this line with the projector B fixes the plan b. The distance of b from xy measures the distance of B from the VP. The point B is 45 mm behind the VP.

10. Draw the projection of a point 30 mm above the HP and in the first quadrant, if its shortest distance from the line of intersection of reference planes is 50 mm. Also find the distance of the point from the VP.

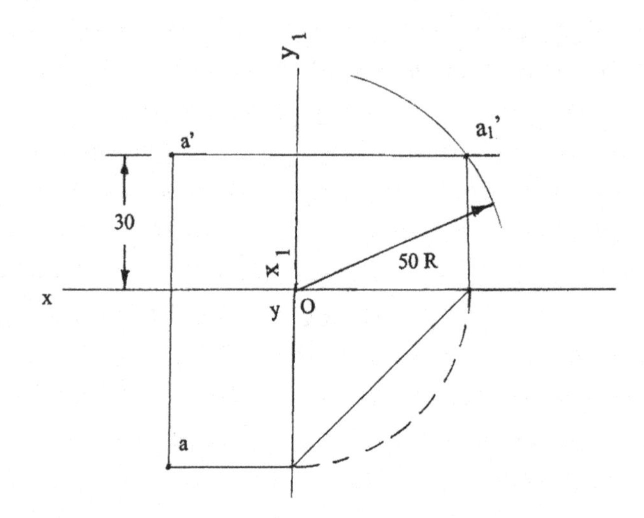

Fig. 6.18 Projections of point A

Draw the reference line xy and locate the elevation a′ of the point above xy as shown in Fig. 6.18. Assume that the projection of this point is to be obtained on the profile plane with reference to the new reference line x_1 y_1, the intersection of the vertical plane and the profile plane as shown in the figure. A projector is drawn from a′ perpendicular to x_1 y_1, and the elevation a_1' on the profile plane should lie along this line from x_1 y_1. The intersection of reference line xy with x_1y_1 is designated as O. An arc of radius 50 mm is drawn with centre O to intersect the projector drawn from a′. The intersecting point gives the projection of this point on the profile plane. The distance of a_1' from x_1 y_1 gives the distance of the point from the VP. The plan a is located on the projector drawn perpendicular to xy from a′ as shown in the figure. The plan a is 40 mm from xy. Hence, the point is 40 mm in front of the VP.

Practice Problems

1. Draw the projections of the following points using the same reference line xy for all projections and keeping the projectors 20 mm apart.

   A, 50 mm above the HP and 60 mm in front of the VP
   B, 50 mm above the HP and 60 mm behind the VP
   C, 40 mm below the HP and 30 mm behind the VP
   D, 30 mm below the HP and 40 mm in front of the VP
   E, 35 mm above the HP and in the VP
   F, on the HP and 40 mm in front of the VP.

2. A point 30 mm above xy is the elevation of two points A and B. The plan of A is 40 mm behind the VP and the plan of B is 40 mm in front of the VP. Draw the projections of the two points and state their positions with reference to the co-ordinate planes and the quadrants in which they are situated.
3. Draw the projections of the points in Table 6.1 at 20 mm intervals along the xy line.
4. State the position of each of the points in Table 6.2 with reference to the HP and the VP and also the quadrant in which the point is situated:
5. Draw the projections of a point lying 25 mm in front of the VP and in the first quadrant if its shortest distance from the line of intersection of the HP and the VP is 65 mm. Also find the distance of the point from the HP.
6. The plan of the point P lies 40 mm above the reference line xy and its elevation 50 mm above the reference line xy. Mention the quadrant in which the point is situated. Draw its projections and find the shortest distance of the point from the intersection of the HP and VP.
7. A point Q is situated in a plane inclined 30° to the HP and passing through the intersection of the HP and VP. Draw the projections of the point if the distance of Q is 30 mm above the HP.
8. A point R is situated in a plane inclined 40° to the VP and passing through the intersection of the HP and VP. Draw the projections of R if the distance of R is 50 mm in front of the VP.
9. A triangle ABC is in the first quadrant. A is in the VP and 50 mm above the HP. B and C are in the HP. The plan of line ab makes 45° with the reference line xy and the plan of the triangle abc is an equilateral triangle of 50 mm side. Draw the plan and elevation of the triangle.

Table 6.1 Positions of points

Point	A	B	C	D	E	F	G
Distance from the HP (mm)	40	20	10	20	−10	−20	0
Distance from the VP (mm)	20	10	0	−10	−20	30	10

Table 6.2 Projections of points

Point	Elevation	Plan
A	50 mm above xy	20 mm below xy
B	10 mm below xy	30 mm above xy
C	40 mm above xy	20 mm above xy
D	30 mm below xy	40 mm below xy

Chapter 7
Projections of Lines

The projections of a straight line in various positions in the first quadrant are grouped based on its orientation with reference to the vertical and horizontal planes. They are:

1. Parallel to both the planes
2. Perpendicular to one plane
3. Parallel to one plane and inclined to the other plane
4. Lying in one plane
5. Lying in the reference line xy (intersection of the VP and the HP)
6. Inclined to both the planes
7. Inclined to both the planes with an end in the reference line xy
8. Inclined to both the planes with ends in both the planes
9. Perpendicular to the reference line xy

The methods of obtaining the projections are classified under three categories.

Type I

The straight line and its projection on a plane have the same length when the line is parallel to that plane. Hence, it is straight-forward to obtain one of the projections. When the straight line is parallel to both the planes, the projections have the same length. After marking the projections of one end of the line, the projections of the line are completed. There are six problems (nos. 1–6) that belong to this category.

Type II

The projections of a straight line inclined to both the planes will be given or stated. The true length, inclinations to the planes of projection and the meeting points of extension of the line with the planes (traces) are obtained by constructing trapeziums on the projections of the line itself. Following this method, the solutions to nine problems (nos. 7–15) are obtained.

K. Rathnam, *A First Course in Engineering Drawing*,
DOI 10.1007/978-981-10-5358-0_7

Type III

The method of obtaining the projections of a straight line inclined to both the planes is explained. This method is reversed to get the true length and inclinations to the reference planes from the projections of the line inclined to both the planes. An easy method of locating the traces alone from the projections of the line is also given with a caution that this simple procedure poses a problem when the projections of the line are perpendicular to the reference line xy. The remaining problems in this chapter are solved by following the method detailed in this type or by some combinations.

Solved Problems

1. Draw the projections of a line 40 mm long when it is parallel to both the HP and the VP. The line is 20 mm from both the planes.

   Consider the pictorial representation of the line AB in the first quadrant as shown in Fig. 7.1. The line is 20 mm above the HP and 20 mm in front of the VP. Horizontal projectors are drawn from A and B to meet the vertical plane at a′ and b′. Since the line is parallel to the VP, a′b′ will be parallel to AB. The elevation a′b′ will be equal to the true length of the line AB (i.e. a′b′ = 40 mm). Vertical projectors are drawn from A and B to meet the horizontal plane at a and b. Since the line is parallel to the HP, the plan ab will be equal to the true length

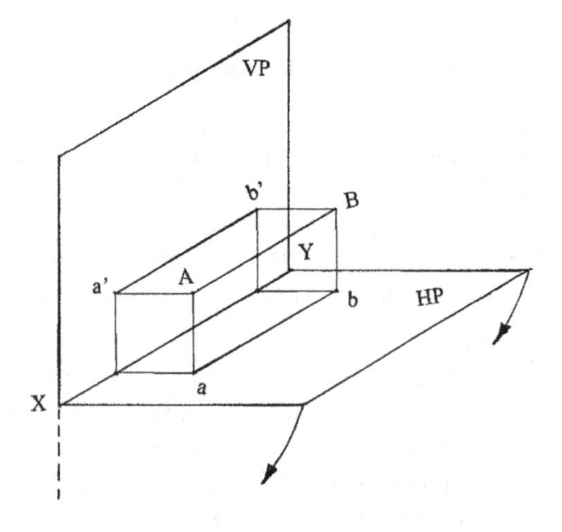

Fig. 7.1 Pictorial representation

of the line (i.e. ab = 40 mm). The horizontal plane is rotated about XY to fall on the vertical plane. The projections of the line AB obtained on the VP and the HP are shown in Fig. 7.2.

2. A vertical line AB 40 mm long is 20 mm in front of the VP, the lower end B being in the HP. Draw its projections.

   The pictorial representation of the line is shown in Fig. 7.3. Horizontal projectors are drawn from A and B to meet the VP at a′ and b′. Since the line is vertical, the elevation a′b′ will be equal to the true length AB. The point B is on the HP and its projection on the VP lies on the reference line xy. As the line is vertical and 20 mm in front of the VP, its projection on the HP will be a point and 20 mm below the reference line xy. The projections of the line a′b′ and ab are shown in Fig. 7.4.

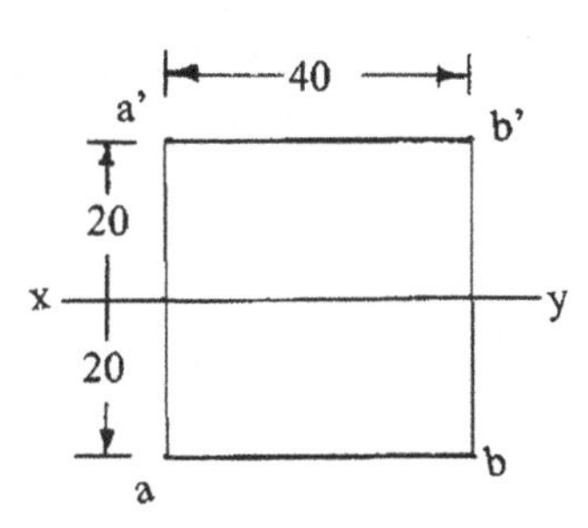

Fig. 7.2 Projections of line AB

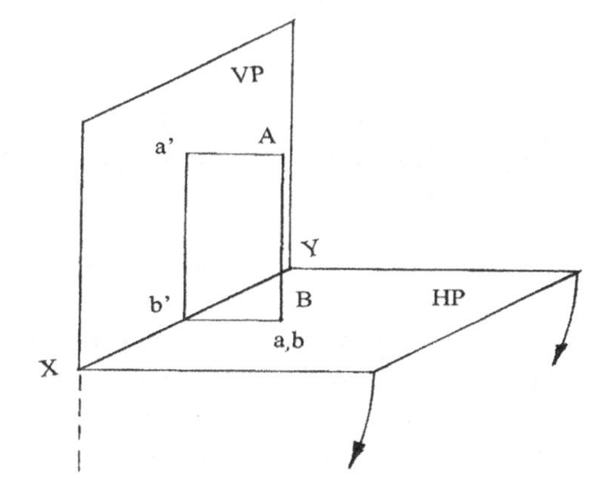

Fig. 7.3 Pictorial representation

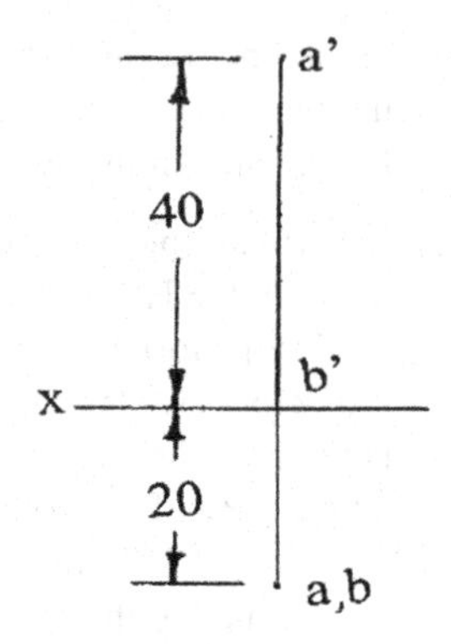

Fig. 7.4 Projections of line AB

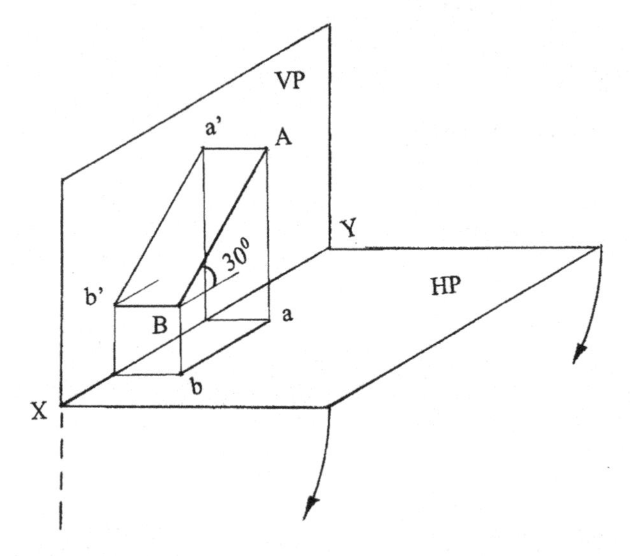

Fig. 7.5 Pictorial representation

Note: The pictorial representation is given for guidance only and is not required to be drawn. The projections of the line are to be drawn as the solution to the problem.

3. A line 40 mm long is parallel to the VP and inclined at 30° to the HP. The end nearer to the HP is 15 mm above the HP and 15 mm in front of the VP. Draw its projections.

The pictorial representation of the line AB is shown in Fig. 7.5. Since the line is parallel to the VP, its elevation a′b′ will be equal to the true length and the plan ab will be parallel to the reference line xy. The elevation a′b′ is inclined at 30° to the HP. The elevation b′ of the point B which is nearer to the HP is 15 mm above xy and the plan b is 15 mm below xy. The projections of B are located on a projector drawn perpendicular to the reference line xy as shown in Fig. 7.6. A horizontal line is drawn at b′ and a line b′d′ inclined at 30° to the horizontal is drawn. The elevation a′ is located on b′d′ 40 mm from b′. A projector is drawn from a′ to locate the plan a at the intersection of the projector with the line drawn

Fig. 7.6 Projections of line AB

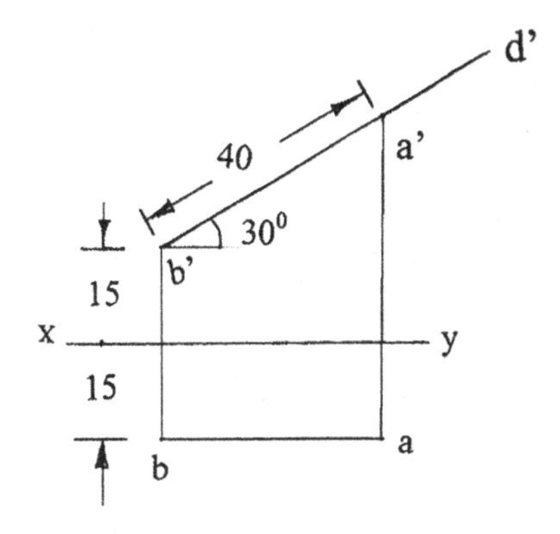

Fig. 7.7 Projections of line AB

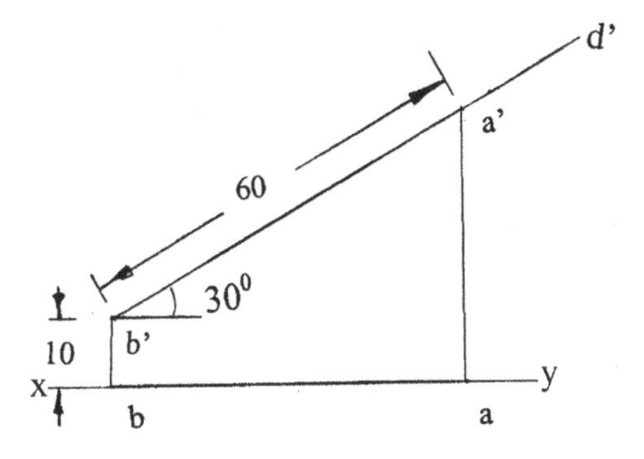

parallel to xy from b. The projections a′b′ and ab of the line AB are shown in Fig. 7.6.

4. Draw the projections of a line AB 60 mm long lying in the VP and inclined to the HP at 30°. The end which is nearer to the HP is 10 mm above it.

   The elevation a′b′ will be 60 mm since the line is lying on the VP. Also, the plan ab will be on the reference line xy. The elevation b′ of the end B, which is nearer to the HP, is 10 mm above xy. Draw the reference line xy and locate the plan b on the reference line xy as shown in Fig. 7.7. A projector is drawn from b and the elevation b′ is located 10 mm above xy. A line b′d′ inclined at 30° to xy is drawn from b′ as shown in the figure. The point a′ is located on b′d′ 60 mm from b′. A projector is drawn from a′ to meet the reference line xy at a. This completes the projections.

5. The plan ab of a line makes an angle of 30° with xy, ab being 45 mm long. The point B is in the HP and the point A is in the VP and 40 mm above the HP. Draw the projections of the line AB.

   The elevation a′ of the line is 40 mm above xy and the elevation b′ of the line is on xy. The plan a of the line is on xy. Draw the reference line xy and locate the plan a as shown in Fig. 7.8. Draw a line ae inclined at 30° with xy as shown in the figure and locate b along this line 45 mm from a. Draw projectors from a and b. The elevation a′ is located 40 mm above xy and b′ is on xy. The elevation points a′ and b′ are connected to complete the projections.

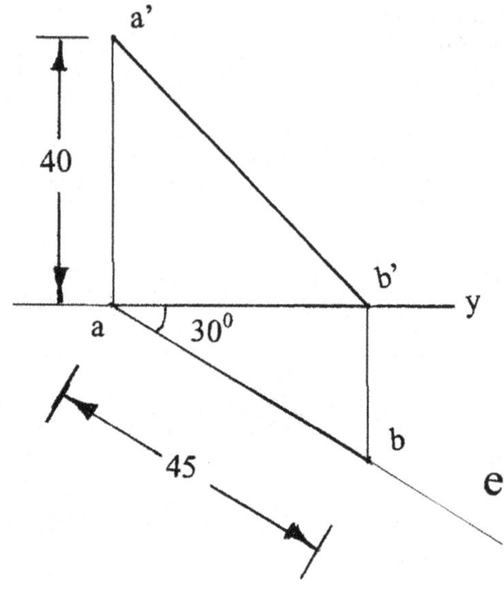

Fig. 7.8 Projections of line AB

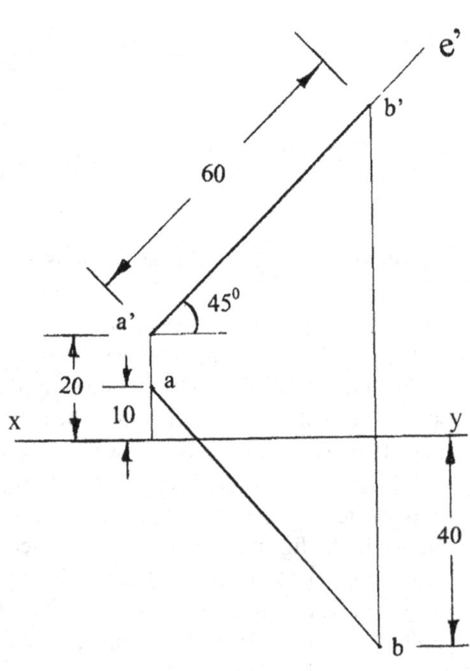

Fig. 7.9 Projections of line AB

6. The elevation a′b′ of a line is inclined at 45° to xy, a′b′ being 60 mm long, a′ is 20 mm above xy and b′ is also above xy. The end A is 10 mm behind VP and the end B is 40 mm in front of the VP. Draw the projections of the line.

   The plan a of the point A is 10 mm above xy and the elevation a′ is 20 mm above xy. Draw the reference line xy and locate the elevation a′ and the plan a along a projector drawn perpendicular to xy as shown in Fig. 7.9. Since the elevation b′ is also above xy, a line a′e′ is drawn inclined at 45° with xy from a′ as shown in the figure. The point b′ is located 60 mm from a′ on this line. A projector is drawn from b′ to locate the plan b 40 mm below the reference line xy and the points a and b are joined. This completes the projections of the line AB. It is to be noted that the end A of the line lies in the second quadrant and the end B lies in the first quadrant.

Trapezium Method

The determination of true length and inclinations to the reference planes are obtained from the projections of a line inclined to both the planes by following the trapezium method. The pictorial representation of a straight line AB and its projections on the reference plane appear in Fig. 7.10. It is obvious from Fig. 7.10 that the line AB, its plan ab and vertical projectors Aa and Bb form a trapezium. Also Aa is equal to the distance of a′ from xy, Bb is equal to the distance of b′ from xy and the angles Aab and Bba are right angles. The trapezium aABb is rotated about the plan ab to fall on the HP as shown in Fig. 7.10. The corresponding projections appear in Fig. 7.11. A line is drawn perpendicular to ab at a. Point A is located on this line taking the distance of a′ from xy as shown in Fig. 7.11. Another line is drawn perpendicular to ab at b. Point B is located on this line taking the distance of b′ from xy (Note: If the points A and B are on opposite sides of the HP, the elevations a′ and b′ appear on opposite sides of the reference line xy. The

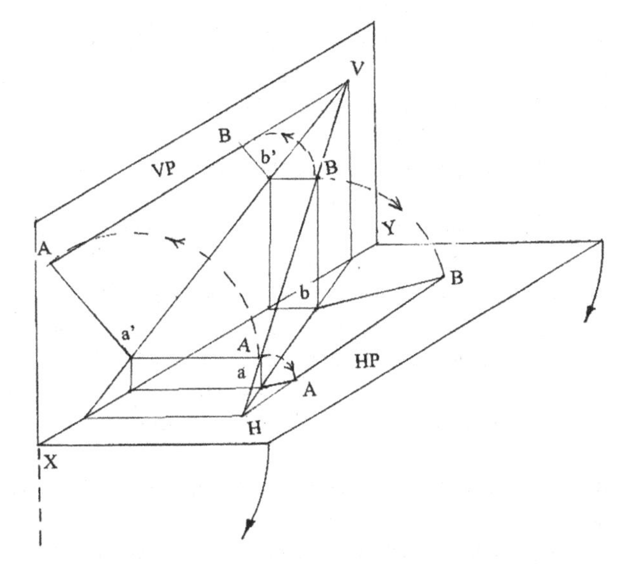

Fig. 7.10 Pictorial representation

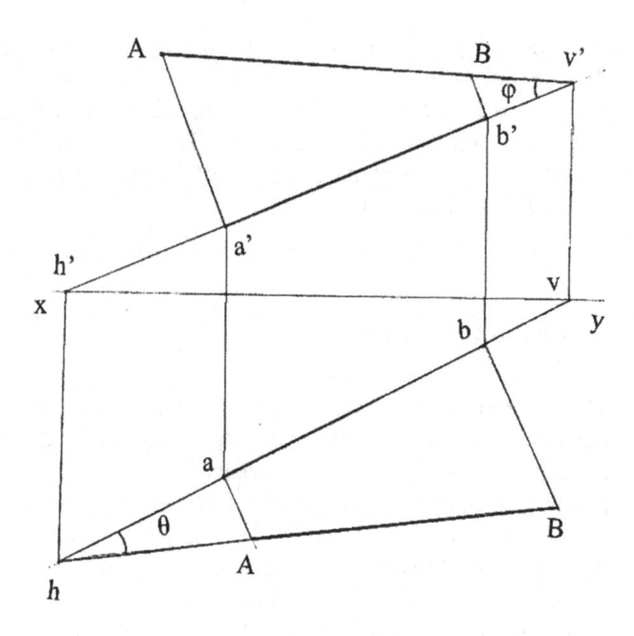

Fig. 7.11 Projections of line AB

perpendiculars aA and Bb must be drawn on opposite sides of the plan ab). Join the points A and B and extend it to meet the extension of the plan ab at h (H in Fig. 7.10). The meeting point is called the horizontal trace of the line, and the angle between ab and AB is the inclination of the line to the horizontal plane. The construction given for finding the true length on the plan of the line also serves for finding its inclination to the horizontal plane and the horizontal trace, h. It is evident that the elevation h′ of the horizontal trace should lie on xy. The inclination of the line to the horizontal plane is denoted by the symbol θ (θ is the alphabet O with a horizontal line through it, i.e. θ).

Refer to the pictorial representation in Fig. 7.10 once again. The line AB, its elevation a′b′ and horizontal projectors Aa′ and Bb′ form a trapezium. Also Aa′ is equal to the distance of a from xy, Bb′ is equal to the distance of b from xy and the angles Aa′b′ and Bb′b are right angles. The trapezium is rotated about the elevation a′b′ to fall on the VP as shown in Fig. 7.10. Refer to the projections of the line in Fig. 7.11. A line is drawn at a′ perpendicular to a′b′ and on this line the point A is located taking the distance of a from xy. Another line is drawn at b′ perpendicular to a′b′ and on this line the point B is located taking the distance of b′ from xy (If the points A and B are on opposite sides of the VP, the plans a and b will appear on opposite sides of the reference line xy. The perpendiculars a′A and b′B must be drawn on opposite sides of the elevation a′b′). Join the points A and B and extend it to meet the extension of the elevation a′b′ at v′ (V in Fig. 7.10). The meeting point is the vertical trace of the line, and the angle between a′b′ and AB is the inclination of the line AB to the vertical plane. The construction given for finding the true length AB from its elevation serves for finding its inclination to the vertical plane and the vertical trace, v′. The inclination of the line to the vertical plane is denoted by the symbol φ (φ is the alphabet O with a vertical line through it, i.e. φ). The plan v of the vertical trace lies on xy.

7. The elevations a′ and c′ of two points A and C in the VP are shown in Fig. 7.12. The plan b of another point B in the HP is also shown in Fig. 7.12. Determine the distances between A&B, B&C and C&A.

   The projections of the points are redrawn in Fig. 7.13. Since the points A and C are in the VP, the plans a and c lie on xy as shown in Fig. 7.13. The elevation b′ of the point B is also located on xy. Join the points a, b and c in pairs and also the points a′, b′ and c′ in pairs. Since the points A and C lie on the VP, the elevation a′ and c′ include the space points A and C also. As the point B is on the HP, its plan b includes the space point B also. A line is drawn perpendicular to ab at

Fig. 7.12 Projections of points

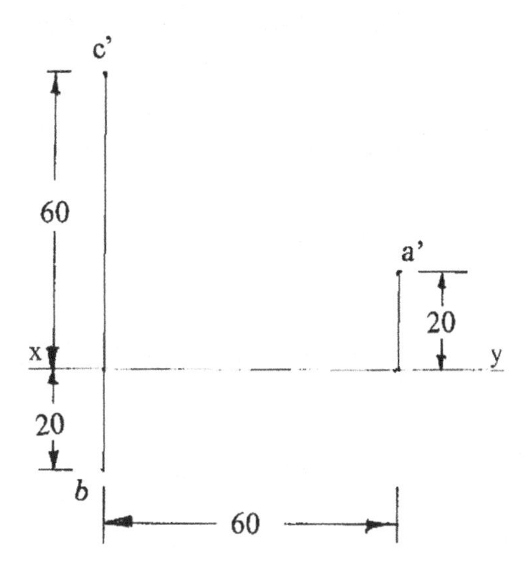

Fig. 7.13 Projections of lines AB, BC and CA

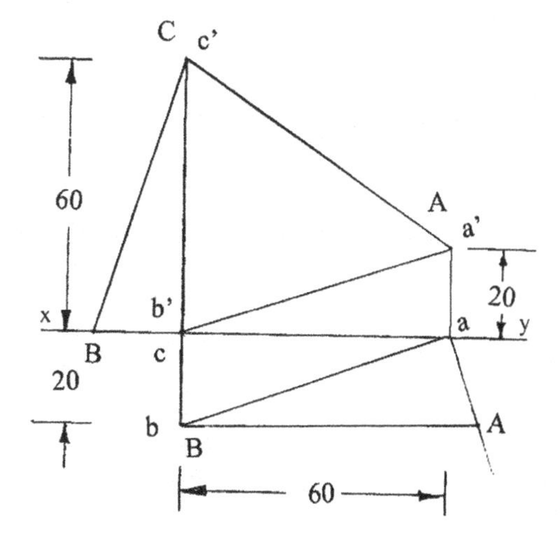

a. The point A is located on this line taking the distance of a′ from xy. Join the points A and B and measure the true length of AB. Draw a line perpendicular to b′c′ at b′. The point B is located on this line taking the distance of b from xy as shown in Fig. 7.13. Join the point B and C and measure the true length of BC. The points C and A coincide with the elevation c′a′ and give directly the true length of CA. Answers: AB = 66 mm; BC = 63 mm; CA = 72 mm.

8. A line AB is shown by its projections a′b′ and ab in Fig. 7.14. Find the length of AB and its inclinations to the HP and the VP.

   The projections of the line AB are redrawn as shown in Fig. 7.15. A line is drawn perpendicular to ab at a, and point A is located on this line taking the distance of a′ from xy. A line is drawn perpendicular to ab at b, and point B is located on this line taking the distance of b′ from xy. Join the points A and B and extend it to meet the extension of ab as shown in Fig. 7.15. The distance between A and B measures its true length, and the angle between AB and ab gives the

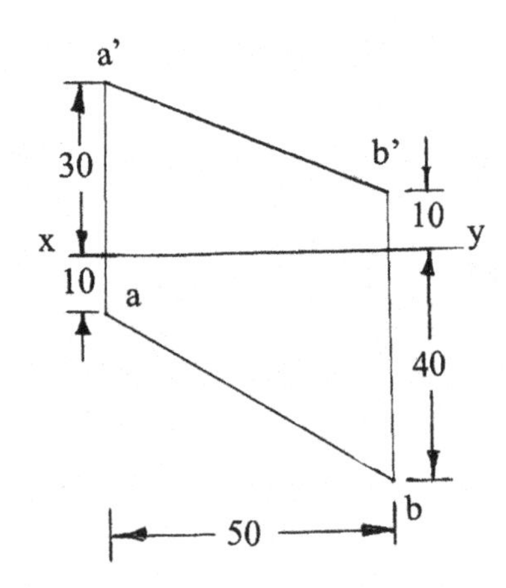

Fig. 7.14 Projections of points

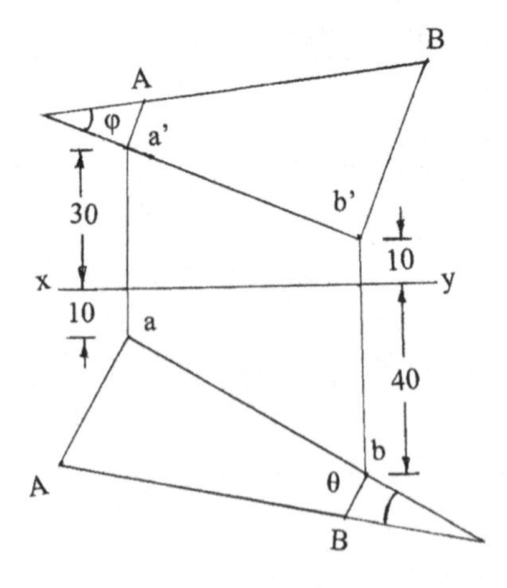

Fig. 7.15 Projections of line AB

inclination of the line to the HP, θ. Two lines are drawn perpendicular to a′b′ at a′ and b′. The points A and B are located on these lines taking, respectively, the distances of a and b from the reference line xy. Join the points A and B and extend it to meet the extension of a′b′ as shown in Fig. 7.15. The angle between AB and a′b′ gives the inclination of the line AB to the VP, φ.

Answers: AB = 61 mm; θ = 19°; φ = 29°.

9. The plan of a line 60 mm long is shown in Fig. 7.16. The elevation of one end is at b′. Complete the elevation and measure the inclinations of the line to the HP and the VP.

   The given projections of the line AB are drawn as shown in Fig. 7.17. Two lines are drawn at a and b perpendicular to ab. The point B is located on the line drawn from b taking the distance of b′ from xy. With B as centre, an arc of radius equal to 60 mm is drawn to intersect the line drawn from a as shown in Fig. 7.17. The intersecting point is A. Join the points A and B. The line AB is extended to meet the extension of ab at h. The angle between AB and ab gives the inclination of the line to the HP. The projector of a is extended above xy, and point a′ is located on it taking the distance of A from a. Join a′ and b′ to show the elevation a′b′ of the line AB. Lines are drawn perpendicular to a′b′ at a′ and b′. Point A and B are located on these lines taking the distances of a and b from xy. Join the points A and B and extend it to meet the extension of a′b′ at v′. The angle between AB and a′b′ gives the inclination of the line to the VP.

   Answers: θ = 42°; φ = 20°.

10. A point P is 20 mm above the HP and 20 mm in front of the VP. Another point Q is 15 mm below the HP and 45 mm behind the VP. The distance between the projectors measured along xy is 70 mm. Draw the projections of the line and determine the true length, traces and inclinations of the line to the co-ordinate points.

    The elevation p′ is 20 mm above xy and the plan p is 20 mm below xy. The elevation q′ is 15 mm below xy and the plan q is 45 mm above xy. The distance between the projectors is 70 mm. The projections of the line PQ are shown in Fig. 7.18. Two lines are drawn perpendicular to pq on both sides of the plan as the elevations lie on opposite sides of xy. Points P and Q are located on the

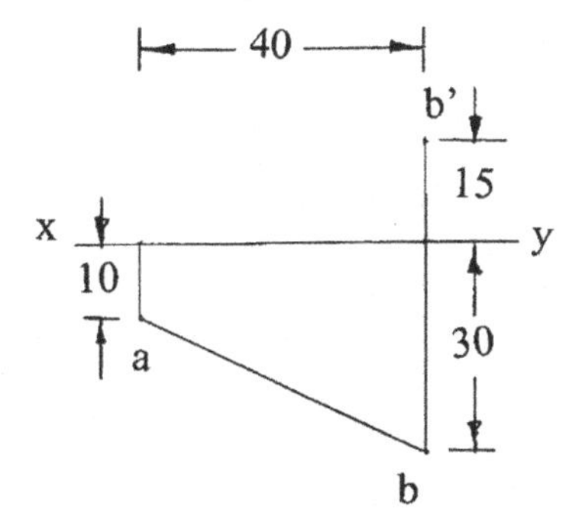

Fig. 7.16 Projections of points

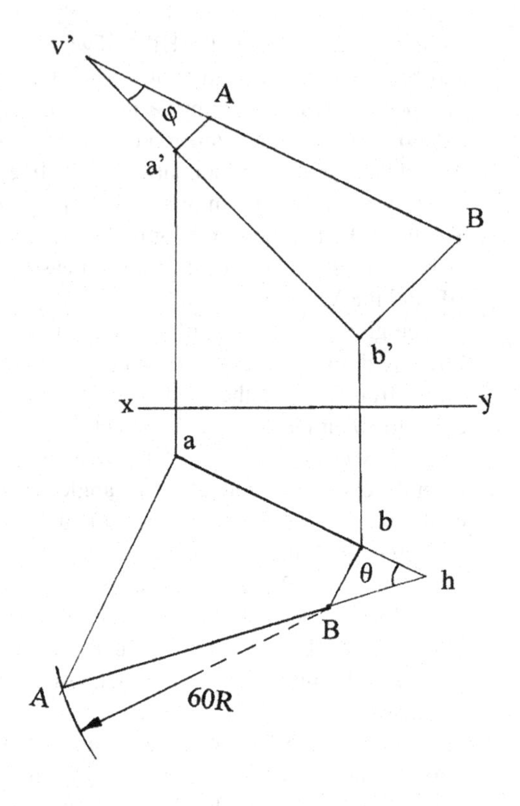

Fig. 7.17 Projections of line AB

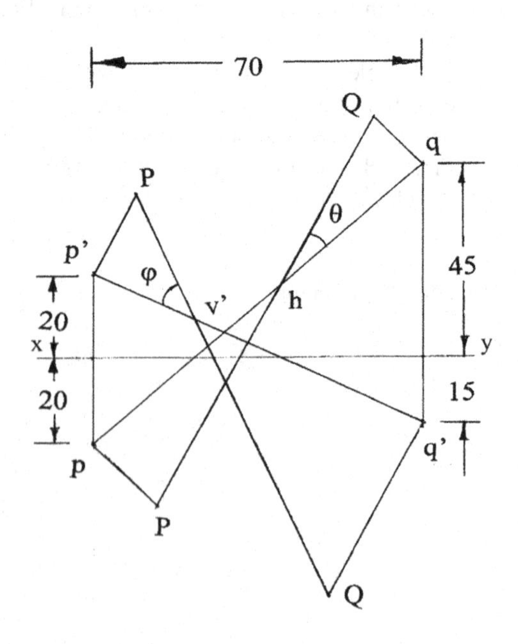

Fig. 7.18 Projections of line PQ

respective lines taking the distances of p′ and q′ from xy as shown in the figure. Join P and Q and measure its length. Since the line PQ intersects the plan pq at h, the angle between the intersecting lines measures the inclination of the line to the HP. The intersecting point locates the horizontal trace of the line. Two lines are drawn perpendicular to p′q′ at p′ and q′ on both sides of the elevation as the plans p and q are on opposite sides of xy. Points P and Q are located on the respective lines taking the distances of p and q from xy. Join P and Q and check with the true length already obtained. Since the line PQ intersects its elevation p′q′ at v′, the angle between the intersecting lines measures the inclination of the line to the VP. The intersecting point locates the vertical trace of the line.

Answers: PQ = 101 mm; $\theta = 20°$; $\varphi = 40°$.

11. The plan of a line is 60 mm long and is inclined to xy at 30°. One end is 20 mm above the HP and 10 mm in front of the VP. The other end is 40 mm above the HP and is in front of the VP. Determine the true length of the line and its inclinations to the HP and the VP. Locate its traces also.

Let the line be designated as AB. The plan ab is 60 mm long and is inclined to xy at 30°. The elevation a′ is 20 mm above xy and the plan a is 10 mm below xy. Draw the reference line xy and locate a′ and a on the projector A as shown in Fig. 7.19. A line ae is drawn from a inclined at 30° to xy as shown in the

Fig. 7.19 Projections of line AB

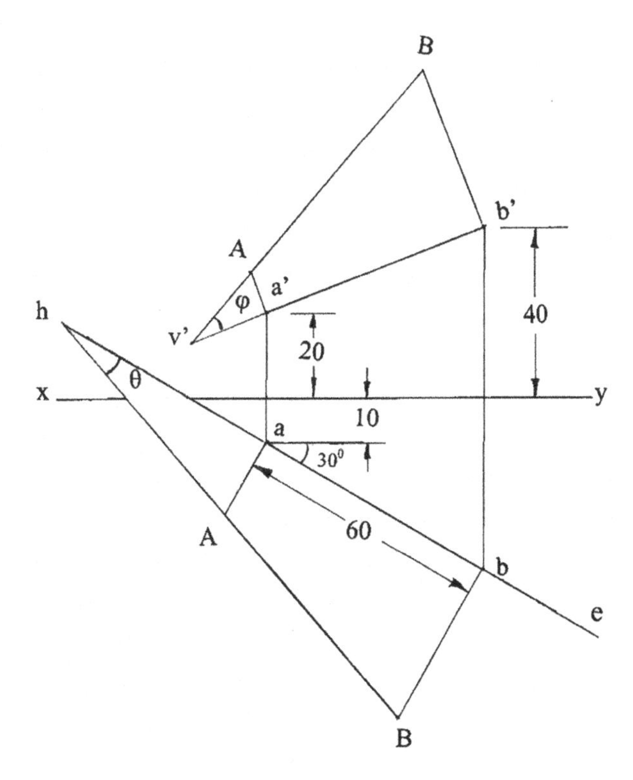

figure. The point b is located on this line 60 mm from a, and the projector B is drawn through b. The elevation b′ is located 40 mm above xy. Complete the elevation a′b′ by joining the points a′ and b′. A trapezium abBA is constructed on the base ab as shown in the figure to determine the true length of the line AB and the inclination of the line to the HP. The horizontal trace is also located by extending the lines ba and BA as shown in the figure. Similarly another trapezium a′b′BA is drawn on the base a′b′ to determine the true length and the inclination of the line to the VP. The vertical trace is also located by extending the lines BA and b′a′.

Answers: AB = 63 mm; $\theta = 19°$; $\varphi = 28°$.

12. AB is a straight line 60 mm long. The end A is 10 mm above the HP and 20 mm in front of the VP. The plan measures 50 mm. The end B is above the HP and 30 mm in front of the VP. Draw the projections of the line AB and find its inclinations to the HP and the VP.

    The elevation a′ is 10 mm above xy and the plan a is 20 mm below xy. The elevation b′ is above xy and the plan b is 30 mm below xy. Draw the reference line xy and locate a′ and a as shown in Fig. 7.20. Draw a line 30 mm below xy and parallel to it. This line represents the locus of b (the plan of point B). With

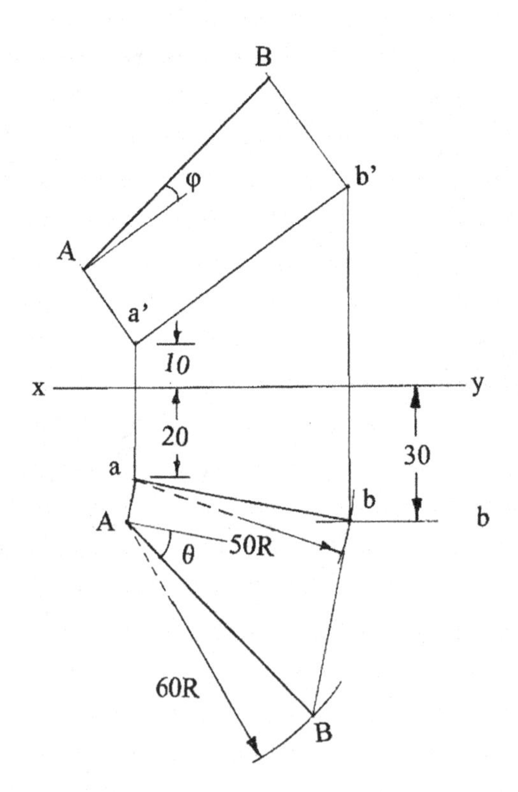

Fig. 7.20 Projections of line AB

centre a and the given plan length ab 50 mm as radius, draw an arc to cut the locus line b drawn below xy at b. Join the points a and b. Two lines are drawn perpendicular to ab at a and b. The point A is located along the line passing through a taking the distance of a′ from xy (10 mm in this case). With centre A and the true length as radius draw an arc to cut the line passing through b to locate the point B. The distance Bb will be equal to the distance of b′ from xy or the distance of B above the HP. The elevation b′ is located on the projector passing through b. Join the points a′ and b′. The trapezium a′b′BA is now constructed on the base a′b′ taking, respectively, the distances of a and b from xy. Draw a line parallel to a′b′ at A. The angle between this line and the line AB gives the inclination of the line to the VP. A line is drawn parallel to ab at A in the trapezium abBA to find the inclination of the line to the HP.

Answers: $\theta = 34°$; $\varphi = 9°$.

13. The plan of a line is 50 mm long and it makes 35° with xy, the elevation makes 45° with xy and the line intersects xy. Draw its projections and find its true length and inclinations to the planes of projections.

Draw a reference line xy and mark a′ and a as the point of intersection of the line AB on xy as shown in Fig. 7.21. At a and a′ draw lines ae and a′e′ inclined, respectively, 35° and 45° with xy. The point b is located on ae 50 mm from a. A projector is drawn perpendicular to xy through b to intersect the line a′e′ at b′. A

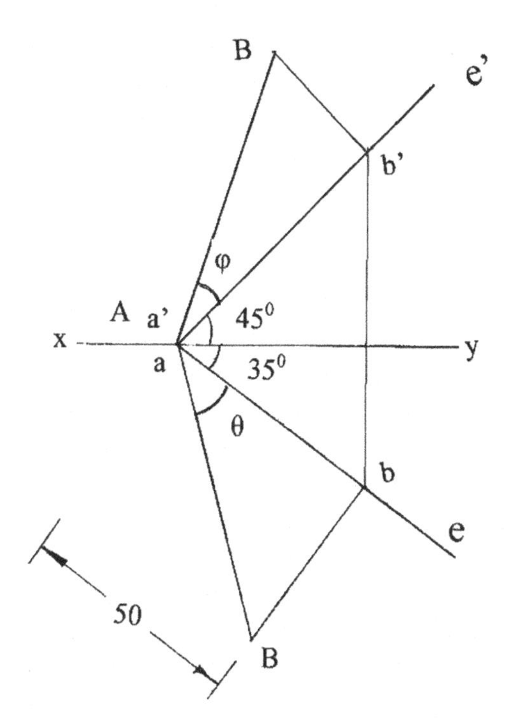

Fig. 7.21 Projections of line AB

line is drawn at b perpendicular to ab, and point B is located taking the distance of b′ from xy. The point of intersection of the line with xy locates the point A also. Join A and B and measure its length AB. The angle between AB and ab gives the inclination of AB to the HP. A line is drawn at b′ perpendicular to a′b′, and point B is located taking distance of b from xy. Join A and B. The angle between AB and a′b′ gives the inclination of AB to the VP. Answers: $\theta = 39°$; $\varphi = 26°$; AB = 64 mm.

14. Line AB is in the first quadrant. Its ends A and B are, respectively, 20 mm and 50 mm in front of the VP. The distance between the end projectors is 45 mm. The line is inclined at 30° to the HP, and its horizontal trace is 10 mm below the xy line. Draw the projections and determine its true length. Locate the vertical trace also.

    The plan a, the plan b and the horizontal trace h are, respectively, 20 mm, 50 mm and 10 mm below xy. Draw the reference line xy, and the projections a and b are marked on the projectors 45 mm apart as shown in Fig. 7.22. Join a and b. Draw a horizontal line 10 mm below xy to indicate the locus of the

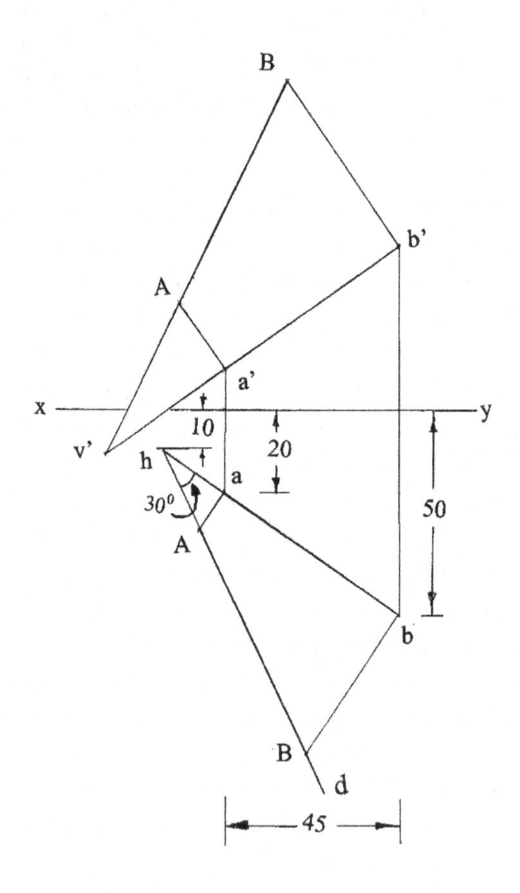

Fig. 7.22 Projections of line AB

horizontal trace h. Extend plan ba to intersect the horizontal line to locate the horizontal trace at h. Draw lines perpendicular to ab at a and b. Draw a line hd inclined at 30° to the extension of the plan ab at h and locate points A and B on the respective perpendicular lines as shown in the figure. The line AB measures its true length. The distance of A from a measures the distance of a′ from xy or the distance of A above the HP. The elevation a′ is marked on the projector A. Similarly, the elevation b′ is also marked on the projector B to complete the elevation a′b′. The trapezium a′b′BA is now constructed on the base a′b′ taking the distances of a and b from xy. The lines BA and b′a′ are extended to meet at the vertical trace v′ of the line.

Answers: AB = 62 mm.

15. The projections of a line AB are perpendicular to xy. The end A is 20 mm above the HP and 20 mm in front of the VP and the end B is 60 mm above the HP and 70 mm in front of the VP. Draw its projections and determine its true length and inclinations with the reference planes. [Answers: AB = 64 mm; $\theta = 39°$; $\varphi = 51°$. Note: $\theta + \varphi = 90°$]

The elevation a′ is 20 mm above xy and the plan a is 20 mm below xy. The elevation b′ is 60 mm above xy and the plan b is 70 mm below xy. Draw the reference line xy and the projector which contains projections of the line, i.e. a′, b′, a and b as shown in Fig. 7.23. The points a′ and b′ are located above xy. The points a and b are located below xy on the same projector. The

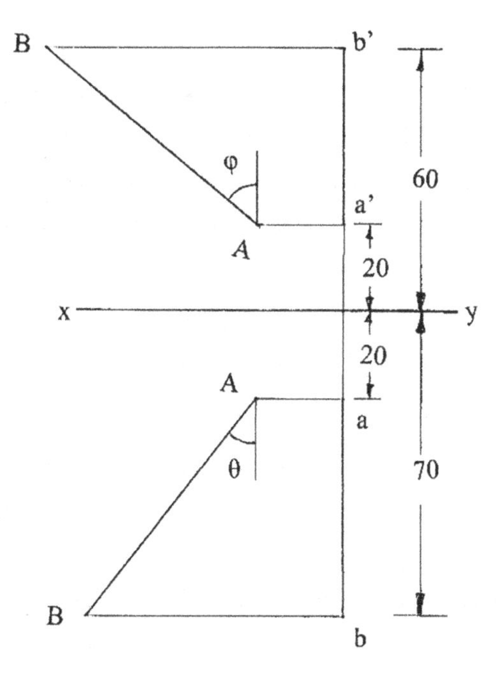

Fig. 7.23 Projections of line AB

elevation a′b′ and the plan ab are perpendicular to xy. The trapezium abBA is drawn on the base ab taking the distances of a′ and b′ from xy. AB measures the true length, and the angle between AB and ab gives the inclination of the line to HP. The trapezium a′b′BA is constructed on a′b′ taking the distances of a and b from xy. The angle between AB and a′b′ gives the inclination of the line to the VP.

Projections of a Line Inclined to Both the Planes

Let a line AB of length TL be inclined at an angle θ to the horizontal plane and at φ to the vertical plane. The position of one end of the line is specified with reference to the planes of projections. The solution for this problem is explained in three steps.

Draw a reference line xy long enough to show three sets of projections of the line AB and also three sets of projections of the point A as shown in Fig. 7.24.

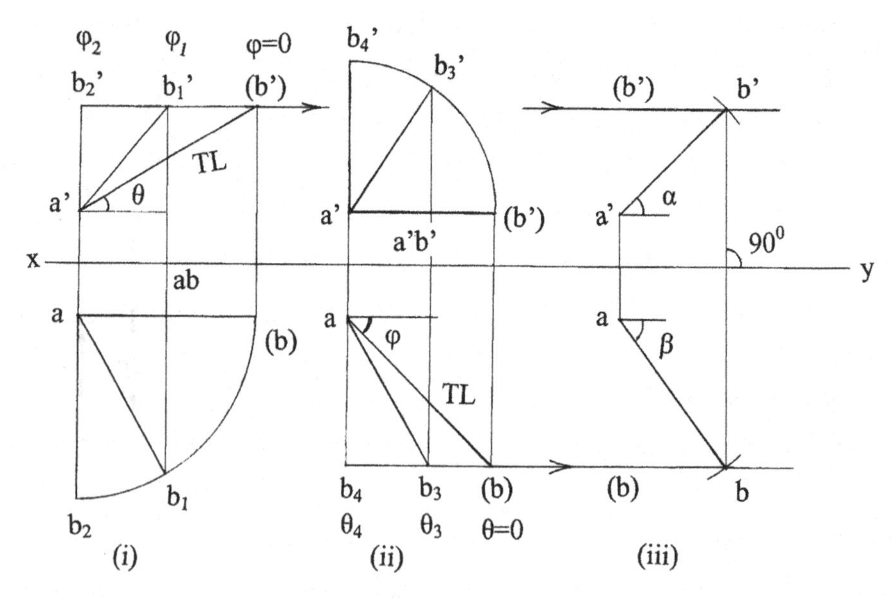

Fig. 7.24 Projections of line AB inclined to both the planes

Step I

Assume the line AB is parallel to the VP and inclined at an angle θ to the HP. The projections of the line are shown in Fig. 7.24(i). The projections of point B are enclosed in parentheses. If the line is made to move about the point A to a new position keeping its inclination θ to the HP the same, the projections of the line $a'b_1'$ and ab_1 appear as shown in Fig. 7.24(i). The elevation b_1' moves along a line parallel to xy. The plan b_1 moves along a circular path. In this position, the line is inclined at an angle φ_1 $(0 < \varphi_1 < \varphi_{max})$ to the VP. The line is moved further till the projections $a'b_2'$ and ab_2 are perpendicular to xy. Let the inclination of the line to the VP be φ_2. It is to be noted that the sum of the inclinations of the line to the reference planes is 90° when the projections of the line are perpendicular to xy. Hence, the inclination φ_2 will be equal to φ_{max} or $\varphi_{max} = (90 - \theta)°$.

Step II

Assume the line AB is parallel to the HP and inclined at an angle φ to the VP. The projections shall appear as shown in Fig. 7.24(ii). If the line is moved about the point A to a new position keeping its inclination to the VP the same, the projections of the line $a'b_3'$ and ab_3 appear as shown in Fig. 7.24(ii). The elevation b_3' moves along a circular path. The plan b_3 moves along a line parallel to xy. In this position the line is inclined to the HP at an angle θ_3 $(0 < \theta_3 < \theta_{max})$. The line is moved further till the projections $a'b_4'$ and ab_4 are perpendicular to xy. The inclination θ_4 will be equal to θ_{max} or $\theta_{max} = (90 - \varphi)°$.

Step III

Assume the line AB is inclined at an angle θ to the HP and at an angle φ to the VP. The projections of point A are already shown in Fig. 7.24(iii). The locus of (b′) when the line is inclined at θ to the HP is extended further (indicated by an arrow) as shown in Fig. 7.24(iii). The locus of (b) when the line is inclined at φ to the VP is also extended as shown in Fig. 7.24(iii). With centre a′ and radius a′b′ [obtained from Fig. 7.24(ii)], draw an arc to cut the locus line (b′) at b′. With centre a and radius ab [obtained from Fig. 7.24(i)], draw an arc to cut the locus line (b) at b. It can be shown the projections b′ and b lie on the projector B perpendicular to xy. The lines a′b′ and ab represent the projections of AB inclined to both the planes. The elevation a′b′ makes an angle α $(\alpha > \theta)$ with xy and the plan ab makes an angle β $(\beta > \varphi)$ with xy. The angles α and β are called apparent inclinations of elevation and plan of a straight line with the reference line xy.

16. A line PQ measuring 50 mm is inclined to the HP at 30° and to the VP at 45° with the end P 20 mm above the HP and 15 mm in front of the VP. Draw its projections.

    Draw a reference line xy long enough to show three sets of projections of the line AB and also three sets of projections of the point P as shown in Fig. 7.25.

    Assume the line PQ is parallel to the VP and inclined at 30° to the HP with the end P as projected in Fig. 7.25(i). Draw a line inclined at 30° to xy at p′ and locate the point (q′) on this line 50 mm from p′. The plan will be parallel to xy. Draw a projector through (q′) to locate the plan (q). The distance between p

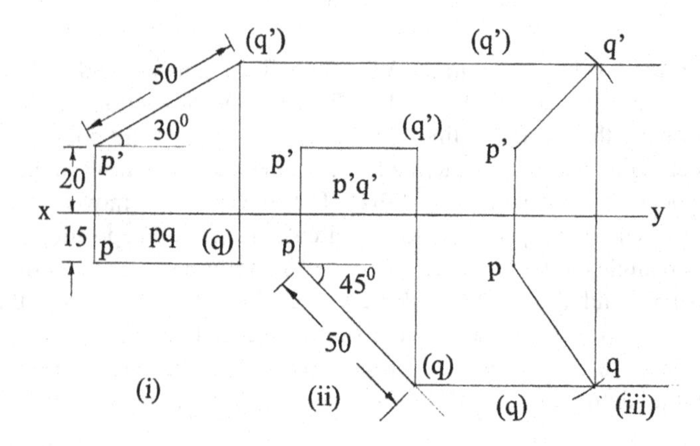

Fig. 7.25 Projections of line PQ

and (q) measures the plan length pq. The locus of elevation of Q is drawn parallel to xy through (q′).

Assume the line PQ is parallel to the HP and inclined at 45° to the VP with the end P as projected in Fig. 7.25(ii). Draw a line inclined at 45° to xy at p and mark on this line (q) 50 mm from p. The line is parallel to the HP. Hence, the elevation p′(q′) will be parallel to xy. Draw a projector through (q) to locate (q′) on the line drawn parallel to xy from p′. The distance between p′ and (q′) measures the elevation length p′q′.

The locus line (q′) when the line is inclined to the HP at 30° is extended as shown in Fig. 7.25(iii). The locus line (q) when the line is inclined at 45° to the VP is extended as shown in Fig. 7.25(iii). With centre p′ and radius p′q′ [obtained from Fig. 7.25(ii)], draw an arc to cut the locus line (q′) at q′. With centre p and radius pq [obtained from Fig. 7.25(i)], draw an arc to cut the locus line (q) at q. The lines p′q′ and pq are the required projections of the line PQ inclined to both the planes.

17. Draw the projections of a straight line AB 80 mm long inclined at 45° to the HP and 30° to the VP with the end A in the HP and the end B in the VP.

The elevation a′ of the end A will be on xy. The plan b of the end B will be on xy. Draw a reference line xy as shown in Fig. 7.26. The elevation a′ is located on xy as shown in Fig. 7.26(i). The line is made parallel to the VP and inclined at 45° to the HP. The projections are obtained as shown in Fig. 7.26(i). The locus of (b′) and the plan length ab are obtained.

The plan b is located on xy. The projections of the line are now obtained keeping it parallel to the HP and inclined at 30° to the VP as shown in Fig. 7.26 (ii). The locus of (a) and the elevation length a′b′ are obtained.

The elevation a′ is marked on xy. The Locus lines (b′) and (a) are extended as shown in Fig. 7.26(iii). With centre a′ and radius a′b′, draw an arc to cut the

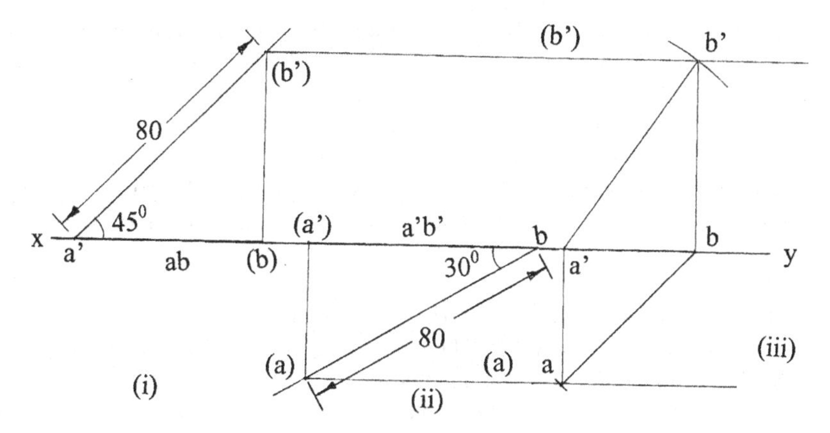

Fig. 7.26 Projections of line AB

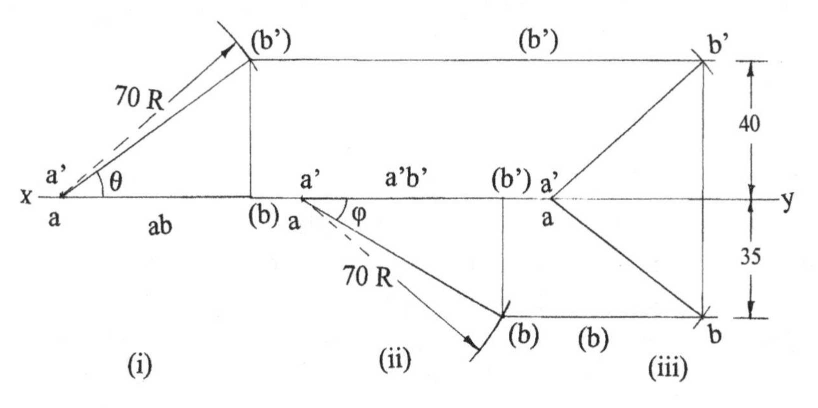

Fig. 7.27 Projections of line AB

locus line passing through (b′). This point locates the elevation of B, i.e. b′. Draw a projector through a′ to intersect the locus line (a). The intersecting point locates the plan of A, i.e. a. Another projector is drawn from b′ to locate the plan b on xy. Join points a′b′ and ab to represent the elevation and plan of the line AB. The plan length ab can be checked with the plan length obtained in Fig. 7.26(i).

18. Draw the projections of a line 70 m long having one of its ends lying both in the HP and the VP. The other end is 40 mm above the HP and 35 mm in front of the VP. Find the inclinations of the line with the HP and the VP.

 Draw a reference line xy as shown in Fig. 7.27. The elevation and plan of the end A of line AB lie on xy itself in this case as shown in three sets of projections. Draw a line parallel to xy and 40 mm above it as shown in the figure. This line represents locus b′ of the point B 40 mm above the HP. Draw another line 35 mm below xy. This line represents the locus b of the point B 35 mm in front of the VP. With centre a′ and radius 70 mm (true length), draw an arc to cut the locus of (b′) at (b′) as shown in Fig. 7.27(i). The angle between a′(b′) and xy measures the inclination of AB to the HP (θ). Draw a projector from (b′) to meet the reference line xy at (b). The plan length ab is obtained on xy.

With centre a and radius 70 mm (true length), draw an arc to cut the locus line (b) at (b) as shown in Fig. 7.27(ii). The angle between a(b) and xy gives the inclination of AB to the VP (φ). Draw a projector from (b) to meet xy at (b′). The elevation length a′b′ is obtained on xy.

With centre a′ and radius a′b′ [obtained from Fig. 7.27(ii)], draw an arc to cut the locus line (b′) at b′ as shown in Fig. 7.27(iii). Join the points a′ and b′. A projector is drawn through b′ to meet the locus line (b) at b. Connect the points a and b. The plan length ab can be checked with the plan length obtained in Fig. 7.27(i). The projections of the line AB are shown in Fig. 7.27(iii).

Answers: $\theta = 35°$; $\varphi = 30°$.

Determination of True Length and Inclinations with the Reference Planes of a Straight Line from Its Projections (Rotating Line Method)

Let the plan ab and elevation a′b′ of a line AB inclined to both the planes be given as shown in Fig. 7.28. Draw the locus line parallel to xy through b′. The plan ab is rotated and made parallel to xy as shown in the figure. A projector is drawn from b_1 to meet the locus line (b′) at B. The line a′B gives the true length AB, and the angle between a′B and xy measures the inclination θ of AB with the HP.

Draw the locus line parallel to xy through b. The elevation a′b′ is rotated and made parallel to xy as shown in the figure. Draw a projector from b_2' to meet the locus line (b) at B. The line aB gives the true length, and the angle between aB and xy measures the inclination φ of AB with the VP.

Traces of a Straight Line

Let the plan ab and elevation a′b′ of a line AB inclined to both the planes be given as shown in Fig. 7.29. The traces can easily be located from the projections of the line.

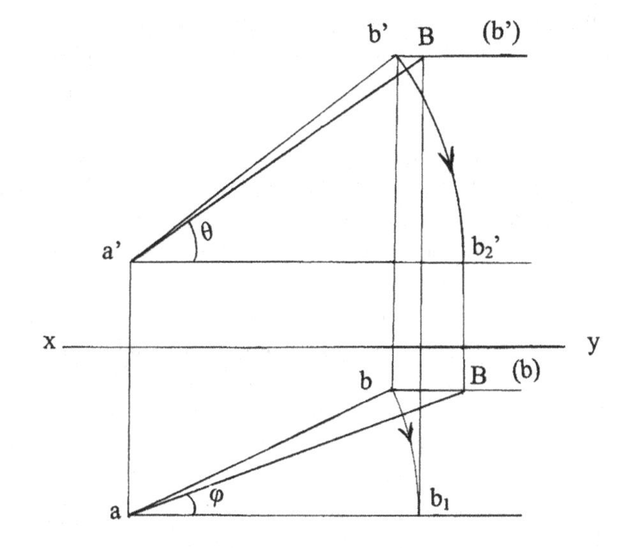

Fig. 7.28 Determination of true length and inclinations of line AB

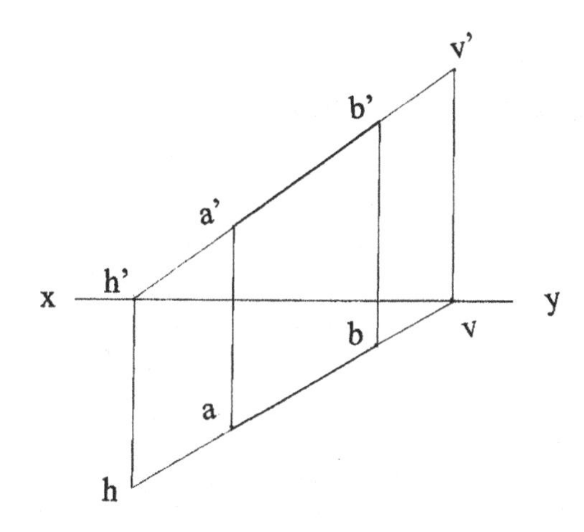

Fig. 7.29 Traces of line AB

Produce the elevation a′b′ to meet xy at h′. The point h′ is the elevation of the horizontal trace. A line perpendicular to xy is drawn from h′ to meet the plan ab produced at h. The point h is the horizontal trace of the line AB. Similarly produce the plan ab to meet xy at v, the plan of the vertical trace. Draw a line perpendicular to xy from v to meet the elevation a′b′ produced at v′. The point v′ is the vertical trace of the line AB. This construction may not locate the traces when the projections of the line are perpendicular to xy. The trapezium method explained previously will apply in this case also provided the plan and elevation of the line are given.

19. The plan and elevation lengths of a line AB 100 mm long are 75 mm and 80 mm, respectively. End A is on the HP and 20 mm in front of the VP. Draw the projections of the line and determine its inclinations with the HP and VP and traces.

    Draw reference line xy and mark the projections a′ and a as shown in Fig. 7.30. Mark the plan length ab (75 mm) along a line parallel to xy and from a. Draw a projector at the extremity of the plan ab. With centre a′ and 100 mm (true length) as radius, draw an arc to cut the projector drawn above xy at (b′). A line drawn parallel to xy through (b′) represents the locus of elevation of point B. The angle between a′(b′) and xy gives the inclination of AB with the HP. With centre a′ and 80 mm (a′b′) as radius, draw an arc to cut the locus line (b′) at b′. Join points a′ and b′. Draw a projector from b′. With centre a and 75 mm (ab) as radius, draw an arc to cut the projector at b. Join the points a and b. Since the point A is on HP, the elevation of the horizontal trace coincides with a′. The horizontal trace h coincides with the plan a as shown in the figure.

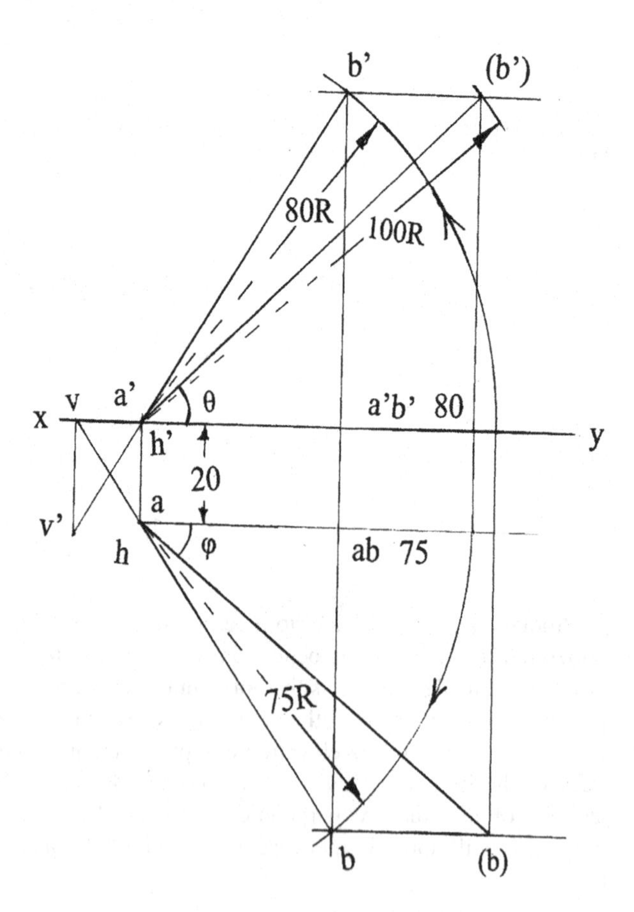

Fig. 7.30 Projections of the line AB

The plan ab is extended to meet xy at v. A projector is drawn from v to meet the extension of a′b′ at v′ as shown in the figure. The point v′ locates the vertical trace of the line AB. The angle between a(b) and xy measures the inclination of AB with the VP.

Answers: $\theta = 41°$; $\varphi = 37°$.

20. One end of a line PQ is at a distance of 25 mm from the VP and 20 mm from the HP. The plan of the line measures 70 mm. The distance between the end projectors when measured parallel to the line of intersection of HP and VP is 50 mm, and the vertical trace is 10 mm above the line of intersection of HP and VP. Draw the projections of the line PQ. Find the true length of PQ and true angles the line makes with HP and VP. Also locate its horizontal trace.

Let the end P of the line PQ be 25 mm in front of the VP and 20 mm above the HP. Draw the reference line xy and also a projector for the point P as shown in Fig. 7.31. The elevation p′ and plan p are marked on the projector P, respectively, 20 mm above xy and 25 mm below xy. Another projector is drawn 50 mm from the projector P as shown in the figure. With p as centre and 70 mm radius, draw an arc to cut the projector Q. The intersecting point locates plan q of the point Q. Extend qp to meet the reference line xy at v, the plan of the vertical trace. A point v′ 10 mm above xy on the projector drawn from v locates the vertical trace. Join v′p′ and extend the line to locate q′ on the projector Q already drawn. Draw the locus line q′. Rotate the plan pq about p to make it parallel to xy. A projector is drawn from the extremity q of the plan pq to locate Q on the locus line q′. The distance p′Q gives the true length PQ, and the angle between p′Q and xy measures the inclination of PQ with the

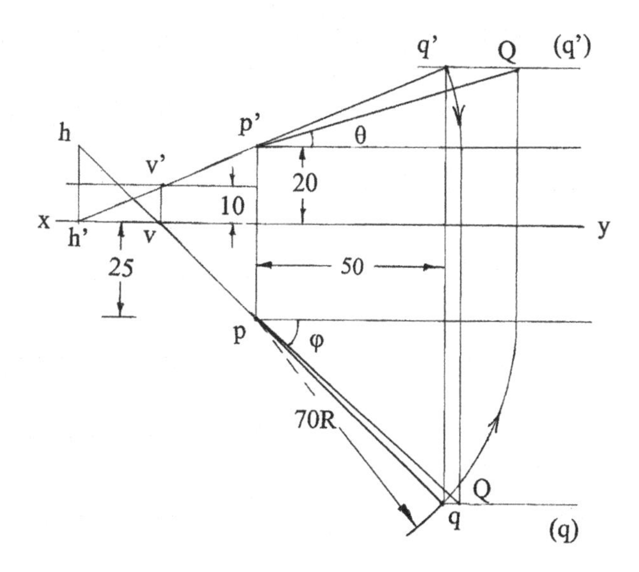

Fig. 7.31 Projections of line PQ

HP. Draw the locus line q. Rotate the elevation p′q′ about p′ to make it parallel to xy. A projector is drawn from the extremity q′ of the elevation p′q′ to locate Q on the locus line q. The distance pQ gives the true length, and the angle between pQ and xy measures the inclination of PQ with the VP. Extend the elevation q′p′ to meet the reference line xy at h′. A projector is drawn from h′ to meet the extension of the plan qp at h as shown in the figure. The point h locates the horizontal trace of the line PQ.

Answers: PQ = 72 mm; $\theta = 16°$; $\varphi = 42°$.

21. The projectors drawn from the HT and VT of a straight line AB are 80 mm apart while those drawn from its ends are 50 mm apart. The HT is 35 mm in front of the VP, the VT is 55 mm above the HP and the end A is 10 mm above the HP. Draw the projections of the line AB. Determine the true length of AB and the inclinations of the line AB with the reference plane.

The vertical trace v′ is 55 mm above xy, the horizontal trace h is 35 mm below xy and the distance between their projectors is 80 mm. The elevation a′ is 10 mm above xy. Draw the reference line xy and mark the traces keeping a distance of 80 mm between their projectors as shown in Fig. 7.32. Join points v′h′ and also vh. The elevation a′b′ and plan ab will lie, respectively, on the lines v′h′ and vh. Draw a line 10 mm above and parallel to xy. The intersecting point on v′h′ locates the elevation a′. A projector drawn through a′ intersects vh and locates the plan a at the intersecting point as shown in the figure. Draw a projector B 50 mm from the projector A and locate the projections b′ and b, respectively, on v′h′ and vh. The projections a′b′ and ab are thus obtained. A trapezium a′b′BA is constructed on the base a′b′ taking the distances of a and b from xy. The line AB gives the true length. The line AB is extended to meet the elevation at v′. The angle between AB and a′b′ gives the inclination of the line AB with the VP. Another trapezium abBA is constructed on the base ab taking the distances of a′ and b′ from xy. The line AB is extended to meet the plan

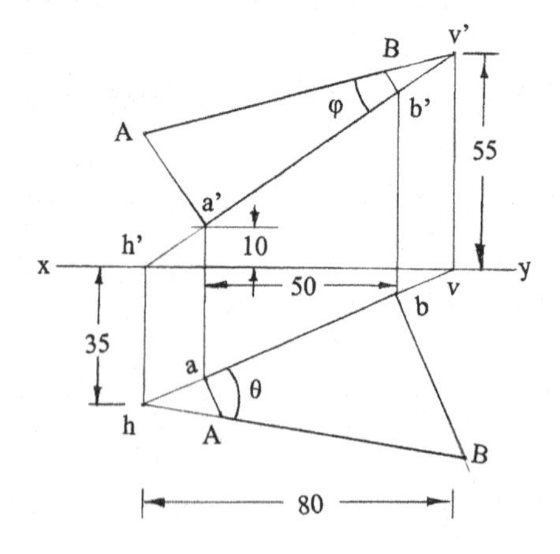

Fig. 7.32 Projections of line AB

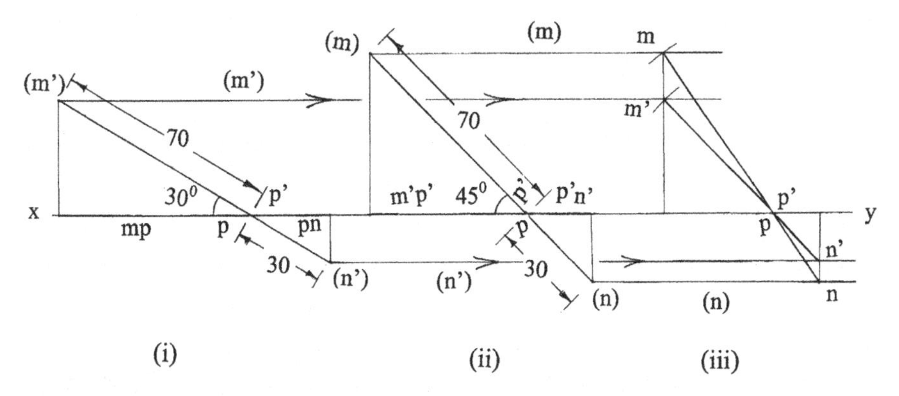

Fig. 7.33 Projections of line MN

h. The angle between AB and ab measures the inclination of the line AB with the HP.

Answers: AB = 65 mm; $\theta = 32°$; $\varphi = 20°$.

22. A line MN 100 mm long is inclined at 45° to the VP and 30° to the HP. Its end M is in the second quadrant and the end N is in the fourth quadrant. A point P on MN, 70 mm from M, lies on both the HP and the VP. Draw the projections of the line MN and locate its traces.

The elevation p′ and plan p lie on xy as the point P is on both the HP and the VP. The elevation m′ and plan m lie above xy and the elevation n′ and plan n lie below xy.

Draw a reference line long enough to show three sets of the projections of line MN as shown in Fig. 7.33. The projections p′ and p are marked on xy at three places as shown in the figure.

The line PM is made parallel to the VP and inclined at 30° to the HP. Its projections are shown in Fig. 7.33(i). The locus line (m′) is drawn parallel to xy through (m′). The plan length mp is obtained on xy. The other portion of the line PN is also projected in the same figure. The locus line (n′) is drawn parallel to xy through (n′). The plan length pn is obtained on xy.

The line PM is now made parallel to the HP and inclined at 45° to the VP. Its projections are shown in Fig. 7.33(ii). The locus line (m) is drawn parallel to xy through (m). The elevation length m′p′ is obtained on xy. The other portion of the line PN is also projected in the same figure. The locus line (n) is drawn parallel to xy through (n). The elevation length p′n′ is obtained on xy.

The locus lines (m′), (m), (n′) and (n) are extended as shown in Fig. 7.33(iii). With centre p′ and radius p′m′, draw an arc to locate the elevation m′ on the locus line (m′). Join m′p′ and extend it to cut the locus line (n′) at n′. With centre p and radius pm, draw an arc to locate the plan m on the locus line (m). Join mp and extend it to cut the locus line (n) at n. A projector is drawn from m

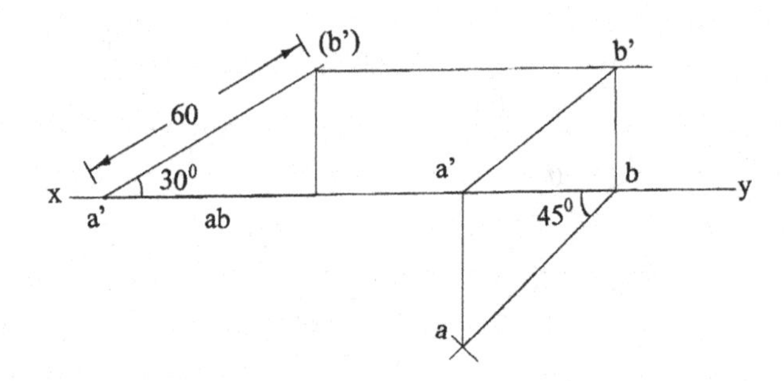

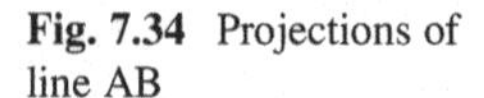
Fig. 7.34 Projections of line AB

to meet xy. Another projector is drawn from n to meet xy. The elevations m′ and n′ will lie on the respective projectors. Since the point P lies on both the HP and the VP, its projections contain both the traces.

23. A straight line is 60 mm long. One end is in the horizontal plane and the other end is in the vertical plane. The line is inclined at 30° to the horizontal plane, and its plan makes an angle of 45° with xy. Draw the projections of the line.

    Let the end A be on the HP and the end B be on the VP. Hence, the elevation a′ and the plan b will lie on xy.

    Draw the reference line xy and mark a′ and b on xy as shown in Fig. 7.34. Assume the line is made parallel to the VP and inclined at 30° to the HP. The elevation of the line is drawn as shown in the figure to get the locus line (b′) and the plan length ab on xy. A projector is drawn through b to locate b′ on the locus line (b′). A line inclined at 45° to xy is drawn at b as shown in the figure. Point a is located on this line taking the distance of plan length ab already obtained. A projector is drawn through a to meet xy at a′. The points a′ and b′ are joined. The lines a′b′ and ab represent the projections of the line AB.

24. The distance between the ends projectors of a straight line AB when measured parallel to the line of intersection of HP and VP is 50 mm. The end B is 40 mm above HP and 50 mm in front of VP. The end A is 10 mm below HP. The line is inclined at 30° to HP. Complete the plan and determine the inclination of the line with VP. Find the true length of AB.

    The elevation a′ is 10 mm below xy. The elevation b′ is 40 mm above xy and the plan b is 50 mm below xy. The distance between the projectors A and B is 50 mm. Draw the reference line xy and locate the projections a′, b′ and b as shown in Fig. 7.35. Join a′b′. The locus line (b′) is drawn parallel to xy. Draw a line a′e′ inclined at 30° to xy. The line a′e′ intersects the locus line (b′) and locates B. The horizontal projection of a′B gives the plan length ab. With centre b and radius ab, draw an arc to intersect the projector A at a as shown in the figure. The point above xy is chosen for convenience. It is to be noted that the position of vertical trace alone will change if the point a is chosen below xy. The plan a and b are opposite sides of

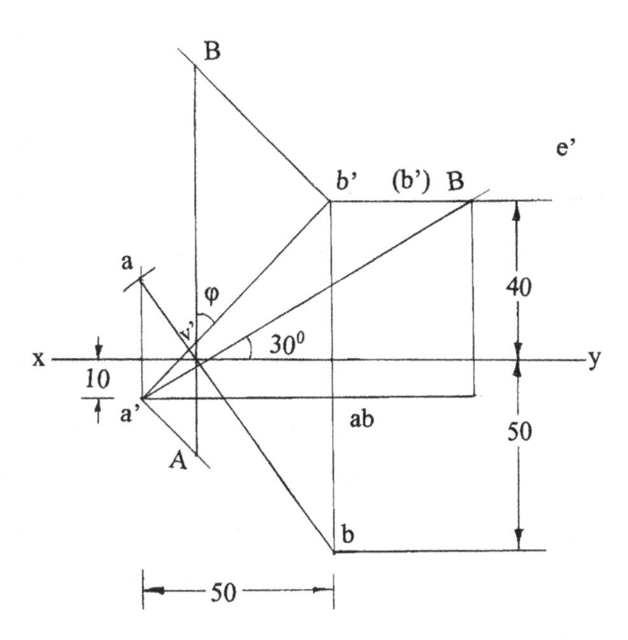

Fig. 7.35 Projections of the line AB

xy. Hence, two lines are drawn perpendicular to a′b′ at a′ and b′ on opposite sides as shown in the figure. Points A and B are located on the lines taking the distances of a and b from xy. Join AB. The intersecting point of AB and a′b′ locates the vertical trace v′, and the angle between Bv′ and b′v′ gives the inclination of AB with the VP.

Answer: AB = 100 mm; $\varphi = 45°$.

25. The plan ab of a straight line is 45 mm long and makes an angle of 30° with xy line. The end A is in the VP and 30 mm from the HP. The end B is 50 mm from the HP. The whole line lies in the first quadrant. Draw the projections of the line AB and determine the length of the elevation, true length, traces and inclinations with the reference planes.

    The whole line lies in the first quadrant. The plan a is on xy and the elevation a′ is 30 mm above xy. The elevation b′ is 50 mm above xy. The plan length ab is 45 mm.

    Draw the reference line xy and mark the projections of A as shown in Fig. 7.36. Draw the locus line (b′) 50 mm above xy. Draw a line inclined 30° to xy at a. Locate point b along this line 45 mm from a. Draw a projector through b to intersect the locus line (b′) at b′. Join a′b′. The trapezium abBA is drawn on the base ab taking the distances of a′ and b′ from xy. Extend the lines BA and ba to meet at h, the horizontal trace of the line. The angle between BA and ba gives the inclination of the line with the HP. Draw a line perpendicular to a′b′ at b′. Locate point B on this line taking the distance of b from xy. Since

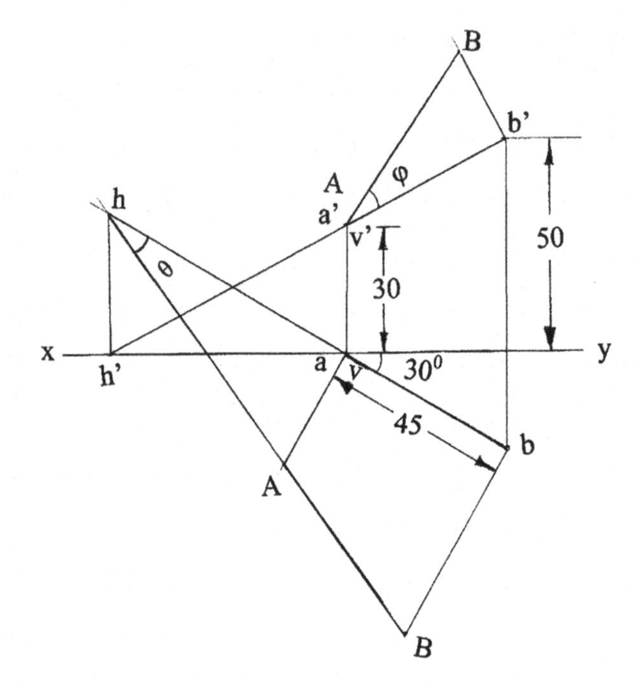

Fig. 7.36 Projections of line AB

the point A lies on the VP, the space point A coincides with the elevation a′. Join AB and find its true length. The angle between AB and a′b′ measures the inclination of the line with the VP. The vertical trace coincides with the elevation a′ in this case.

Answers: a′b′ = 44 mm; AB = 49 mm; $\theta = 24°$; $\varphi = 27°$.

26. The end A of a straight line AB is in front of the VP and 30 mm above the HP. The end B is 20 mm in front of the VP and above the HP. The line is 70 mm long and is inclined at 45° to the HP and 30° to the VP. Draw the projections of the line.

    The elevation a′ is 30 mm above xy and the plan a is below xy. The elevation b′ is above xy and the plan b is 20 mm below xy.

    Draw the reference line xy as shown in Fig. 7.37. The projections a′ and b are marked leaving sufficient space between their projectors to show the final projections in the mid-portion of the figure.

    The line AB is made parallel to the VP and inclined at 45° to the HP. The projections are shown with reference to the elevation a′. This gives the locus line (b′) and the plan length ab.

    The line is made parallel to the HP and inclined at 30° to the VP. The projections are shown with reference to the plan b. This gives the locus line (a) and the elevation length a′b′.

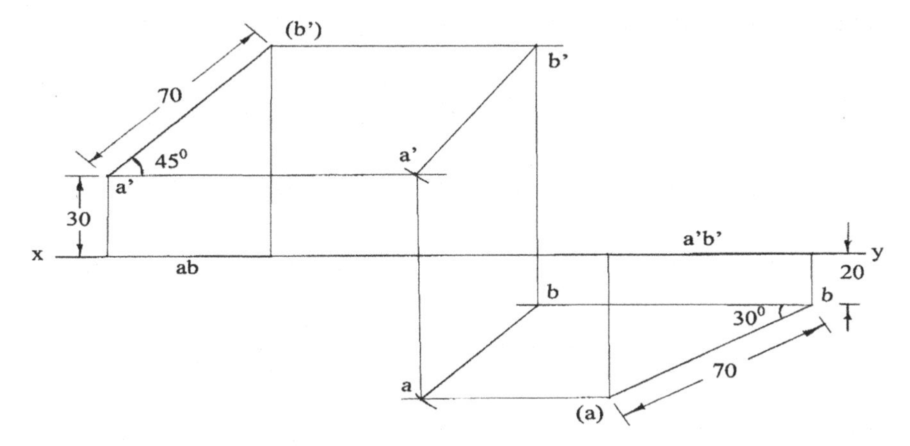

Fig. 7.37 Projections of line AB

Draw the projector B and mark b on it as shown in the figure and also locate b′ at the intersection the projector with locus line (b′). With centre b′ and radius a′b′, draw an arc to cut the locus line (a′). The intersecting point locates the elevation a′. Join a′b′. Draw the projector passing through a′. With centre b and radius ab, draw an arc to cut the projector at a. Join ab. The lines a′b′ and ab represent the projections of the line AB.

27. The plan of certain triangle is an equilateral one abc of side 40 mm, one side ab coinciding with xy. The corners A, B and C of the triangle are, respectively, 20 mm, 30 mm and 45 mm above the HP. Draw the projections of the triangle ABC and determine the true lengths of its sides.

    An equilateral triangle abc of side 40 mm with side ab on xy represents the plan of certain triangle. The elevation a′ is 20 mm above xy. The elevation b′ is 30 mm above xy. The elevation c′ is 45 mm above xy. Draw the reference line xy and locate the plans a, b and c as shown in Fig. 7.38. Draw projectors through a, b and c. The elevations of the corners of the triangle are marked on the respective projectors. Since the plans a and b lie on xy, the corners A and B lie on the VP. The elevations a′ and b′ represent the space points A and B. The elevation length gives directly the true length of side AB. The trapezium acCA is constructed on the base ac taking the distances of a′ and c′ from xy. The line AC gives the true length of the side AC. The trapezium bcCB is constructed on the base bc taking the distances of b′ and c′ from xy. The line BC measures the true length of the side BC.

    Answers: AB = 41 mm; BC = 43 mm; CA = 47 mm.

28. Three vertical poles AB, CD and EF are, respectively, 2.5 m, 3 m and 4 m long. Their ends B, D and F are on the ground and form the corners of an equilateral triangle of 4 m sides. Determine graphically the distances between the top ends

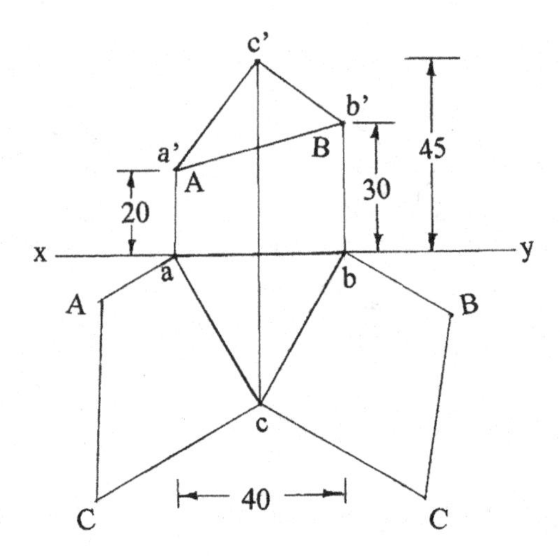

Fig. 7.38 Projections of triangle ABC

of the pole, namely AC, CE and AE and the inclinations of these with the ground.

The vertical poles form an equilateral triangle on the ground (horizontal plane). In the scale specified 50 units on the actual size are represented as 1 unit on the drawing. The equilateral triangle of side 80 mm (4 m) represents the poles in the plan. The top ends of the poles are, respectively, 50 mm (2.5 m), 60 mm (3 m) and 80 mm (4 m) above the horizontal plane (ground). The elevations a′, c′ and e′ are, respectively, 50 mm, 60 mm and 80 mm above xy.

Draw the reference line xy as shown in Fig. 7.39. The equilateral triangle is drawn with one side parallel to xy, and the corner points are marked in the plan. Projectors are drawn from the corner points in the plan. The elevations a′, b′, c′, d′, e′ and f′ are also marked on the respective projectors. As the ends B, D and F are on the ground, the elevations b′, d′ and f′ lie on xy. Join the points a′, c′ and e′ in pairs. The plan ac is parallel to xy and the elevation a′c′ gives the true length AC. The angle between a′c′ and xy gives the inclination θ_1 of the top ends AC with the ground. The trapezium aeEA is drawn on the base ae taking the distances of a′ and e′ from xy. The line AE gives the distance between A and E. The angle between AE and ae measures the inclination θ_2 of the top ends A and E with the ground. Another trapezium ceEC is drawn on the base ce taking the distances of c′ and e′ from xy. The line CE gives the distance between C and E. The angle between CE and ce measures the inclination θ_3 of the top ends C and E with the ground.

Answers: AC = 4.03 m; $\theta_1 = 7°$; AE = 4.25 m; $\theta_2 = 20°$; CE = 4.1 m; $\theta_3 = 13°$.

29. A room is 5 m long, 3 m wide and 4 m high. Determine graphically the distance between the top corner and the bottom corner diagonally opposite to it.

The pictorial representation of the room is shown in Fig. 7.40. Choose the line connecting the points AG whose true length is required graphically. The vertical plane is chosen passing through the inner surface ABFE of the room.

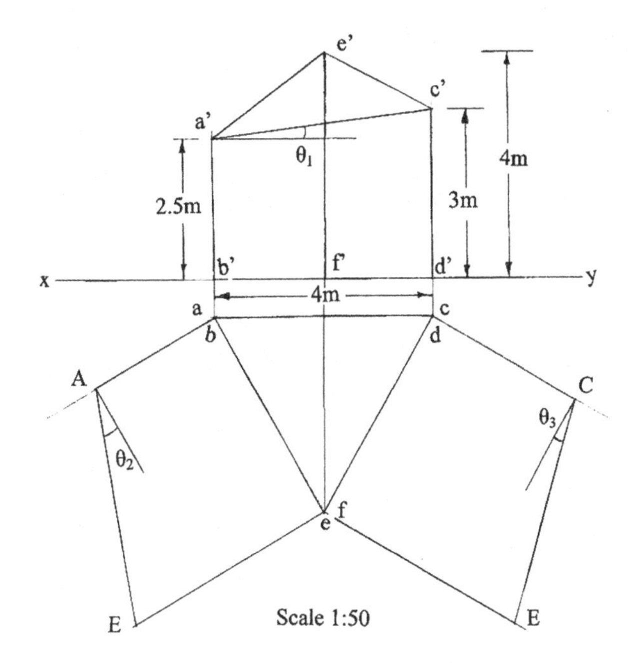

Fig. 7.39 Projections of poles

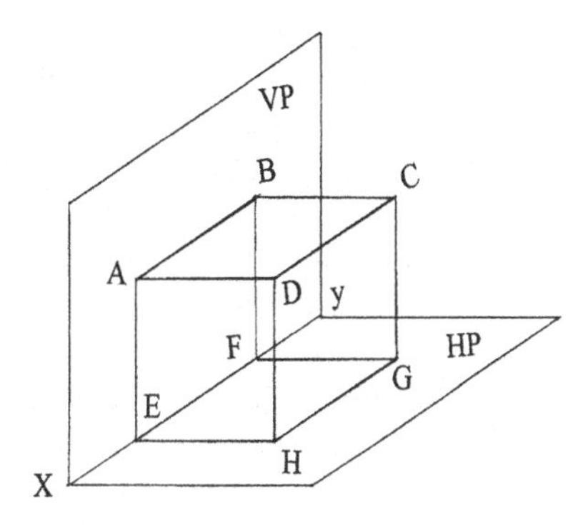

Fig. 7.40 Pictorial view of the Room

The horizontal plane contains the floor of the room. The positions of the points A and G are stated with reference to the VP and the HP. The point A is in the VP and 4 m above the HP. The point G is in the HP and 3 m in front of the VP. Adopt a scale 1:50 for this problem. The elevation a′ is 80 mm (4 m) above xy and the plan a is on xy. The elevation g′ is on xy and the plan g is 60 mm (3 m) below xy. The distance between the projectors A and G is 100 mm (5 m).

Draw the reference line xy and mark the projections of AG as shown in Fig. 7.41. Draw a line perpendicular to a′g′ at g′. Locate point G on this line taking the distance of g from xy. Since the plan a is on xy, the space point A coincides with the elevation a′. The line AG measures the true length of the diagonal.

Answer: AG = 7.05 m.

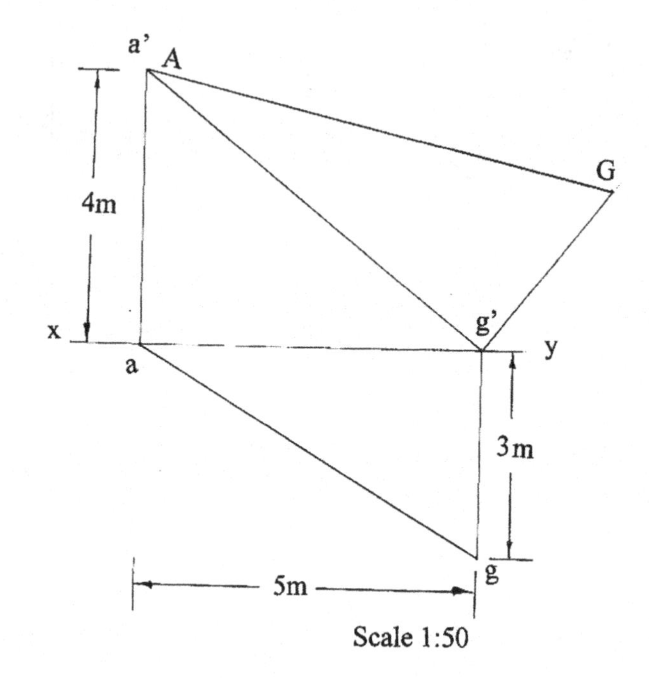

Fig. 7.41 Projections of line AG

30. Two mangoes on a tree are, respectively, 5 m and 3 m above the ground and 1.5 m and 2.5 m from the wall surface, but on opposite sides of the wall. The wall thickness is 0.3 m. The distance of mangoes measured along the ground and parallel to the wall is 5 m. Determine the true distance between the mangoes and the angles of inclinations of the line joining the mangoes (i) with the ground and (ii) with the wall.

    A scale of 1:50 is adopted for this problem. Let A and B be designated as the two mangoes on the tree. The vertical plane is oriented on one face of the wall. The ground is chosen as the horizontal plane. The positions of the mangoes are stated with reference to the planes of projections. The point A is 30 mm (1.5 m from the wall) in front of the VP and 100 mm (5 m above the ground) above the HP. The point B is 56 mm (2.5 m + 0.3 m from the wall including its thickness) behind the VP and 60 mm (3 m above the ground) above the HP. The elevation a′ is 100 mm above xy and the plan a is 30 mm below xy. The elevation b′ is 60 mm above xy and the plan b is 56 mm above xy. The distance between the projectors A and B is 100 mm (5 m).

    Draw the reference line xy and mark the projections a′, a, b′ and b on the projectors A and B, 100 mm apart, as shown in Fig. 7.42. The trapezium abBA is constructed on the base ab taking the distances of a′ and b′ from xy. The line AB gives the distance between the mangoes, and the angle between AB and ab measures the inclination of the line joining the mangoes with the ground (Θ). The plans a and b are on opposite sides of xy. Hence, two lines perpendicular to a′b′ are drawn at a′ and b′ on opposite sides of a′b′ as shown in the figure. Points

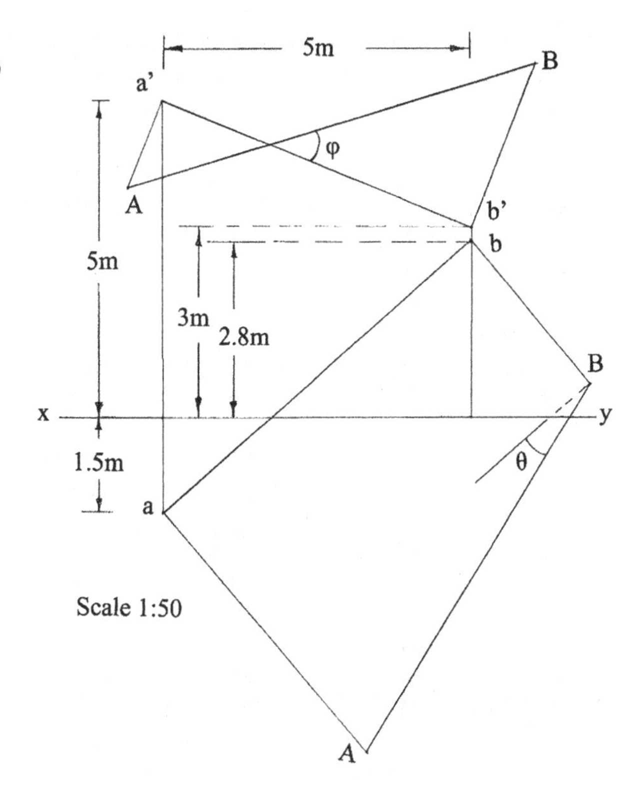

Fig. 7.42 Projections of line AB (mangoes A and B)

A and B are located on the respective lines taking the distances of a and b from xy. The angle between the intersecting lines AB and a′b′ measures the inclination of the line joining the mangoes with the wall (Φ).

Answers: AB = 6.9 m; Θ = 18°; Φ = 39°.

Practice Problems

1. Draw the projections of the following lines:
   (a) AB, 50 mm long, parallel to XY, 20 mm above the HP and 30 mm in front of the VP.
   (b) CD, 60 mm long, parallel to the HP, inclined at 30° to the VP, the end C to be in the VP and 30 mm above the HP.
   (c) EF, 40 mm long, perpendicular to the HP, the end E to be 15 mm above the HP and 25 mm in front of the VP.
   (d) GH, 50 mm long, perpendicular to the VP, 20 mm above the HP and the end G to be 15 mm in front of the VP.
2. A line PQ 90 mm long is parallel to the HP and inclined 30° to the VP. End P is 30 mm above the HP and 20 mm in front of the VP. Draw the projections of the line.

3. A line MN 70 mm long has its end M 30 mm above the HP and end N 20 mm below the HP. If the line is 30 mm in front of and parallel to the VP, draw its projections and measure its inclination to the HP.
4. A line GH 90 mm long is parallel to and 30 mm above the HP. Its ends G and H are, respectively, 30 mm and 40 mm in front of the VP. Find its inclination with the VP.
5. The elevation of a line inclined at 30° to the VP is 65 mm long. Draw the projections of the line when it is parallel to and 30 mm above the HP, its one end being 30 mm in front of the VP.
6. a′b′, 60 mm long, is the elevation of a straight line which is parallel to the VP. The end A is in the HP and 20 mm in front of the VP. The end B is 30 mm above the HP. Draw the plan and elevation of AB.
7. The projections of a line AB are perpendicular to xy. The end A is in the HP and 50 mm in front of the VP and the end B is in the VP and 40 mm above the HP. Draw its projections, determine its true length and the inclinations with the reference planes.
8. The plan and elevation of a line are inclined at 35° and 50° to the xy line. One end of the line is touching both the HP and the VP. The other end is 50 mm above the HP and in front of the VP. Find its true length and inclinations of the line with HP and VP.
9. The distance between the end projectors passing through the end points of line AB is 40 mm. The end A is 15 mm above the HP and 10 mm in front of the VP. The line AB appears as 55 mm long in the elevation and the end B is 20 mm in front of the VP. Complete the projections. Find the true length of the line and its inclinations with the HP and VP. Locate the traces also.
10. A line AB, 65 mm long, has its end A in the HP and 15 mm behind the VP. The end B is in the first quadrant. The line is inclined at 30° to the HP and 60° to the VP. Draw its projections.
11. The plan of line PQ is 70 mm long while its elevation is 60 mm long. Q is nearer to the HP and the VP than P. The distance between the end projectors of the line when measured parallel to the line of intersection of the HP and the VP is 50 mm. Determine the true length PQ, traces of PQ and the inclinations of PQ with the reference planes.
12. The elevation of line AB is 100 mm long and makes 40° with the reference line xy. Its mid-point is 50 mm in front of the VP and 50 mm above the HP. The point A is 40 mm in front of the VP. Draw the projections of the line and determine its true length.
13. The middle point of a straight line 100 mm long is 20 mm above the HP and 30 mm in front of the VP. The line is inclined at 30° to the HP and 40° to the VP. Draw the projections of the line.
14. A straight line inclined at 50° to the HP has one end 50 mm above the HP and 15 mm in front of the VP. The other end is on the HP and 50 mm in front of the VP. Draw the projections of the line.
15. Figure 7.43 shows the plan cd of a straight line whose true length is 80 mm. The end C is 10 mm below the HP and the end D is above the HP. Draw the elevation of the line.

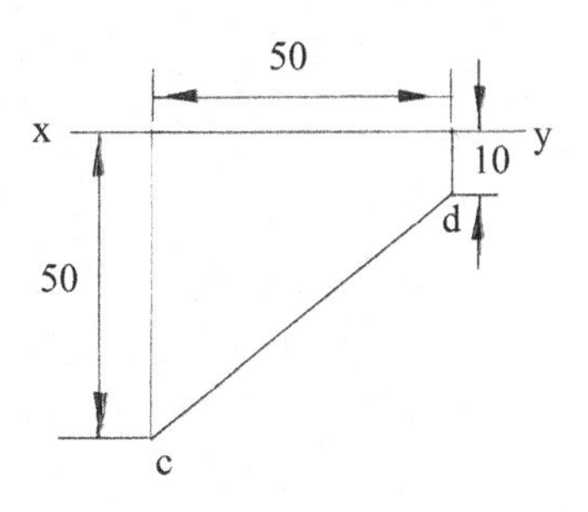

Fig. 7.43 Plan cd of a line

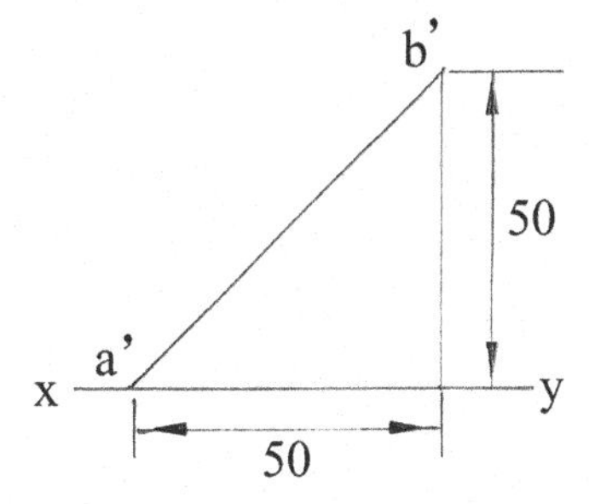

Fig. 7.44 Elevation a′b′ of a line

16. Figure 7.44 shows the elevation a′b′ of a straight line. B is in the VP. The line is inclined at 35° to the HP. Draw the plan of the line.
17. A straight line AB is 140 mm long. Its elevation a′b′ measures 90 mm and plan ab measures 100 mm. Its end B and the mid-point M are in the first quadrant. The mid-point M is 30 mm from both the planes. Draw the projections of line AB and find inclinations of the line with the HP and the VP.
18. A chimney of boiler is 20 m high and 1 m in diameter. This chimney is supported by three guy wires which appear in plan at 120° to each other. The ends of the wires are pegged to the ground at distance 3 m, 4 m and 5 m from the centre of the chimney. The other ends of the wires are connected to the top of the chimney. Find the length of the three guy wires.
19. A pipe is to be fixed on a vertical wall. One end of the pipe is touching on the floor and the other end is at a height of 3 m. If the distance between the ends of the pipe measured along the floor is 7 m, find graphically the length of the pipe and its inclination to the floor. Adopt a scale 1:100.
20. Three sticks AB, AC and AD of equal length are joined at A, the lower ends B, C and D resting on the ground. The lower ends occupy the corners of an equilateral triangle of side 80 mm. The common end point A is 80 mm above the ground. Find the length of the sticks.

Chapter 8
Projections of Plane Figures

A plane surface enclosed or bounded by straight lines or a curved line is called a plane figure. The least number of straight lines which can enclose a space is three. When three straight lines intersect, the plane surface enclosed is a triangle. With four straight lines, the figure obtained is a quadrilateral. The most common plane figures are set squares (30°–60° and 45°). Other common plane figures include square, rectangle, pentagon, hexagon, octagon, circle, ellipse, etc.

Projections of Plane Figure

If a plane figure is perpendicular to the HP and parallel to the VP, its plan will be a straight line parallel to xy. The elevation will have the exact size and shape of the plane figure. If a plane figure is perpendicular to the VP and parallel to the HP, its elevation will be a straight line parallel to xy. The plan will have the exact size and shape of the plane figure. If the plane figure is perpendicular to both the reference planes, its projections on both the reference planes will be straight lines and perpendicular to xy. The projection on the profile plane will have the exact size and shape of the plane figure. The plane figure on extension will intersect the two reference planes (HP and VP) at two different lines. These lines of intersections are called traces of the plane figure. The line of intersection of plane figure with horizontal plane is called horizontal trace (HT). The line of intersection of plane figure with vertical plane is called vertical trace (VT). If the plane figure is parallel to one of the reference plane, it has no trace on that reference plane.

K. Rathnam, *A First Course in Engineering Drawing*,
DOI 10.1007/978-981-10-5358-0_8

Position of Plane Figure with Respect to the Reference Planes

The position of plane figures can be classified as follows:

(i) Parallel to one of the reference plane and perpendicular to the other plane.
(ii) Inclined to one of the reference plane and perpendicular to the other plane.
(iii) Perpendicular to both the reference planes.
(iv) Inclined to both the reference planes.

Solved Problems

1. A 45° set square has its hypotenuse 100 mm long and is kept on the HP in such a way that its hypotenuse is touching the HP and perpendicular to the VP. Draw the projections of the set square if it is inclined at 40° to the HP.

   When a 45° set square is kept on the HP, its plan will be an isosceles triangle and the elevation will be a straight line on the reference line. The angle opposite to hypotenuse is 90°. The hypotenuse is to be kept perpendicular to xy.

   Draw the reference line xy and the plan abc of the set square lying on the HP with its hypotenuse perpendicular to the reference line as shown in Fig. 8.1. The elevation of the set square is a straight line on the reference line. The elevation a′b′c′ is projected from the plan abc. The elevation is rotated about a′b′ at an angle of 40° to xy, and the elevation $a_1'b_1'c_1'$ is redrawn as shown in the figure. The corner point C of the set square moves parallel to the VP during the rotation. Hence, the distance of the corner point C from the VP remains the

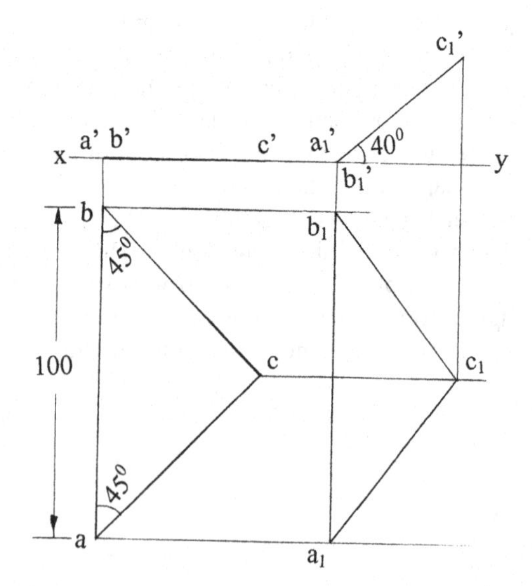

Fig. 8.1 Projections of 45° set square

same. Projectors are drawn from a_1', b_1' and c_1'. Locus lines are drawn from a, b and c parallel to xy. The intersections of the points with the corresponding projectors drawn from a_1', b_1' and c_1' locate the plan a_1, b_1 and c_1 of the corner points of the set square. Join the points in the plan to complete the projections $a_1'b_1'c_1'$ and $a_1b_1c_1$ of the set square.

2. The plan of a 45° set square ABC with the side BC on the HP and the side AB in the VP is a triangle abc. The side bc is 100 mm and perpendicular to xy. The angle bca is 35°. Draw the plan and elevation of the set square and measure the inclination of the set square with the HP.

   The side BC of the set square is on the HP and the side AB is in the VP. Hence, the side AC will be hypotenuse of the set square. The set square is kept on the HP with the side AB touching the VP. The plan abc will be an isosceles and the elevation a'b'c' will be a straight line on the reference line. The equal sides of the set square are 100 mm long.

   Draw the reference line xy and the projections of the set square as shown in Fig. 8.2. The set square is rotated about the side BC so that the point A moves on the VP. The side BC stays on the HP. The plan b_1c_1 is drawn as shown in the figure. Draw a line inclined at 35° to b_1c_1 at c_1. This line intersects xy to locate the plan a_1. Draw a projector through a_1. With centre b_1' and radius a'b' (i.e. $a_1'b_1'$), draw an arc to cut the projector at a_1'. Join the point a_1' with b_1' to complete the elevation of the set square. The angle between $a_1'b_1'$ and xy measures the inclination of the set square with the HP.

   Answer: $\theta = 46°$.

3. A 30°–60° set square of hypotenuse 100 mm long is placed such that its hypotenuse is in the VP and inclined at 30° to the HP. The surface of the set square is perpendicular to the VP. Draw its projections.

   Let ABC represent the 30°–60° set square with the side BC opposite to 30° angle and side AC the hypotenuse. Since the hypotenuse is 100 mm, the side BC will be 50 mm. The set square is kept initially on the HP with the

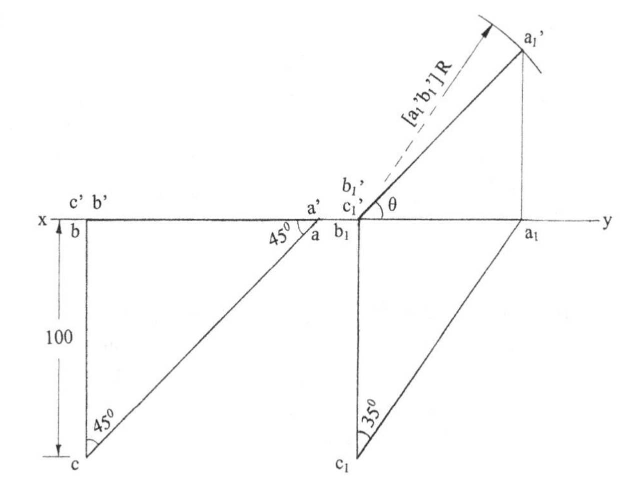

Fig. 8.2 Projections of 45° set square

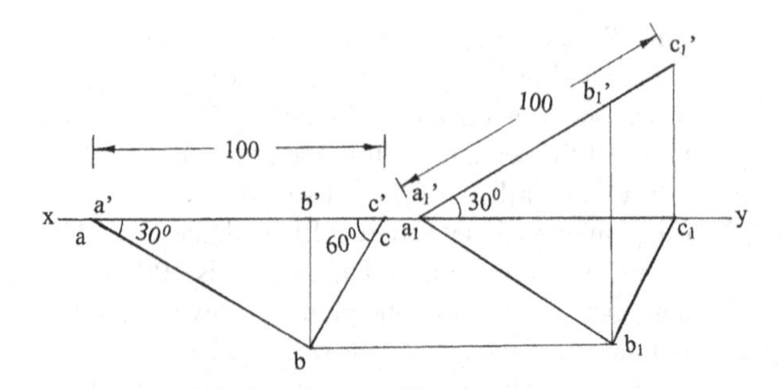

Fig. 8.3 Projections of 30°–60° set square

hypotenuse AC touching the VP. In this position, the plan will be the exact size of the set square and the elevation will be a straight line on xy.

Draw the reference line xy and also the projections abc and a′b′c′ of the set square as shown in Fig. 8.3. The set square is moved over the VP such that side AC (hypotenuse) makes an angle of 30° with xy and the set square is perpendicular to the VP. Draw a line inclined at 30° to xy and redraw the elevation $a_1'b_1'c_1'$ on it as shown in the figure. Since the point B moves parallel to the VP, the distance of b_1 from xy remains the same as the distance of b from xy. Projectors are drawn from a_1' and c_1'. Since the points A and C lie on the VP, the plans a_1 and c_1 will lie on xy. A projector is drawn through b_1'. Locus line is drawn parallel to xy from b as shown in the figure. The intersecting point of the projector with the locus line locates the plan b_1. Join the points of the plan a_1, b_1 and c_1 to complete the projections of the set square.

4. A thin triangular lamina having sides 40 mm, 60 mm and 80 mm is held in such a way that the smallest side is parallel to the HP and perpendicular to the VP. The plane of the lamina is inclined at 60° to the HP. Draw the projections of the lamina.

   Let ABC be the triangular lamina with the sides AB, BC and CA of 40 mm, 60 mm and 80 mm, respectively. The smallest side AB is parallel to the HP and perpendicular to the VP. The lamina ABC is kept initially on the HP with the side AB perpendicular to the VP. The plan abc will be the triangle with the sides 40 mm, 60 mm and 80 mm. The elevation a′b′c′ will be a straight line on xy.

   Draw the reference line xy and the projections of the lamina ABC as shown in Fig. 8.4. The lamina is rotated about the side AB such that it is inclined at 60° to the HP. The elevation of the lamina is rotated about a′b′ such that the elevation a′b′c′ makes an angle of 60° with xy as shown in the figure. The point C moves parallel to the VP and the distance of c_1 from xy is unaltered. The points A and B stay as such. Hence, locus lines are drawn parallel to xy from a, b and c. Projectors are drawn from a_1', b_1' and c_1'. The intersecting points of the corresponding projectors locate the plan a_1, b_1 and c_1. Join the points a_1, b_1 and c_1 to complete the projections of the lamina.
5. Draw the projections of a square 40 mm side. Its centre is 30 mm above the HP with its surface parallel to the VP and 20 mm in front of it.

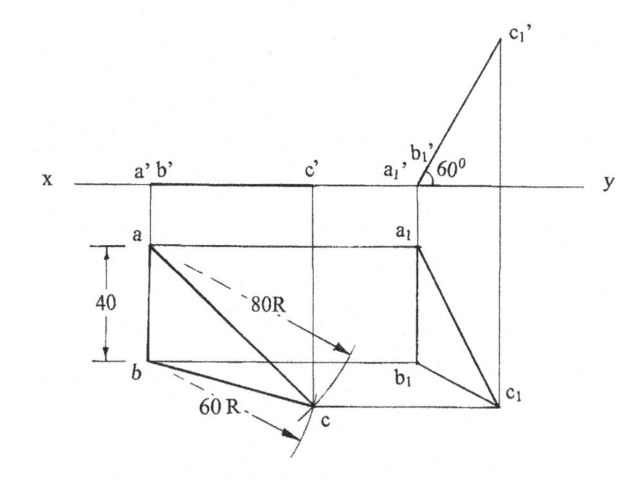

Fig. 8.4 Projections of triangular lamina

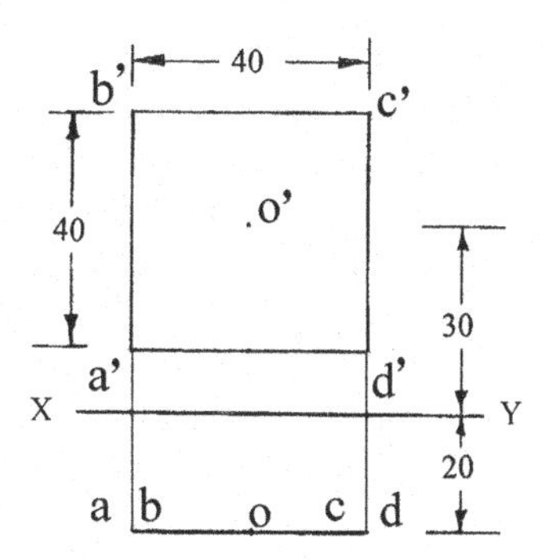

Fig. 8.5 Projections of square

Let the square be designated as ABCD. The square is parallel to the VP. Hence, the projections of ABCD on the VP will be a square of 40 mm side. The centre of the square is 30 mm above xy. Draw a square of side 40 mm as the elevation a′b′c′d′ of the square and locate its centre O′ as shown in Fig. 8.5. Draw the reference line xy 30 mm below O′. The square is parallel to the VP and 20 mm in front of it. The plan will be a straight line parallel to xy and 20 mm below xy. The plan abcd of the square is drawn as shown in the figure to complete the projections of the square.

6. A rectangular lamina of 40 mm × 60 mm rests on the HP on one of its longer edges. The lamina is tilted about the edge on which it rests till the plane of the lamina is inclined at 40° to the HP. The edge on which it rests is perpendicular to the VP. Draw the projections of the lamina.

   Let the lamina be designated as ABCD with longer sides as AB and CD. The lamina is kept initially on the HP with the edge AB perpendicular to the VP.

   Draw the reference line xy and the plan abcd of size 60 mm × 40 mm as shown in Fig. 8.6.

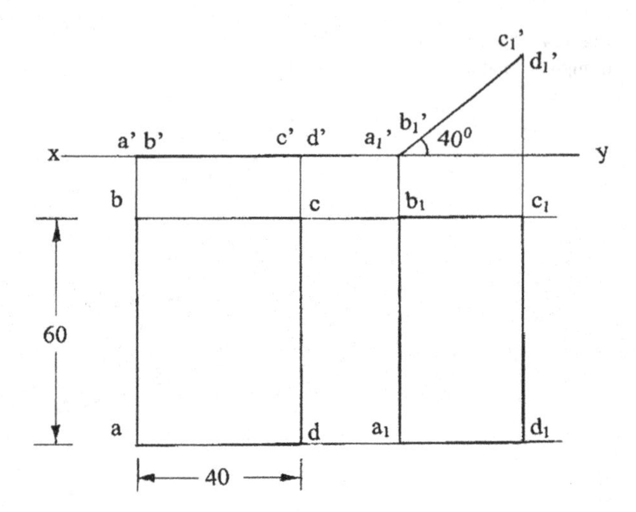

Fig. 8.6 Projections of rectangular lamina

The elevation a′b′c′d′ is obtained on the reference line xy. The lamina is tilted about a′b′ till the elevation a′b′c′d′ makes an angle of 40° to xy. The new elevation is marked as $a_1'b_1'c_1'd_1'$. The points C and D move parallel to the VP during tilting. Hence, the distances of c and d from xy remain the same. Locus lines are drawn parallel to xy from a, b, c and d. Projectors are drawn through a_1', b_1', c_1' and d_1'. The intersections of the projectors with locus lines of the corresponding points locate the plans a_1, b_1, c_1 and d_1. Join the points to complete the plan $a_1b_1c_1d_1$ of the lamina.

7. A rectangular lamina ABCD of 80 mm × 50 mm has its plane perpendicular to the VP and inclined at 45° to the HP. The longer edge AB is 20 mm above the HP and the corner B is 15 mm in front of the VP. Draw the projections of the lamina and locate its traces.

   The lamina is kept initially parallel to the HP and 20 mm above it. The longer edge AB is perpendicular to the VP with the corner B 15 mm in front of the VP.

   Draw the reference line xy, and the projections of the lamina in the initial position are also drawn as shown in Fig. 8.7. The elevation is rotated about the longer edge a′b′ such that the elevation is inclined at 45° to xy. The elevation $a_1'b_1'c_1'd_1'$ is drawn as shown in the figure. The points C and D move parallel to the VP during the rotation of the plane. Locus lines are drawn parallel to xy from a, b, c and d. Projectors are drawn through a_1', b_1', c_1' and d_1'. The intersections of the projectors with the locus lines of the corresponding points locate the plan a_1, b_1, c_1 and d_1. Join the points to complete the plan $a_1b_1c_1d_1$ of the lamina. Since the lamina is perpendicular to the VP, the elevation $a_1'b_1'c_1'd_1'$ is the vertical trace (VT) of the lamina. Extend the elevation to cut the reference line xy at h′. Draw a projector through h′ to get the horizontal trace (HT) of the lamina as shown in the figure.

Fig. 8.7 Projections of rectangular lamina

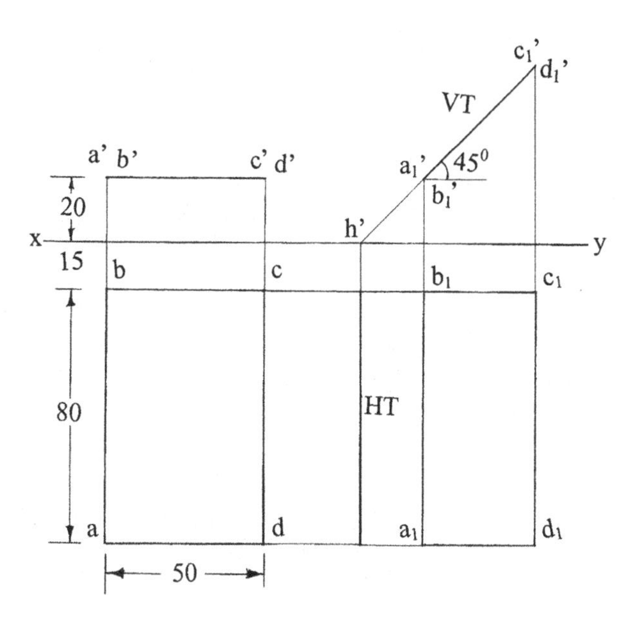

8. A circular lamina of 50 mm diameter has its plane vertical and inclined at 30° to the VP. Its centre is 40 mm above the HP and 40 mm in front of the VP. Draw its projections.

   The circular lamina is kept parallel to the VP initially. Its centre is 40 mm above the HP and 40 mm in front of the VP.

   Draw the reference line xy as shown in Fig. 8.8. Locate the elevation of circle centre 40 mm above xy and the plan of circle centre 40 mm below xy. The elevation of the lamina is a circle of 50 mm in diameter and the plan is a straight line. With centre O′ and radius 25 mm, draw a circle which represents the elevation of the lamina. The plan is projected from the extremities of the circle. Eight points are chosen equidistant on the circumference of the circle as shown in the figure. The points in the plan are projected from the elevation. The lamina is tilted about the diameter 1–5 so that the plane of the lamina is inclined at 30° to the reference line xy. The plan is redrawn and numbered as 1_1, 2_1, 3_1,...etc. Since the points other than 1 and 5 move parallel to the HP, the heights of points 1, 2, 3, 4, 5, etc. above the HP remain the same. Draw locus lines parallel to xy from 1′, 2′, 3′, 4′, 5′, etc. Projectors are drawn through 1_1, 2_1, 3_1, 4_1, 5_1, etc. The intersections of the projectors with the locus lines of the corresponding points locate the elevation $1_1'$, $2_1'$, $3_1'$, $4_1'$, $5_1'$, $6_1'$, $7_1'$ and $8_1'$, and a smooth curve is drawn through these points to complete the elevation of the circular lamina.

9. Draw the projections of a regular pentagon of 30 mm side, having its surface inclined at 50° to the HP and a side perpendicular to the VP. Show the traces.

   Let the pentagon be designated as ABCDE. The pentagon is kept initially parallel to the HP with side AB perpendicular to the VP. The plan of the pentagon will be a regular pentagon of side 30 mm, and the elevation a′b′c′d′e′

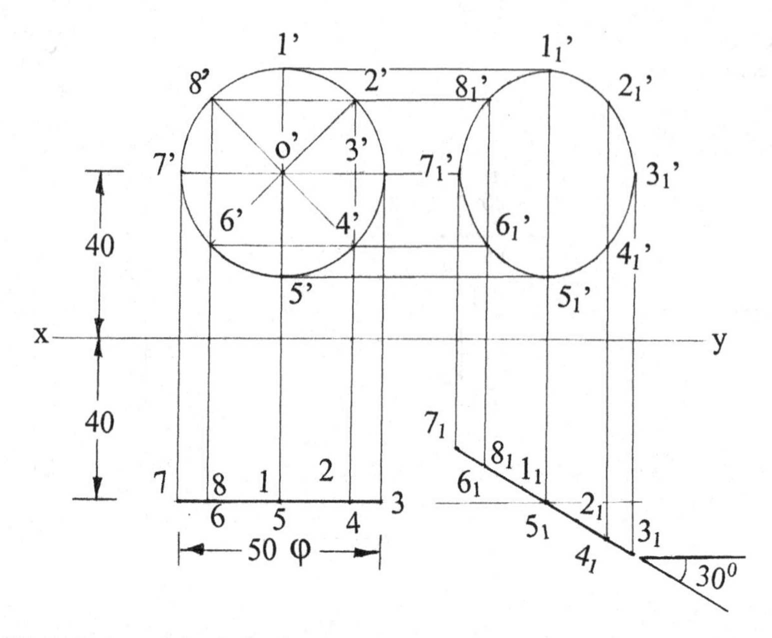

Fig. 8.8 Projections of circular lamina

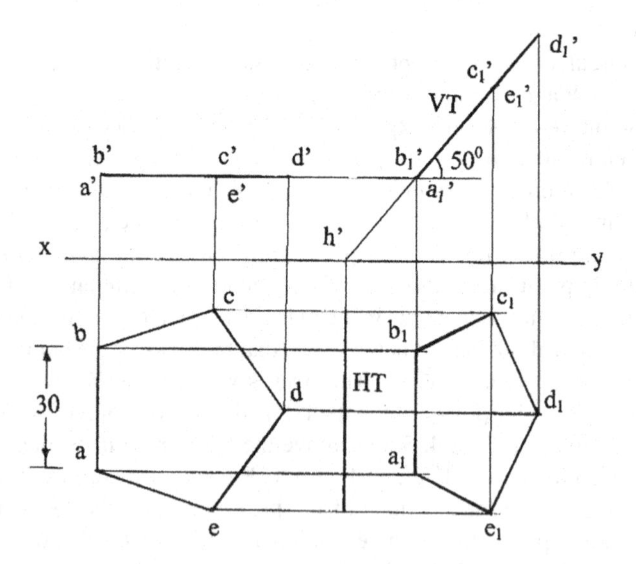

Fig. 8.9 Projections of pentagon

will be a straight line. Draw the reference line xy and the plan of the pentagon abcde as shown in Fig. 8.9. The elevation is projected from the plan. The pentagon is tilted about AB till it makes an angle of 50° with the HP. The elevation is rotated about a′b′ till the elevation makes an angle of 50° with xy. The elevation $a_1'b_1'c_1'd_1'e_1'$ is redrawn making an angle of 50° with xy. The angular points C, D and E move parallel to the VP during tilting. Locus lines are drawn parallel to xy from a, b, c, d and e. Projectors are drawn from a_1', b_1', c_1', $d_1'$and e_1' to intersect the locus lines. The intersecting points are marked as a_1, b_1, c_1, d_1 and e_1. Join the points to complete the plan $a_1b_1c_1d_1e_1$. Since the pentagon is perpendicular to the VP, the elevation $a_1'b_1'c_1'd_1'e_1'$ is the vertical trace (VT) of the pentagon. Extend the elevation to cut the reference line xy at h′. Draw a projector through h′ to get the horizontal trace (HT) of the pentagon as shown in the figure.

10. A regular hexagonal lamina of 30 mm side has one side on the VP. If the lamina is inclined at 40° to the VP and perpendicular to the HP, draw its projections. Also show its traces.

    Let the hexagonal lamina be designated as ABCDEF. It is kept initially on the VP with side AB perpendicular to the HP. The elevation will be a regular hexagon. The plan will be a straight line on xy.

    Draw the reference line xy and the elevation a′b′c′d′e′f′ and the plan abcdef of the hexagonal lamina as shown in Fig. 8.10. The lamina is tilted at an angle of 40° to the VP about the side AB. The plan abcdef is rotated about ab such that it makes an angle 40° with xy. The new plan $a_1b_1c_1d_1e_1f_1$ is drawn as shown in the figure. The angular points C, D, E and F move parallel to the HP during tilting. Locus lines are drawn parallel to xy from a′, b′, c′, d′, e′ and f′. Projectors are drawn through a_1, b_1, c_1, d_1, e_1 and f_1 to intersect the locus lines. The intersecting points are marked as a_1', b_1', c_1', d_1', e_1' and f_1'. Join the points to complete the elevation $a_1'b_1'c_1'd_1'e_1'f_1'$. Since the hexagon is perpendicular to the HP, the plan $a_1b_1c_1d_1e_1f_1$ is the horizontal trace (HT) of the hexagon. Extend the plan to cut the reference line xy at v. Draw a projector through v to get the vertical trace (VT) of the hexagon as shown in the figure.

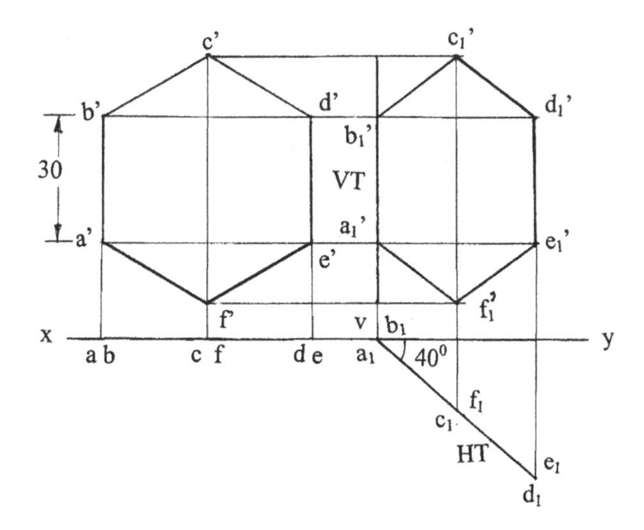

Fig. 8.10 Projections of hexagon

11. An isosceles triangular plate has its base 50 mm and altitude 70 mm. It is so placed that in the elevation it is seen as an equilateral triangle of 50 mm sides and one side is inclined at 45° to the reference line xy. Draw its projections.

    Draw the reference line xy and the elevation a′b′c′ of the isosceles triangle lying on the VP with its base perpendicular to the HP as shown in Fig. 8.11. The plan abc is a straight line on xy. The plate is tilted parallel to the HP about the side ab such that the projection on the VP will be an equilateral triangle of sides 50 mm. The elevation is stated as an equilateral triangle. This triangle is drawn and a projector is drawn from c_1'. With a as centre and ac as radius, draw an arc to cut the projector at c_1. The plan of the plate is abc_1 and the elevation is $a'b'c_1'$ (an equilateral triangle). The plate is moved parallel to the VP so that the side a′b′ is inclined at 45° to xy. The elevation is redrawn with the side $a_2'b_2'$ inclined at 45° to xy as shown in the figure. Projectors are drawn from a_2', b_2' and c_2'. Locus lines are drawn from a, b and c_1. The intersections of points with the corresponding projectors locate the plans a_2, b_2 and c_2. These points are joined to show the required plan of the plate.

12. An equilateral triangular plate of 60 mm edge lies with one of its edges on the HP, and the surface of the plate is inclined at 40° to the HP. The edge on which it rests is inclined to the VP at 50°. Draw its projections.

    Draw the reference line xy and the equilateral triangle abc, the plan of the plate lying on the HP with edge ab perpendicular to xy as shown in Fig. 8.12. The elevation a′b′c′ is a straight line on xy. The elevation is tilted parallel to the VP such that its surface is inclined at 40° to the HP. The elevation $a_1'b_1'c_1'$ is redrawn as shown in the figure. As the point C moves parallel to the VP, its distance from the VP remains the same. Projectors are drawn from a_1', b_1' and c_1'. Locus lines are drawn from a, b and c. The intersections of the points with the corresponding projectors locate the plans a_1, b_1 and c_1. Join the points to get the plan of the plate. The plate is moved parallel to the HP with the edge a_1b_1

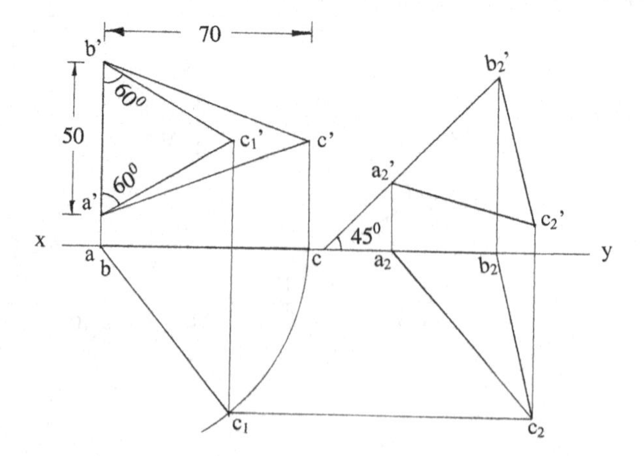

Fig. 8.11 Projections of isosceles triangle

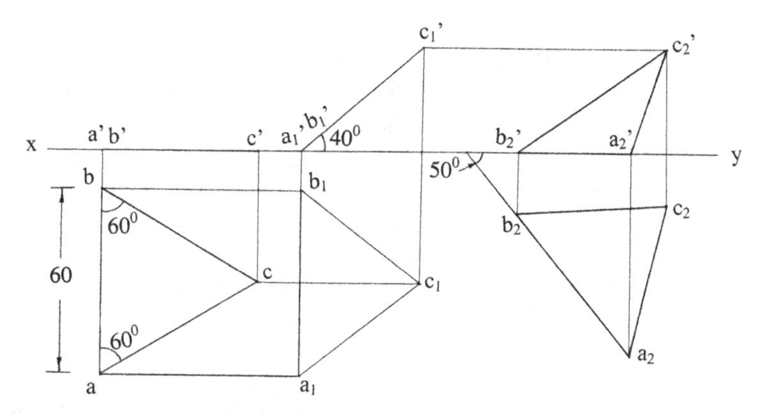

Fig. 8.12 Projections of equilateral triangle

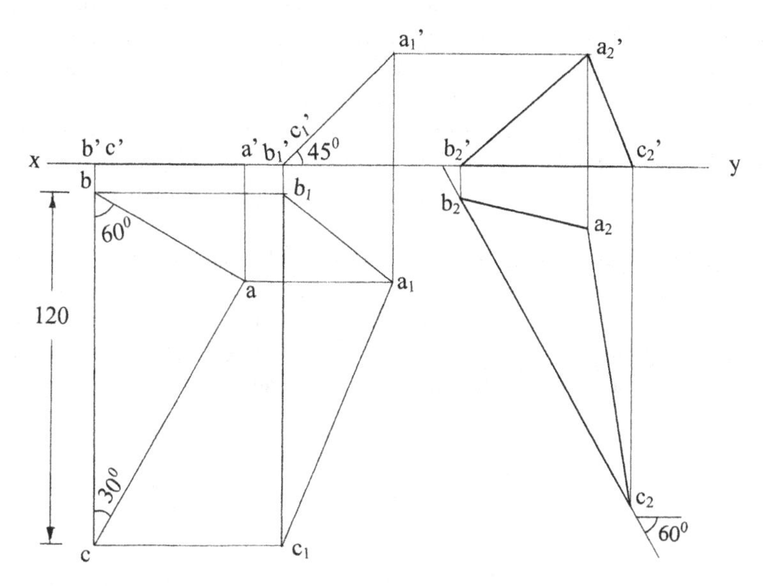

Fig. 8.13 Projections of 30°–60° set square

making an angle of 50° with xy. The plan is redrawn with the edge a_2b_2 making an angle of 50° with xy. Projectors are drawn from a_2, b_2 and c_2. Locus lines are drawn from a_1', b_1' and c_1'. The intersections of the points with the corresponding projectors locate the elevations a_2', b_2' and c_2'. These points are joined to complete the required elevation of the plate.

13. A 30°–60° set square of 120 mm longest side is so kept that the longest side is in the HP making an angle of 60° with the VP. The set square itself is inclined at 45° to the HP. Draw the projections of the set square.

Draw the reference line xy and the plan abc of the set square on the HP with the longest side bc perpendicular to xy as shown in Fig. 8.13. The elevation a′b′c′ is a straight line on xy. The elevation is rotated about b′c′ such that the set square is inclined at 45° to the HP. The point A moves parallel to the VP. The elevation $a_1'b_1'c_1'$ is redrawn as shown in the figure. Projectors are drawn from a_1', b_1' and c_1'. Locus lines are drawn from a, b and c. The intersections of the points with the corresponding projectors locate the plans a_1, b_1 and c_1. These points are joined to get the plan of the set square. The set square is moved parallel to the HP such that the edge b_1c_1 makes an angle of 60° with the VP. The plan $a_2b_2c_2$ is redrawn with the side b_2c_2 making an angle of 60° with xy. Projectors are drawn from a_2, b_2 and c_2. Locus lines are drawn from a_1', b_1' and c_1'. The intersections of the points with the corresponding projectors locate the elevations a_2', b_2' and c_2'. Join the points to complete the elevation of the set square.

14. A square plate of 40 mm side rests on the HP such that one of the diagonals is inclined at 30° to the HP and 45° to the VP. Draw its projections.

    Draw the reference line xy and the plan abcd of the square plate with its sides equally inclined to xy as shown in Fig. 8.14. The diagonal ac is parallel to both the HP and the VP. The elevation a′b′c′d′ lies on xy. The plate is tilted parallel to the VP about the corner a′ such that the diagonal a′c′ makes an angle of 30° with the HP. The elevation is redrawn with the diagonal making an angle of 30° with xy. Projectors are drawn from a_1', b_1', c_1' and d_1'. Locus lines are drawn from a, b, c and d. The intersections of the points with the corresponding projectors locate the plans a_1, b_1, c_1 and d_1. These points are joined to show the plan of the plate. The length a_1c_1 provides the plan length of the diagonal. The apparent angle of inclination of the plan of the diagonal is to be found out when the diagonal is inclined at 45° to the VP. A line x_1y_1 is drawn parallel to xy. Locate point a on the line x_1y_1 as shown in the figure. A line inclined at 45° with x_1y_1 is drawn from a. A point c′ is marked on this line such that the length

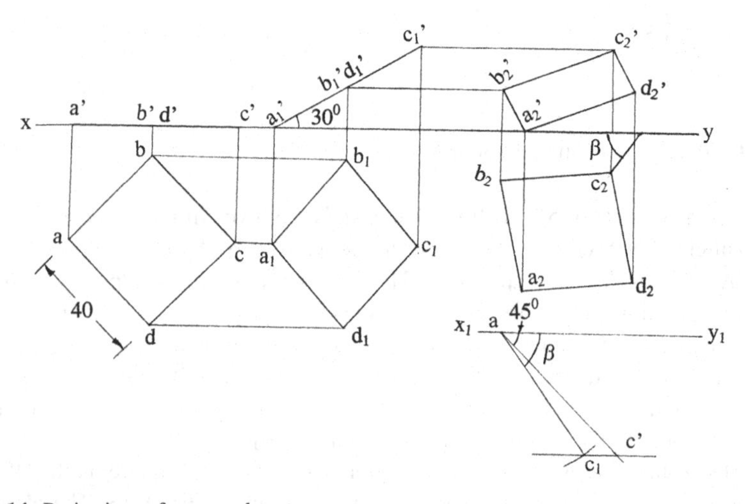

Fig. 8.14 Projections of square plate

of ac′ is equal to the diagonal length of the square plate. Draw locus line parallel to xy through c′. With centre a and the plan length of the diagonal a_1c_1 as radius, draw an arc to cut the locus line at c_1. The inclination of ac_1 furnishes the apparent angle (β) of inclination of the plan of the diagonal with xy. The square plate is moved parallel to the HP such that the plan of the diagonal makes an angle of β with xy. The plan is redrawn such that the plan of the diagonal a_2c_2 makes an angle of β with xy. Projectors are drawn from a_2, b_2, c_2 and d_2. Locus lines are drawn from a_1', b_1', c_1' and d_1'. The intersections of the points with the corresponding projectors locate the elevations a_2', b_2', c_2' and d_2'. These points are joined to get the required elevation of the plate.

15. Draw the projections of an octagonal lamina of sides 30 mm when the lamina rests on one of its sides, the side on which it rests being parallel to the VP and the surface of the lamina inclined to the HP at an angle of 50°.

    Draw the reference line xy and the plan of the octagonal lamina lying on the HP with one side perpendicular to the VP as shown in Fig. 8.15. The elevation is a straight line on xy. The lamina is tilted parallel to the VP about the edge ab such that the lamina surface makes an angle of 50° with the HP. The elevation is redrawn as shown in the figure. Projectors are drawn from a_1', b_1', c_1', d_1', e_1', f_1', g_1' and h_1'. Locus lines are drawn from a, b, c, d, e, f, g and h. The intersections of the points with the corresponding projectors locate the plans a_1, b_1, c_1, d_1, e_1, f_1, g_1 and h_1. These points are joined to get the plan of the lamina. The lamina is moved parallel to the HP such that the edge a_1b_1 is parallel to the VP. The plan is redrawn with a_2b_2 being parallel to xy as shown in the figure. Projectors are drawn from a_2, b_2, c_2, d_2, e_2, f_2, g_2 and h_2. Locus lines are drawn from a_1', b_1', c_1', d_1', e_1', f_1', g_1' and h_1'. The intersections of

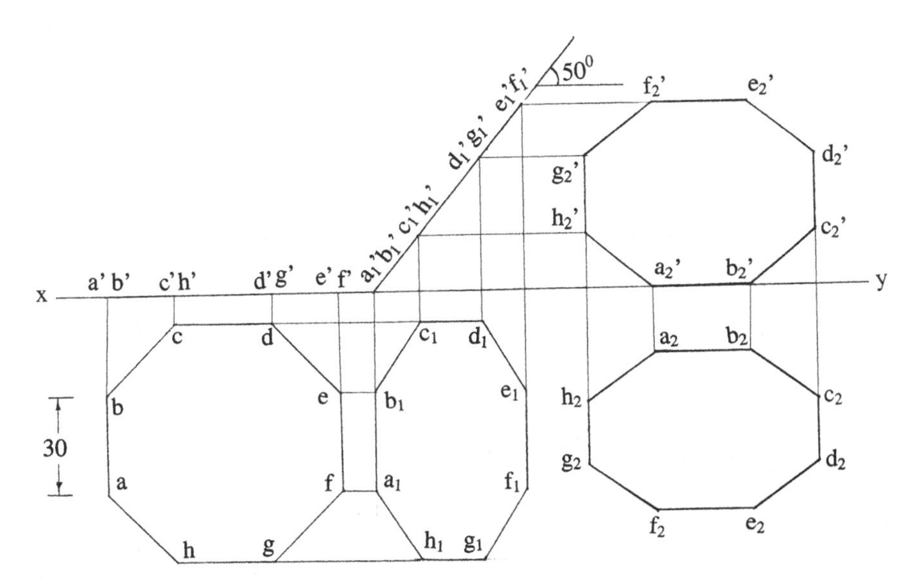

Fig. 8.15 Projections of octagonal lamina

points with the corresponding projectors locate the elevations a_2', b_2', c_2', d_2', e_2', f_2', g_2' and h_2'. These points are joined to show the sides of the lamina in the elevation.

16. Draw the projections of a square lamina of side 40 mm when its centre is 40 mm from the HP and VP and one of its diagonals makes an angle of 30° to the HP and the other 45° to the VP.

Let the square lamina be designated as ABCD. The square lamina is kept parallel to the HP with its centre 40 mm in front of the VP and 40 mm above the HP. The plan is drawn with its diagonal ac parallel to the VP as shown in Fig. 8.16. The elevation is projected from the plan. The lamina is tilted about the diagonal bd till the diagonal ac makes an angle of 30° to the HP. The elevation $a_1'b_1'c_1'd_1'$ is redrawn as shown in the figure. The plan $a_1b_1c_1d_1$ is to be projected from the elevation $a_1'b_1'c_1'd_1'$. Locus lines are drawn parallel to xy from a, b, c and d. Projectors are drawn from a_1', b_1', c_1' and d_1' to intersect the locus lines. The intersecting points are marked as a_1, b_1, c_1 and d_1. Join the points to complete the plan $a_1b_1c_1d_1$. It should be noted that the centre of lamina is 40 mm above xy and 40 mm below xy. The lamina in this position is moved parallel to the HP such that the diagonal b_1d_1 makes 45° to the VP. The plan $a_2b_2c_2d_2$ is redrawn such that the centre of the lamina is 40 mm below xy and the diagonal b_2d_2 makes an angle of 45° to xy. Locus lines are drawn parallel to xy from a_1', b_1', c_1' and d_1'. Projectors are drawn from a_2, b_2, c_2 and d_2 to intersect the locus lines. The intersecting points are marked a_2', b_2', c_2' and d_2'. Join the points to complete the final elevation $a_2'b_2'c_2'd_2'$.

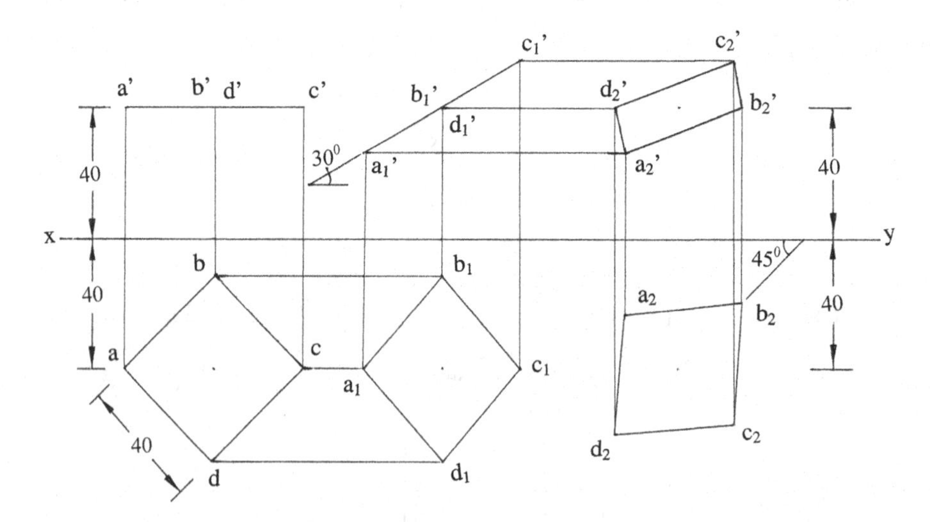

Fig. 8.16 Projections of square lamina

Practice Problems

1. A thin triangular lamina having sides 50 mm, 70 m and 100 mm is held in such a way that the largest side is vertical and perpendicular to the HP. The plane of the lamina is inclined at 40° to the VP. Draw the projections of the lamina.
2. The elevation of a rectangular lamina of sides 80 mm × 60 mm is a square of 60 mm side. Draw the plan and elevation of the lamina. Determine the inclination of the surface of the lamina with the HP and the VP.
3. A square lamina of 25 mm side rests on one of its sides on the VP. The lamina makes an angle of 60° to the VP. Draw its projections.
4. A regular pentagonal plate with edge 30 mm length is held on the VP on one of its edges. This edge is perpendicular to the HP and the plate surface makes 30° to the VP. Draw its projections.
5. Draw the projections of a regular hexagon of 40 mm side standing vertically on the HP on one edge with its plane inclined at 40° to the VP, and one angular point 30 mm from the VP. Show its traces.
6. A circular plate of 60 mm diameter has its plane perpendicular to the VP and inclined 45° to the HP. The centre of the plate is 40 mm in front of the VP and 35 mm above the HP. Show the projections of the plate. Also mark the traces.
7. A regular pentagon of 30 mm edges has an edge on the HP and perpendicular to the VP. The corner opposite to the edge on the HP is 30 mm above the HP. Show the projections of the pentagon. Determine the inclination of the surface with the HP.

Chapter 9
Projections of Solids

Classification of Solids

A solid may be defined as an object having dimensions like length, breadth and thickness. Solids generally used in the study of Engineering Drawing may be classified as:

1. Polyhedra
2. Solids of revolution.

Solids may also be classified as:

1. Right solids
2. Oblique solids.

If the axis of a solid is perpendicular to its base or end faces, that solid is called a right solid. If all the edges of the base or of the end faces of a right solid are of equal lengths and they form a plane figure, that right solid is called a right regular solid.

If the axis of a solid is inclined to its base or end faces, that solid is called an oblique solid. If all the edges of the base or of the end faces of an oblique solid are of equal lengths and they form a plane figure, that oblique solid is called an oblique regular solid.

Polyhedron

A polyhedron is defined as a solid bounded by planes called faces. Regular polyhedra, prisms and pyramids are some of the solids coming under this group.

K. Rathnam, *A First Course in Engineering Drawing*,
DOI 10.1007/978-981-10-5358-0_9

Fig. 9.1 Regular polyhedra

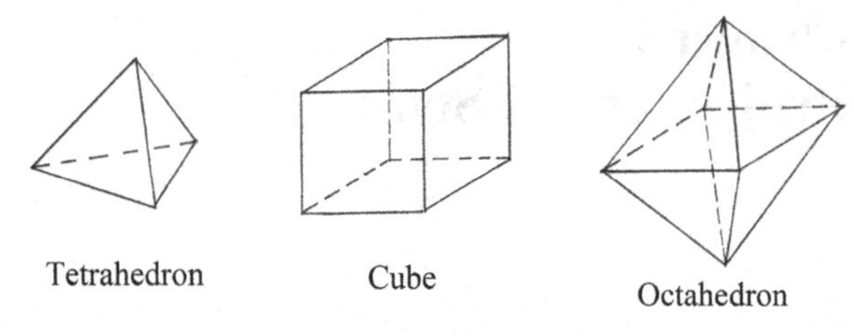

Regular Polyhedra

A regular polyhedron is defined as a solid bounded by regular planes called faces whose edges are equal in length. The three regular polyhedra encountered in our course are shown in Fig. 9.1.

Prism

A prism is a polyhedron having two equal and similar end faces parallel to each other and joined by side faces which are either rectangles or parallelograms. Prisms are named according to the shape of the end faces such as triangular, square, pentagonal, hexagonal, etc. Some common types of prisms are shown in Fig. 9.2.

The axis of a prism is an imaginary line joining the centre of the end faces. In a right prism, its axis is perpendicular to its base; but in an oblique prism, its axis is inclined to its base. Figure 9.3 shows the nomenclature of a square prism.

Pyramid

A pyramid is a polyhedron having a polygon as base and a number of equal isosceles triangular faces equal to the number of sides of the polygon, meeting at a point called apex or vertex. Pyramids are named according to the shape of its base such as triangular, square, pentagonal, hexagonal, etc. Some important types of pyramids are shown in Fig. 9.4. Axis of pyramid is an imaginary line joining the apex to the centre of its base. The perpendicular distance between the apex and the base of the pyramid is called altitude or height of the pyramid. In a right pyramid, the axis is perpendicular to its base, but in an oblique pyramid, the axis is inclined to its base. Figure 9.5 shows the nomenclature of a square pyramid. In the figure shown, the distance OP represents the height of the pyramid.

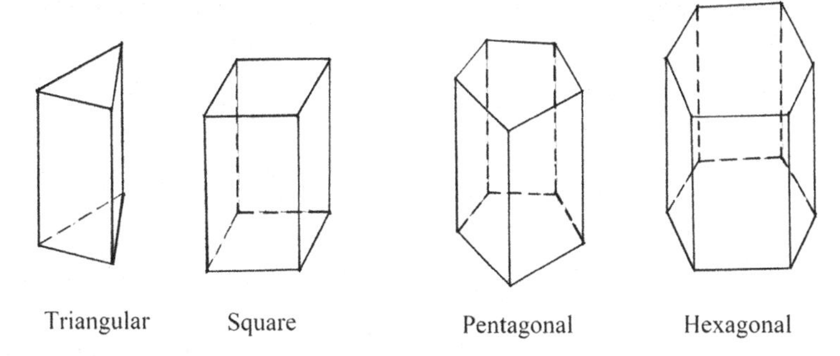

Fig. 9.2 Prisms

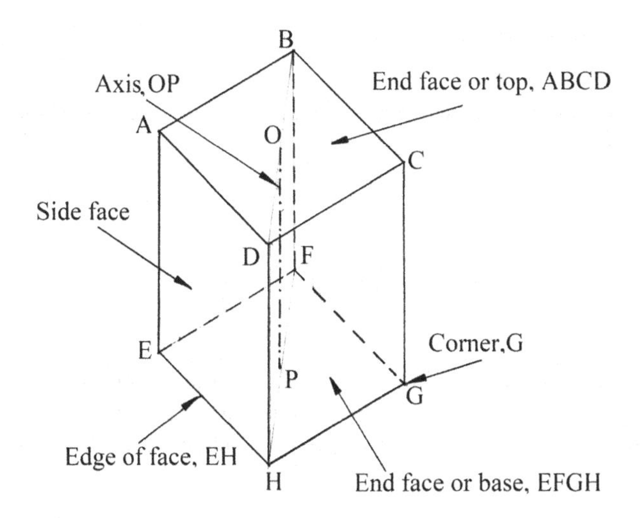

Fig. 9.3 Nomenclature of a square prism

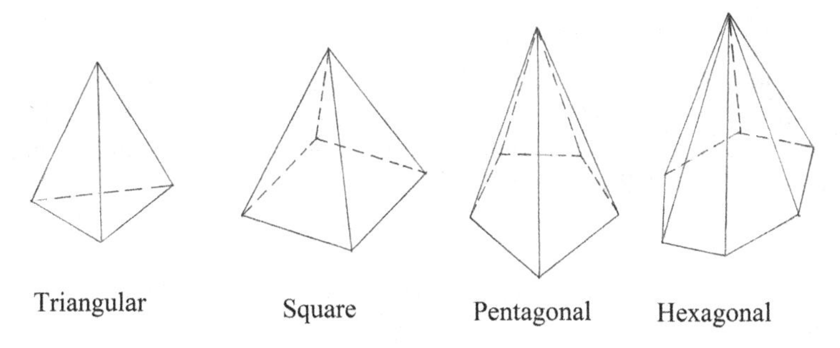

Fig. 9.4 Pyramids

Fig. 9.5 Nomenclature of square pyramid

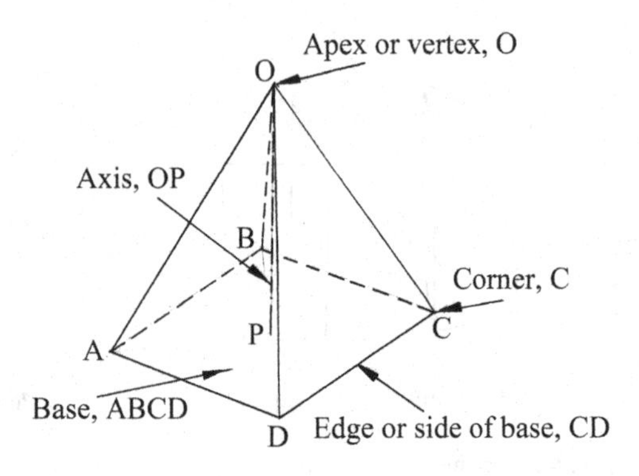

Solids of Revolution

A solid of revolution is defined as a solid generated by the revolution of a plane figure about a line called axis. Cylinder, cone and sphere are some of the solids coming under this group and are shown in Fig. 9.6.

Cylinder

A solid generated by the revolution of a rectangle about one of its sides which remains stationary is called a right circular cylinder. The right circular cylinder is generated by the revolution of the rectangle ABPO about OP as shown in Fig. 9.6(i).

Cone

A solid generated by the revolution of a right angled triangle about one of its sides which remains stationary and containing the right angle is called right circular cone. The right circular cone as shown in Fig. 9.6(ii) is generated by the revolution of the right angled triangle OPA about OP.

Sphere

A solid generated by the revolution of a semi-circle about its diameter which remains stationary is called a sphere. The sphere shown in Fig. 9.6(iii) is generated by the revolution of the semi-circle about AB.

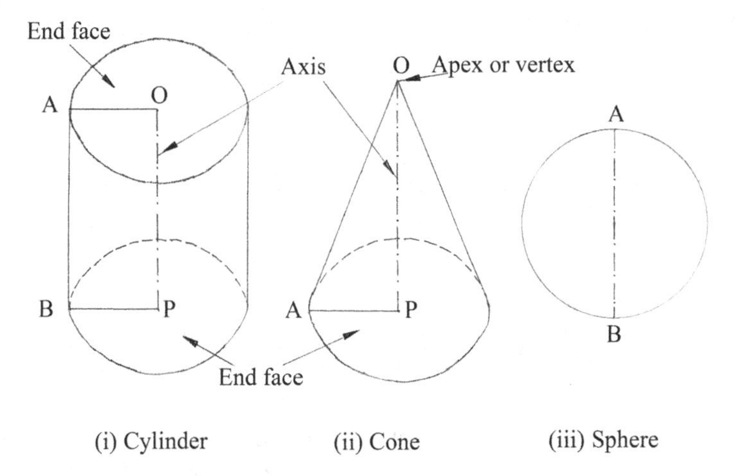

Fig. 9.6 Solids of revolution

Position of a Solid to Get Its Projections

The solid is generally placed in any one of the following ways, based on the inclination of its axis with the reference planes:

1. Axis perpendicular to one of the reference planes (simple position)
2. Axis parallel to both the reference planes (simple position)
3. Axis parallel to one of the reference planes and inclined to the other plane.

The position of a solid with reference to the reference planes can also be grouped as under:

1. Solid resting on its base (simple position in general except the solid Octahedron)
2. Solid resting on any one of its faces/an edge of face/an edge of base/a generator/a slant edge.
3. Solid suspended freely from one of its corners, etc.

Axis perpendicular to one of the reference planes

When the axis is perpendicular to one of the reference planes, its projection on this plane is obtained first. After drawing this projection, the projection on the other plane is obtained easily.

Axis parallel to both the reference planes

If the axis of a solid is parallel to both the VP and HP, its projections (elevation and plan) can be easily obtained by drawing one of its end views first. End view of the solid is the projection of the solid on a plane perpendicular to both the VP and HP. This perpendicular plane is called profile plane and is abbreviated by PP (not to be confused with the picture plane). These three mutually perpendicular planes HP, VP and PP are shown in Fig. 9.7. The projection obtained on the profile plane kept on the right hand side is called end view, side view or profile view from the left (left

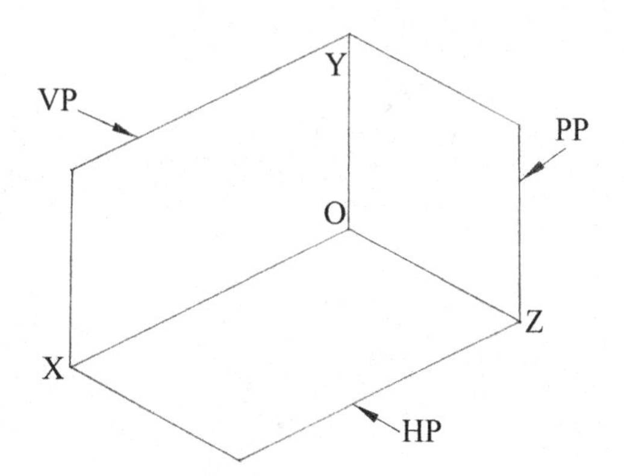

Fig. 9.7 Pictorial representation of three reference planes

side elevation to be drawn on the right side). The lines of intersection of HP with VP, VP with PP and HP with PP are marked by OX, OY and OZ, respectively. To show the projections of the solid on a sheet of paper, the planes HP and PP are rotated through 90° and are represented as shown in Fig. 9.8.

Axis parallel to one of the reference planes and inclined to the other
In the preparation of the orthographic projections of a solid whose axis is parallel to one of the reference planes and inclined to the other plane, the method adopted is as follows. Assume the axis is initially perpendicular to the plane to which its inclination is given in the problem. The projections of the solid are obtained. Based upon to which plane the solid is inclined, one of the above views is rotated to obtain the required view.

In this chapter, the projections of solids in simple position are given. Accordingly, the projections of a prism is taken up first and explained with the help of the pictorial representations.

Projections of a Prism

The projections of a prism are simple when the prism is placed such that its ends are parallel to one of the reference planes. If the prism is a right prism with its ends parallel to one of the planes, its projection on that plane will show its true form. Also the projections of its ends will coincide on the plane of projection. The projections of the ends on the other plane of projection will be straight lines parallel to the reference line xy.

The projections of a square prism are illustrated in Figs. 9.9 and 9.10. Figure 9.9 shows the pictorial representation of the prism and the planes of projection. Figure 9.10 shows the plan and elevation of a square prism when its ends are horizontal. The plan abcd of the top end is drawn first; its position with reference to xy will be specified in the problem. The plan of the lower end efgh coincides with the top end. The elevation is projected from the plan as shown in Fig. 9.10. The

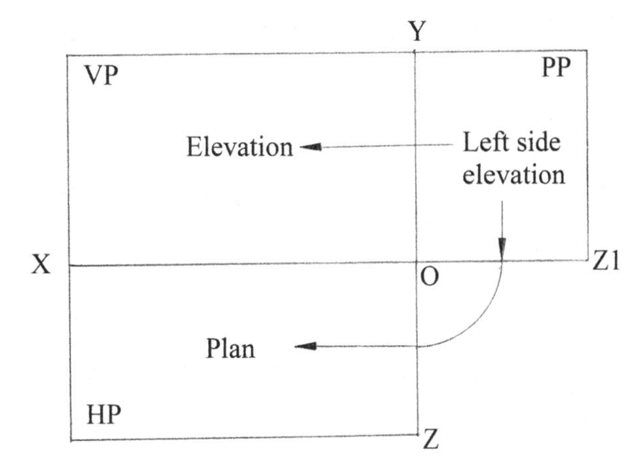

Fig. 9.8 Orthographic views of an object (axis parallel to both the planes)

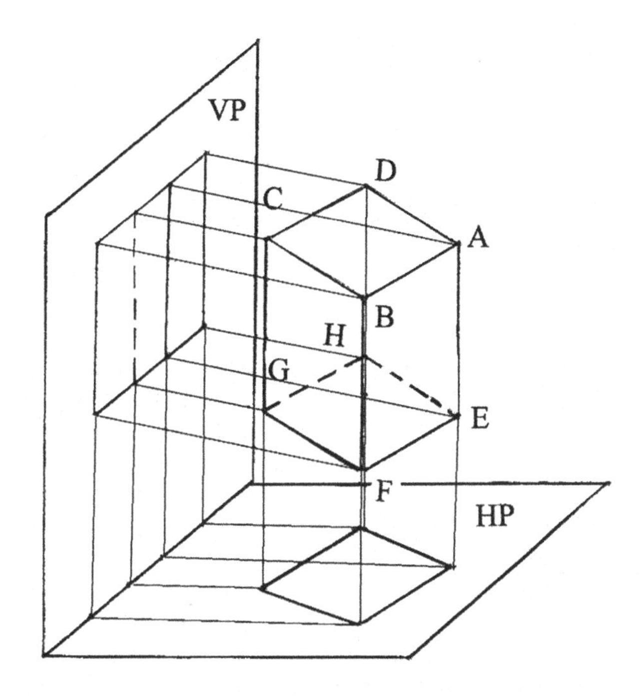

Fig. 9.9 Pictorial view of planes and prism

height of the elevation of lower end above xy will be specified. The distance between the horizontal lines a′b′c′d′ and e′f′g′h′ is equal to the altitude/axis of the prism. It will be noticed in Fig. 9.10 that the elevation c′g′ of the edge CG is shown as a dotted line. This edge is hidden by the solid when viewed/projected from the front. Horizontal edges b′c′, c′d′, f′g′ and g′h′ are also invisible in the elevation. When an invisible edge coincides fully with the visible edge in any projection, the visible edge alone is shown by full line ignoring the invisible edge. When a part of

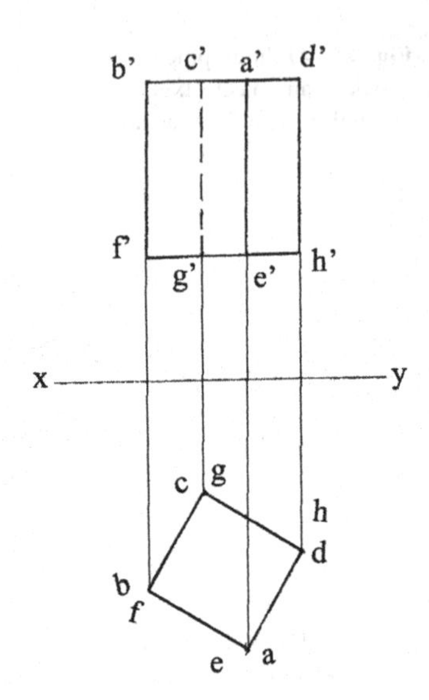

Fig. 9.10 Projections of prism

the invisible edge is distinct in any projection, that part of invisible portion is shown by a dotted line.

Solved Problems

1. Draw the projections of a cube of 40 mm side when resting on the HP on one of its faces with a vertical face inclined at 30° to the VP.

   When the cube rests on the HP, its projection on the HP is a square of 40 mm side. The four sides of the square represent four vertical faces of the cube. The projections of the top face and the bottom face coincide. The plan of the cube is drawn first with a side of square inclined at 30° to the VP. The elevation is projected from the plan.

   Draw the reference line xy and the plan of the cube with one side inclined 30° to xy as shown in Fig. 9.11. The plan of the bottom face efgh coincides with the top face abcd. Draw projectors through a, b, c, d, e, f, g and h. The bottom face EFGH lies on the HP. Hence, the elevation of the bottom face lies on xy. The intersecting points of the projectors with reference line locate the elevation e′f′g′h′ of the bottom face. The elevation of the top face will be 40 mm above the bottom face. Draw a horizontal line 40 mm above xy. The intersecting points of the projectors with this horizontal line locate the elevation a′b′c′d′ of the top face of the cube. Join the points a′e′, b′f′, c′g′ and d′h′ to complete the elevation of the cube. The edge BF is hidden by the solid when viewed from the front. Hence, the elevation b′f′ is shown by dotted line.

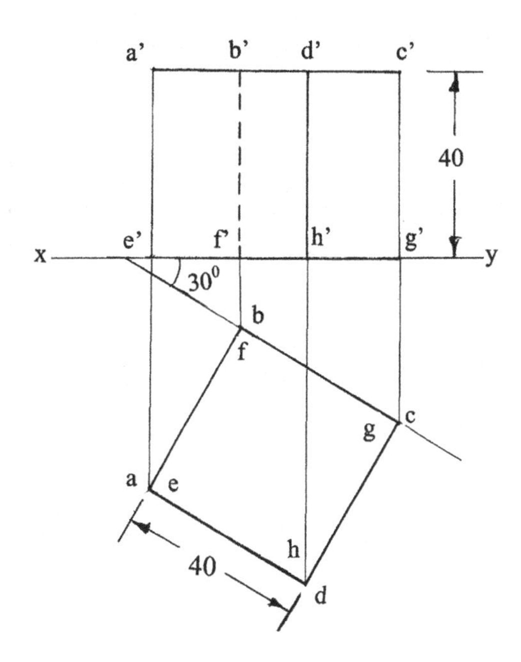

Fig. 9.11 Projections of a cube

2. A triangular prism 40 mm side of base and 60 mm long rests on the HP on one of its ends with an edge of the base inclined at 30° to the VP. Draw its projections.

   When the triangular prism rests on the HP on its end, its projection on the HP is an equilateral triangle of side 40 mm. The plans of its ends coincide. The plan of the prism is drawn first with a side of the triangle inclined at 30° to the VP. The elevation is projected from the plan.

   Draw the reference line xy and the plan of the prism keeping one of its sides inclined at 30° to xy as shown in Fig. 9.12. The plan of the bottom face def coincides with the top face abc. Draw projectors through a, b, c, d, e and f. The bottom face DEF lies on the HP. Hence, the elevation of the bottom face lies on xy. The intersecting points of the projectors with xy locate the elevation d′e′f′ on xy. The elevation of the top face will be 60 mm above the bottom face. Draw a horizontal line 60 mm above xy. The intersecting points of the projectors with the horizontal line locate the elevation a′b′c′. Join a′d′, b′e′ and c′f′ to complete the elevation of the prism. The invisible edge b′e′ coincides with the visible edge a′d′.

3. Draw the projections of a pentagonal prism 30 mm side of base and axis 50 mm long resting on the HP on one of its ends such that two of its rectangular faces are equally inclined to the VP.

   When the pentagonal prism rests on the HP, its projection on the HP is a pentagon of side 30 mm. The plans of its ends coincide. The pentagon is drawn keeping one side either parallel to xy or perpendicular to xy. In both cases, two of its rectangular faces are equally inclined to the VP. The elevation is projected from the plan.

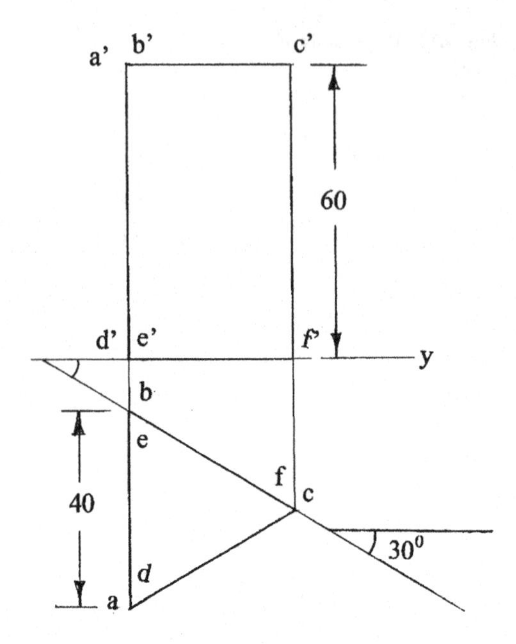

Fig. 9.12 Projections of triangular prism

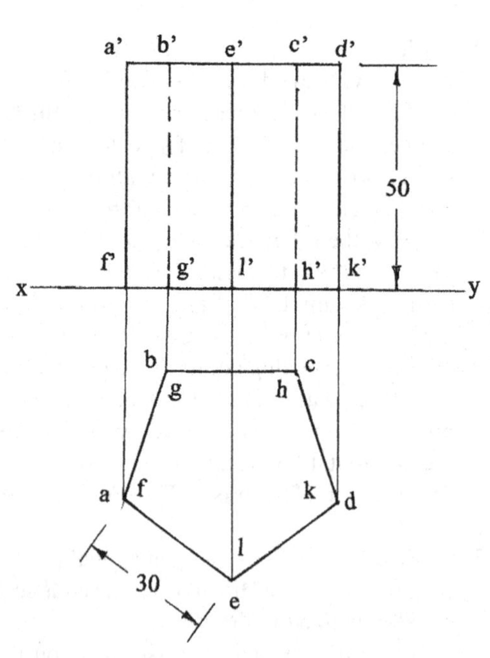

Fig. 9.13 Projections of pentagonal prism

Draw the reference line xy and the plan of the pentagonal prism keeping one of its sides parallel to xy as shown in Fig. 9.13. In this position, two faces AELF and DELK are equally inclined to the VP. The plan of the bottom end coincides with the plan of the top end. Draw projectors through a, b, c, d, e, f, g, h, k and l. The elevation of the bottom end lies on xy. The intersecting points of the projectors with xy locate the elevation f′g′h′k′l′ of the bottom end. The elevation of the top end will be 50 mm above xy. Draw a horizontal line 50 mm above xy. The intersecting points of the projectors with this horizontal line locate the elevation a′b′c′d′e′ of the top end of prism.

The vertical edges are joined to complete the elevation of the prism. The edges b′g′ and c′h′ are hidden by the solid when viewed from the front. Hence, these two edges are shown by dotted lines.

4. A hexagonal prism, base 30 mm side and axis 70 mm long, is placed with its base on the VP such that one of its base edges is parallel to and 20 mm above the HP. Draw its projections.

   When the hexagonal prism has its base on the VP, its projection on the VP is a hexagon with a base edge parallel to and 20 mm above xy. The elevations of its ends coincide. The plan is projected from the elevation. The plan of base of the prism is a straight line on xy. The plan of top of the prism is also a straight line 70 mm below xy.

   Draw the reference xy and the elevation of the hexagonal prism with a side parallel to and 20 mm above xy as shown in Fig. 9.14. The elevation g′h′k′l′m′n′ of the base of the prism coincides with the elevation a′b′c′d′e′f′ of the top of the prism. Projectors are drawn through a′, b′, c′, etc. The base of the prism rests on the VP. Hence, the plan ghklmn of the prism lies on xy. The plan of the front/top end of the prism will be 70 mm below xy. Draw a horizontal line 70 mm below xy. The intersecting points of the projectors with this horizontal line locate the

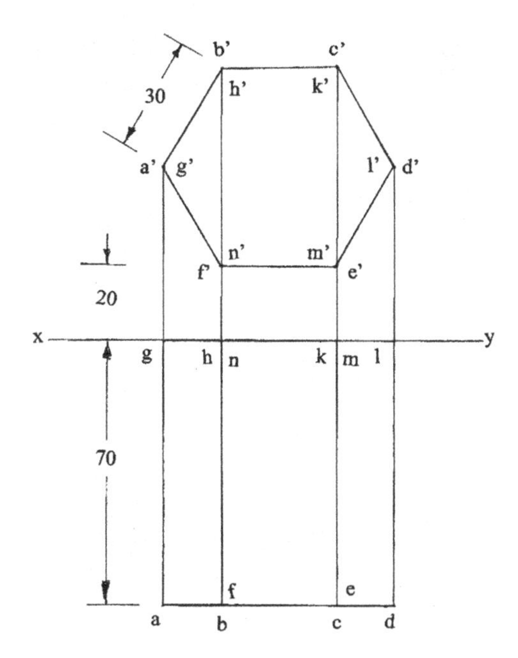

Fig. 9.14 Projections of hexagonal prism

plan of the front end abcdef. The horizontal edges of the prism are joined to complete the plan. The invisible edges em and fn coincide with the visible edges ck and bh in the plan.

Projections of a Pyramid

The projections of a pyramid are simple when the base of the pyramid is parallel to one of the reference planes. When the base of the pyramid is parallel to one of the reference planes, its projection of the base on that plane will show its true form. This projection is drawn first. For a right pyramid, the projection of its vertex will be at the centre of the base. The projection of the base on the other plane will be straight line.

The projections of a square pyramid are illustrated in Figs. 9.15 and 9.16. Figure 9.15 shows the pictorial representation of the pyramid and the planes of projection. Figure 9.16 shows the plan and elevation of a square pyramid when its base is parallel to the horizontal plane. The plan is a square abcd with its diagonals representing the slant edges. The intersection of the diagonals locates the plan v of the vertex/apex of the pyramid. The elevation is projected from the plan as shown in Fig. 9.16. The elevation of base of the pyramid will be a straight line parallel to the reference line xy. Its position will be specified in the problem. The elevation v′ of the vertex is located on the projector drawn through v. The distance of v′ from the elevation of base of the pyramid is the height/altitude of the pyramid. The invisible slant edge v′c′ is shown by dotted line.

5. A square pyramid of 40 mm side of base and 60 mm height rests on the HP on its base with a base edge inclined at 40° to the VP. Draw its projections.

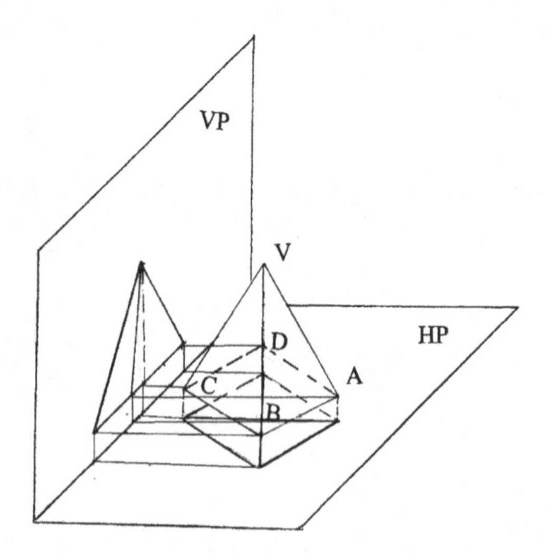

Fig. 9.15 Pictorial view of planes and pyramid

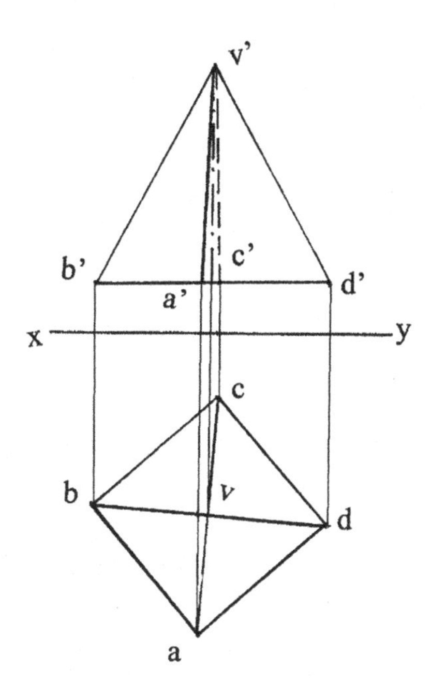

Fig. 9.16 Projections of pyramid

The square pyramid rests on the HP. The projection of the pyramid on the HP will be a square with its diagonals representing the slant edges. The diagonals intersect at v, the plan of the vertex/apex of the pyramid. The plan is drawn first. The elevation is projected from the plan.

Draw the reference xy and the plan of the pyramid as shown in Fig. 9.17. The side ad of the base is inclined at 40° to xy. The diagonals of the square intersect at v and locate the plan of the vertex/apex of the pyramid. The elevation is projected from the plan. The base of the pyramid rests on the HP. Hence, the elevations of base points lie on xy. Projectors are drawn through a, b, c and d.

The intersecting points of projectors with xy locate the elevation a′b′c′d′. The elevation v′ of the vertex lies on the projector drawn through v. The distance of v′ from the elevation of the base is the height/altitude of the pyramid. Hence, v′ is 60 mm from the elevation of base of the pyramid. The slant edges are joined. The slant edge VB is hidden by the pyramid when viewed from the front. The elevation of edge v′b′ is shown as dotted line. The axis of the pyramid is also shown in the elevation.

6. Draw the projections of a pentagonal pyramid, side of base 30 mm and height 60 mm, resting with its base on the VP such that one of the corners of the base is on the HP and the adjacent corner is 15 mm above the HP.

   The hexagonal pyramid rests with its base on the VP. The projection on the VP will be a pentagon. The elevation of an angular point lies on xy, and the elevation of the adjacent angular point will be 15 mm above xy. The centre of the pentagon represents the vertex and is joined to the angular points to show the elevations of slant edges of the pyramid. The elevation is drawn first. The plan of

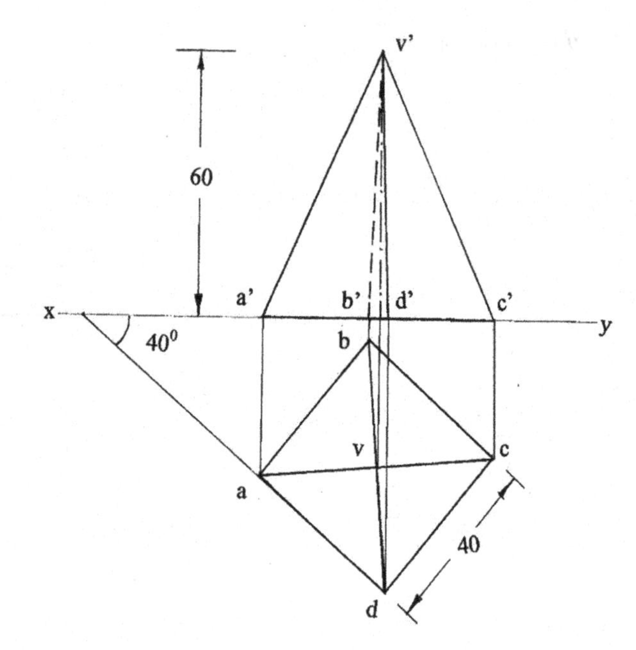

Fig. 9.17 Projections of square pyramid

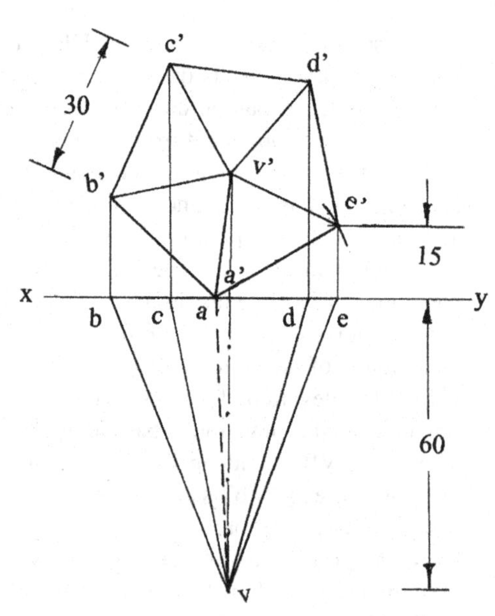

Fig. 9.18 Projections of pentagonal pyramid

base of the pyramid lies on xy. The plan of the vertex/apex will be 60 mm from the plan of the base of the pyramid.

Draw the reference line xy as shown in Fig. 9.18. The elevation a′ of an angular point is located on xy, and locus line of the adjacent angular point is drawn 15 mm above xy. With centre a′ and radius 30 mm, draw an arc cutting the locus line at e′. Join a′e′ and construct the pentagon a′b′c′d′e′ as shown in the figure. Locate the elevation of the vertex at the centre of the pentagon. The vertex is connected to the angular points to show the elevations of slant edges. Projectors are drawn through a′, b′, c′, d′ and e′ to intersect the reference line xy to locate the plan of base points of the pyramid. A projector is drawn through v′. The plan v is located on this projector 60 mm from the plan of base of the pyramid. The slant edges are connected. The invisible edge av is shown as a dotted line.

7. A hexagonal pyramid, side of base 30 mm and height 70 mm, rests on the HP with a base edge inclined at 45° to the VP. Draw its projections and also its end elevation.

The hexagonal pyramid rests on the HP. The plan of the pyramid will be a hexagon with a side of the base inclined at 45° to xy. The elevation is projected from the plan. The end elevation (left side elevation) is projected from the plan and elevation of the pyramid.

Draw the reference line xy and the plan of the pyramid as shown in Fig. 9.19. The side af of the hexagon is inclined at 45° to xy. The diagonals of the hexagon represent the slant edges of the pyramid. The intersecting point of the diagonals locates v, the vertex of the pyramid in the plan. Projectors are drawn through a, b, c, d, e and f. The base of the pyramid rests on the HP. Hence, the intersecting points of the projectors with the reference line xy locate the elevations of the base points a′, b′, c′, d′, e′ and f′. A projector is drawn through v in the plan. The

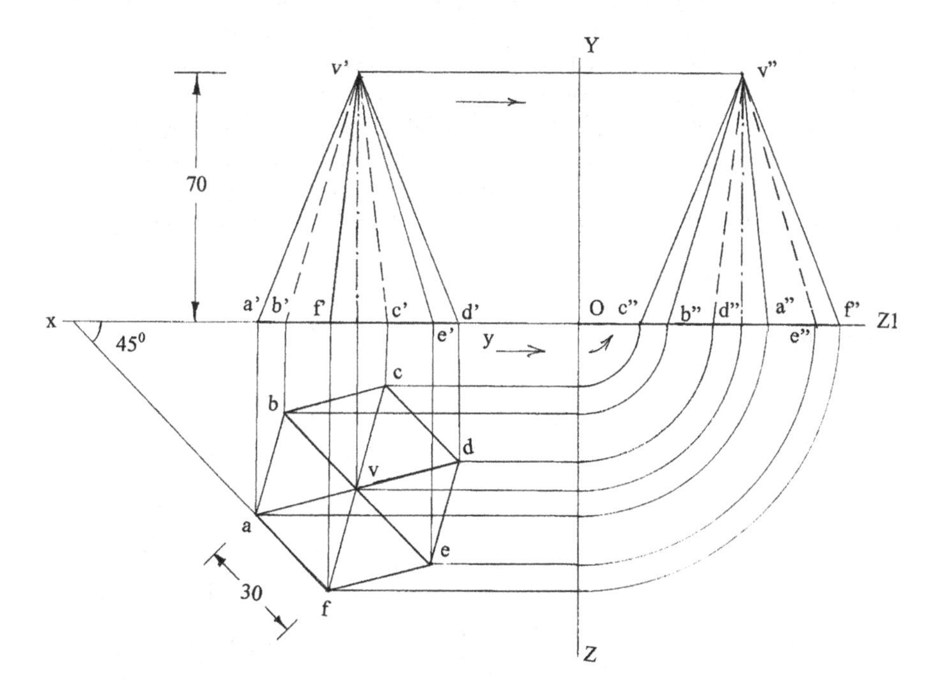

Fig. 9.19 Projections of hexagonal pyramid

elevation v′ is located on this line 70 mm from the elevation of the pyramid a′b′c′d′e′f′. The elevations of the slant edges are connected. The invisible edges are shown as dotted lines. The side elevation can be projected by drawing locus lines through the angular points in the plan and elevation of the pyramid. The locus lines drawn through the angular points in the plan are terminated at OZ. The locus lines are turned through 90° by drawing circular arcs with centre O so that the distances of the angular points and the vertex from the VP remain the same. The base points are on the HP and the side elevations of the base points lie on OZ1, the intersection of the HP with the PP (profile plane). These are designated as a″, b″, c″, d″, e″ and f″. The locus line through v in the plan is drawn and turned through 90°' and a projector is drawn upwards as shown in the figure. The locus line through v′ intersects this projector and locates the vertex v″ in the side elevation. The slant edges are connected in the side elevation. The hidden edges v″d″ and v″e″ when viewed from the left are shown as dotted lines. The side elevation is obtained following the first angle projection (left side elevation drawn on the right side).

8. Draw the projections of a tetrahedron of 40 mm edge resting on the HP on one of its faces such that an edge of that face is parallel to and 20 mm in front of the VP.

   The tetrahedron stands on the HP. All four faces of the tetrahedron are equilateral triangles. The plan of the tetrahedron will be an equilateral triangle with an edge parallel to and 20 mm below xy. The elevation is projected from the plan.

   Draw the reference line xy and the plan of the tetrahedron abc as shown in Fig. 9.20. The plan d of the top angular point is at the centre of triangle. The projections of slant edges ad, bd and cd are shown in the figure. Projectors are drawn through a, b and c. The base of the tetrahedron is on the HP. Hence, the intersecting points of the projectors with the reference line xy locate the elevations of base points a′, b′ and c′. The height of point D from the base ABC of the tetrahedron is found as follows. The true length of the edge will be equal to ac. Draw a line perpendicular to ad at d. With a as centre and ac as radius, draw

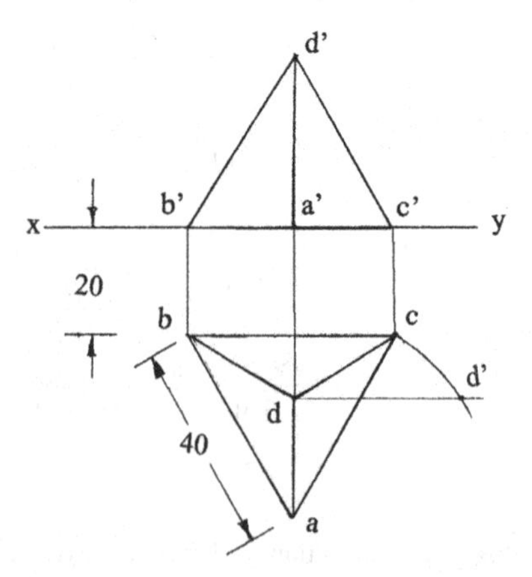

Fig. 9.20 Projections of tetrahedron

an arc to cut the perpendicular line at d′. The distance d′d provides the height (or altitude of the tetrahedron) of D from the base ABC. A projector is drawn through d, and point d′ is located on this projector taking the height of point D from ABC. The slant edges are connected to complete the elevation a′b′d′c′ of the tetrahedron.

9. An octahedron of 40 mm edges rests on the HP on one of its corners such that the solid diagonal through that corner is perpendicular to the HP. Draw the projections when one of its horizontal edges is parallel to the VP.

   An octahedron has 8 faces, 6 angular points (6 corners) and 12 edges. The lines joining opposite angular points are its solid diagonals equal in length and bisecting one another at right angles at the centre of the solid. When the octahedron is placed with a solid diagonal perpendicular to the HP, the projection on the HP will be a square with its two diagonals representing its slant edges. The plan is drawn first and the elevation is projected from the plan.

   Draw the reference line xy and the plan of the octahedron as shown in Fig. 9.21. The square abcd represents the outline of the plan with side bc parallel to xy. The two diagonals ac and bd represent the projections of eight slant edges. The four visible edges ae, be, ce and de have the common topmost point e. The four invisible edges af, bf, cf and df coincide with the four visible edges and have the common bottommost point f. The bottommost point f lies on the HP. A projector is drawn through f to locate the elevation f′ on xy. The elevation e′f′ of the solid diagonal is perpendicular to xy and has length equal to ac or bd. The elevation e′ of the topmost point is located on the projector drawn through e and f. Projectors are drawn through a, b, c and d. Draw a line parallel to xy and bisecting e′f′. The intersecting points of the projectors with the bisecting line locate the elevations a′, b′, c′ and d′. The points are joined with e′ and f′ to complete the elevation of the octahedron. The solid diagonal e′f′ lies inside the solid. The other two solid diagonals a′c′ and b′d′ coincide with visible horizontal edge ad of the solid.

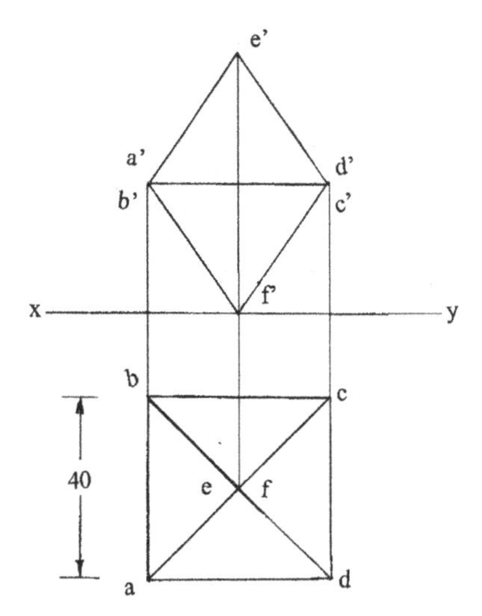

Fig. 9.21 Projections of octahedron

Projections of Solids of Revolution

Cylinder, cone and sphere are some of the solids coming under this group.

When the axis of the cylinder is perpendicular to one of the planes of projection, the projection of the cylinder on that plane will be a circle of diameter equal to the diameter of the cylinder. The projection on the other plane will be a rectangle.

When the axis of the cone is perpendicular to one of the planes of projection, the projection of the cone on that plane will be a circle having a diameter equal to the diameter of the base of cone. The projection on the other plane will be an isosceles triangle.

All the projections of a sphere are circles having diameter equal to that of the sphere.

10. Draw the plan and elevation of a right circular cylinder of base 40 mm diameter and axis 50 mm long lying on the HP on one of its generators such that one of its ends is parallel to and 20 mm in front of the VP.

    Draw a circle of diameter 40 mm with o′ as centre that represents the axis of the cylinder as shown in Fig. 9.22. Draw a tangent at the bottom of the circle. This tangent is the reference line xy. Draw a projector through o′. Projectors are also drawn through the extremities of the circle as shown in the figure. A line is drawn parallel to and 20 mm below xy. The intersections of projectors with this line drawn parallel to xy locate the plan of the rear end of the cylinder. Another line is drawn parallel to and 70 mm (50 + 20) below xy. The intersections of the projectors with this line locate the plan of the front end of the cylinder. The rectangle represents the plan of the cylinder.
11. A cone of base 60 mm diameter and height 80 mm is resting on the HP on its base such that the apex is 40 mm in front of the VP. Draw its projections.

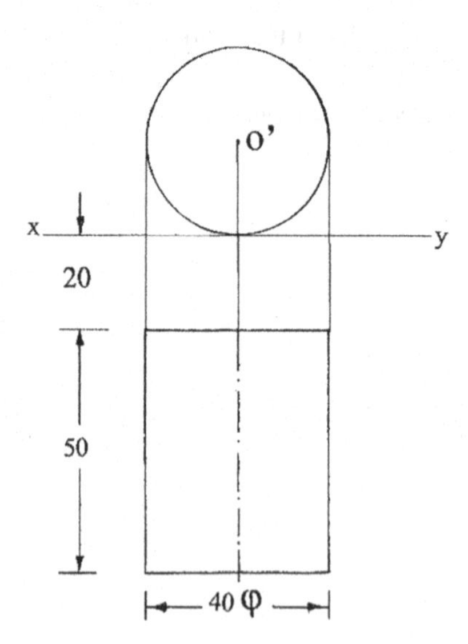

Fig. 9.22 Projections of cylinder

Draw a circle of diameter 60 mm, plan of the cone, as shown in Fig. 9.23. The plan a of apex of the cone coincides with the centre of circle. Draw a horizontal line 40 mm above a. This represents the reference line xy. Draw projectors at the extremities of circle as shown in the figure. The base of the cone is on the HP. Hence, the intersection of the projectors with xy locates the elevation of base of the cone. The elevation a′ is located at 80 mm from xy on the projector drawn through the plan a of apex of the cone. The isosceles triangle is completed to represent the elevation of the cone.

12. Draw the plan and elevation of a sphere of 60 mm diameter lying on the HP. The centre of sphere is 40 mm in front of the VP.

Draw the plan of sphere as a circle of 60 mm diameter as shown in Fig. 9.24. Locate the plan of centre of the sphere at o. The reference line xy is 40 mm from o. The plan b of base point of the sphere coincides with the plan o of the centre of sphere. Draw a projector through b. Locate the elevation b′ at the intersection of the projector with xy. The elevation o′ of the centre of sphere is located 30 mm above the elevation b′. With centre o′ and radius 30 mm, draw a circle to represent the elevation of the sphere.

13. A triangular prism, side of base 35 mm and axis 70 mm long, lies on the HP on one of its longer edges such that its axis is parallel to the VP. Draw its projections.

The axis of the prism is parallel to the VP. One of its longer edges is on the HP. Hence, the axis of the prism is parallel to both the VP and HP. The side elevation of the prism is an equilateral triangle with a corner of the triangle on

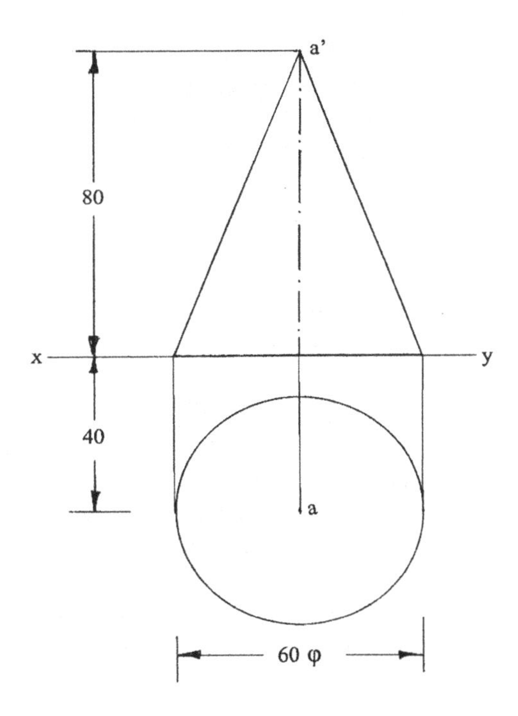

Fig. 9.23 Projections of cone

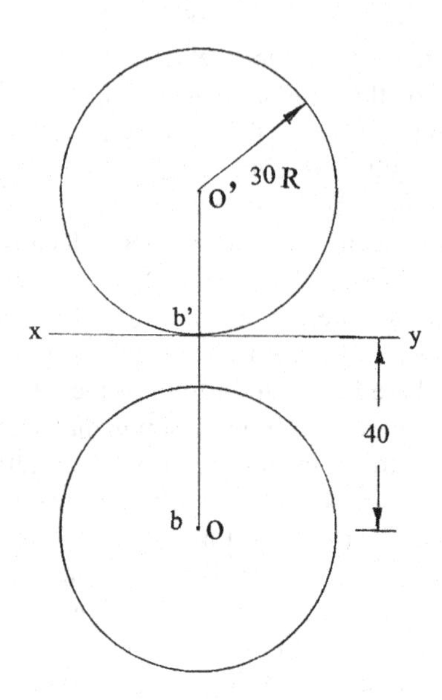

Fig. 9.24 Projections of sphere

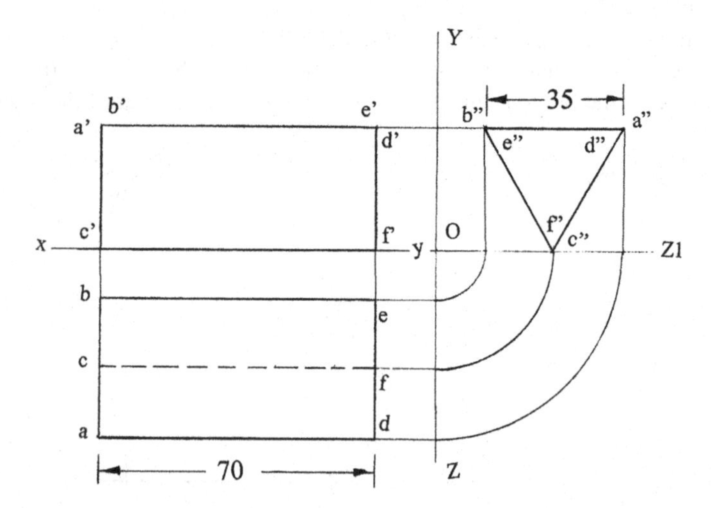

Fig. 9.25 Projections of triangular prism

the intersection of the profile plane and the horizontal plane. The plan and elevation are projected from the side elevation. The triangular ends of the prism are perpendicular to the VP and HP.

The reference lines xy, OZ, OZ1 and OY are drawn as shown in Fig. 9.25. The side elevation is drawn first on OZ1. Locus lines are drawn parallel to xy from the angular points a″, b″ and c″. Projectors are drawn through a″, b″ and c″ to meet OZ1 as shown in the figure. These projectors are turned through 90° by drawing arcs with centre O. These arcs intersect OZ as indicated. Locus lines are drawn through the intersecting points. The ends of the prism are perpendicular to the HP and VP. Hence, the projections of the ends are perpendicular to xy. Draw the elevation d′e′f′ and plan def by drawing straight lines perpendicular to xy as shown in the figure. The elevation and plan of the other end will be 70 mm to the left of the end DEF. The invisible edge fc is shown by dotted line in the plan.

14. A pentagonal prism, side of base 30 mm and axis 60 mm long, lies with one of its rectangular faces on the HP such that its axis is parallel to the VP. Draw its projections.

    The prism lies on the HP on one of its rectangular faces. The axis of the prism is parallel to the VP. The ends of the prism will be perpendicular to the VP and HP. The side elevation will be a pentagon of side 30 mm with one side coinciding on the intersection of the profile plane with the horizontal plane. The side elevation is drawn first. The elevation and plan are projected from the side elevation by drawing locus lines from the angular points of side elevation.

    Draw the reference lines xy, OY, OZ and OZ1 as shown in Fig. 9.26. The pentagon is drawn with side c″d″ on OZ1 (the rectangular face CDKH lies on the HP). Locus lines are drawn parallel to xy to locate elevations of angular points from the angular points of side elevation. Projectors are drawn through a″, b″, c″, d″ and e″ to meet OZ1. These projectors are turned through 90° by drawing circular arcs with centre O as shown in the figure. Locus lines are drawn below xy to locate plans of the angular points. The plan and elevation of

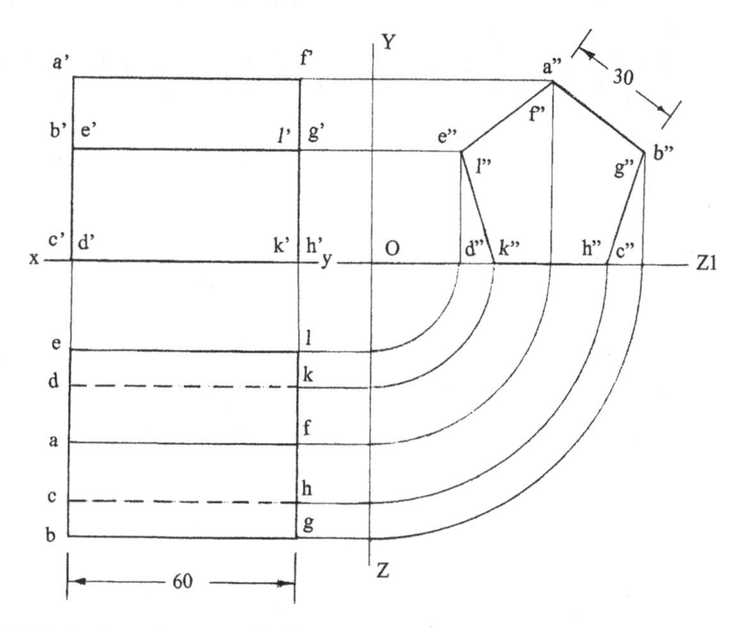

Fig. 9.26 Projections of pentagonal prism

the ends of the prism will be straight lines perpendicular to xy. A straight line is drawn perpendicular to xy to show the elevation and plan of the right end of the prism. Another line is drawn perpendicular to xy 60 mm in the left to show the elevation and plan of the other end of the prism. The angular points are also named in the elevation and plan of the prism. The invisible edges ch and dk that are distinct in the plan are shown by dotted lines.

15. Draw the plan and elevation of a tetrahedron whose edges are 50 mm long when one face is in the VP and an edge of that face is inclined at 20° to the HP.

    The tetrahedron has one face on the VP with an edge of that face inclined at 20° to xy. The elevation is drawn first. The plan is projected from the elevation.

    Draw the reference line xy and the elevation a′b′c′ of the face ABC of the tetrahedron with side b′c′ on a line inclined at 20° to xy as shown in Fig. 9.27. The centre of the equilateral triangle represents the elevation d′ of the angular point D. The elevations of the edges a′d′, b′d′ and c′d′ are joined. Projectors are drawn through a′, b′ and c′ to intersect the line xy at a, b and c giving the plan abc of the face in the VP. The distance of point D from the face ABC is found as follows. a′d′ is the elevation of slant edge AD and the true length of the edge will be equal to a′b′ or b′c′ or c′a′. Draw a line perpendicular to a′d′ at d′. With a′ as centre and a′b′ as radius, draw an arc to cut the perpendicular line at d. The distance dd′ provides the height (or altitude) of the tetrahedron. A projector is drawn through d′, and point d is located keeping the height of the tetrahedron from the plan abc as shown in the figure. The slant edges da, db and dc in the plan are connected to complete the plan of the tetrahedron.

16. Draw the plan and elevation of an octahedron whose edges are 40 mm long when one axis (solid diagonal) is perpendicular to the VP and the elevation of another axis (solid diagonal) is inclined at 60° to xy.

    The elevation of octahedron is a square when an axis (solid diagonal) is perpendicular to the VP. The other two axes (solid diagonals) are parallel to the

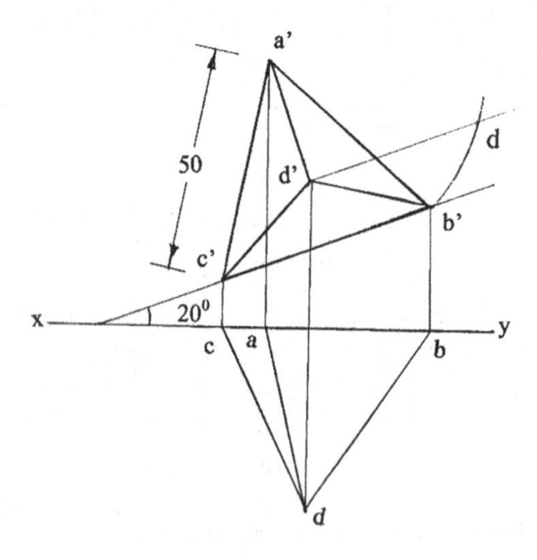

Fig. 9.27 Projections of tetrahedron

VP. When an axis which is parallel to the VP is inclined at 60° to the xy, the inclination of one side of the square (elevation of the octahedron) will be 15° (60° − 45°) to xy.

Draw the reference line xy and the elevation of octahedron a′b′c′d′ with side d′c′ inclined at 15° to xy as shown in Fig. 9.28. The sides of the square represent elevations of four edges and the diagonals of the square represent eight remaining edges. The solid diagonal EF is perpendicular to the VP (parallel to the HP) and the solid diagonal BD is inclined at 60° to the xy. The length of solid diagonal is a′c′ or b′d′. Draw a projector through e′. The plan of points e and f are located on this projector keeping the distance between them equal to a′c′ or b′d′. A horizontal line is drawn through the middle point of ef. Projectors are drawn through a′, b′, c′ and d′. The intersecting points of the projectors with the horizontal line locate points a, b, c and d in the plan. The edges are connected to complete the plan of the octahedron. The invisible edges ed and fd are shown as dotted lines.

17. Lateral faces of a pentagonal pyramid are isosceles triangles of base 40 mm and equal sides 70 mm. The pyramid rests on the HP. Draw its projections.

    The base of the pyramid is a pentagon of 40 mm sides and the slant edges are 70 mm. The plan of the pyramid is drawn keeping one of its sides perpendicular to the reference line. The elevation is projected from the plan.

    Draw the reference line xy and the plan abcde of the pyramid with side ab perpendicular to xy as shown in Fig. 9.29. The centre of the pentagon represents plan of the vertex/apex of the pyramid and is joined to the angular points to

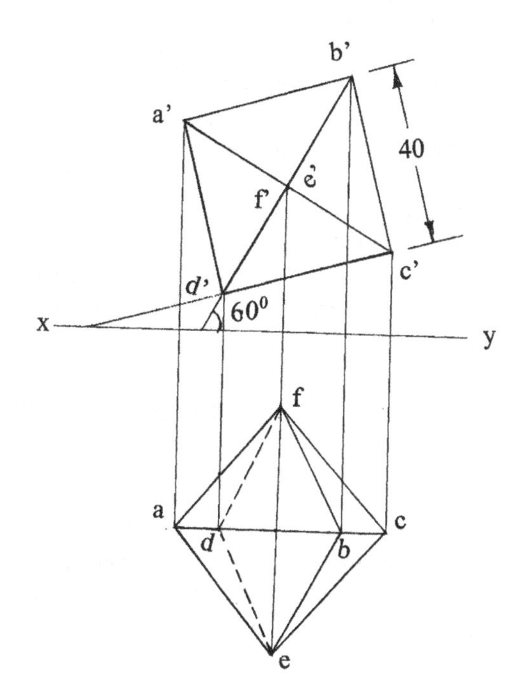

Fig. 9.28 Projections of octahedron

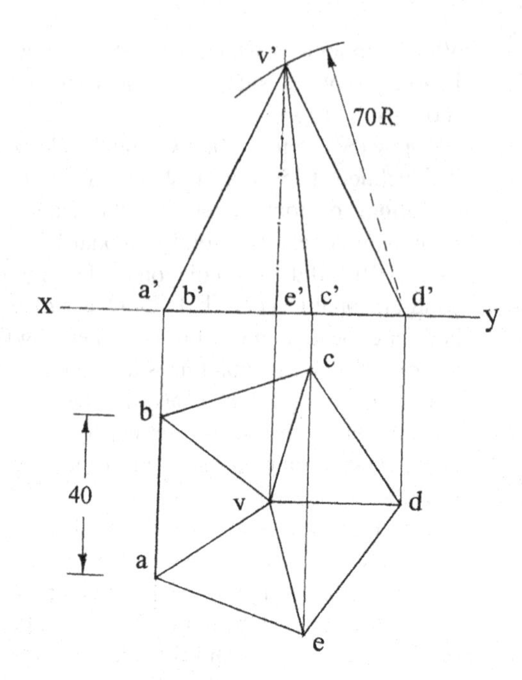

Fig. 9.29 Projections of pentagonal pyramid

show the plan of the slant edges. Projectors are drawn through a, b, c, d, and e to intersect the reference line to locate the elevation of base points a′, b′, c′, d′ and e′. A projector is drawn through v. The equal side of the isosceles triangle provides the slant edge length of the pyramid. The plan vd of the slant edge is parallel to xy. With d′ as centre, draw an arc of radius 70 mm. This arc intersects the projector and locates v′, elevation of the vertex/apex of the pyramid. The vertex is joined to the base points a′, b′, c′, d′ and e′ to complete the elevation of the pyramid.

18. A square pyramid is placed over a cube of edge 50 mm such that the base corners of pyramid touch the middle points of top end of the cube. The axis of pyramid is 60 mm. Draw the projections of the solids.

    Draw the reference line xy and a square of 50 mm side as shown in Fig. 9.30. The square represents the plan of the cube. The middle points of its sides are located. A square is drawn connecting the middle points. This square represents the plan of base of the square pyramid. The diagonals the smaller square represent the plan of slant edges of the pyramid. The intersecting point of the diagonals locates the plan v of vertex/apex of the pyramid. The elevation is projected from the plan. A square of 50 mm side on xy represents the elevation of the cube. The base points of pyramid are located on the top end of the cube. A projector is drawn through v to locate the elevation v′ of the pyramid 60 mm from the base of pyramid. The slant edges are connected as shown in the figure to complete the projections of the solids.

Fig. 9.30 Projections of solids

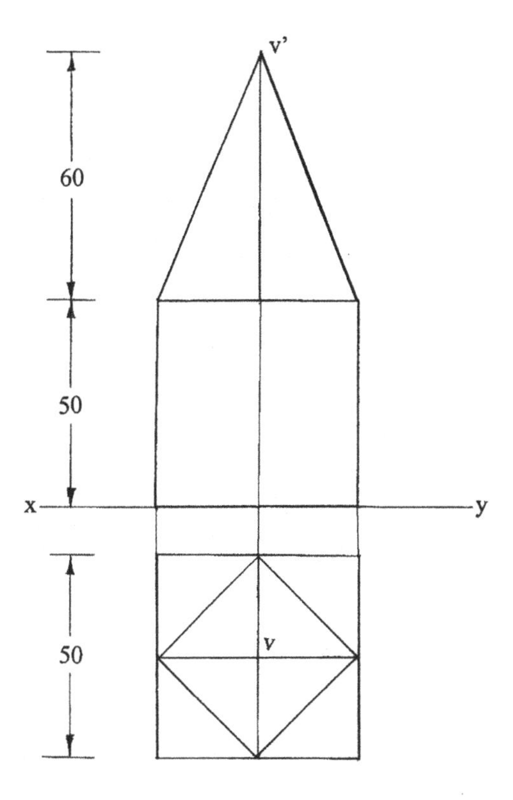

19. A sphere is placed centrally over a vertical hexagonal prism of edge of base 40 mm and height 60 mm such that the plan of the sphere touches the sides of the hexagon. Draw the projections of the solids.

    Draw the reference line xy and a hexagon of 40 mm side as shown in Fig. 9.31. The hexagon represents the plan of the prism. A circle which touches all the sides of the hexagon is to be drawn. The centre of the hexagon is located by drawing perpendicular bisectors of the sides. The length op of any one of these perpendiculars from the centre of the hexagon to one of the mid-points of its sides gives the radius of the circle. With centre o and radius op, draw a circle which touches all sides of the hexagon. This inscribed circle is the plan of the sphere resting centrally over the prism. The elevation of the prism is projected first. A projector is drawn through o. The elevation of the centre of the sphere is located on this projector at o′ as shown in the figure using the radius of the circle op. Draw a circle of radius op to show the elevation of the sphere.
20. The semicircle a′b′c′ (Fig. 9.32) is the elevation of the half of a right circular cone, the base being in the VP. The semicircle a′d′c′ is the elevation of the half of a right circular cylinder, one end being in the VP. The altitude of the cone is the same as that of the cylinder, 70 mm. Draw the given elevation and from it project the plan.

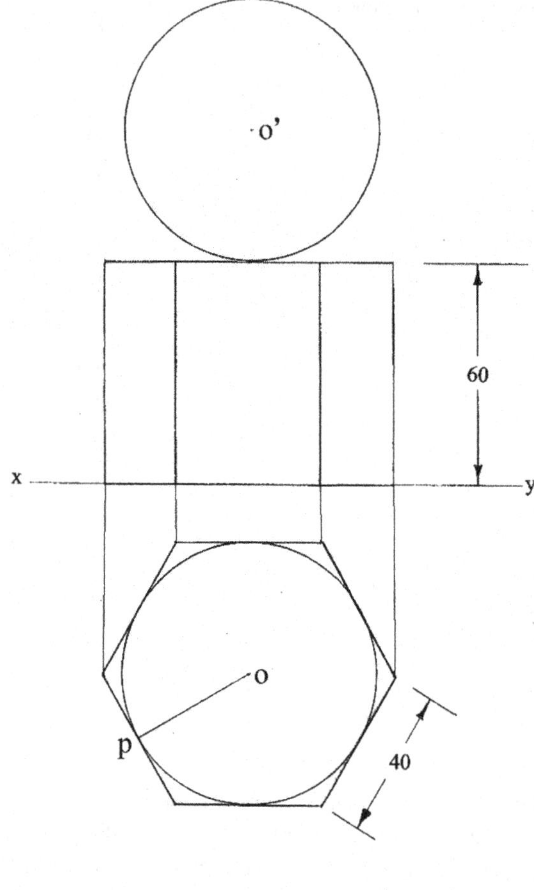

Fig. 9.31 Projections of solids

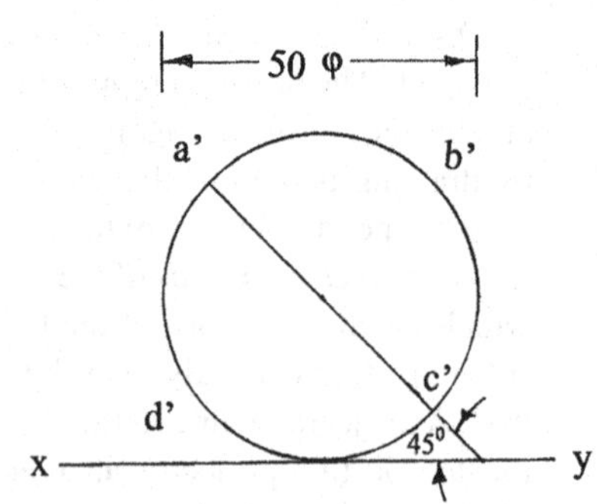

Fig. 9.32 Elevations of cylinder and cone

The elevation of the half of the cylinder is drawn as shown in Fig. 9.33(i). The plan is projected from the elevation.

The elevation of the half of the cone is drawn as shown in Fig. 9.33(ii). The plan is projected from the elevation.

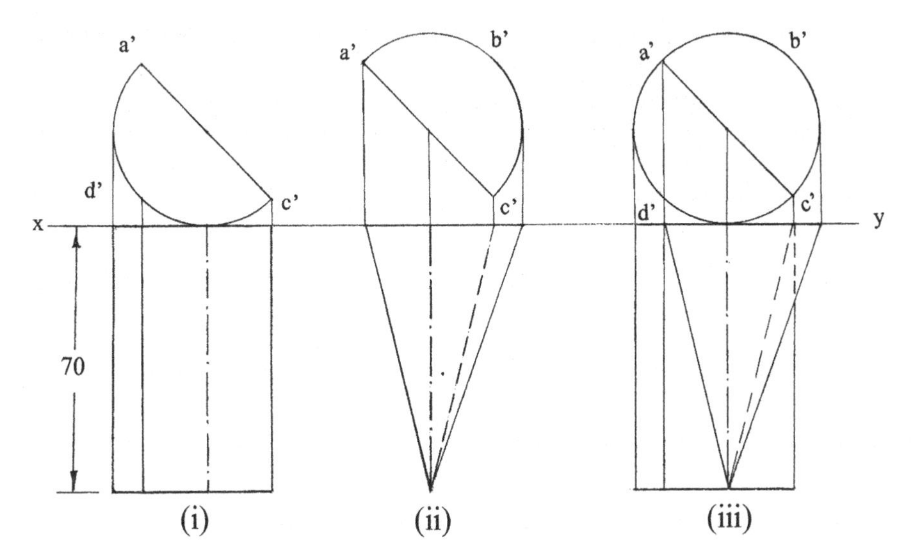

Fig. 9.33 Projections of half-cylinder and half-cone

Figure 9.33(iii) shows the given elevation and the plan projected from the elevation.

Practice Problems

1. Draw the projections of a cube 40 mm side when resting on the HP on one of its faces with its vertical faces equally inclined to the VP.
2. A square prism of 40 mm × 40 mm × 70 mm is on the HP with one of its rectangular faces parallel to and 20 mm in front of the VP. Draw its projections.
3. A rectangular prism 30 mm × 40 mm × 70 mm is on the HP on one of its largest faces with its axis parallel to the VP. Draw its projections.
4. A right pentagonal prism of base edge 40 mm and axis 70 mm long rests with its base in the VP and axis parallel to the HP. Draw its projections when one of the base edges of the prism is parallel to and 20 mm above the HP.
5. A triangular pyramid of base edges 40 mm and axis 60 mm long has its base in the VP with an edge of base perpendicular to the HP. Draw its projections.
6. A square pyramid of base edges 40 mm and axis 60 mm long has its base on the HP with the base edges equally inclined to the VP. Draw its projections.
7. A right hexagonal pyramid of base edges 30 mm and axis 60 mm long rests on its base in the HP with an edge of base parallel to and 20 mm in front of the VP. Draw its projections.
8. A tetrahedron of 40 mm edge rests on one of its faces in the HP such that an edge is perpendicular to the VP. Draw its projections.

9. An octahedron of 40 mm edge stands on the HP on one of its corners such that the solid diagonal through that corner is perpendicular to the HP. Draw its projections when the horizontal edges are equally inclined to the VP.
10. A square pyramid of 40 mm side of base and 60 mm slant length rests on the HP on its base with an edge of base inclined at 30° to the VP. Draw its projections and also its end elevation.

Chapter 10
Auxiliary Projections

The projections of an object when its base is inclined to one of the principal planes (HP/VP) are difficult to obtain in one step. Hence, the object is kept initially in simple position to get its projections. Thereafter, the position of the object is changed to satisfy the given condition with reference to one of the principal planes (VP or HP) to which its base/axis is inclined. In this method, the angular points of the object move parallel to the other principal plane (HP or VP). One of the principal views (plan or elevation) will be common to the other two views. The common view and the latest view obtained complete the projections of the object. This method of projection is called change of position method. In the other method, an auxiliary plane is chosen to satisfy the given condition instead of changing the position of the object. The projection on the auxiliary plane furnishes the solution of the problem.

Auxiliary planes are of two types, perpendicular or oblique. Perpendicular planes are perpendicular to one or both of the principal planes. These are called auxiliary vertical planes (AVP) perpendicular to the horizontal plane or auxiliary inclined planes (AIP) perpendicular to the vertical plane. Oblique planes are inclined to both the principal planes. The auxiliary vertical plane and the auxiliary inclined plane are represented by lines in which they meet the principal planes. These lines are called its traces. The line in which the plane meets the HP is called the horizontal trace, HT, and the line in which the plane meets the VP is called the vertical trace, VT. A feature common to all perpendicular planes is that one trace represents an edge view of the plane.

An auxiliary vertical plane is shown in Fig. 10.1. The position of the auxiliary vertical plane with reference to the principal planes is shown in the left side of the figure. The principal planes and the auxiliary vertical plane are shown after rotation about the reference line xy and the horizontal trace in the middle of the figure. The conventional representation of the principal planes and the auxiliary vertical plane are shown in the right side of the figure. An auxiliary elevation is a projection on any auxiliary vertical plane (AVP) not parallel to the VP.

K. Rathnam, *A First Course in Engineering Drawing*,
DOI 10.1007/978-981-10-5358-0_10

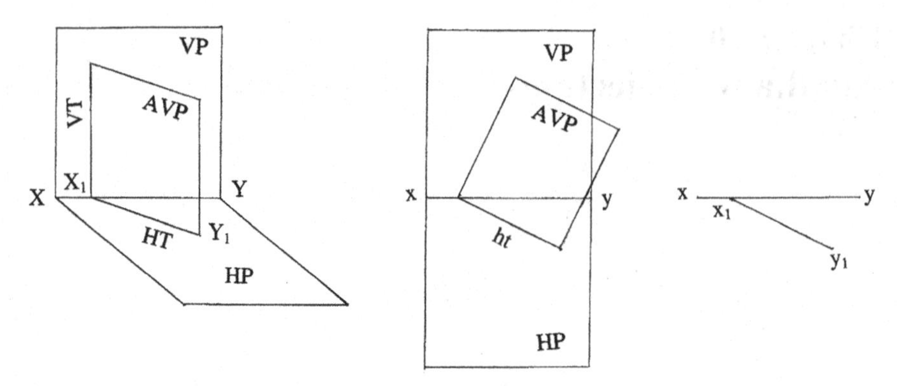

Fig. 10.1 Representation of an auxiliary vertical plane

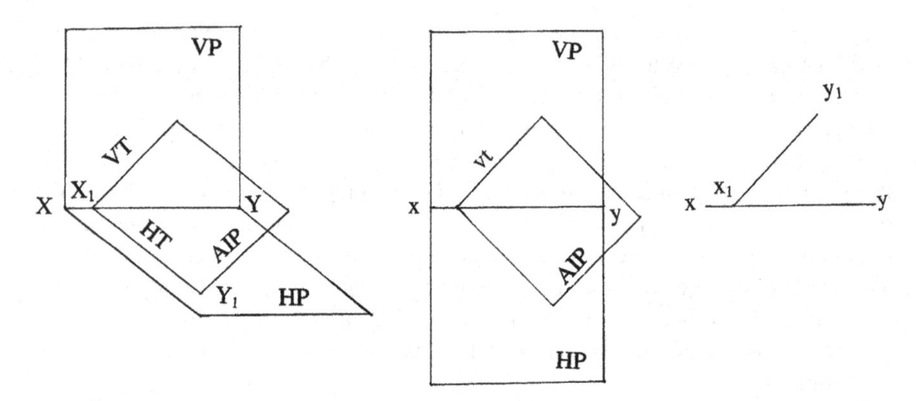

Fig. 10.2 Representation of an auxiliary inclined plane

An auxiliary inclined plane is shown in Fig. 10.2. The position of the auxiliary inclined plane with reference to the principal planes is shown in the left side of the figure. The principal planes and the auxiliary inclined plane are shown after rotation about the reference line xy and the vertical trace in the middle of the figure. The conventional representation of the principal planes and the auxiliary inclined plane are shown in the right side of the figure. An auxiliary plan is a projection on any auxiliary inclined plane (AIP).

Projections of a Point on an Auxiliary Vertical Plane

Refer to Fig. 10.3. The point P, the plane inclined to the VP, AVP and the trace of AVP on the HP, X_1Y_1 are shown in the left side of the figure. The projections of the point on the horizontal plane and the vertical plane are given by p and p′. The projection on the auxiliary vertical plane is given by p_1'. The rotation of the

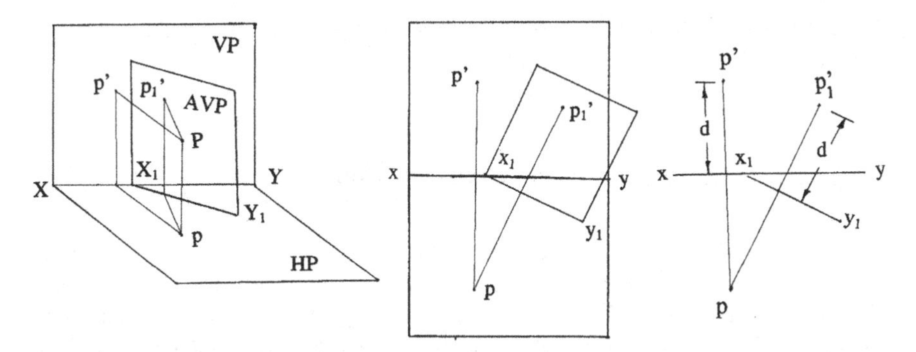

Fig. 10.3 Projections of point P on AVP

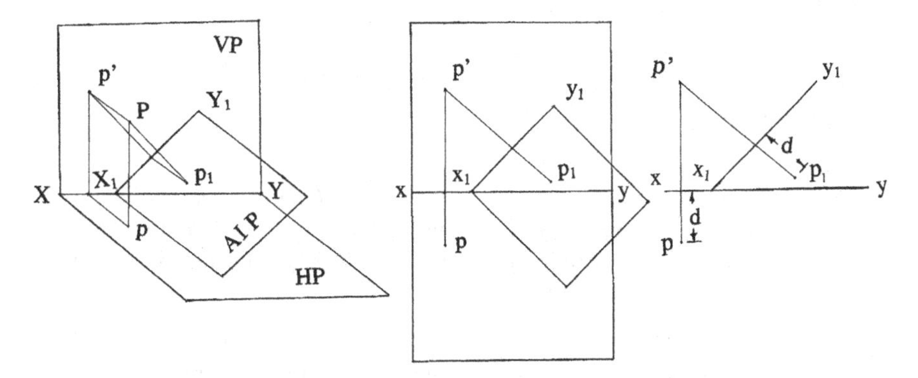

Fig. 10.4 Projections of point P on AIP

auxiliary vertical plane about x_1y_1 shows the projection p_1' as given in the middle of the figure. The actual reference lines necessary for showing the projections are shown in the right side of the figure. The plan is common to the elevations p′ and p_1'. The height of the point P from the HP remains the same. Hence, the distance of p_1' from x_1y_1 will be equal to that of p′ from xy.

Projections of a Point on an Auxiliary Inclined Plane

Refer to Fig. 10.4. The point P, the plane inclined to the HP, AIP and the trace of AIP on the VP, X_1Y_1 are shown in the left side of the figure. The projections of the point on the horizontal plane and the vertical plane are given by p and p′. The projection on the auxiliary inclined plane is given by p_1. The rotation of the auxiliary inclined plane about x_1y_1 shows the projection p_1 as given in the middle of the figure. The actual reference lines necessary for showing the projections are shown in the right side of the figure. The elevation is common to the plans p and p_1.

The distance of the point P from the VP remains the same. Hence, the distance of p_1 from x_1y_1 will be equal to that of p from xy.

Projections of a Line on an Auxiliary Plane

The elevation a′b′ and plan ab of a straight line are shown in Fig. 10.5. An auxiliary inclined plane (AIP) parallel to a′b′ is chosen at x_1y_1. Projectors are drawn perpendicular to x_1y_1 from a′ and b′. An auxiliary plan on x_1y_1 is given by a_1b_1 as shown in the figure. The distance of a_1 from x_1y_1 will be equal to the distance of a from xy. The distance of b_1 from x_1y_1 will be equal to the distance of b from xy. The projection a_1b_1 gives the true length of the line, and the angle between a_1b_1 and x_1y_1 gives the inclination of the line with the vertical plane.

Solved Problems

1. The projections of a line a′b′, ab are given in Fig. 10.6. Determine auxiliary plans on x_1y_1 and x_3y_3 and auxiliary elevations on x_2y_2 and x_4y_4. Find the length of each projection. x_1y_1 is parallel to a′b′ and x_3y_3 is perpendicular to a′b′. x_2y_2 is parallel to ab and x_4y_4 is perpendicular to ab.

   (i) The projections of the line are drawn as shown in Fig. 10.7. The auxiliary inclined planes are chosen at x_1y_1 and x_3y_3, the distances of x_1y_1 and x_3y_3 from a′b′ being arbitrary. Projectors are drawn perpendicular to x_1y_1 from a′ and b′. The auxiliary plans a_1 and b_1 are obtained using the distances of a and b from xy. Projectors are drawn perpendicular to x_3y_3 from a′ and b′. The auxiliary plans a_3 and b_3 are obtained using the distances of a and b from xy.

   Answers: $a_1b_1 = 45$ mm;
   $a_3b_3 = 15$ mm.

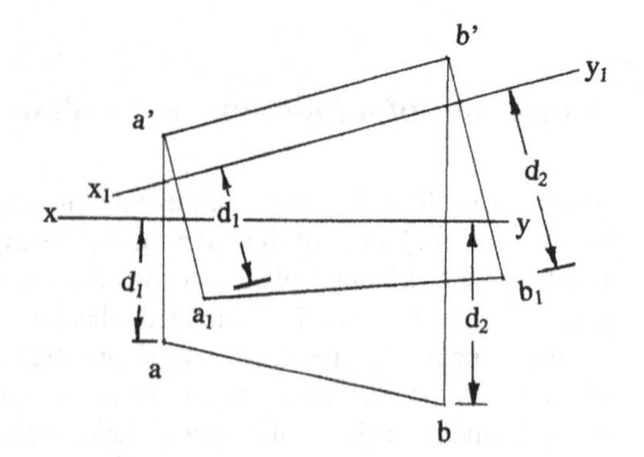

Fig. 10.5 Projections of line AB on an auxiliary plane

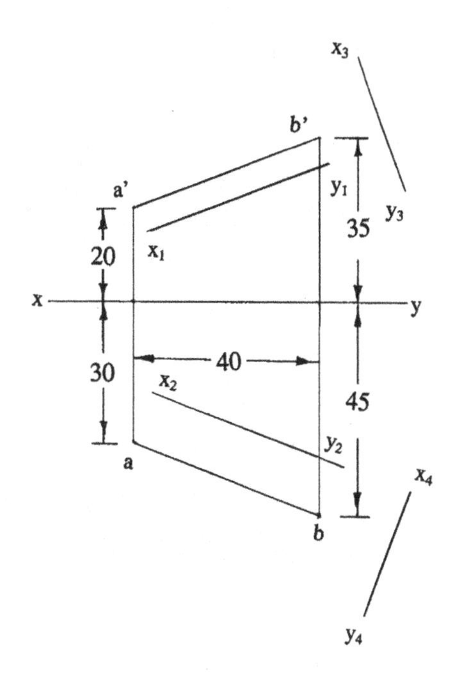

Fig. 10.6 Projections of line AB

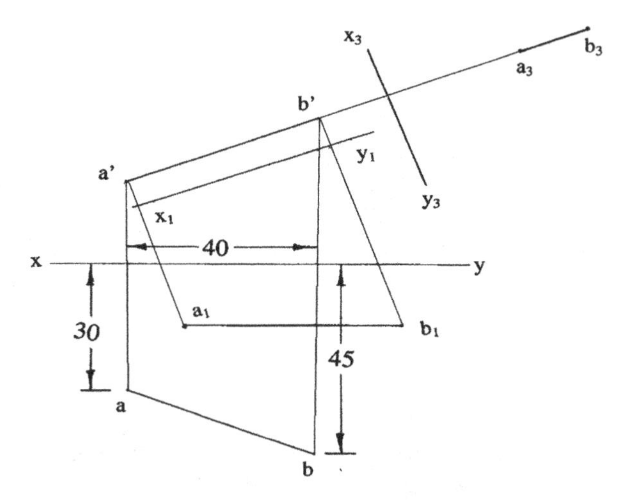

Fig. 10.7 Auxiliary plans of line AB

(ii) The projections of the line are drawn as shown in Fig. 10.8. The auxiliary vertical planes are chosen at x_2y_2 and x_4y_4, the distances of x_2y_2 and x_4y_4 from ab being arbitrary. Projectors are drawn perpendicular to x_2y_2 from a and b. The auxiliary elevations a_2' and b_2' are obtained using the distances of a′ and b′ from xy. Projectors are drawn perpendicular to x_4y_4 from a′ and b′. The auxiliary elevations a_4' and b_4' are obtained using the distances of a′ and b′ from xy.

Answers; $a_2'b_2' = 45$ mm;

$a_4'b_4' = 15$ mm.

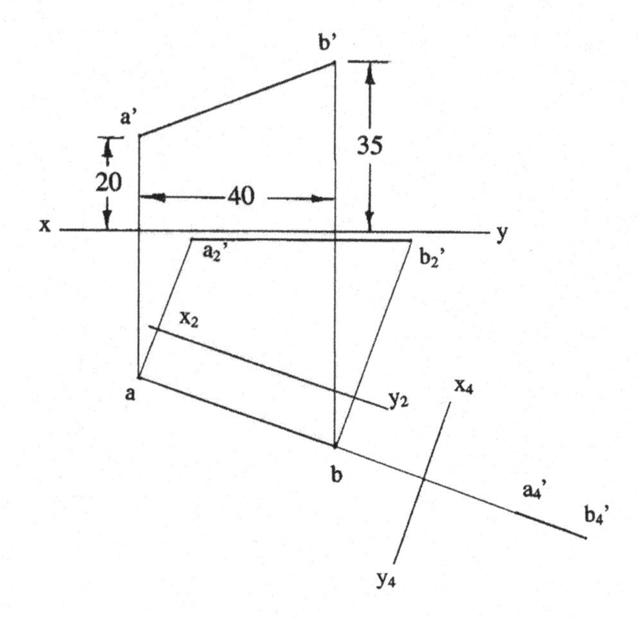

Fig. 10.8 Auxiliary elevations of line AB

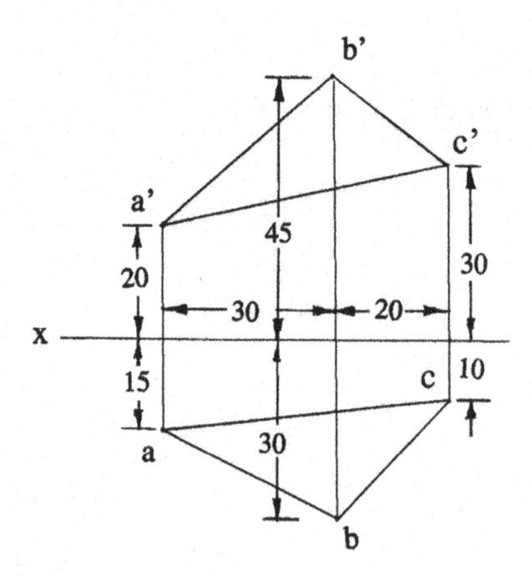

Fig. 10.9 Projections of triangle ABC

2. Determine the true shape of the triangle whose projections are given in Fig. 10.9. Measure the sides of the triangle.

   The elevation a′b′c′ and plan abc of the triangle are drawn as shown in Fig. 10.10. A horizontal line c′d′ lying in the elevation is drawn as shown in the figure. The plan cd of the line is also drawn. The line cd is produced. An auxiliary vertical plane perpendicular to cd is chosen with its reference line x_1y_1. Projectors are drawn perpendicular to x_1y_1 from a, b and c. Project an auxiliary elevation $a_1'b_1'c_1'$ from the plan abc. The distance of a_1' from x_1y_1 is equal to the distance of a′ from xy. The auxiliary projection of the triangle $a_1'b_1'c_1'$ is now a straight line. Choose a reference line x_2y_2 parallel to $a_1'b_1'c_1'$ and project an auxiliary plan $a_2b_2c_2$. The distance of a_2 from x_2y_2 is equal to the distance of a

Fig. 10.10 Auxiliary projections of triangle ABC

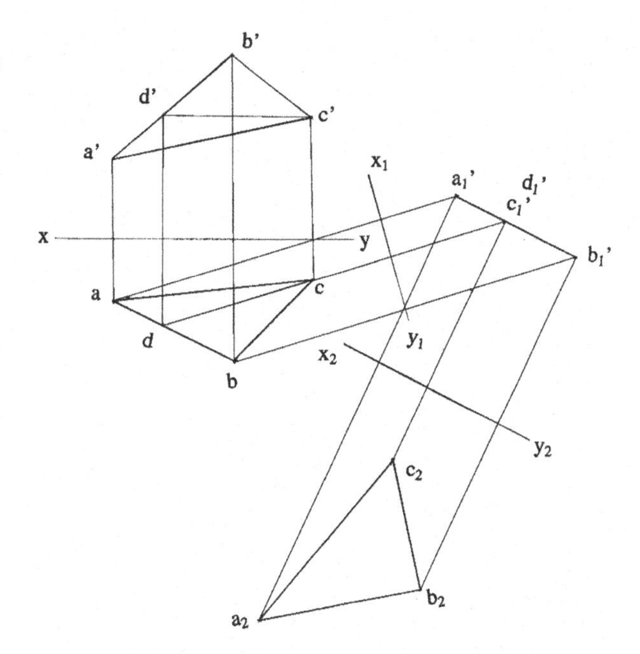

from x_1y_1. The final auxiliary plan $a_2b_2c_2$ gives the true shape of the triangle. The sides of Δ are AB = 41 mm, BC = 33 mm and CA = 51 mm.

3. A triangular prism of edge of base 40 mm and axis 70 mm long rests with one of its faces on the HP, and axis is inclined at 30° to the VP. Draw its projections.

(i) Auxiliary plane method

The axis of the prism is kept perpendicular to the VP initially. The elevation is drawn with a side of triangle on xy. The plan is projected from the elevation. An auxiliary vertical plane (perpendicular to the HP) inclined 30° to the axis of the prism is chosen. The projection on this plane is obtained from the plan of the prism. The auxiliary elevation and the plan furnish the solution of the problem.

Draw the reference line xy and the elevation of the prism with a side on xy as shown in Fig. 10.11. The plan is projected from the elevation. An auxiliary vertical plane is chosen at convenient distance from the plan of the prism such that its inclination with the axis of the prism is 30°. The reference line x_1y_1 represents the position of the auxiliary vertical plane. Projectors are drawn perpendicular to x_1y_1 from the angular points of the prism in the plan to get the auxiliary elevations of the angular points of the prism. The auxiliary elevation b_1' is located on the projector drawn from b. The distance of b_1' from x_1y_1 is equal to the distance of b′ from xy since the plan is common to the elevation and the auxiliary elevation of the prism. The elevations of the remaining angular points are located following the same procedure. The auxiliary elevation is completed with the invisible edge $b_1'c_1'$ shown as a dotted line.

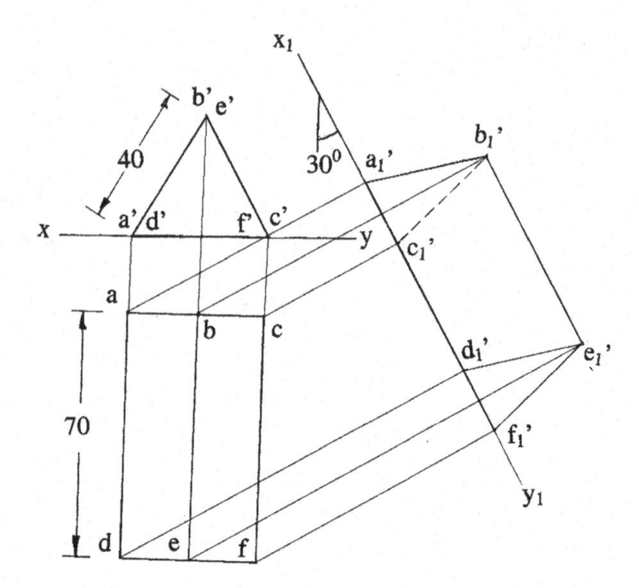

Fig. 10.11 Auxiliary projections of Δ prism

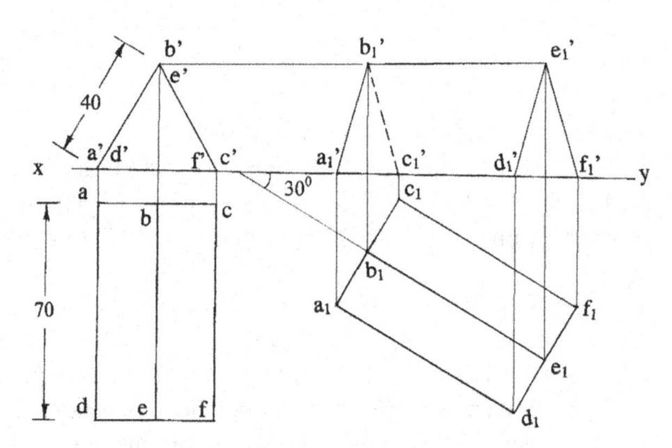

Fig. 10.12 Auxiliary projections of Δ prism

(ii) Change of position method

The prism is kept initially in simple position with the axis perpendicular to the VP. The elevation and plan of the prism are dawn as shown in Fig. 10.12. The prism is moved over the HP such that its axis is inclined at 30° to the VP. In this process, all the angular points of the prism move parallel to the HP. The plan of the prism is redrawn as shown in the figure. A projector is drawn from b_1. Since the angular point moves parallel to the HP, locus line is drawn parallel to xy from b′. The intersection of the projector with the locus line locates the auxiliary elevation b_1'. The same procedure is repeated to get the auxiliary elevations of the remaining points. The auxiliary elevation of the prism is completed with the distinct invisible edge $b_1'c_1'$ shown as a dotted line.

Note: The change of position method does not require an auxiliary plane. The drawing of projectors perpendicular to the new reference line x_1y_1 and transferring the distances of projections of points become cumbersome in an auxiliary plane method. It will be easy to redraw one view from its projections in simple position in the change of position method. The drawing of locus lines parallel to the reference line from the other view will eliminate the measurement of the distances of the points from the reference line. Hence, the change of position method is recommended for the problems in auxiliary projections. If the given conditions require two stages in the projections, the change of position is adopted in the first stage. An auxiliary plane (AVP/AIP) is used in the second stage.

4. A pentagonal prism of edge of base 30 mm and axis 70 mm long rests with one of its rectangular faces on the HP and the ends inclined at 30° to the VP. Draw its projections.

The prism is kept initially in simple position with the axis perpendicular to the VP. The elevation and plan of the prism are drawn as shown in Fig. 10.13. The prism is moved over the HP such that its end is inclined at 30° to the VP. In this process, all the angular points of the prism move parallel to the HP. The plan of the prism is redrawn as shown in the figure. A projector is drawn perpendicular to xy from b_1. Since the angular point moves parallel to the HP, locus line is drawn parallel to xy from b'. The intersection of the projector with the locus line locates the auxiliary elevation b_1'. Projectors are drawn perpendicular to xy from the remaining angular points in the plan of the prism redrawn. Locus lines are drawn parallel to xy from the corresponding points in the elevation of the prism. The intersections of the projectors with the corresponding locus lines locate the auxiliary elevations of the remaining angular points, and the auxiliary elevation

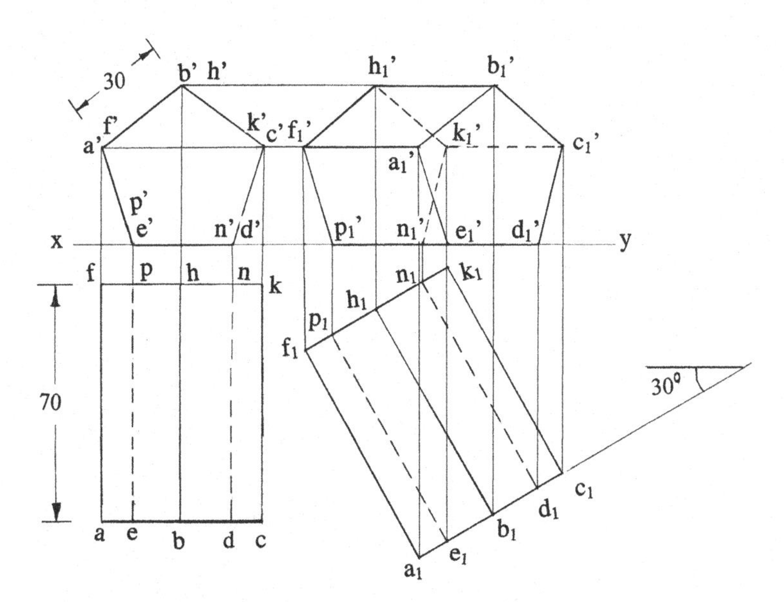

Fig. 10.13 Auxiliary projections of pentagonal prism

of the prism is completed with the distinct invisible edges $c_1'k_1'$, $h_1'k_1'$ and $k_1'n_1'$ shown as dotted lines.

5. Draw the projections of a square prism of edge of base 40 mm and axis 70 mm long when one of its solid diagonals is vertical.

   There are four solid diagonals connecting the angular points of the base with the opposite angular points of the top of a square prism. The square prism is kept on the HP with its base edges equally inclined to the VP. Two of its solid diagonals will be parallel to the VP. The prism is tilted parallel to the VP about the right extreme base point so that the solid diagonal passing through this base point will be perpendicular to the HP. The elevation is redrawn, and the plan is projected from the elevation.

   Draw the reference line xy and the plan of the square prism with its base edges equally inclined to xy as shown in Fig. 10.14. The elevation is projected from the plan. The solid diagonals a′h′ and c′e′ are parallel to the VP. The prism is tilted parallel to the VP about the base point H. The elevation is redrawn such that the solid diagonal a′h′ is perpendicular to xy. The plan is to be projected from the elevation. A projector is drawn perpendicular to xy from a_1'. The locus line is drawn parallel to xy from a. The intersection of the projector with the locus line locates the plan a_1. The same procedure is repeated to get the plans of the remaining angular points. These points are connected to show the edges of the prism in the plan. The distinct invisible edges c_1h_1, f_1h_1 and n_1h_1 are shown as dotted lines. The redrawn elevation and projection of the plan from it complete the solution of the problem.
6. Draw the projections of a hexagonal prism of edge of base 30 mm and axis 80 mm long resting on a corner of its base with the base inclined at 45° to the HP.

   The prism is kept on the HP with a side of base parallel to the VP. The prism can be tilted parallel to the VP about the right extreme corner such that its base is

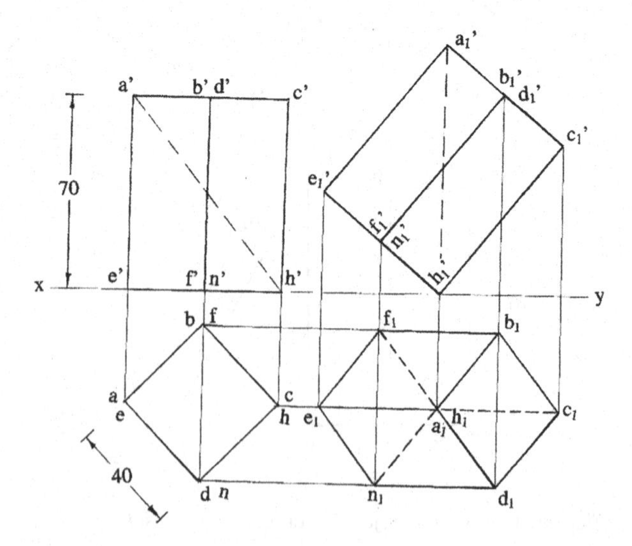

Fig. 10.14 Auxiliary projections of square prism

inclined at 45° to the HP. The elevation is redrawn and the plan is projected from the elevation.

Draw the reference line xy and the plan of the hexagonal prism with a side of base parallel to xy as shown in Fig. 10.15. The elevation is projected from the plan. The prism is tilted parallel to the VP about the right extreme corner N such that the base is inclined at 45° to the HP. The elevation is redrawn so that prism base is inclined at 45° with xy. The plan is to be projected from the elevation. A projector is drawn perpendicular to xy from a_1'. The locus line is drawn parallel to xy from a. The intersection of the projector with the locus line locates the plan a_1. The same procedure is repeated to get the plans of the remaining points. These points are connected to show the edges of the prism in the plan. The distinct invisible edges m_1n_1 and n_1p_1 and a portion of the invisible edge d_1n_1 are shown as dotted lines. The redrawn elevation and projection of the plan from it complete the solution of the problem.

7. A hexagonal pyramid edge of base 30 mm and axis 70 mm stands with a corner of its base on the HP. Draw the projections of the pyramid when the slanting edge containing that corner is vertical.

   The hexagonal pyramid is kept on the HP with a base edge parallel to the VP. Two of its slanting edges containing the two extreme corners are parallel to the VP in this position. The pyramid is tilted parallel to the VP about one extreme corner so that the slanting edge containing that corner is made perpendicular to the HP. In this process, all the points of the prism move parallel to the VP. The elevation is redrawn with the slanting edge perpendicular to the HP. The plan is projected from the elevation.

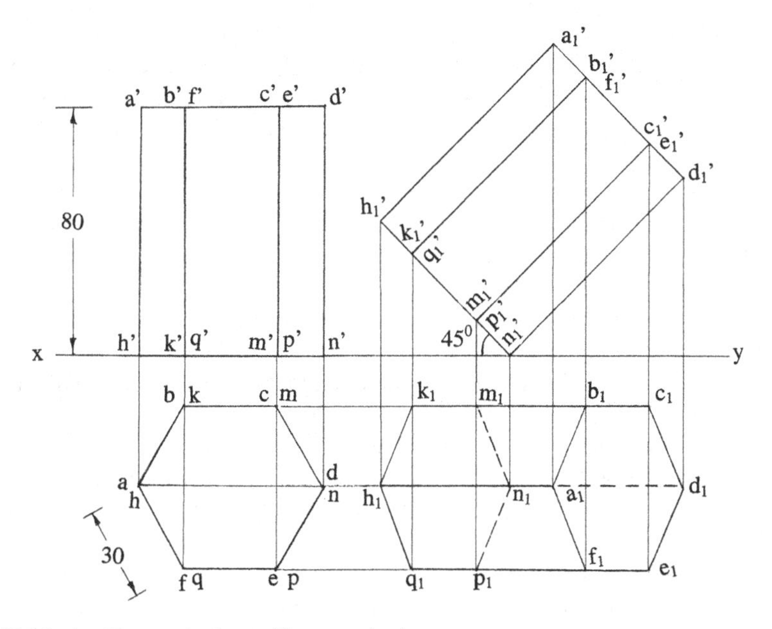

Fig. 10.15 Auxiliary projections of hexagonal prism

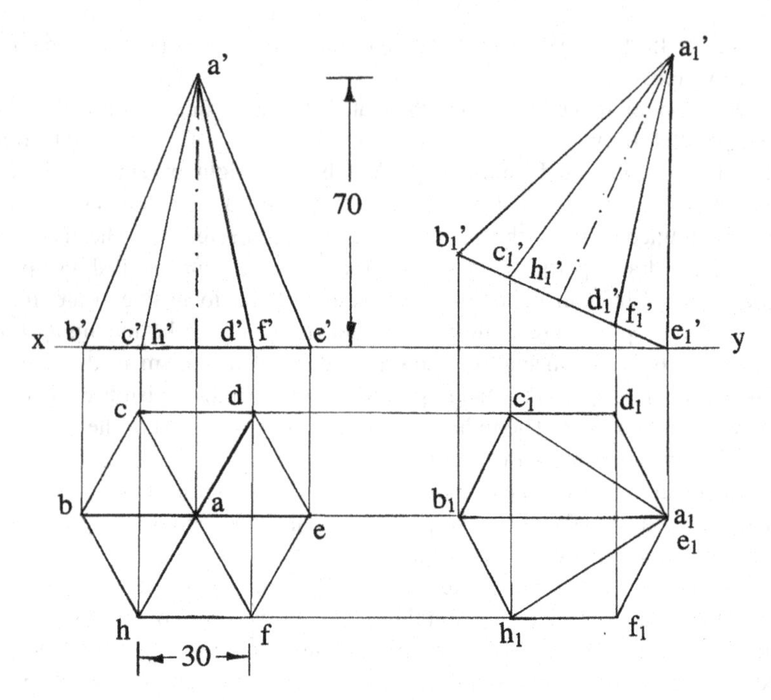

Fig. 10.16 Auxiliary projections of hexagonal pyramid

Draw the reference line xy and the plan of the hexagonal pyramid with a side of base parallel to xy as shown in Fig. 10.16. The elevation is projected from the plan. The pyramid is tilted parallel to the VP about the right extreme corner E such that the slant edge EA is perpendicular to the HP. The elevation is redrawn so that the slant edge $e_1'a_1'$ is perpendicular to the xy. The plan is to be projected from the elevation. A projector is drawn perpendicular to xy from b_1'. The locus line is drawn parallel to xy from b. The intersection of the projector with the locus line locates the plan b_1. The same procedure is repeated to get the plans of the remaining points. These points are connected to show the edges of the prism in the plan. The redrawn elevation and projection of the plan from it complete the solution of the problem.

8. Draw the projections of an octahedron of edge 40 mm when it rests on the HP on one of its faces.

   The solid diagonal of the octahedron is kept perpendicular to the HP with the end point of the solid diagonal on it to get its projection in its simple position. An edge of the octahedron is kept parallel to the VP. The outline of the plan will be a square and the outline of the elevation will be a rhombus. The octahedron is tilted parallel to the VP about the end point till one of its faces rests on the HP. The elevation is redrawn, and the plan is projected from it.

   Draw the reference line xy and the plan of the octahedron with an edge parallel to xy and a solid diagonal perpendicular to the HP as shown in Fig. 10.17. The elevation is projected from the plan. The octahedron is tilted

Fig. 10.17 Auxiliary projections of octahedron

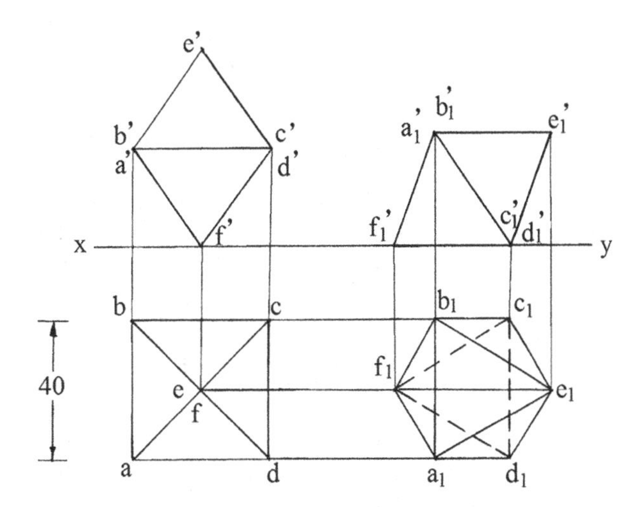

parallel to the VP about the end point F of the diagonal till the face FCD rests on the HP. The elevation is redrawn with the face $f_1'c_1'd_1'$ on the xy. The plan is to be projected from the elevation. A projector is drawn perpendicular to xy from b_1'. The locus line is drawn parallel to xy from b. The intersection of the projector with the locus line locates the plan b_1. The same procedure is repeated to get the plans of the remaining points. These points are connected to show the edges of the octahedron in the plan. The invisible edges f_1c_1, c_1d_1 and d_1f_1 are shown as dotted lines. The redrawn elevation and projection of the plan from it complete the solution of the problem.

9. A rectangular prism 50 mm × 30 mm × 70 mm stands on its shorter edge of its base such that this edge is inclined at 40° to the VP, and the face containing that edge is inclined at 60° to the HP. Draw its projections.

   This problem needs two conditions to be satisfied. The prism is kept on the HP on its base with a shorter edge perpendicular to the VP. The prism is tilted parallel to the VP about this shorter edge such that the face containing this edge is inclined at 60° to the HP. The elevation is redrawn and the plan is projected from it. The outline of the plan will be a rectangle. The prism is moved parallel to the HP to satisfy the other condition. The new plan is redrawn with the shorter edge inclined at 40° to the VP. The final elevation is obtained from this new plan.

   Draw the reference line xy and the plan of the rectangular prism resting on its base on the HP with a shorter edge perpendicular to the reference line xy as shown in Fig. 10.18. The elevation is projected from the plan. The prism is tilted parallel to the VP about the shorter edge NH such that the face NDCH is inclined at 60° to the HP. The elevation is redrawn with the face n′d′c′h′ inclined at 60° to xy as shown in the figure. The plan is to be projected from the elevation. A projector is drawn perpendicular to xy from a_1'. The locus line is drawn parallel to xy from a in the plan. The intersection of the projector with the locus line locates the plan a_1. The same procedure is repeated to get the plans of the remaining points. The plan is completed with the distinct invisible edge h_1n_1 shown by a dotted line. Since the

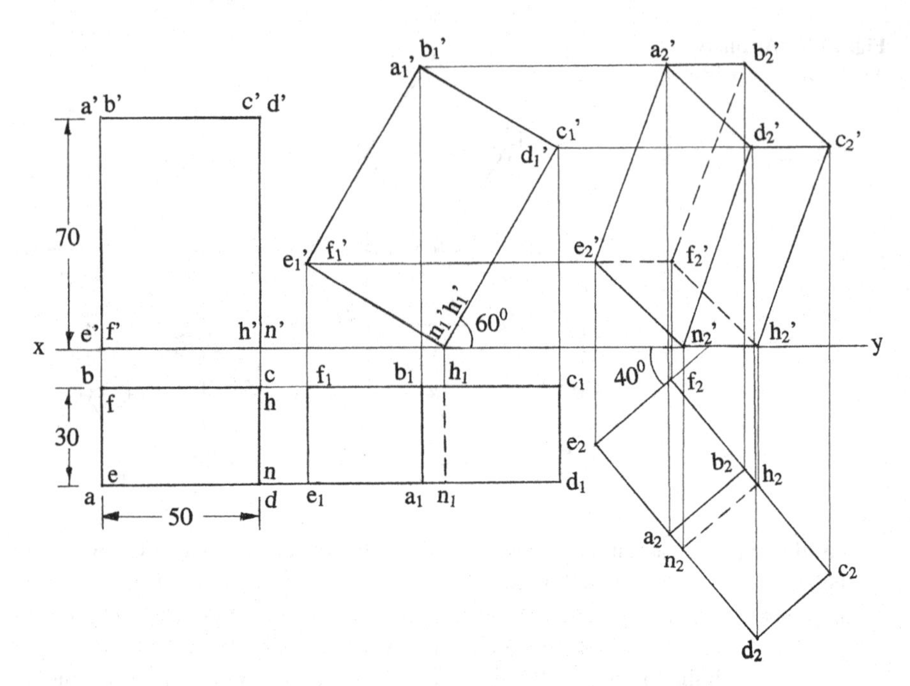

Fig. 10.18 Auxiliary projections of rectangular prism

outline of the plan $c_1d_1e_1f_1$ is a rectangle, the prism can be moved parallel to the HP with its shorter edge inclined at 40° to the VP. The plan is redrawn with its shorter edge inclined at 40° to xy as shown in the figure. The elevation is to be projected from the plan. A projector is drawn perpendicular to xy from a_2. The locus line is drawn parallel to xy from a_1'. The intersection of the projector with the locus line locates the elevation a_2'. The same procedure is repeated to get the elevations of the remaining points. These points are connected to show the elevation of the edges of the prism. The invisible edges $b_2'f_2'$, $e_2'f_2'$ and $h_2'f_2'$ are shown by dotted lines. The outline of elevation $a_2'b_2'c_2'h_2'n_2'e_2'$ and the outline of plan $c_2d_2e_2f_2$ furnish the solution of the problem.

10. Draw the projections of a cube of edge 40 mm standing on one of its corners with its solid diagonal perpendicular to the VP.

    The cube is kept on the HP with its base edges equally inclined to the VP. In this position two of its solid diagonals will be parallel to the VP. The cube is tilted parallel to the VP about one of its extreme base corners till the solid diagonal passing through the other extreme corner is parallel to the HP. The elevation is redrawn with a corner touching the HP. The plan is projected from the elevation. An auxiliary vertical plane is chosen at convenient distance from the plan such that the plane is perpendicular to solid diagonal parallel to the HP. An auxiliary elevation completes the solution.

    Draw the reference line xy and the plan of the cube with its base edges equally inclined to the VP as shown in Fig. 10.19. The elevation is projected from the plan. The solid diagonals $a'h'$ and $c'e'$ are parallel to the VP. The

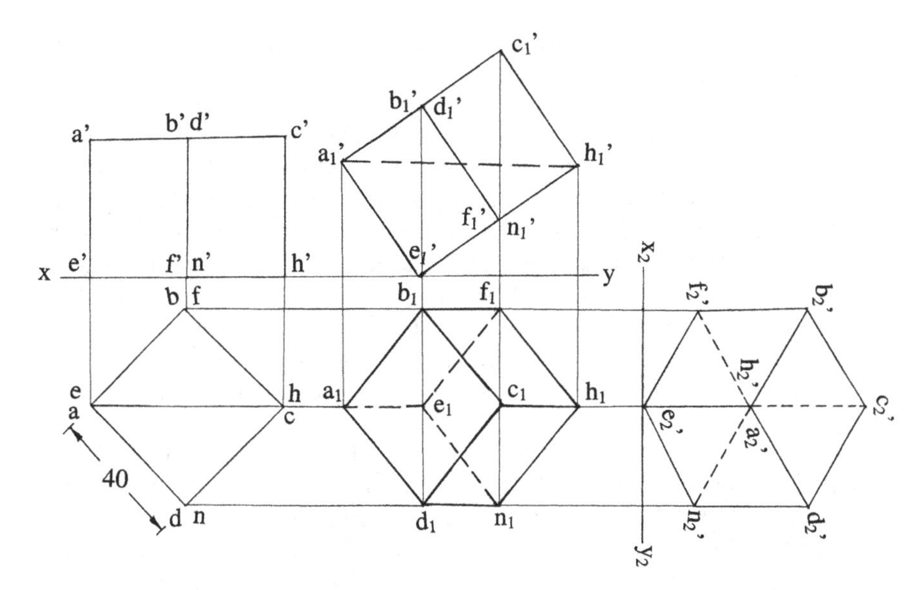

Fig. 10.19 Auxiliary projections of cube

elevation is tilted about the corner e′ so that the solid diagonal a′h′ is parallel to xy. The elevation is redrawn as shown in the figure. The plan is to be projected from the elevation. A projector is drawn perpendicular to xy from a_1'. The locus line is drawn parallel to xy from a. The intersection of the projector with the locus line locates the plan a_1. The same procedure is followed to get the plans of the remaining points. These points are joined to show the plans of the edges of the cube. The invisible edges a_1e_1, f_1e_1 and n_1e_1 are shown by dotted lines. An auxiliary vertical plane is chosen perpendicular to a_1h_1 at x_2y_2 as shown in the figure. The auxiliary elevation is to be projected from the plan. A projector is drawn perpendicular to x_2y_2 from c_1. Since the plan c_1 is common, the distance of c_1' from xy will be equal to the distance of c_2' from x_2y_2. The distance of c_1' from xy is measured and making use of this distance the point c_2' is located on the projector drawn through c_1. The same procedure is followed to get the elevations of other points. These points are joined to show the elevations of the edges of the cube. The invisible edges $c_2'h_2'$, $f_2'h_2'$ and $n_2'h_2'$ are shown by dotted lines.

11. Draw the projections of a tetrahedron of edge 40 mm resting on one of its angular points with the axis inclined at 60° to the HP.

    A face of the tetrahedron is kept parallel to the HP, and the angular point opposite to this face is on the HP. The solid is tilted parallel to the VP about the base angular point such that the axis passing through the base point is inclined at 60° to the HP. The elevation is redrawn and the plan is projected from the elevation.

    Draw the reference line xy and the equilateral triangle abc representing the outline of the plan of the inverted tetrahedron as shown in Fig. 10.20. The invisible edges ad, bd and cd are shown by dotted lines. The elevation is projected from the plan. The elevation of the angular point d′ lies on xy, and the axis passing through d′ is also shown in the elevation. The tetrahedron is tilted parallel to the VP about d_1' such that the axis makes an angle of 60° with the HP. As the angular points move parallel to the VP, the elevation is redrawn with the axis inclined at 60° to xy as shown in the figure. The plan is to be projected from the elevation. A projector is drawn perpendicular to xy from a_1'. The locus line is drawn parallel to xy from a. The intersection of the projector with the locus line locates the plan a_1. The same procedure is followed to get b_1, c_1 and d_1. These points are joined to get the plan of the tetrahedron. The invisible edge c_1d_1 is shown as a dotted line.

12. Draw the projections of a tetrahedron of edge 40 mm resting on an edge inclined at 30° to the VP and a face contained the edge is inclined at 40° to the HP.

    The tetrahedron is kept with a face on the HP, and an edge of this face is perpendicular to the VP. The projections are obtained. The tetrahedron is tilted parallel to the VP about the base edge perpendicular to the VP such that one face containing the edge is inclined at 40° to the HP. The elevation is redrawn and the plan is obtained from the elevation. An auxiliary vertical plane inclined at 30° to the base edge is chosen and the elevation is obtained on this plane.

    Draw the reference line xy and the plan of the tetrahedron as shown in Fig. 10.21 with the edge ac perpendicular to xy. The elevation is projected

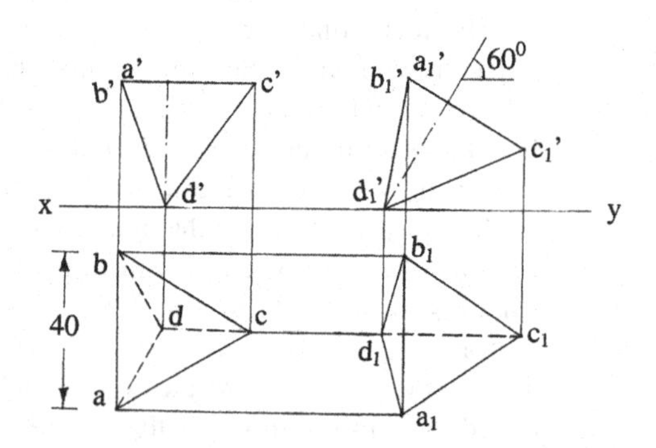

Fig. 10.20 Auxiliary projections of tetrahedron

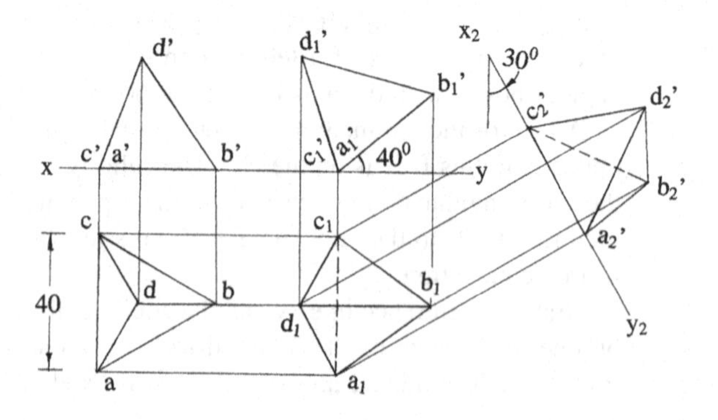

Fig. 10.21 Auxiliary projections of tetrahedron

from the plan. The tetrahedron is tilted about the edge AC till the face ABC is inclined at 40° to the HP. As the angular points move parallel to the VP, the elevation is redrawn with the edge $a_1'c_1'$ on xy and the face $a_1'b_1'c_1'$ is inclined at 40° to xy. The plan is to be projected from the elevation. A projector is drawn perpendicular to xy from a_1'. The locus line is drawn parallel to xy from a. The intersection of the projector with the locus line locates the plan a_1. The same procedure is followed to get b_1, c_1 and d_1. These points are joined to show the edges of the tetrahedron in the plan. The invisible edge a_1c_1 is shown as a dotted line. An auxiliary vertical plane inclined at 30° to the base edge a_1c_1 is chosen at x_2y_2 as shown in the figure. The elevation is to be projected on this plane. Since the plan is common, the heights of the angular points of the tetrahedron from the HP remain the same. A projector is drawn perpendicular to x_2y_2 from d_1. The point d_2' is located on this projector such that the distance of d_2' from x_2y_2 is equal to the distance of d_1' from xy. The same procedure is followed to get the points a_2', b_2' and c_2'. These points are joined to show the edges of tetrahedron in the elevation. The invisible edge $b_2'c_2'$ is shown as a dotted line.

13. Draw the projections of an octahedron of edge 40 mm when it rests on the HP on one of its faces, such that an edge of that face is inclined at 40° to the VP.

    The octahedron is kept on the HP on an angular point in such a way that its plan is a square with an edge parallel to the VP. In this position, four faces of the solid are perpendicular to the VP. The octahedron is tilted parallel to the VP about the base angular point till one of the four faces lies on the HP and the projections are obtained. An auxiliary vertical plane inclined at 40° to the base edge is chosen to get the elevation.

    Draw the reference line xy and the plan of the octahedron as shown in Fig. 10.22. The elevation is projected from the plan. The octahedron is tilted parallel to the VP about the angular point F till one of the adjacent faces lies

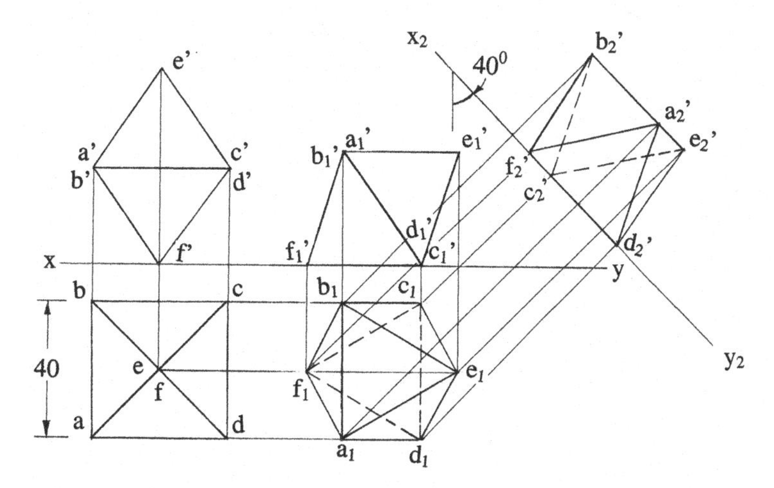

Fig. 10.22 Auxiliary projections of octahedron

on the HP. The elevation is redrawn with the face $c_1'd_1'f_1'$ on xy. The plan is to be projected from the elevation. A projector is drawn perpendicular to xy from a_1'. The locus line is drawn parallel to xy from a. The intersection of the projector with the locus line locates plan a_1. The same procedure is followed to get b_1, c_1, d_1, e_1 and f_1. These points are joined to show the plan of the octahedron. The invisible edges c_1d_1, d_1f_1 and f_1c_1 are shown by dotted lines. An auxiliary vertical plane inclined at 40° to c_1d_1 is chosen at x_2y_2. The elevation is to be obtained on this plane. A projector is drawn perpendicular to x_2y_2 from a_1. The elevation a_2' is marked on this projector such that the distance of a_2' from x_2y_2 is equal to the distance of a_1' from xy. The same procedure is followed to get the remaining points b_2', c_2', d_2', e_2' and f_2'. These points are joined to complete the required elevation. The invisible edges $b_2'c_2'$ and $c_2'e_2'$ are shown by dotted lines.

14. A pentagonal pyramid of side of base 40 mm and axis 70 mm has one face on the VP with the base edge of this face inclined at 70° to the HP. Draw its projections.

    The pentagonal pyramid base is kept on the VP with an edge of base perpendicular to the HP. The projections are obtained. The pyramid is tilted parallel to the HP about the base edge till the face containing the base edge lies on the VP. The plan is redrawn and the elevation is projected from the plan. An auxiliary plane inclined at 70° to the base edge is chosen at x_2y_2. The plan is projected on this plane.

    Draw the reference line xy and the elevation of the pyramid with an edge of base perpendicular to xy as shown in Fig. 10.23. The plan is projected from the elevation. The pyramid is tilted parallel to the HP about bc till the face abc coincides with the reference line xy. The plan is redrawn with the face $a_1b_1c_1$

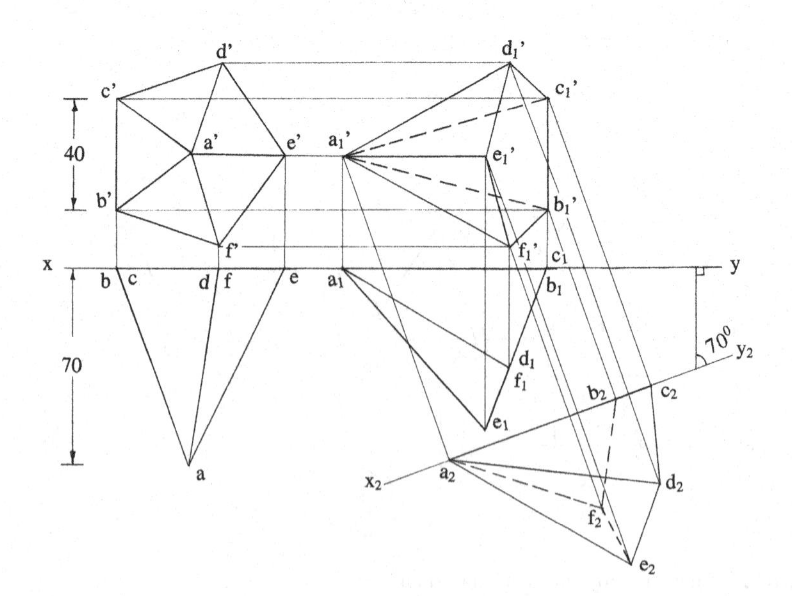

Fig. 10.23 Auxiliary projections of pentagonal pyramid

on xy as shown in the figure. The elevation is to be projected from the plan. A projector is drawn perpendicular to xy from e_1. The locus line is drawn parallel to xy from e'. The intersection of the projector with the locus line locates e_1'. The same procedure is followed to get a_1', b_1', c_1', d_1' and f_1'. These points are joined to get the elevation of the pyramid. The invisible edges $a_1'b_1'$ and $a_1'c_1'$ are shown as dotted lines. An auxiliary plane inclined at 70° to $b_1'c_1'$ is chosen at x_2y_2. The plan is to be projected from the elevation. A projector is drawn perpendicular to x_2y_2 from e_1'. The point e_2 is marked on this projector such that the distance of e_2 from x_2y_2 is equal to the distance of e_1 from xy. The same procedure is followed to get a_2, b_2, c_2, d_2 and f_2. These points are joined to complete the plan of the pyramid. The invisible edges b_2f_2, a_2f_2 and e_2f_2 are shown as dotted lines.

15. An octahedron of 40 mm edge rests with an edge on the HP such that the two faces containing that edge make equal inclinations with the HP. The edge on which it rests should be inclined at 30° to the VP. Draw its projections.

The octahedron is kept initially on an angular point on the HP such that the solid diagonal passing through this angular point and another solid diagonal are parallel to the VP. The projections are obtained. The octahedron is tilted parallel to the VP about the base point till the edge passing through the base point lies on the HP. In this position, the two faces containing the base edge are equally inclined to the HP. The plan is projected from the elevation. An auxiliary vertical plane inclined at 30° to the plan of base edge of the octahedron is chosen. The elevation is projected on this plane.

Draw the reference line xy and the plan and elevation of the octahedron resting on an angular point on the HP as shown in Fig. 10.24. The solid diagonals ef and ac are parallel to the VP. The octahedron is tilted parallel to the VP about the base point F till the edge FC lies on the HP. The angular

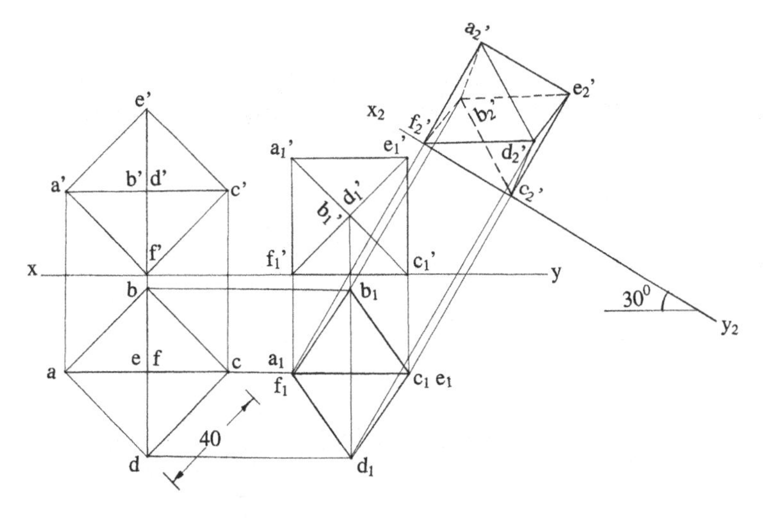

Fig. 10.24 Auxiliary projections of octahedron

points move parallel to the VP. The elevation is redrawn with the edge $f_1'c_1'$ on xy. The faces $f_1'b_1'c_1'$ and $f_1'd_1'c_1'$ are equally inclined to the HP. The plan is to be projected from the elevation. A projector is drawn perpendicular to xy from e_1'. The locus line is drawn parallel to xy from e. The intersection of the projector with the locus line locates the point e_1 in the plan. The same procedure is followed to get a_1, b_1, c_1, d_1 and f_1. These points are joined to complete the plan of octahedron. An auxiliary vertical plane inclined at 30° to the base edge f_1c_1 in the plan is chosen at x_2y_2. The elevation is to be obtained on this plane. A projector is drawn perpendicular to x_2y_2 from e_1. The point e_2' is marked on this projector such that the distance of e_2' from x_2y_2 is equal to the distance of e_1' from xy. The same procedure is followed to get a_2', b_2', c_2', d_2' and f_2'. These points are joined to show the elevation of the octahedron. The invisible edges $b_2'a_2'$, $b_2'e_2'$, $b_2'c_2'$ and $b_2'f_2'$ are shown by dotted lines. (Note: All 12 edges of octahedron are distinct in the final elevation and 8 edges are visible.)

16. A regular octagon abcdefgh of 20 mm side is the plan of a plane figure. The heights of A, B, C, D, E, F, G and H are, respectively, 10 mm, 10 mm, 20 mm, 30 mm, 40 mm, 40 mm, 30 mm and 20 mm. Determine the true form of the figure.

    The plan of the octagon is drawn with a side parallel to the VP as shown in Fig. 10.25. The left extreme points are designated as a and b. The elevations of

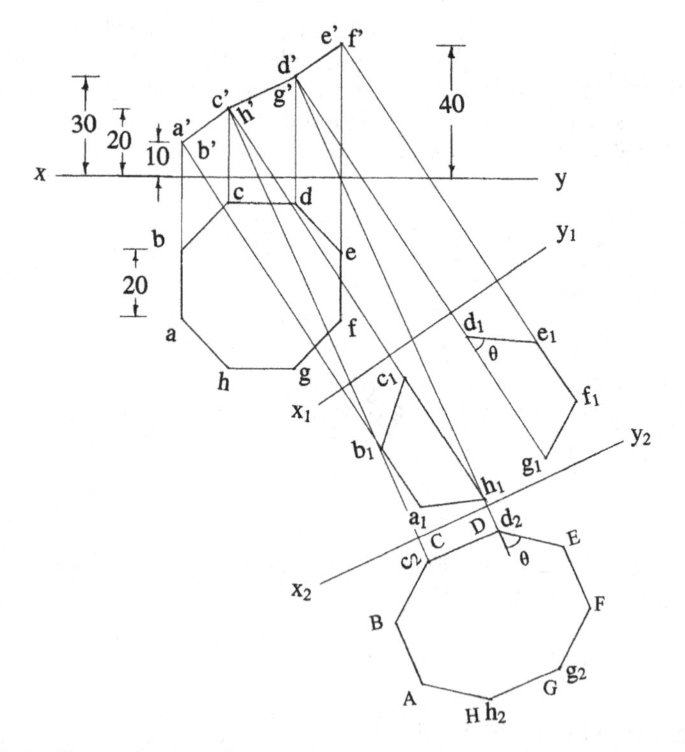

Fig. 10.25 Auxiliary projections of octagon

the points are shown on the respective projectors. It is to be noted that the elevations of the lines a′b′, c′h′, d′g′ and e′f′ are points and are parallel to the HP. Hence, ab, ch, dg and ef give the respective true lengths. Elevations a′h′, b′c′, d′e′ and g′f′ are parallel. An auxiliary plane is chosen parallel to the elevations of these lines at x_1y_1. The projections $h_1a_1b_1c_1$ and $g_1f_1e_1d_1$ on this auxiliary plane give true forms of the two portions of the octagon. The two portions are identical. Another auxiliary plane is chosen parallel to c′d′ at x_2y_2. The projection $c_2d_2g_2h_2$ on this plane gives the true form of the middle portion of the octagon. The two portions $h_1a_1b_1c_1$ and $g_1f_1e_1d_1$ are attached to the middle portion using the angle θ and the true lengths of the sides. ABCDEFGH represents the true form of the octagon.

17. A right square pyramid, edge of base 30 mm and height 60 mm, is suspended freely from a corner of its base. Draw its projections.

The centre of gravity of a pyramid/cone lies on the axis at a distance equal to ¼ axis from the base. The line connecting the corner of suspension with the centre of gravity will be vertical when suspended freely. The base of the square pyramid is kept parallel to the HP and above the apex/vertex with the base edges equally inclined to the VP. The projections are obtained. The pyramid is suspended from one of its extreme corners. The pyramid moves parallel to the VP so that the line joining the extreme corner and the centre of gravity is perpendicular to the HP. The plan is projected from the elevation.

Draw the reference line xy and the plan of the square pyramid with its base above the apex and parallel to the HP as shown in Fig. 10.26. The base edges are equally inclined to the VP so that the square pyramid can be suspended from one of its extreme corner. The elevation is projected from the plan. The centre of gravity g′ is located on the axis 15 mm from the base in the elevation.

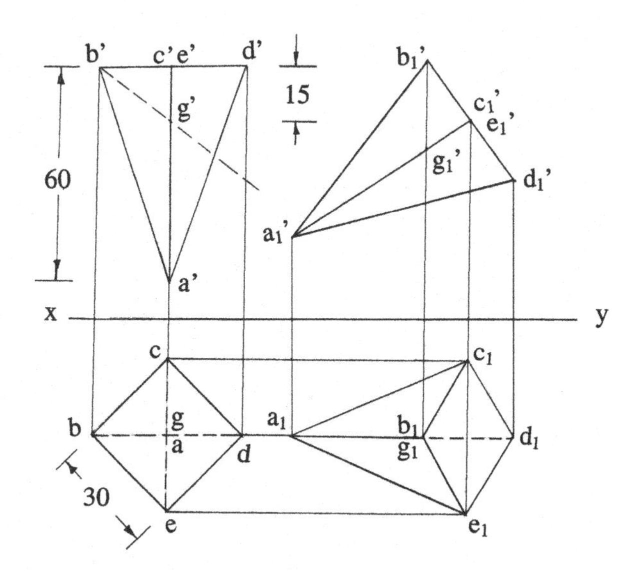

Fig. 10.26 Auxiliary projections of square pyramid

The plan of centre of gravity g coincides with the axis of the pyramid. The pyramid is suspended from b′ so that the line passing through b′ and g′ is perpendicular to xy. The pyramid moves parallel to the VP. The elevation is redrawn as shown in the figure. The plan is to be projected from the elevation. A projector is drawn perpendicular to xy from d_1'. The locus line is drawn parallel to xy from d. The intersection of the projector with the locus line locates d_1. The same procedure is followed to get a_1, b_1, c_1 and e_1. These points are joined to show the edges of the pyramid in the plan with a part of invisible edge a_1d_1 shown as dotted.

18. A tetrahedron of edge 50 mm stands on one of its edges on the HP such that the vertical plane containing that edge and the centre of gravity of the solid makes 30° with the VP. Draw its projections.

The tetrahedron is kept on the HP on one of its faces with a side of that face perpendicular to the VP. The solid is tilted parallel to the VP about the extreme angular point such that the edge containing the extreme angular point is on the HP. The centre of gravity lies on the vertical plane containing the edge on which the solid stands on the HP. An auxiliary vertical plane inclined at 30° to the base edge is chosen, and the projection on this plane is obtained.

Draw the reference line xy and the plan of the tetrahedron resting on the HP on one of its faces with an edge of that face perpendicular to the VP as shown in Fig. 10.27. The elevation is projected from the plan. The centre of gravity g coincides with the point d in the plan. It is to be noted that the centre of gravity g′ is one-fourth the height of d′ from the face a′b′c′. The tetrahedron is tilted parallel to the VP about the extreme point c′ so that the edge c′d′ lies on xy. The elevation is redrawn such that the edge $c_1'd_1'$ lies on xy. All points move parallel to the VP. The plan is to be projected from the elevation. A

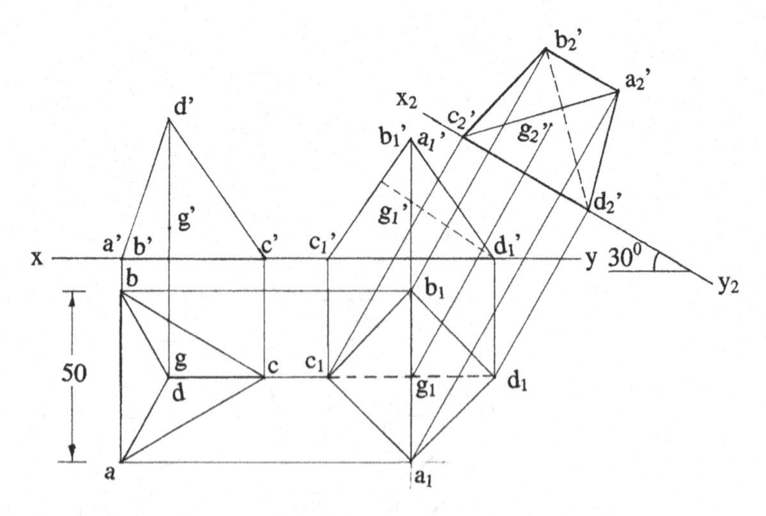

Fig. 10.27 Auxiliary projections of tetrahedron

projector is drawn perpendicular to xy from a_1'. The locus line is drawn parallel to xy from a. The intersection of the projector with the locus line locates the plan a_1. The same procedure is followed to get b_1, c_1, g_1 and d_1. It is to be noted that the centre of gravity lies on the plane containing the edge c_1d_1. An auxiliary vertical plane inclined at 30° to the base edge c_1d_1 is chosen at x_2y_2 as shown in the figure. The elevation is to be projected from the plan. A projector is drawn perpendicular to x_2y_2 from a_1. The elevation a_2' is marked on this projector such that the distance of a_2' from x_2y_2 is equal to the distance of a_1' from xy. The same procedure is followed to get b_2', c_2', d_2' and g_2'. The angular points are joined to complete the elevation. The invisible edge $b_2'd_2'$ is shown by a dotted line.

19. A pentagonal pyramid, side of base 30 mm and axis 60 mm long, stands on the HP with an edge of its base parallel to and 20 mm in front of the VP such that the apex touches the VP. Draw its projections.

The projections of the pentagonal pyramid are obtained keeping its base on the HP with a side of base perpendicular to the VP. An auxiliary vertical plane is chosen at x_2y_2. The vertical trace of the auxiliary plane is also drawn. The pyramid is moved parallel to the VP till the perpendicular base edge is 20 mm from the vertical trace and then tilted about this base edge such that the apex touches the auxiliary plane. The elevation is redrawn and the plan is projected from the elevation. The projection on the auxiliary plane kept at x_2y_2 furnishes the required elevation.

Draw the reference line xy and the projections of the pentagonal pyramid resting on the HP on its base with a side of base perpendicular to the VP as shown in Fig. 10.28. The side ed is perpendicular to xy. An auxiliary vertical plane is chosen at x_2y_2 and its vertical trace is drawn above xy. The pyramid is moved parallel to the VP initially till the edge e′d′ is 20 mm from the vertical

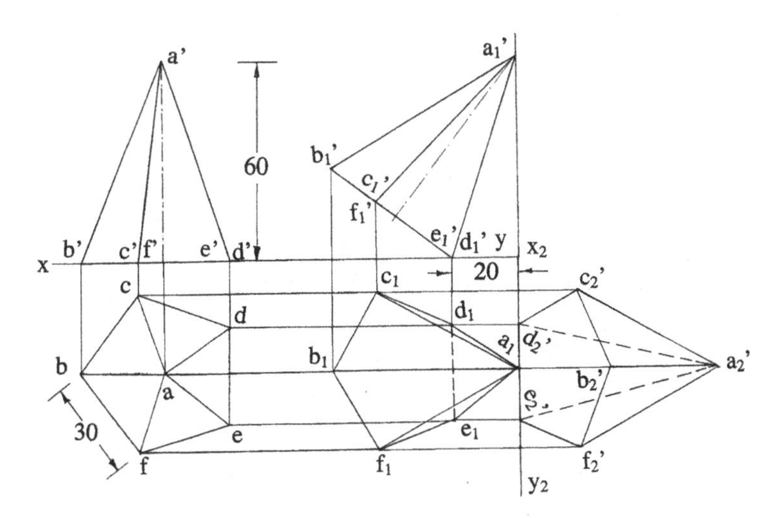

Fig. 10.28 Auxiliary projections of pentagonal pyramid

trace. Thereafter, the pyramid is tilted parallel to the VP about the edge e′d′ till the apex touches the auxiliary plane. The elevation is redrawn with the edge $e_1'd_1'$ on xy and the apex a_1' touching the vertical trace as shown in the figure. The plan is to be projected from the elevation. A projector is drawn perpendicular to xy from b_1'. The locus line is drawn parallel to xy from b. The intersection of the projector with the locus line locates the plan b_1. The same procedure is followed to get a_1, c_1, d_1, e_1 and f_1. These points are joined to complete the required plan of the pyramid. The elevation is to be projected on the auxiliary plane at x_2y_2 from the plan. A projector is drawn perpendicular to x_2y_2 from b_1. The point b_2' is marked on this projector such that the distance of b_2' from x_2y_2 is equal to the distance of b_1' from xy. The same procedure is followed to get a_2', c_2', d_2', e_2' and f_2'. These points are joined to complete the required elevation of the pyramid. The invisible edges $a_2'd_2'$ and $a_2'e_2'$ are shown by dotted lines.

20. A hexagonal pyramid, edge of base 30 mm and axis 50 mm long, stands on an edge of base in the HP and inclined at 45° to the VP and the face containing the base edge inclined at 50° to the HP. Draw its projections.

    The hexagonal pyramid is kept on the HP on its base with a base edge perpendicular to the VP. The pyramid is tilted parallel to the VP about the edge perpendicular to the VP till the face containing this edge makes an angle of 50° with the HP. An auxiliary vertical plane inclined at 45° to the base edge is chosen at x_2y_2. The projection on this plane provides the elevation of the pyramid.

    Draw the reference line xy and the plan of the pyramid with a base edge perpendicular to xy as shown in Fig. 10.29. The elevation is projected from the

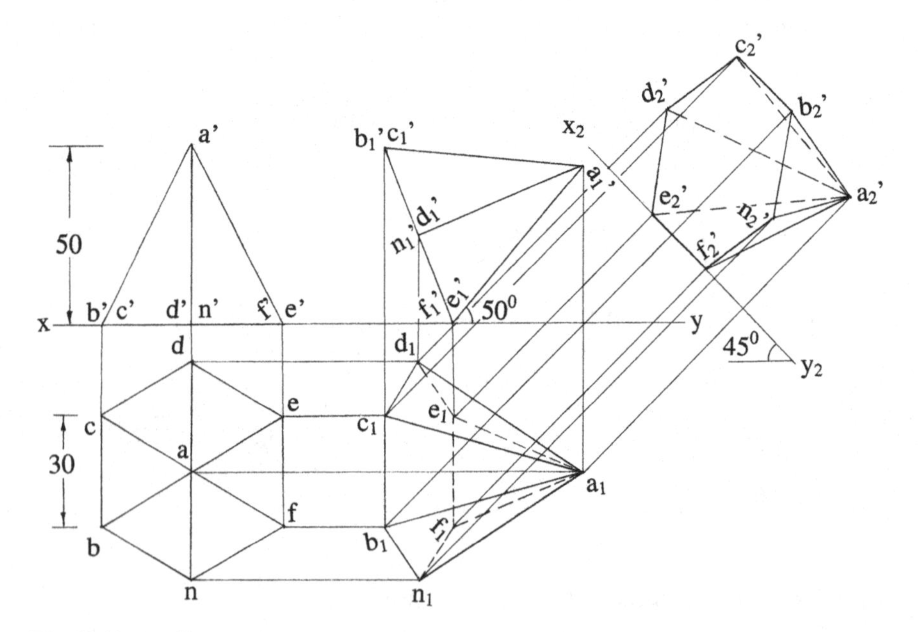

Fig. 10.29 Auxiliary projections of hexagonal pyramid

plan. The pyramid is tilted parallel to the VP about the base edge e′f′ till the face a′e′f′ is inclined at 50° to the HP. The elevation is redrawn such that the face $a_1'e_1'f_1'$ is inclined at 50° to xy. The plan is to be projected from the elevation. A projector is drawn perpendicular to xy from a_1'. The locus line is drawn parallel to xy from a. The intersection of the projector with the locus line locates the plan a_1. The same procedure is followed to get b_1, c_1, d_1, e_1, f_1 and n_1. These points are joined to complete the required plan of the pyramid. The invisible edges a_1e_1, a_1f_1, e_1f_1, d_1e_1 and f_1n_1 are shown by dotted lines. An auxiliary vertical plane inclined at 45° to the base edge is chosen at x_2y_2 as shown in the figure. The elevation is to be projected from the plan. A projector is drawn perpendicular to x_2y_2 from a_1. The point a_2' is marked on this projector such that the distance of a_2' from x_2y_2 is equal to the distance of a_1' from xy. The same procedure is followed to get b_2', c_2', d_2', e_2', f_2' and n_2'. These points are joined to get the required elevation of the pyramid. The invisible edges $a_2'c_2'$, $a_2'd_2'$ and $a_2'e_2'$ are shown by dotted lines.

21. A square pyramid, side of base 40 mm and axis 60 mm long, lies with one of the base diagonals inclined to the HP at 60° and the other diagonal parallel to the HP. Draw its projections.

    The square pyramid is kept on the HP on its base with base edges equally inclined to the VP. The pyramid is tilted parallel to the VP such that one of the base diagonals is inclined at 60° to the HP. The other base diagonal is parallel to the HP. The elevation is redrawn and the plan is projected from the elevation.

    Draw the reference line xy and plan of the square pyramid with its base edges equally inclined to xy as shown in Fig. 10.30. The elevation is projected from the plan. The pyramid is tilted parallel to the VP about the base diagonal c′e′ such that the other base diagonal b′d′ is inclined at 60° to the HP. The elevation is redrawn with the base diagonal $b_2'd_2'$ inclined at 60° to xy and the

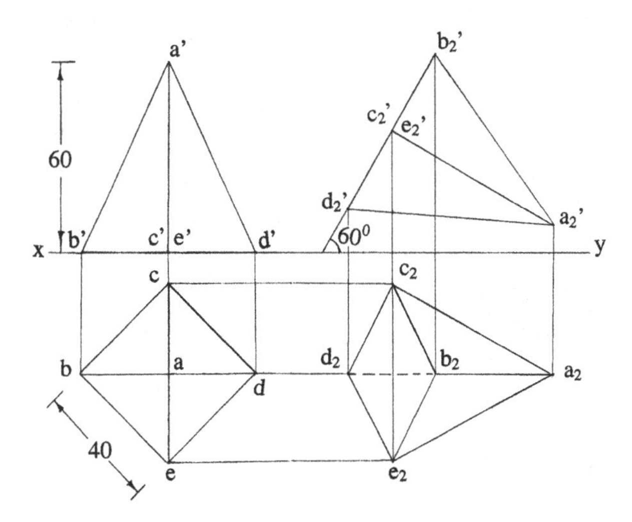

Fig. 10.30 Auxiliary projections of square pyramid

base diagonal $c_2'e_2'$ parallel to the HP as shown in the figure. The plan is to be projected from the elevation. A projector is drawn perpendicular to xy from a_2'. The locus line is drawn parallel to xy from a. The intersection of the projector with the locus line locates the plan a_2. The same procedure is followed to get b_2, c_2, d_2, and e_2. These points are joined to complete the required plan with a portion of the invisible edge a_2d_2 shown as dotted.

22. Draw the projections of a cylinder of diameter of base 50 mm and axis 70 mm long when the base is inclined at 60° to the HP and the plan of the axis is inclined at 45° to xy.

    The cylinder is kept on the HP on its base and the projections are obtained. The cylinder is tilted parallel to the VP about one of its extreme base point so that the base is inclined at 60° to the HP. The elevation is redrawn and the plan is projected from the elevation. An auxiliary vertical plane inclined at 45° to the plan of the axis is chosen at x_2y_2. The elevation is obtained on this plane.

    Draw the reference line xy and the plan of the cylinder with its base on the HP as shown in Fig. 10.31. The alternate generators on the lateral surface of the cylinder are designated as ab, cd, ef and hn in the plan. The elevation is projected from the plan. The cylinder is tilted parallel to the VP about the base point f′ such that the base of the cylinder is inclined at 60° to the HP. The elevation is redrawn with f_1' on xy and the base $b_1'd_1'f_1'n_1'$ inclined at 60° to xy. The plan is to be projected from the elevation. A projector is drawn perpendicular to xy from a_1'. The locus line is drawn parallel to xy from a. The intersection of the projector with the locus line locates the plan a_1. The same procedure is followed to get remaining points in the plan. These points are joined by smooth curves to show the plans of top and bottom ends of

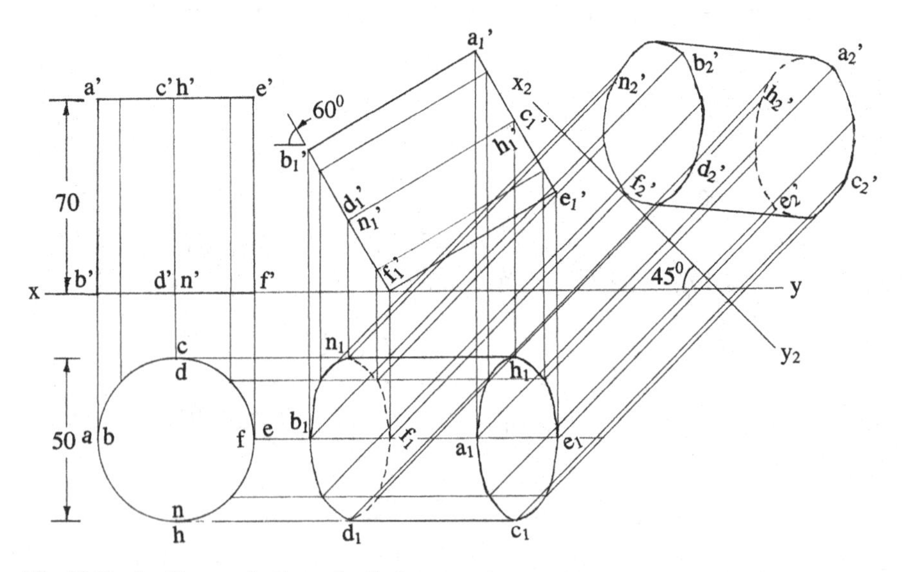

Fig. 10.31 Auxiliary projections of cylinder

the cylinder. Two tangents are drawn to the ellipses to show the extreme generators h_1n_1 and c_1d_1 as shown in the figure. The portion of the invisible curve $d_1f_1n_1$ of the cylinder is shown as dotted in the plan. An auxiliary vertical plane inclined at 45° to the plan of axis is chosen at x_2y_2. The elevation is to be projected from the plan. A projector is drawn perpendicular to x_2y_2 from a_1. The point a_2' is marked on this projector such that the distance of a_2' from x_2y_2 is equal to the distance of a_1' from xy. The same procedure is followed to get the remaining points in the elevation. These points are joined by smooth curves to show the top and bottom ends of the cylinder. Two outside parallel tangents are drawn to the ellipses to show the extreme generators in the elevation. The invisible portion of the top end between the tangents is shown as dotted.

23. A hexagonal pyramid of edge of base 30 mm and axis 50 mm has one of its triangular faces in the VP with the axis parallel to the HP. Draw the three views of the pyramid.

 The hexagonal pyramid base is kept in the VP with a side of base perpendicular to the HP. The pyramid is tilted parallel to the HP about one of its extreme base edges till the triangular face containing the base edge lies in the VP. The three projections are obtained for this position of the pyramid.

 Draw the reference line xy and elevation of the pyramid when its base is in the VP as shown in Fig. 10.32. The plan is projected from the elevation. The pyramid is tilted parallel to the HP about the edge EF till the face AEF lies on the VP. The plan is redrawn with the face $a_1e_1f_1$ on xy. The elevation is to be projected from the plan. A projector is drawn perpendicular to xy from a_1. The locus line is drawn parallel to xy from a′. The intersection of the projector with the locus line locates the elevation a_1'. The same procedure is followed to get b_1', c_1', d_1', e_1', f_1' and h_1'. These points are joined to show the base and slanting edges of the pyramid in the elevation. The invisible edges $a_1'e_1'$ and $a_1'f_1'$ are shown by dotted lines. The left side elevation is drawn from the elevation and

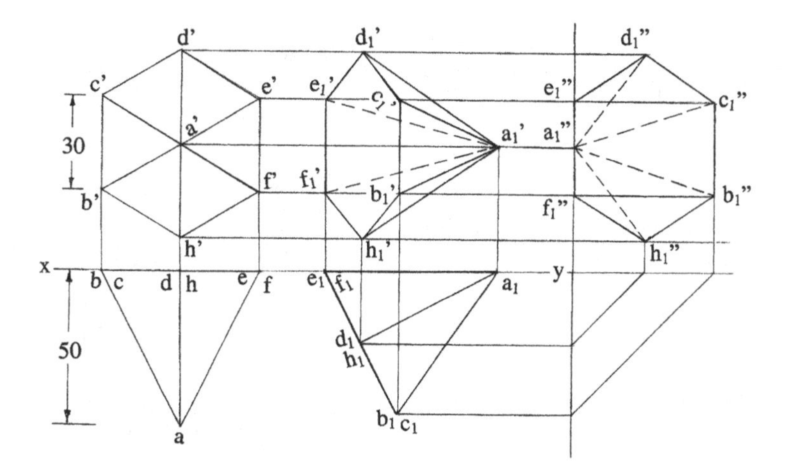

Fig. 10.32 Auxiliary projections of hexagonal pyramid

plan of the pyramid following the principle of projection by drawing locus lines and projectors as shown in the figure. The invisible slanting edges $a_1''d_1''$, $a_1''c_1''$, $a_1''b_1''$ and $a_1''h_1''$ are shown by dotted lines in the side elevation.

24. The diameter of base of a cone is 50 mm and the height is 70 mm. Draw its projections when it is resting on the HP on one of its generators with the axis inclined at 35° to the VP.

    The cone is kept on the HP on its base and the projections are drawn. The cone is tilted parallel to the VP about one of its extreme base point till the generator connecting the base point with the apex lies on the HP. The elevation is redrawn and the plan is projected from the elevation. The next task will be to find out the apparent angle of inclination (β) of the plan axis of the cone with xy. An auxiliary vertical plane inclined at β to the plan axis of the cone is chosen, and the required elevation is obtained on this plane.

    Draw the reference line xy and the plan of cone with alternate generators designated as ab, ac, ad and ae as shown in Fig. 10.33. The elevation is projected with the generators from the plan. The cone is tilted parallel to the VP about the base point D till the generator AD lies on the HP. The elevation is redrawn with $a_1'd_1'$ on xy. The plan is to be projected from the elevation. A projector is drawn perpendicular to xy from b_1'. The locus line is drawn parallel to xy from b. The intersection of the projector with the locus line locates the plan b_1. The same procedure is followed to get the plans of base points and the apex. The base points are joined by a smooth curve. Two tangents are drawn from the apex a_1 to the ellipse to show the extreme generators of the cone. To find the apparent angle of inclination, a line x_1y_1 is drawn parallel to xy. The point a_1 is located on x_1y_1 and another line is drawn from a_1 inclined at 35°. A point n' is located on the inclined line taking its distance from a_1 equal to the

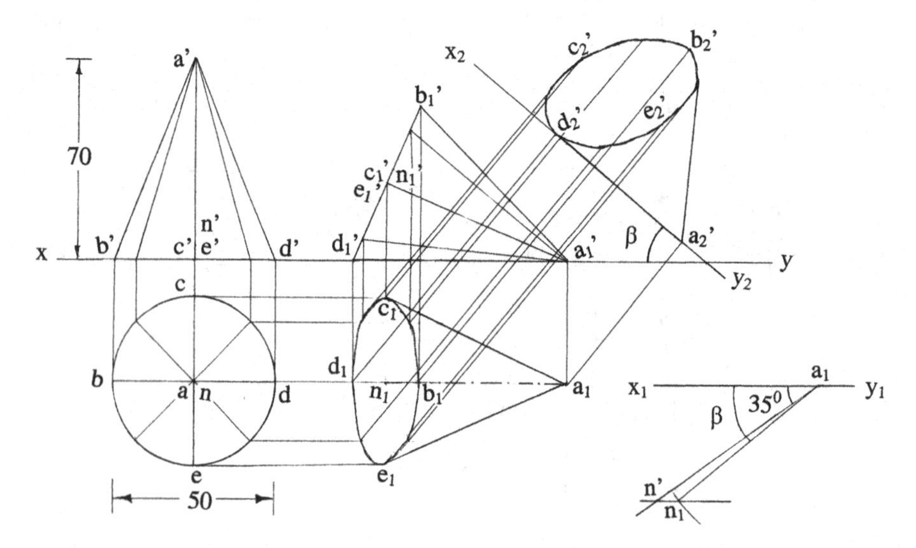

Fig. 10.33 Auxiliary Projections of cone

axis of the cone. Draw the locus line passing through n′. With centre a_1 and the plan axis length a_1n_1 as radius draw an arc to cut the locus line at n_1. The inclination of the line a_1n_1 with x_1y_1 gives the apparent angle β. An auxiliary vertical plane inclined at β to xy is chosen at x_2y_2. The projection on this plane gives the elevation of the cone. A projector is drawn perpendicular to x_2y_2 from b_1. The point b_2' is marked on this projector such that the distance of b_2' from x_2y_2 is equal to the distance of b_1' from xy. The same procedure is followed to get the elevations of the remaining points. The base points are joined by a smooth curve. A tangent is drawn from the apex to the base curve to represent the upper extreme generator of the cone in elevation.

25. The diameter of base of a cone is 50 mm and the height is 60 mm. Draw its projections when it is suspended freely from a point on the circumference of its base. Assume that the axis is parallel to the VP.

    The cone base is kept parallel to the HP and above the apex. The projections are drawn. The centre of gravity is located along the axis in the elevation. The line joining the extreme base point and the centre of gravity will be vertical when suspended freely from the extreme base point. The cone moves parallel to the VP during the suspension. The elevation is redrawn keeping the line joining the base point and the centre of gravity vertical and the plan is projected from the elevation.

    Draw the reference line xy and the projections of the cone as shown in Fig. 10.34. The centre of gravity g′ is located 15 mm from the base of the cone in the elevation. The centre of gravity g coincides with the apex of the cone in the plan. When the cone is suspended freely from b′, the line joining b′ and g′ will be vertical. The cone moves parallel to the VP. The elevation is redrawn as shown in the figure. The plan is to be projected from the elevation. A projector is drawn perpendicular to xy from a_1'. The locus line is drawn parallel to xy from a. The intersection of the projector with the locus line locates the plan a_1.

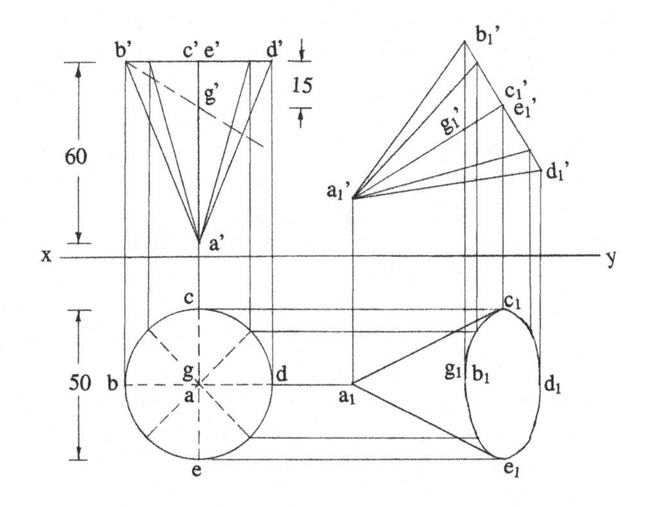

Fig. 10.34 Auxiliary projections of cone

The same procedure is followed to get the plans of other points. The base points of the cone are joined by a smooth curve. Two tangents are drawn from a_1 to the base curve of the cone in the plan.

Practice Problems

1. A square prism, edge of base 40 mm and axis 60 mm long, stands on the HP with a face inclined 30° to the VP. Draw its projections.
2. A hexagonal prism, edge of base 30 mm and axis 70 mm long, rests with one of its rectangular faces on the HP and the ends inclined at 40° to the VP. Draw its projections.
3. A pentagonal prism, edge of base 30 mm and axis 60 mm long, rests with one of its longer edges on the HP. The two rectangular faces containing the longer edge on which it rests make equal inclination with the HP, and the longer edges are inclined at 50° to the VP. Draw its projections.
4. A square pyramid, edge of base 30 mm and axis 60 mm long, stands on the HP on one of its base edges. Draw its projections when the face containing that base edge is perpendicular to the HP.
5. A pentagonal pyramid, edge of base 40 mm and axis 60 mm long, stands with a corner of its base on the HP. Draw the projections of the pyramid when the slanting edge containing that corner is vertical.
6. Draw the projections of a tetrahedron of edge 40 mm when suspended freely from one of its angular points.
7. An octahedron of 50 mm edge rests on one of its angular points on the HP. Draw its projections when one of its solid diagonals is parallel to both the HP and VP.
8. A square pyramid, edge of base 40 mm and axis 60 mm long, has one face on the VP with the base edge of this face inclined at 50° to the HP. Draw its projections.
9. Draw the projections of a hexagonal pyramid, side of base 30 mm and axis 70 mm long, when it is resting on one of its base edges with the axis making an angle of 40° with the HP and its plan axis making an angle of 30° with the reference line xy.
10. An octahedron of edge 40 mm has one of its faces in the VP and an edge of that face is making 30° with the HP. Draw its projections.
11. A tetrahedron of edge 50 mm has one of its edges parallel to the HP and inclined at 40° to the VP while one of the faces containing that edge is vertical. Draw its projections.
12. A cone, base 50 mm diameter and axis 70 mm long, has one of its generators in the HP and perpendicular to the VP. Draw its projections keeping the base nearer to the VP than the apex.
13. A pentagonal pyramid, side of base 30 mm and axis 60 mm long, rests with one of its base edges in the HP, the axis parallel to the VP and the base making an angle of 40° to the HP. Draw the plan, elevation and the left side elevation of the pyramid.

14. Draw the projections of a cylinder of diameter of base 50 mm and axis 70 mm long when the base is inclined to the HP such that the plans of the top and bottom ends touch each other.
15. An octagonal pyramid, base 20 mm side and axis 60 mm long, has one of its triangular faces on the HP. A vertical plane containing the axis is inclined at 30° to the VP. Draw the projections of the pyramid if the base is nearer to the VP than the apex.
16. A rectangular prism of base 50 mm × 20 mm and height 60 mm rests with one of its longer base edges on the HP. The axis of the prism is inclined to the HP at 30° and the plan of the axis is inclined at 40° to the VP. Draw its projections.
17. A pentagonal prism of 30 mm side of base and height 50 mm rests with its base on the HP such that one of its rectangular faces is inclined at 30° to the VP to the left. Draw the plan, elevation and left side elevation of the prism.
18. A hexagonal prism side of base of 30 mm and axis 50 mm long rests on one of its largest edges on the HP and the axis is inclined at 40° to the VP. One of the rectangular faces containing the longer edge on which the prism rests is inclined at 30° to the HP. Draw the projections of the prism.
19. Draw the projections of a right circular cone of 60 mm base diameter and 80 mm height when a generator lies in the VP making an angle of 30° to the HP.
20. A rectangular pyramid of 30 mm × 40 mm sides of base and 70 m high rests with one of its shorter edges of the base on the HP such that the axis is inclined at 30° to the HP and 40° to the VP. Draw its projections.

Chapter 11
Sections of Solids

The working parts of a machine are generally made of a number of geometric shapes assembled to produce the desired form. The details of such parts are hidden in an outside view of the machine. The machine is assumed to be cut through by a section plane. Sectional views are obtained to show the true forms of hidden or internal parts. In solid geometry, the solid is cut through by a section plane. The portion of the solid between the section plane and the observer is removed. If the portion of solid between the section plane and the plane of projection is projected, as well as the section, the projection is called a sectional plan or sectional elevation. The true shape of the section is obtained in sectional plan if the section plane is parallel to the horizontal plane. The sectional elevation shows the true shape of the section if the section plane is parallel to the vertical plane. If the section plane is not parallel to any one of the principal planes, an auxiliary projection is necessary on a plane parallel to the section plane to show the true shape of section. The projection of section is distinguished by drawing evenly spaced thin lines, called section lines, drawn at 45° to the horizontal or the reference line xy.

The section planes are classified as under:

(a) Section plane parallel to the HP and perpendicular to the VP
(b) Section plane inclined to the HP and perpendicular to the VP
(c) Section plane perpendicular to the HP and parallel to the VP
(d) Section plane perpendicular to the HP and inclined to the VP
(e) Section plane perpendicular to both the HP and VP
(f) Section plane inclined to both the HP and VP

(a) Section plane parallel to the HP and perpendicular to the VP

The section plane is shown pictorially in Fig. 11.1. The projection of the section plane on the VP will be a line coinciding with its vertical trace. The orthographic projection of the section plane is shown in Fig. 11.2. The intersections of vertical trace with the edges of the solid locate the section points in the elevation of the solid. These points are then projected in the plan of the solid.

K. Rathnam, *A First Course in Engineering Drawing*,
DOI 10.1007/978-981-10-5358-0_11

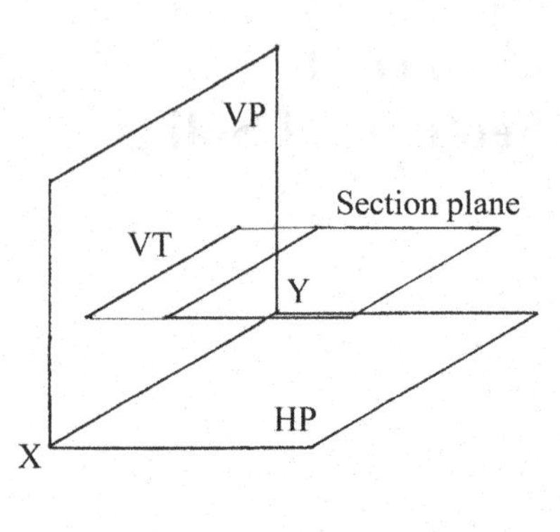

Fig. 11.1 Pictorial representation

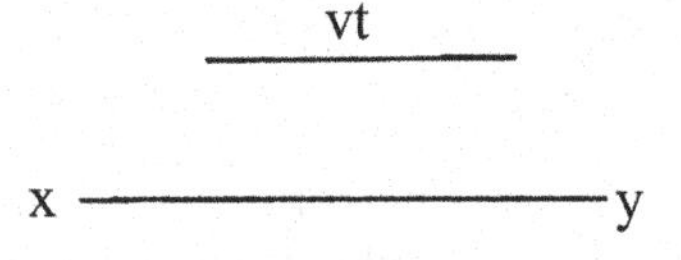

Fig. 11.2 Orthographic projection

(b) Section plane inclined to the HP and perpendicular to the VP

The section plane inclined at θ to the HP is shown pictorially in Fig. 11.3. The orthographic projection of the section plane is shown in Fig. 11.4. The vertical trace is inclined at θ to the reference line xy as shown in Fig. 11.4. The intersections of the vertical trace with edges of the solid locate the section points in the elevation of the solid. These points are then projected in the plan of the solid.

(c) Section plane perpendicular to the HP and parallel to the VP

The section plane is shown pictorially in Fig. 11.5. The orthographic projection of the section plane is shown in Fig. 11.6. The intersections of the horizontal trace with the edges of the solid locate the section points in the plan of the solid. These points are then projected in the elevation of the solid.

(d) Section plane perpendicular to the HP and inclined to the VP

The section plane is shown pictorially in Fig. 11.7. The section plane is inclined at Φ to the VP. The orthographic projection is shown in Fig. 11.8. The horizontal trace is inclined at Φ to the reference line xy as shown in Fig. 11.8. The intersections of the horizontal trace with the plan of the solid locate the section points in the plan of the solid. These points are then projected in the elevation of the solid.

(e) Section plane perpendicular to both the HP and VP

The section plane is shown pictorially in Fig. 11.9. The orthographic projection of the section plane is shown in Fig. 11.10. The vertical trace and the horizontal trace will be perpendicular to the reference line xy. The traces intersect the edges of the solid, and the side elevation shows the true shape of the section.

(f) Section plane inclined to both the HP and VP

Section planes inclined to both the principal planes are called oblique planes and are beyond the scope of this book.

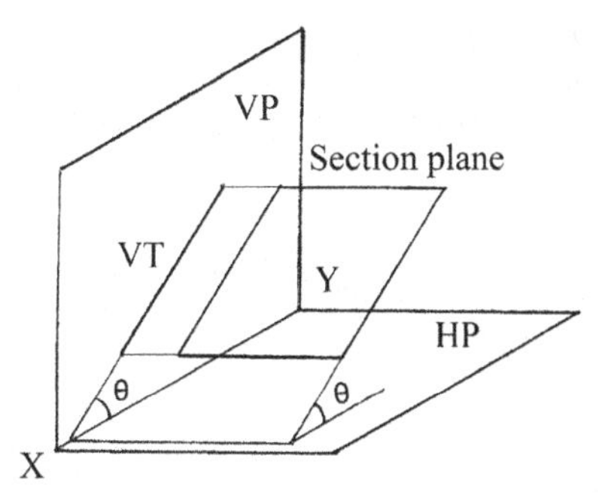

Fig. 11.3 Pictorial representation

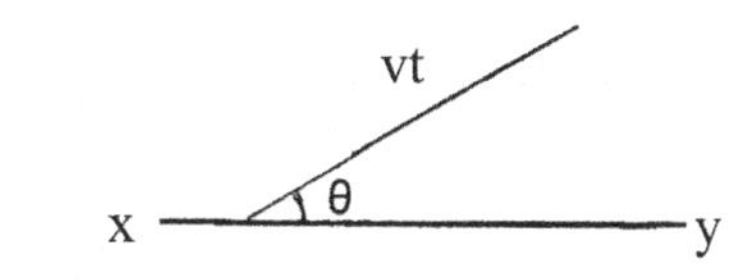

Fig. 11.4 Orthographic projection

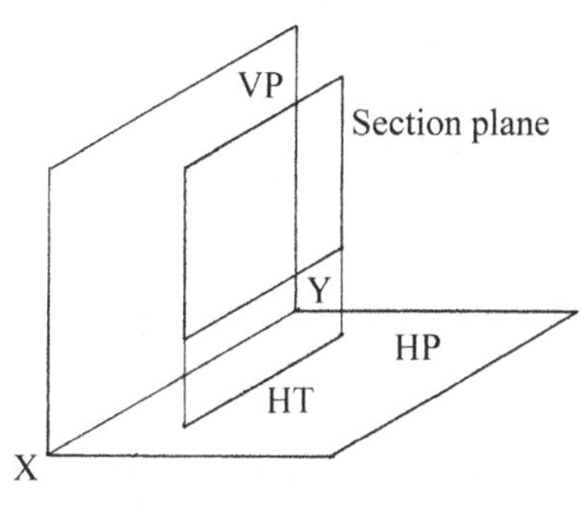

Fig. 11.5 Pictorial representation

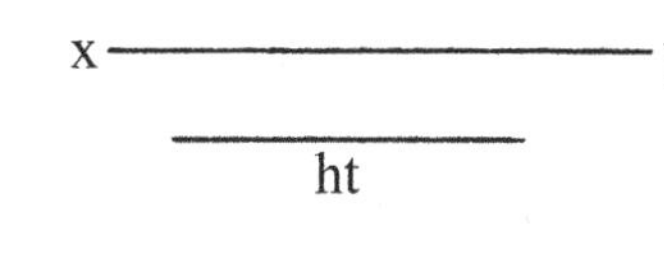

Fig. 11.6 Orthographic projection

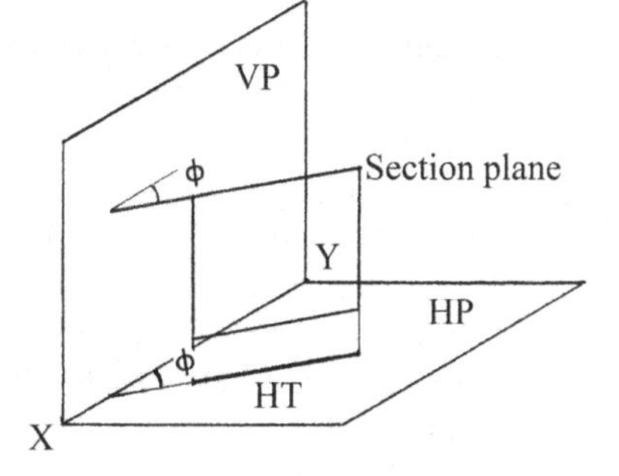

Fig. 11.7 Pictorial representation

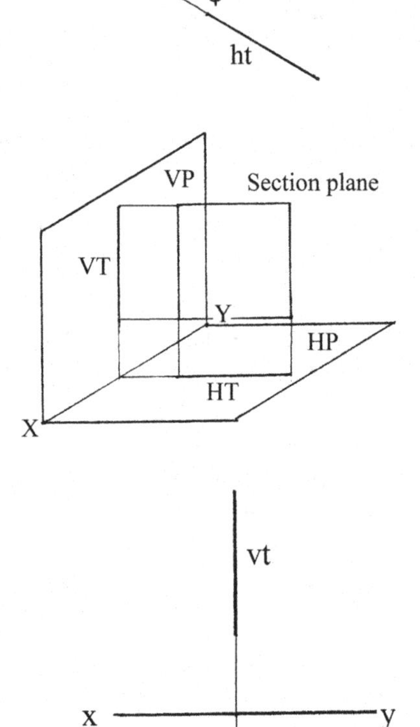

Fig. 11.8 Orthographic projection

Fig. 11.9 Pictorial representation

Fig. 11.10 Orthographic projection

Solved Problems

1. A triangular prism edge of base 30 mm and axis 60 mm long is lying on the HP on one of its rectangular faces with its axis inclined at 30° to the VP. It is cut by a section plane parallel to the VP passing through the mid-point of the axis. Draw its projections and the true shape of the section.

   The plan of the triangular prism will be a rectangle of sides 30 mm × 60 mm when it rests on its face on the HP. The axis is inclined at 30° to the VP. The plan is drawn with the longer sides of the rectangle inclined at 30° to xy as shown in Fig. 11.11. The line connecting the mid-points of the short edges represents the top edge of the prism. The height of this edge above xy is obtained from the equilateral triangle of sides 30 mm. The section plane, which is parallel to the VP, is drawn passing through the mid-point of the axis or the edge be in the plan. The section plane cuts the three longer edges of

Fig. 11.11 Sectional projections of a triangular prism

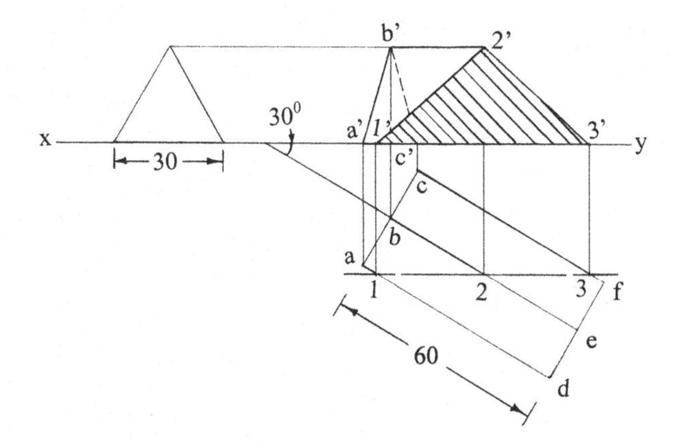

the prism. The intersection points are marked as 1, 2 and 3. These points are then projected in the elevation to get 1′, 2′ and 3′ on the respective edges. The front portion of the solid is removed. The sectioned surface is got by connecting the points 1′-2′-3′-1′. The enclosing area of the sectioned surface is hatched by drawing evenly spaced thin lines inclined at 45° to xy as shown in the figure. Since the section plane is parallel to the VP, the hatched area itself shows the true shape of the section in the elevation.

2. A cube of edge 40 mm stands on the HP with one of its faces inclined at 30° to the VP. It is cut by a section plane parallel to the VP and 10 mm in front of the axis. Draw its projections and the true shape of the section.

   The plan of the cube will be a square when it stands on the HP. Since one of the faces is inclined at 30° to the VP, the plan is drawn with a side of square inclined at 30° to xy as shown in Fig. 11.12. The elevation is projected from the plan. The section plane, which is parallel to the VP, is drawn 10 mm in front of the vertical axis of the cube. The section plane cuts the top end and bottom end of the cube. Hence, there will be four intersection points on the cube edges. Point 2 lies on the edge ad, point 3 lies on the edge cd, point 4 lies on the edge gh and point 1 lies on the edge eh. These points are then projected in the elevation to get 2′, 3′, 4′ and 1′ on the respective edges. The front portion of the cube is removed. The sectioned surface is obtained by connecting the points 1′-2′-3′-4′-1′. The sectioned surface is hatched by drawing evenly spaced thin lines inclined at 45° to the reference line xy. Since the section plane is parallel to the VP, the hatched area itself gives the true shape of the section.

3. A cube of edge 40 mm stands on the HP with one of its faces inclined at 30° to the VP. It is cut by a section plane inclined at 60° to the HP and perpendicular to the VP passing through a point on the axis 5 mm below the top end. Draw its projections and the true shape of the section.

   The plan of the cube will be a square when it stands on the HP. Since one of the faces is inclined at 30° to the VP, the plan is drawn with a side of square inclined at 30° to xy as shown in Fig. 11.13. The elevation is projected from the plan. The section plane is shown by its vertical trace. The vertical trace is

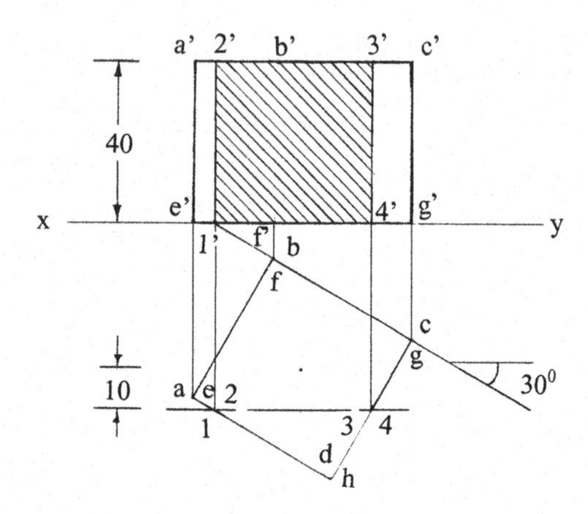

Fig. 11.12 Sectional projections of a cube

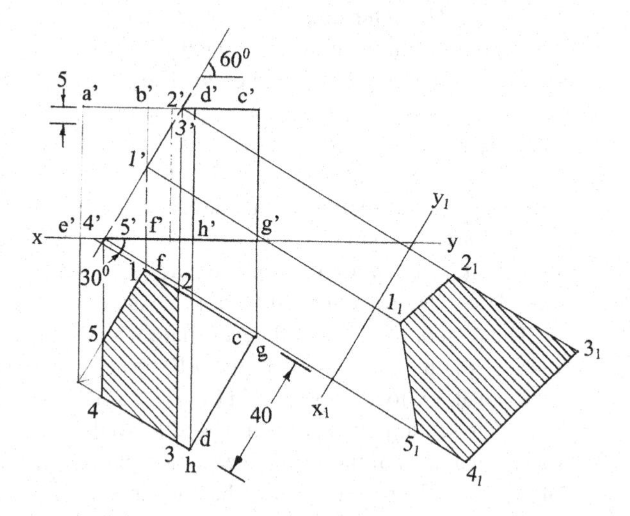

Fig. 11.13 Sectional projections of a cube

inclined at 60° to the reference line xy and passing through a point on the axis 5 mm below the top end. The section plane cuts the top end and bottom end besides the vertical edge b′f′. There are five intersection points on the cube. Point 1 lies on the edge b′f′, point 2′ lies on the edge b′c′, point 3′ lies on the edge d′a′, point 4′ lies on the edge e′h′ and point 5′ lies on the edge e′f′. These points are then projected to get 1, 2, 3, 4 and 5 on the respective edges in the plan. The top portion of the cube is removed. The sectioned surface is obtained by connecting the points 1-2-3-4-5-1. The sectioned surface is hatched by drawing thin lines inclined at 45° to xy. An auxiliary plane parallel to the section plane is chosen at x_1y_1. Projectors are drawn from 1′, 2′, 3′, 4′ and 5′. Point 1_1 is located on the projector drawn from 1′ such that the distance of 1_1 from x_1y_1 is equal to the distance of 1 from xy. The same procedure is followed to get the points 2_1, 3_1, 4_1 and 5_1. These points are then joined to show the true shape of the section. The sectioned surface is hatched by drawing thin lines inclined at 45° to the reference line xy.

4. A square prism 45 mm edge of base and 90 mm high rests with its base on the HP and a face inclined at 30° to the VP. A section plane perpendicular to the VP and inclined at 60° to the HP passes through a point 65 mm above the HP along the axis. Draw its sectional plan and true shape of the section.

The plan of the square prism is a square of 45 mm side with a side of base inclined at 30° to the VP. The plan is drawn with side bc inclined at 30° to xy as shown in Fig. 11.14. The elevation is projected from the plan. The section plane is represented by its vertical trace. The vertical trace is inclined at 60° to xy and passes through a point 65 mm above the base of the prism along the axis. The section plane cuts three vertical edges a′e′, b′f′ and d′h′ and passes through the top end of the prism. Hence, there will be three points on the vertical edges of the prism and two points on the top end of the prism. The intersection points are marked as 1′, 2′, 3′, 4′ and 5′ in the elevation. Points 3′ and 4′ lie on the top edges b′c′ and c′d′. These points are then projected to get points 1, 2, 3, 4 and 5 on the respective edges in the plan. The top portion of the prism is removed. The sectioned surface is got by connecting the points 1-2-3-4-5-1 in the plan. The enclosing area of the sectioned surface is hatched by drawing thin lines

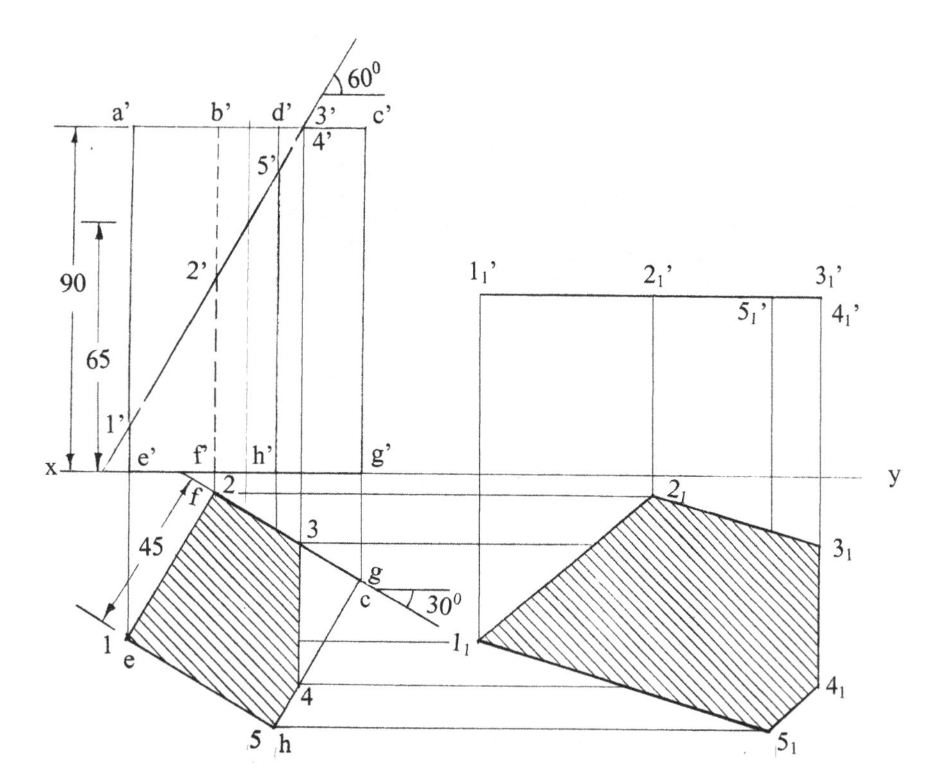

Fig. 11.14 Sectional projections of a square prism

inclined at 45° to xy. The elevation of the sectioned surface is kept horizontal by moving the section points parallel to the VP. The elevation $1_1'$-$2_1'$-$3_1'$-$4_1'$-$5_1'$-$1_1'$ is redrawn as shown in the figure. A projector is drawn perpendicular to xy from $1_1'$. The locus line is drawn parallel to xy from 1. The intersection of the projector with the locus line locates the point 1_1. The same procedure is followed to get the remaining points 2_1, 3_1, 4_1 and 5_1. These points are joined to get the true shape of the section 1_1-2_1-3_1-4_1-5_1-1_1 as shown in the figure. The true shape of the section is hatched by drawing thin lines inclined at 45° to xy.

5. A pentagonal prism, edge of base 40 mm and axis 70 mm long, lies with one face on the HP and the axis perpendicular to the VP. It is cut into two halves by a vertical section plane inclined at 60° to the axis. Draw the true shape of the section and a sectional elevation on a plane parallel to the cutting plane.

   The elevation of the prism will be a pentagon with a side on the reference line xy. The elevation is drawn as shown in Fig. 11.15. The plan is projected

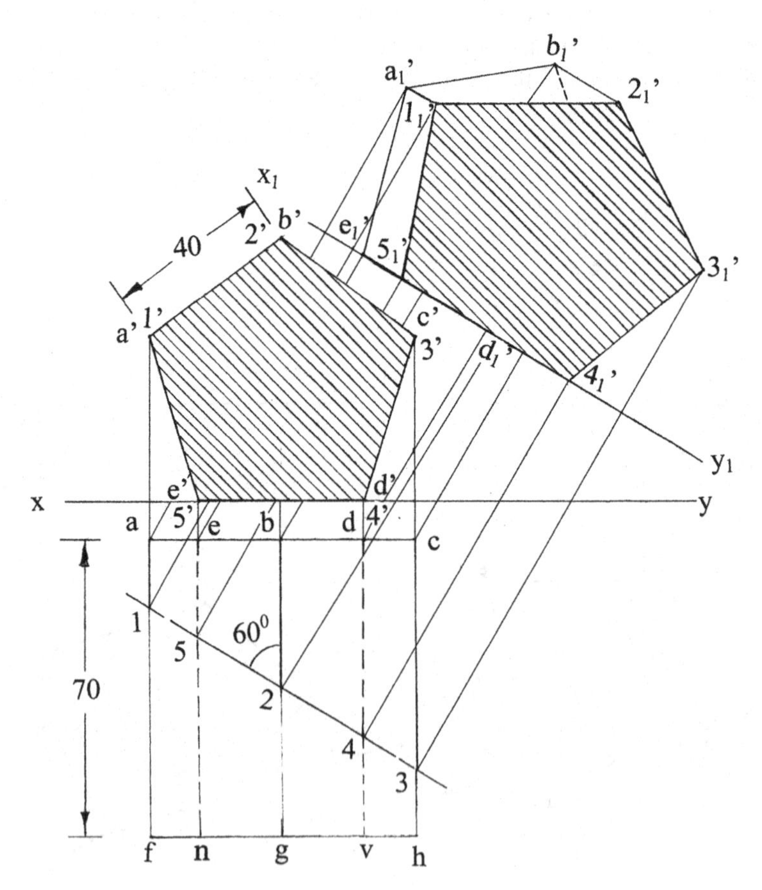

Fig. 11.15 Sectional projections of a pentagonal prism

from the elevation. The section plane is represented by its horizontal trace. The horizontal trace is inclined at 60° to the axis of the prism and passes through the mid-point of the axis. The section plane cuts the five horizontal edges as shown in the figure. The intersection points are marked as 1, 2, 3, 4 and 5 in the plan. These points are then projected to get 1′, 2′, 3′ 4′ and 5′ on the respective edges of the prism in the elevation. The front portion of the prism is removed. The apparent sectioned surface is obtained by joining the points 1′-2′-3′-4′-5′-1′. The sectioned surface is hatched by drawing thin lines inclined at 45° to xy. An auxiliary plane parallel to the section plane is chosen at x_1y_1 as shown in the figure. A projector is drawn perpendicular to x_1y_1 from point 1. The point $1_1'$ is located on this projector such that the distance of $1_1'$ from x_1y_1 is equal to the distance of 1′ from xy. Similarly, points $2_1'$, $3_1'$, $4_1'$, $5_1'$, a_1', b_1', c_1', d_1' and e_1' are located following the same procedure. The sectioned surface is obtained by connecting the points $1_1'$-$2_1'$-$3_1'$-$4_1'$-$5_1'$-$1_1'$. The sectioned surface is hatched by drawing thin lines inclined at 45° to xy. The end edges and the horizontal edges are joined to complete the auxiliary projection of the prism. The invisible edges are shown by dotted lines avoiding the sectioned surface as shown in the figure.

6. A pentagonal prism, edge of base 30 mm and axis 60 mm long, rests with one of its longer edges on the HP. The two rectangular faces containing the longer edge on which it rests make equal inclinations with the HP, and the longer edges are inclined at 50° to the VP. The horizontal trace of the section plane parallel to the VP passes through the mid-point of the axis. Draw the sectional view of the prism and the true shape of the section.

   The elevation and the plan of the prism are drawn keeping the axis perpendicular to the VP with a longer edge on the HP and the two rectangular faces containing this edge equally inclined to the HP as shown in Fig. 11.16. The plan is redrawn keeping the longer edges inclined at 50° to xy. The elevation is projected from the plan. The mid-point of the edge en is located, and the section plane parallel to the VP passes through this mid-point. The horizontal trace is drawn parallel to xy and passing through the mid-point of the edge en. The section plane cuts five longer edges. The section points are marked as 1, 2, 3, 4 and 5 in the plan. These points are then projected to get 1′, 2′, 3′, 4′ and 5′ on the respective edges in the elevation. The front portion of the prism is removed. The sectioned surface is obtained by connecting the points 1′-2′-3′-4′-5′-1′ and is hatched by drawing thin lines inclined at 45° to xy. Since the section plane is parallel to the VP, the sectioned surface in the elevation itself gives the true shape of the section.

7. A hexagonal prism, edge of base 25 mm and axis 65 mm long, has a face on the HP and the axis parallel to the VP. It is cut by a section plane whose horizontal trace makes an angle of 45° with xy and passes through a point 20 mm along the axis from one of its ends. Draw its sectional elevation and true shape of the section.

   The axis of the prism is parallel to the VP with a face on the HP. Hence, its left side elevation is drawn first as shown in Fig. 11.17. The elevation and the plan are projected from the side elevation. The section plane is represented by

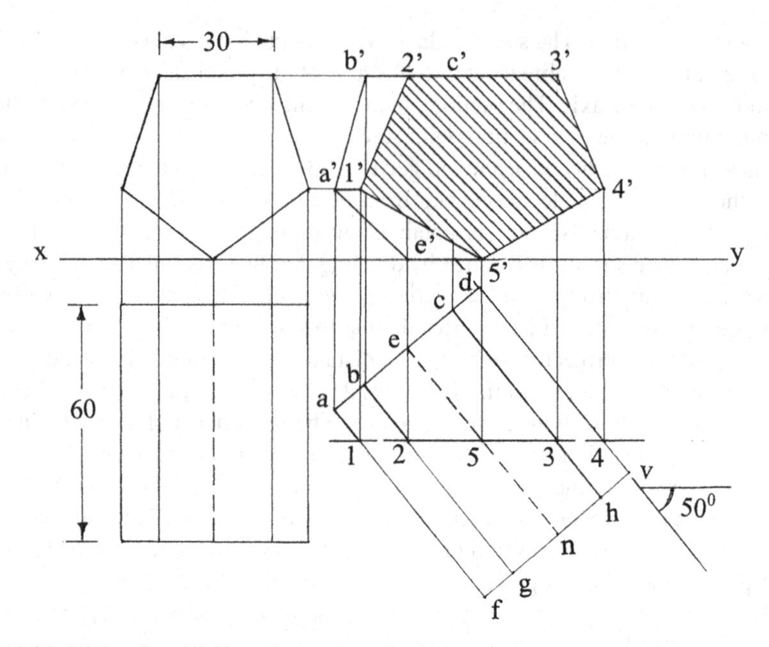

Fig. 11.16 Sectional projections of a pentagonal prism

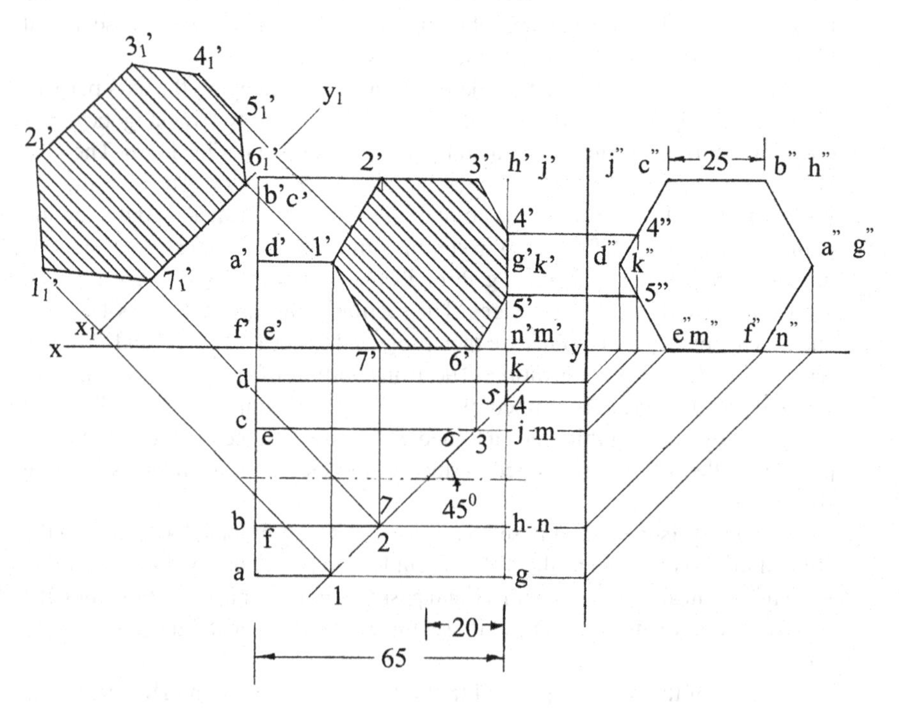

Fig. 11.17 Sectional projections of a hexagonal prism

its horizontal trace which makes an angle of 45° with xy and passes through a point 20 mm from its right end on its axis as shown in the figure. The section plane cuts five horizontal longer edges and two shorter edges on the right end of the prism. The intersection points are marked as 1, 2, 3, 4, 5, 6 and 7 in the plan. Points 4 and 5 lie on the edges jk and km. The points 1′, 2′, 3′, 6′ and 7′ are located directly on the respective longer edges of the prism in the elevation. Since the plan of the edges jk and km are perpendicular to xy, points 4′ and 5′ cannot be projected directly from the plan. These points are projected on the side elevation and are marked as 4″ and 5″. Locus lines are drawn parallel to xy from 4″ and 5″. These lines intersect the edges j′k′ and k′m′ and locate the points 4′ and 5′ in the elevation of the prism. The portion of the prism lying in front of the section plane is removed so that the sectioned surface will be visible in the elevation of the prism. The sectioned surface is obtained by connecting the points 1′-2′-3′-4′-5′-6′-7′-1′. The enclosing area is hatched by drawing thin lines inclined at 45° to xy. An auxiliary plane parallel to the section plane is chosen at x_1y_1 as shown in the figure. A projector is drawn perpendicular to x_1y_1 from point 1. Point $1_1'$ is located on this projector such that the distance of $1_1'$ from x_1y_1 is equal to the distance of 1′ from xy. The remaining points $2_1'$, $3_1'$, $4_1'$, $5_1'$, $6_1'$ and $7_1'$ are located following the same procedure. These points are connected to show the true shape of the section formed. The sectioned surface is hatched by drawing thin lines inclined at 45° to xy.

8. A tetrahedron of edge 60 mm is placed on the HP with one of its edges parallel to the VP. It is cut by a vertical section plane whose horizontal trace is inclined at 60° to xy and 10 mm from the axis. Draw its projections and true shape of the section.

   The plan of the tetrahedron is drawn as shown in Fig. 11.18. The elevation is projected from the plan. An arc of radius 10 mm is drawn with centre d. The horizontal trace of the section plane inclined at 60° to xy is drawn tangential to the arc as shown in the figure. The section plane cuts four edges of the tetrahedron. The intersection points are marked as 1, 2, 3 and 4 in the plan. Points 1′, 2′ and 4′ are projected on the respective edges in the elevation. Point 3 lies on the edge dc which is perpendicular to xy. With centre d and radius d3, an arc is drawn to cut the edge db at e as shown in the figure. A projector is drawn perpendicular to xy from e, and the intersection of the projector with d′b′ locates e′ in the elevation. A horizontal line is drawn from e′ to locate 3′ on the edge d′c′. The portion of the solid in front of the section plane is removed. The sectioned surface is obtained by connecting the points 1′-2′-3′-4′-1′ in the elevation. The sectioned surface is hatched by drawing thin lines inclined at 45° to xy. An auxiliary plane parallel to the section plane is chosen at x_1y_1. A projector is drawn perpendicular to x_1y_1 from point 2. Point $2_1'$ is located on this projector such that the distance of $2_1'$ from x_1y_1 is equal to the distance of 2′ from xy. The remaining points $1_1'$, $3_1'$ and $4_1'$ are located following the same procedure. These points are connected to show the true shape of the section formed. The sectioned surface is hatched by drawing thin lines inclined at 45° to xy.

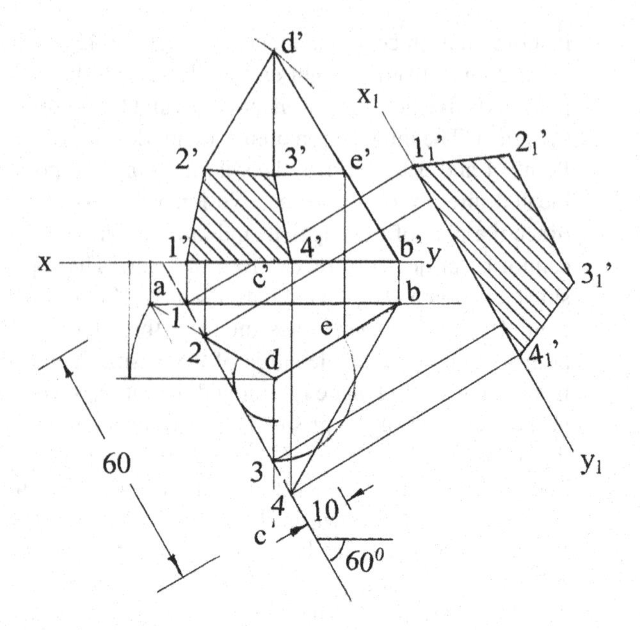

Fig. 11.18 Sectional projections of a tetrahedron

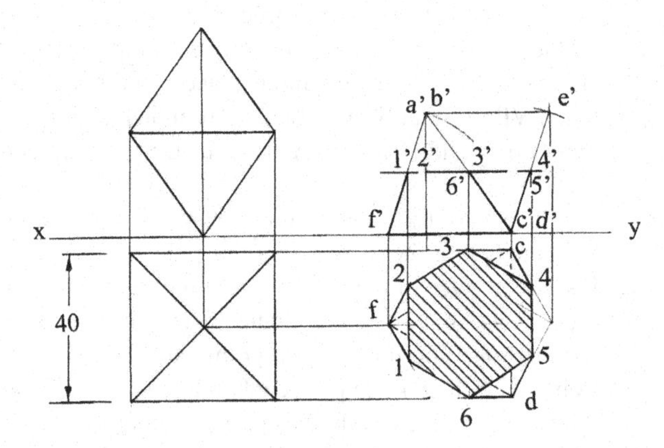

Fig. 11.19 Sectional projections of an octahedron

9. An octahedron of edge 40 mm rests with one of its faces on the HP. It is cut by a section plane parallel to the HP passing through the mid-point of its solid diagonal/axis. Draw the sectional plan and true shape of the section.

The plan of the octahedron will be a square if one of its axis or solid diagonal is vertical. Since the octahedron is to rest on one of its faces, the plan is drawn with a side of square parallel to xy as shown in Fig. 11.19. The elevation is projected from the plan. The elevation is redrawn such that the face c′d′f′ lies on xy. The plan is projected from the elevation. The horizontal section plane passes through the mid-point of the axis or the solid diagonal. The vertical trace of the section plane, which passes through the mid-point of e′f′, is drawn. The section plane cuts six edges of the octahedron as shown in the figure. These points are marked as 1′, 2′, 3′, 4′, 5′ and 6′ (1′ lies on a′f′, 2′ on bf′, etc.) in the elevation. These points are then projected to locate 1, 2, 3, 4, 5 and 6 on the respective edges in the plan. The top portion of the solid is removed. The sectioned surface is obtained by connecting the points 1-2-3-4-5-6-1. The sectioned surface is hatched by drawing thin lines inclined at 45° to xy. Since the section plane is horizontal, the sectioned surface in the plan itself shows the true shape of the section formed.

10. A square pyramid base 40 mm side and axis 60 mm long has its base on the HP and base edges equally inclined to the VP. It is cut by a section plane inclined at 30° to the HP and bisecting the axis. Draw its sectional plan, sectional end elevation and true shape of the section.

The plan of the square pyramid resting on the HP with its base is a square and is drawn with the base edges equally inclined to xy (inclined at 45° to xy) as shown in Fig. 11.20. The elevation is projected from the plan. The side elevation is drawn from the plan and elevation of the pyramid by drawing locus lines parallel to xy and the 45° inclined lines as shown in the figure. The section plane is represented by its vertical trace. The vertical trace is inclined at 30° to xy and passes through the mid-point of the axis in the elevation. The section plane cuts the four slant edges. The intersection points are marked as 1′, 2′, 3′ and 4′ in the elevation. Points 1′ and 3′ are projected to get points 1 and 3 on the respective edges in the plan. Points 2′ and 4′ lie on the slant edges whose projections are perpendicular to xy. A horizontal line is drawn from 2′ and 4′ to intersect the slant edge e′c′ at g′. A projector is drawn perpendicular to

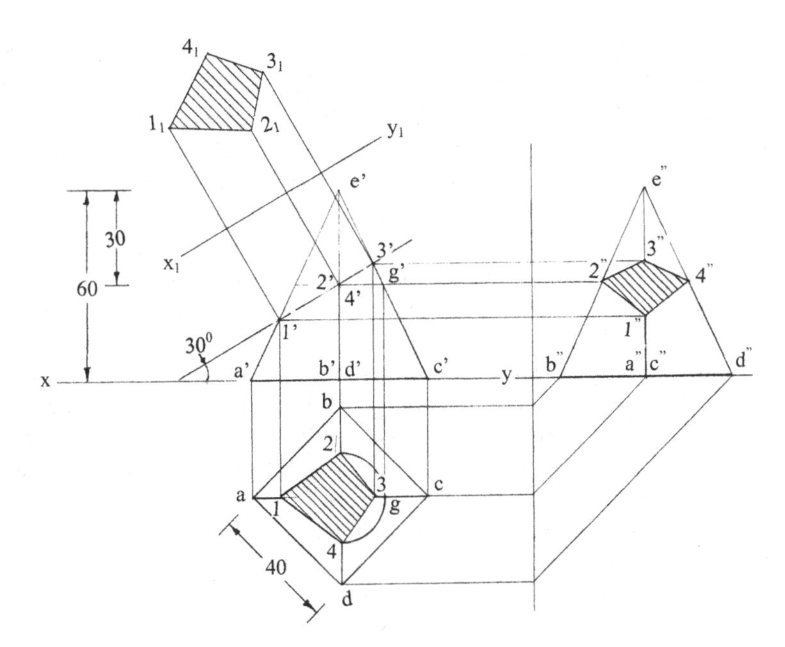

Fig. 11.20 Sectional projections of square pyramid

xy from g' to intersect the edge ec at g. With centre e and eg as radius, draw an arc to cut the edges eb and ed to locate the points 2 and 4 as shown in the figure. The top portion of the pyramid is removed. The sectioned surface is obtained by connecting the points 1-2-3-4-1 in the plan. The sectioned surface is hatched by drawing thin lines inclined at 45° to xy. The section points are projected in the side elevation by drawing locus lines as shown in the figure. The sectioned surface in the side elevation is obtained by connecting the points $1''$-$2''$-$3''$-$4''$-$1''$. The sectioned surface in the side elevation is hatched by drawing thin lines inclined at 45° to xy. An auxiliary plane parallel to the section plane is chosen at x_1y_1. A projector is drawn perpendicular to x_1y_1 from $1'$. Point 1_1 is located on this projector such that the distance of 1_1 from x_1y_1 is equal to the distance of 1 from xy. The remaining points 2_1, 3_1 and 4_1 are obtained following the same procedure. These points are joined to show the true shape of the section. The sectioned surface is hatched by drawing thin lines inclined at 45° to xy.

11. A right square pyramid, side of base 40 mm and axis 80 mm long, is resting on the HP on one of its longer edges such that the sides of the base are equally inclined to the HP and the axis is parallel to the VP. It is cut by a vertical section plane parallel to the axis and at a distance of 10 mm in front of it. Draw the plan and sectional elevation.

    The square pyramid is kept initially on the HP on its base with the base edges equally inclined to the VP. Then the pyramid is tilted about the right extreme corner of its base such that the slant longer edge containing this corner lies on the HP. In this position, the base edges are equally inclined to the HP and the axis of the pyramid is parallel to the VP.

    The plan of the square pyramid is drawn in its simple position with the base edges equally inclined to xy as shown in Fig. 11.21. The elevation is projected from the plan. The pyramid is tilted about the right extreme base corner so that the longer edge containing this corner touches the HP. All the points move parallel to the VP. The elevation is redrawn such that the edge $b'v'$ coincides with the reference line xy. The plan is projected from the elevation. The horizontal trace of the vertical section plane is drawn 10 mm in front of the axis of the pyramid. The section plane cuts two base edges and one longer slant edge of the pyramid. The intersection points are marked as 1, 2 and 3 as shown in the figure. These points are projected to get $1'$, $2'$ and $3'$ on the respective edges in the elevation. The portion of the pyramid in front of the section plane is removed. The sectioned surface is obtained by connecting the points $1'$-$2'$-$3'$-$1'$. The sectioned surface is hatched by drawing thin lines inclined at 45° to xy. A portion of the invisible edge $c'v'$, which lies inside the sectioned surface, is not shown as dotted.

12. A pyramid has for its base a square of 50 mm side which is on the horizontal plane. The altitude of the pyramid is 70 mm and the plan of the vertex is at the middle point of one side of the plan of the base. This pyramid is cut into two portions by a horizontal section plane which is 25 mm above the base. Draw the plan of the lower portion.

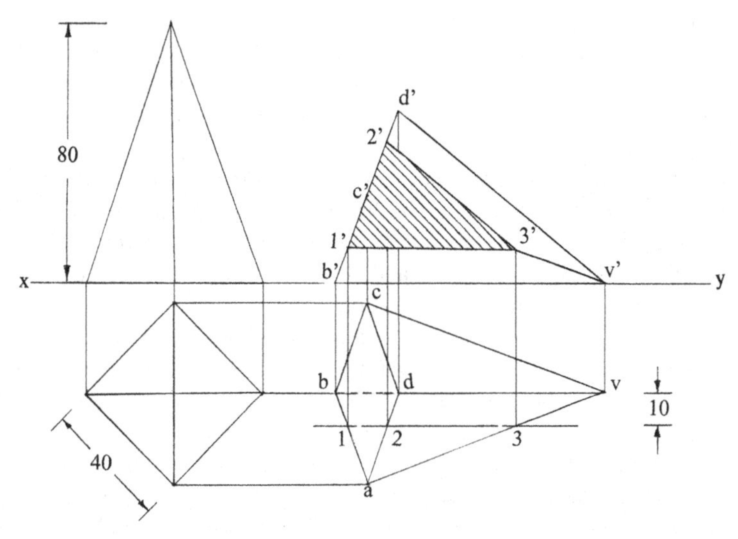

Fig. 11.21 Sectional projections of a square pyramid

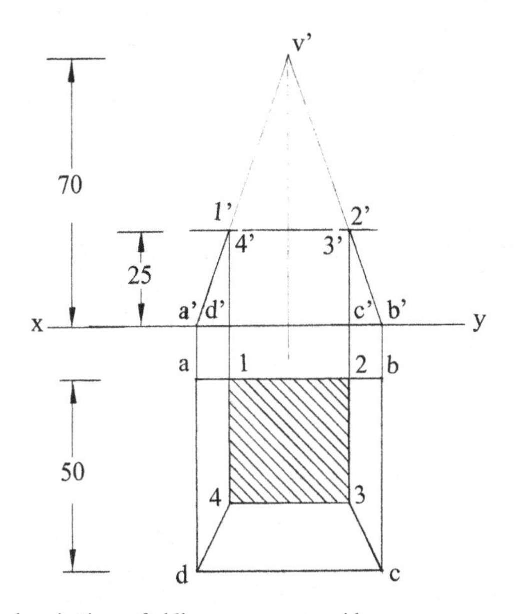

Fig. 11.22 Sectional projections of oblique square pyramid

The plan of the oblique pyramid is drawn as a square of side 50 mm as shown in Fig. 11.22. The plan of the vertex is chosen at the middle point of the side ab. The elevation is projected from the plan. The vertical trace of horizontal section plane is drawn 25 mm above the base in the elevation as shown in the figure. The section plane cuts the four slant edges. The intersection points are marked as 1′, 2′, 3′ and 4′ in the elevation. These points are projected to get 1, 2, 3 and 4 on the respective edges in the plan. The top portion of the pyramid is removed. The sectioned surface is obtained by connecting the points 1-2-3-4-1 in the plan. The sectioned surface is hatched by drawing thin lines inclined at 45° to xy.

13. A pentagonal pyramid edge of base 30 mm and axis 60 mm stands on the HP with an edge of base parallel to the VP. It is cut by a section plane inclined at 45° to the HP passing through one of the extreme angular points of the base of pyramid. Draw the sectional plan and also the plan of the truncated pyramid on an auxiliary plane parallel to the section plane.

    The plan of the pentagonal pyramid is drawn with a side of base parallel to xy as shown in Fig. 11.23. The elevation is projected from the plan. The vertical trace of the section plane is drawn inclined at 45° to xy and passing through the left extreme corner a′. The section plane cuts four slant edges and passes through the angular point a′. The intersection points are marked as 1′, 2′, 3′, 4′ and 5′ in the elevation. These points are projected to get 1, 2, 3, 4 and 5 on the respective edges in the plan (construction is shown for point 5). The top portion of the pyramid is removed. The sectioned surface is obtained by connecting the points 1-2-3-4-5-1 in the plan. The sectioned surface is hatched by drawing thin lines inclined at 45° to xy. An auxiliary plane parallel to the section plane is chosen at x_1y_1. A projector is drawn perpendicular to x_1y_1 from 1′. Point 1_1 is located on this projector such that the distance of 1_1 from x_1y_1 is equal to the distance of 1 from xy. Points 2_1, 3_1, 4_1, 5_1, b_1, c_1, d_1 and e_1 are located on the

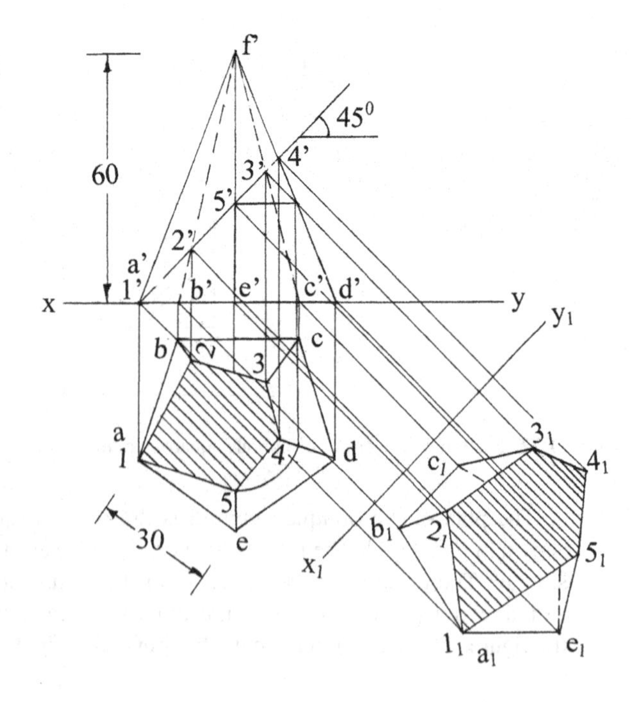

Fig. 11.23 Sectional projections of a pentagonal pyramid

respective projectors following the same procedure. The sectioned surface is obtained by connecting the points 1_1-2_1-3_1-4_1-5_1-1_1. The sectioned surface is hatched by drawing thin lines inclined at 45° to xy. The remaining edges of the truncated pyramid are also joined. The invisible edges c_1d_1 and d_1e_1 are shown as dotted avoiding the sectioned surface.

14. A hexagonal pyramid edge of base 30 mm and axis 50 mm rests with a triangular face on the HP and the axis is parallel to the VP. It is cut by a section plane parallel to the HP and passing through the centre of base of the pyramid. Draw the sectional plan.

    The plan of the hexagonal pyramid resting on the HP on its base is drawn keeping a side of base perpendicular to xy as shown in Fig. 11.24. The elevation is projected from the plan. The pyramid is tilted about the right extreme base edge such that the triangular face containing this edge lies on xy. All points move parallel to the VP. The elevation is redrawn as shown in the figure. The plan is projected from the elevation. The vertical trace of the horizontal section plane is drawn passing through the centre of the base of the pyramid. The section plane cuts two slant edges a′v′ and b′v′ and passes through the angular points c′ and f′. The intersection points are marked as 1′, 2′, 3′ and 4′ in the elevation. These points are projected to get 1, 2, 3 and 4 on the respective edges and the angular points in the plan as shown in the figure. The top portion of the pyramid is removed. The sectioned surface is obtained by connecting the points 1-2-3-4-1 in the plan. The sectioned surface is hatched by drawing thin lines inclined at 45° to xy. The invisible edges dv and ev are shown as dotted avoiding the sectioned surface.

15. A regular hexagonal pyramid, edge of base 40 mm and axis 70 mm, is standing on a base edge on the HP with the triangular face containing that edge

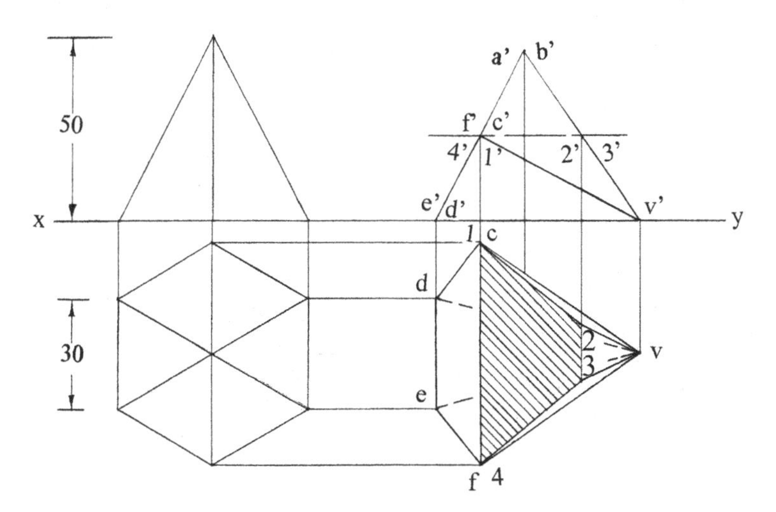

Fig. 11.24 Sectional projections of a hexagonal pyramid

perpendicular to both the HP and VP. A horizontal section plane cuts the solid at a height of 40 mm above the HP. Draw the sectional plan and elevation.

The plan of the hexagonal pyramid resting on the HP on its base is drawn keeping a side of base perpendicular to xy as shown in Fig. 11.25. The elevation is projected from the plan. The pyramid is tilted about the right extreme base edge such that the triangular face containing this edge is perpendicular to both the HP and VP. The elevation is redrawn as shown in the figure. The plan is projected from the elevation. The vertical trace of the horizontal section plane is drawn 40 mm above xy. The section plane cuts the six slant edges of the pyramid. The intersection points are marked as $1'$, $2'$, $3'$, $4'$, $5'$ and $6'$. Projectors are drawn from $1'$, $2'$, $3'$ and $6'$ to locate the points 1, 2, 3 and 6 on the respective edges in the plan. Since the elevations of edges v′d′ and v′e′ are perpendicular to xy, the points $4'$ and $5'$ lying on these edges are located in the elevation previously drawn as shown in the figure. Projectors are drawn from $4'$ and $5'$ to locate the points 4 and 5 on the edges vd and ve in the plan previously drawn. Locus lines are drawn from 4 and 5 to locate the same points in the sectional plan of the pyramid. The top portion of the solid is removed. The sectioned surface is obtained by connecting the points 1-2-3-4-5-6-1. The sectioned surface is hatched by drawing thin lines inclined at 45° to xy.

16. A vertical cylinder 40 mm diameter is cut by an auxiliary vertical plane making an angle of 30° to the VP in such a way that the true shape of the section is a rectangle of sides 25 mm × 65 mm. Draw its projections and the true shape of the section.

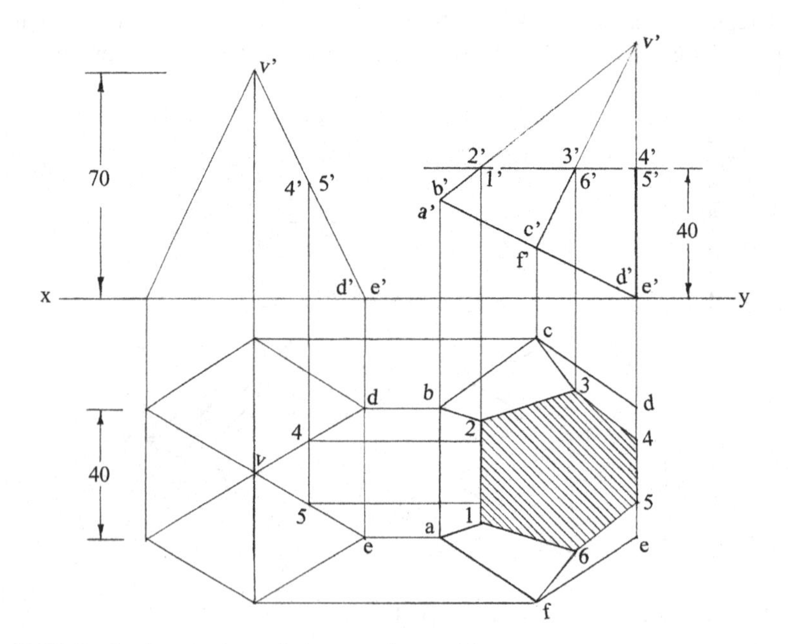

Fig. 11.25 Sectional projections of a hexagonal pyramid

When a vertical cylinder is cut by an auxiliary vertical plane parallel to the axis of the cylinder, the true shape of the section obtained will always be a rectangle. The diameter of the cylinder is 40 mm. The given dimensions of the rectangle are 25 mm × 65 mm. Hence, the height of the cylinder is 65 mm. Draw a line inclined at 30° to xy and locate the chord of the circle 25 mm long as shown in Fig. 11.26. Draw the perpendicular bisector of the chord and locate the centre of the circle of diameter 40 mm. The plan of the cylinder is a circle of diameter 40 mm. The plan is drawn knowing the circle centre. The elevation is projected from the plan. The inclined line represents the horizontal trace of the section plane. The section plane cuts the top and bottom ends of the cylinder. Hence, there are two points on the circumference of top end and two points on the bottom end of the cylinder. These points are marked as 1, 2, 3 and 4. Projectors are drawn from the points 1, 2, 3 and 4. The intersections of the projectors with top and bottom ends of the cylinder locate the points 1′, 2′, 3′ and 4′ in the elevation. The portion of the cylinder lying in front of the section plane is removed. The sectioned surface is obtained by connecting the points 1′-2′-3′-4′-1′ in the elevation. The sectioned surface is hatched by drawing thin lines inclined at 45° to xy. An auxiliary plane parallel to the section plane is chosen at x_1y_1. A projector perpendicular to x_1y_1 is drawn from point 2. Point $2_1'$ is located on this projector such that the distance of $2_1'$ from x_1y_1 is equal to the distance of 2′ from xy. The remaining points are located following the same procedure. The sectioned true shape is obtained by connecting the points

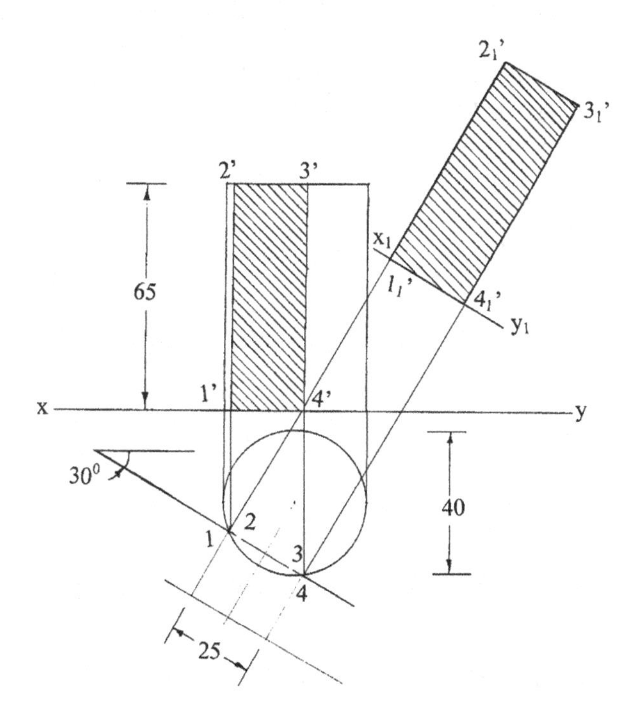

Fig. 11.26 Sectional projections of a cylinder

$1_1'$-$2_1'$-$3_1'$-$4_1'$-$1_1'$. The sectioned surface is hatched by drawing thin lines inclined at 45° to xy.

17. A cylinder, base 50 mm diameter and axis 80 mm long, has a square hole of 25 mm side cut through it so that the axis of the hole coincides with the axis of the cylinder. The cylinder is lying on the HP with the axis perpendicular to the VP and the faces of the hole equally inclined to the HP. The cylinder is cut in halves by a vertical plane which is inclined at 30° to the VP. Draw the elevation of the cylinder on an auxiliary plane parallel to the section plane.

    The elevation of the cylinder is drawn on xy with the sides of square hole equally inclined to the HP as shown in Fig. 11.27. The plan is projected from the elevation. Eight generators are shown on the lateral surface of the cylinder by dividing the circle into eight equal parts in the elevation and projecting them in the plan (only the extreme two generators alone are shown in the diagram). The horizontal trace of the vertical section plane inclined at 30° to xy is drawn passing through the mid-point of the cylinder. The section plane cuts eight generators on the surface of the cylinder and four longer edges of the hole. The 12 points are marked in the plan as shown in the figure. The intersection points 1, 2, 3 and 4 lie on the edges of square hole, and the remaining points lie on the outer surface of the cylinder. These points are projected to get 1′, 2′, 3′, etc. on the respective edges/generators in the elevation. The portion of the solid lying in front of the section plane is removed. The area lying between the surfaces 1′-2′-3′-4′-1′ and 5′-6′-7′-8′-9′-10′-11′-12′-5′ is the required sectioned surface. An auxiliary projection of the cut cylinder is required on a plane parallel to vertical section plane. The cut cylinder is moved parallel to the HP such the sectioned surface is parallel to the VP instead of choosing an auxiliary plane parallel to the section plane. The sectional plan of the cut cylinder is redrawn keeping the sectioned surface parallel to xy as shown in the figure. Projectors are drawn (perpendicular to xy) from the intersection points and also the end

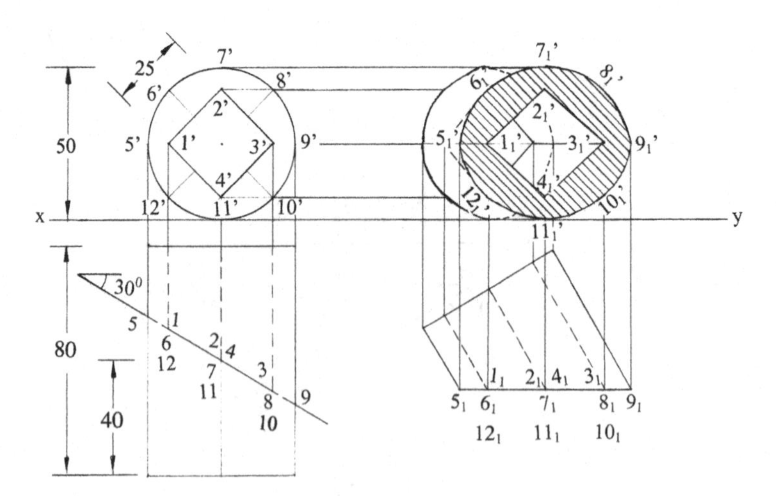

Fig. 11.27 Sectional projections of a cylinder with a square hole

points of the cylinder in the plan redrawn. Locus lines are drawn parallel to xy from the intersection points and the end points of the cylinder in the elevation. The intersections of the projectors with the locus lines locate the points $1_1'$, $2_1'$, $3_1'$, $4_1'$, $5_1'$, etc. The intersection points are joined in the new elevation, and the sectioned surface lying in between the surfaces 1′-2′-3′-4′-1′ and 5′-6′-7′-8′-9′-10′-11′-12′-5′ is hatched by drawing thin lines inclined at 45° to xy. The end points are also joined in the elevation, and the invisible portion of the curve and edges of the hole are shown as dotted avoiding the sectioned surface.

18. A right circular cone of 40 mm diameter of the base and 60 mm altitude stands on the HP. A plane normal to the HP and inclined at 45° to the VP cuts the cone at a distance of 10 mm from its axis. Draw the sectional elevation and true shape of the section.

    The plan and the elevation of the cone resting on the HP on its base are drawn as shown in Fig. 11.28. The horizontal trace of section plane inclined at 45° to xy and perpendicular to the HP is drawn at a distance of 10 mm from the axis of the cone. Five generators fa, fb, fc, fd and fe are chosen on the surface of cone as shown in its plan. The generator fc is kept perpendicular to the horizontal trace of the section plane. The elevations of the generators f′a′, f′b′, f′c′, f′d′ and f′e′ are projected from the plan. The section plane cuts these generators as shown in the plan. The intersection points 1, 2, 3, 4 and 5 are marked in the plan. These points are projected to get 1′, 2′, 3′, 4′ and 5′ on the respective generators in the elevation. The portion of the cone lying in front of the section plane is removed. The sectioned surface is obtained by connecting the points 1′-2′-3′-4′-5′-1′ in the elevation. The sectioned surface is hatched by drawing thin lines inclined at 45° to xy. An auxiliary plane parallel to the section plane is chosen at x_1y_1. A projector is drawn perpendicular to x_1y_1 from the point 2 in the plan. Point $2_1'$ is located on this projector such that the

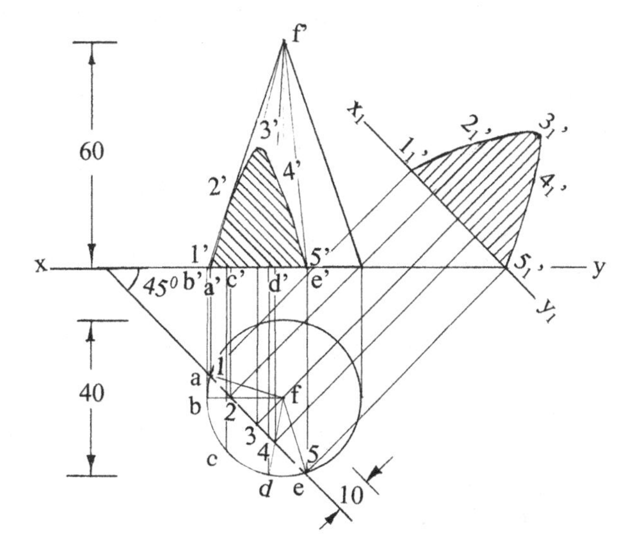

Fig. 11.28 Sectional projections of a cone

distance of $2_1'$ from x_1y_1 is equal to the distance of 2′ from xy. Points $1_1'$, $3_1'$, $4_1'$ and $5_1'$ are located following the same procedure. These points are connected to show the true shape of the section. The sectioned surface is hatched by drawing thin lines inclined at 45° to xy.

19. A right circular cone, 60 mm of base diameter and 80 mm in altitude, is resting with its base on the HP. It is cut by a plane parallel to one of the generators of the cone bisecting the axis. Draw the sectional plan and true shape of the section.

    The plan and elevation of the cone resting on the HP on its base are drawn as shown in Fig. 11.29. Eight generators at equal intervals are shown in the plan naming alternate generators as fa, fb, fc and fd. These generators are then projected in the elevation. The vertical trace of the section plane parallel to the left extreme generator f′a′ is drawn passing through the mid-point of the axis. The section plane cuts five generators and passes through the base of the cone. There are seven intersection points on the cone. The intersection points are marked as 1′, 2′, 3′, 4′, 5′, 6′ and 7′ in the elevation. These points are to be projected in the plan. Projectors are drawn from 1′ and 7′ to intersect the base of the cone at points 1 and 7 in the plan. Projectors are drawn from 3′, 4′ and 5′ to intersect the corresponding generators at 3, 4 and 5 in the plan. Since the points 2′ and 6′ lie on the generators f′b′ and f′d′, whose projections are perpendicular to the reference line xy, horizontal lines are drawn from these two points to intersect the right extreme generator f′c′ at v′. A projector is drawn from v′ to

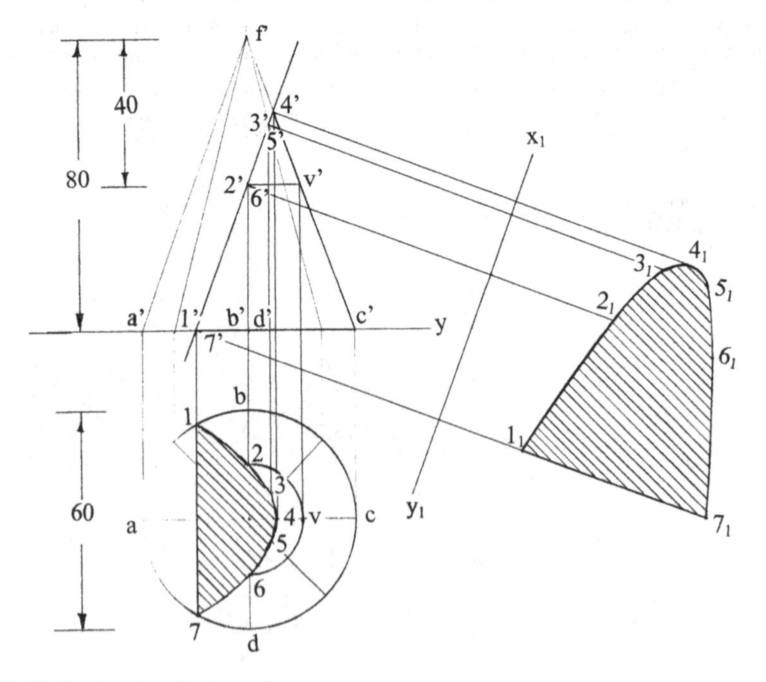

Fig. 11.29 Sectional projections of a cone

intersect the projector fc at v in the plan. With centre f and fv as radius, an arc is drawn to intersect the generators fb and fd at 2 and 6 as shown in the figure. The top portion of the cone is removed. The sectioned surface is obtained by connecting the points 1-2-3-4-5-6-7-1 in the plan. The sectioned surface is hatched by drawing thin lines inclined at 45° to xy. An auxiliary plane parallel to the section plane is chosen at x_1y_1. A projector perpendicular to x_1y_1 is drawn from point $1'$. Point 1_1 is located on this projector such that the distance of 1_1 from x_1y_1 is equal to the distance of 1 from xy. Points $2_1, 3_1, 4_1, 5_1, 6_1$ and 7_1 are located following the same procedure. The true shape of the section is obtained by connecting points 1_1-2_1-3_1-4_1-5_1-6_1-7_1-1_1. The sectioned surface is hatched by drawing thin lines inclined at 45° to xy.

20. A plastic bucket of height 4000 mm, bottom and top diameters of 2000 mm and 3000 mm, respectively, contains certain quantity of water. When the bucket is tilted about the base rim through an angle of 45°, the liquid is about to ooze out from the bucket. Draw the projections of the bucket in this inclined position and also the outline of water surface. Adopt a scale 1:50.

    The plan of the bucket is drawn adopting a scale of 1:50 as shown in Fig. 11.30. Eight generators are chosen by dividing the circumference of the outer circle which represents the top edge of the bucket into eight equal parts. The eight lines connecting the outer circle with the inner circle represent the eight generators on the lateral conical surface of the bucket. The elevation of the bucket is projected from the plan with the eight generators. The bucket is

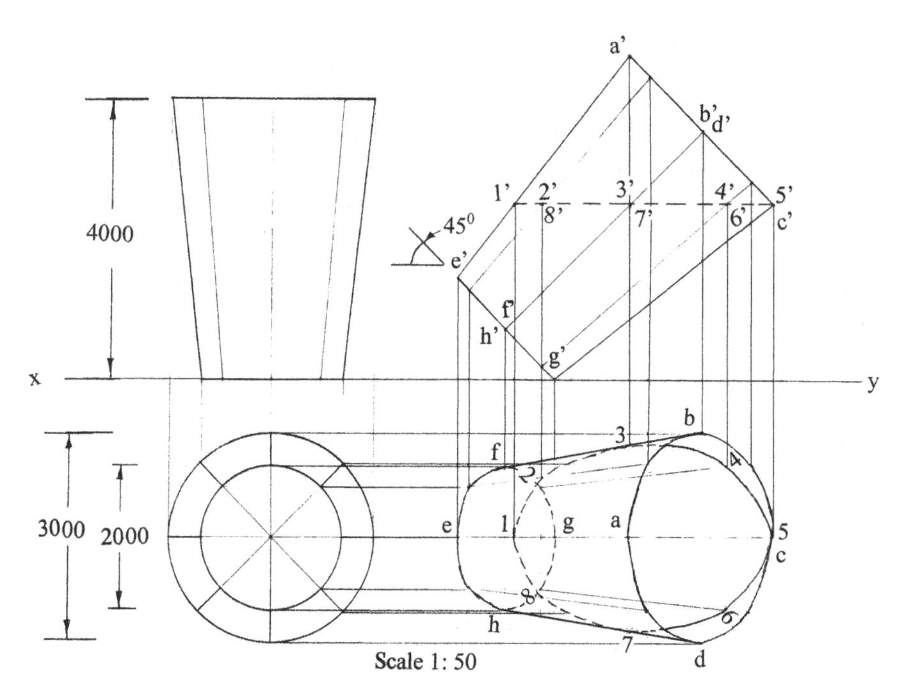

Fig. 11.30 Projections of a bucket

tilted about the right extreme base point such that the base is inclined at 45° to the horizontal. All the points move parallel to the VP. The elevation is redrawn with the base inclined at 45° to xy, and the alternate generators are named as a′e′, b′f′, c′g′ and d′h′. The points a′, b′, c′ and d′ lie on the top edge of the bucket. The points e′, f′, g′ and h′ lie on the bottom edge of the bucket. Projectors are drawn perpendicular to xy from the generator ends on the top of the bucket. Locus lines are drawn parallel to xy from the plan of generator ends on the outer circle. The intersection points locate the points on the top rim of the bucket. The ellipse is drawn passing through the intersection points. The projection of the base of the bucket is obtained following the same procedure. The eight generating lines are drawn connecting the respective points on the top and bottom ends of the bucket. The alternate generating lines are identified as a-e, b-f, c-g and d-h in the plan. A horizontal line passing through the point c′ is drawn in the elevation. This line represents the water surface on the point of oozing out from the bucket. The horizontal line cuts the eight generators. The intersection points are marked as 1′, 2′, 3′, 4′, 5′, 6′, 7′ and 8′. Point 1′ lies on the generator a′e′, point 3′ lies on the generator b′f′, point 5′ lies on the generator c′g′ and point 7′ lies on the generator d′h′. A projector is drawn from 1′ to intersect the generator ae in the plan and locates point 1. Points 2, 3, 4, 5, 6, 7 and 8 are located on the respective generators following the same procedure. The outline of water surface is obtained by connecting the points 1-2-3-4-5-6-7-8-1 in the plan. The invisible portion of the ellipse representing the outline of the water surface is shown as dotted.

21. A cone diameter of base 60 mm and axis 70 mm long is resting on its base on the HP. It is cut by an auxiliary plane inclined such that the true shape of section is an isosceles triangle of 50 mm base. Draw the sectional plan and elevation of the cone. Also show the true shape of the section.

    The plan and elevation of the cone resting on the HP on its base are drawn as shown in Fig. 11.31. When the cone is cut by a section plane passing through its apex, the true shape of the section formed shall always be an isosceles triangle.

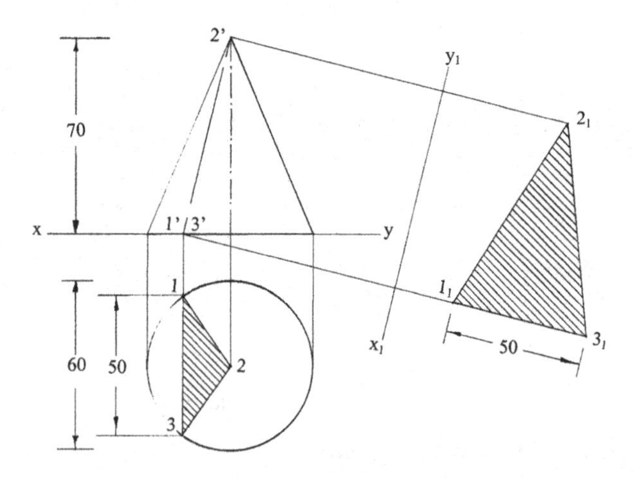

Fig. 11.31 Sectional projections of a cone

A chord of length 50 mm is drawn for the circle as shown in the figure. This chord is projected in the elevation as a point, and the vertical trace of the section plane passes through this point and the apex of the cone. The section plane cuts the base of the cone and passes through its apex. There are three points on the sectioned surface. The intersection points are marked as $1'$, $2'$ and $3'$ in the elevation. These points are then projected to get 1, 2 and 3 in the plan. The left portion of the cone is removed. The sectioned surface is obtained by connecting the points 1-2-3-1 in the plan. The sectioned surface is hatched by drawing thin lines inclined at 45° to xy. An auxiliary plane parallel to the section plane is chosen at x_1y_1. A projector perpendicular to x_1y_1 is drawn from point $2'$ in the elevation. Point 2_1 is located on this projector such that the distance of 2_1 from x_1y_1 is equal to the distance of 2 from xy. Points 1_1 and 3_1 are also located following the same procedure. The isosceles triangle representing the true shape of the section is obtained by connecting the points 1_1-2_1-3_1-1_1 as shown in the figure. The true shape of the section is hatched by drawing thin lines inclined at 45° to xy.

22. A sphere of 50 mm diameter is on the HP. A vertical section plane inclined at 45° to the VP cuts the sphere. If the cutting plane is 10 mm from the centre of the sphere, draw the sectional front elevation of the sphere.

    The plan and elevation of the sphere are drawn as shown in Fig. 11.32. The horizontal trace of the section plane inclined at 45° to xy is drawn 10 mm in front of the centre of sphere (i.e. the centre of the circle) in the plan. The true shape of sections obtained on a sphere will always be circles by any plane. An

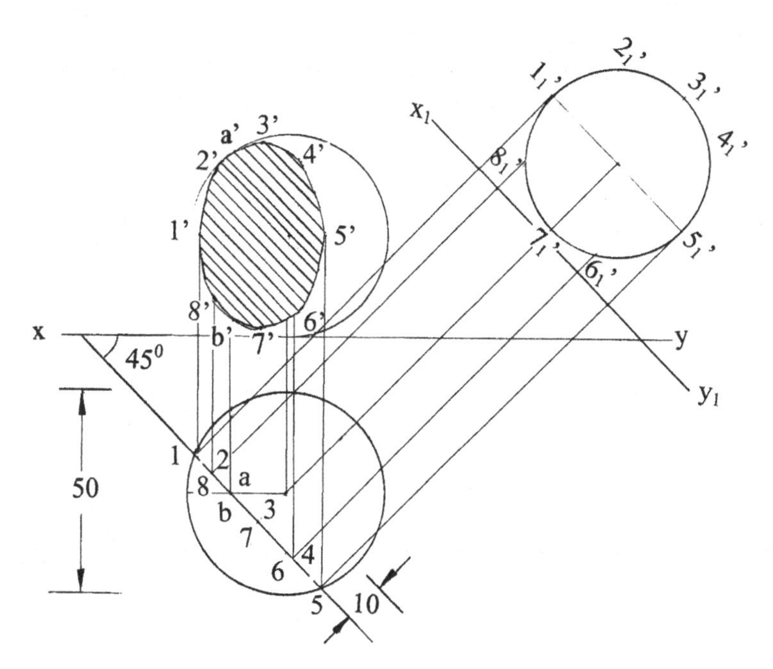

Fig. 11.32 Sectional projections of a sphere

auxiliary plane parallel to the section plane is chosen at x_1y_1. A projector is drawn perpendicular to x_1y_1 from the centre of sphere in the plan. The centre of sphere is located on this projector. A circle of diameter equal to the chord made by the section plane in the plan is drawn as shown in the figure. This circle is the true shape of the section obtained. The circle is divided into eight equal parts, and eight points are located on the circumference of the circle. The points are marked as $1_1'$, $2_1'$, $3_1'$, $4_1'$, $5_1'$, $6_1'$, $7_1'$ and $8_1'$. Projectors are drawn from these points to intersect the horizontal trace of the section plane at 1, 2, 3, 4, 5, 6, 7 and 8 in the plan. A projector is drawn perpendicular to xy from point 1. Point 1′ is located on this projector such that the distance of 1′ from xy is equal to the distance of $1_1'$ from x_1y_1. Points 2′, 3′, 4′, 5′, 6′, 7′ and 8′ are located on the respective projectors following the same procedure. A line is drawn parallel to xy from the centre of the sphere in the plan. This line intersects the horizontal trace of the section plane. There will be two points lying on the surface of the sphere and also the sectioned surface. These points are marked as b and a in the plan. These points are then projected to get b′ and a′ in the elevation of the sphere. The portion of the sphere lying in front of the section plane is removed. The sectioned surface is obtained by drawing an ellipse connecting the 1′-2′-a′-3′-4′-5′-6′-7′-b′-8′-1′. The circle of the sphere connecting the points a′ and b′ is drawn thick to show the projection of the cut-sphere in the elevation. The sectioned surface is hatched by drawing thin lines inclined at 45° to xy.

23. A tetrahedron of edge 60 mm is lying on the HP with one of its edges perpendicular to the VP. It is cut by a section plane perpendicular to the VP and inclined to the HP such that the true shape of section is an isosceles triangle of base 50 mm and altitude 40 mm. Find the inclination of the section plane with the HP and draw the sectional plan and the true shape of the section.

    The plan and elevation of the tetrahedron lying on the HP with an edge perpendicular to the VP are drawn as shown in Fig. 11.33. The points 1 and

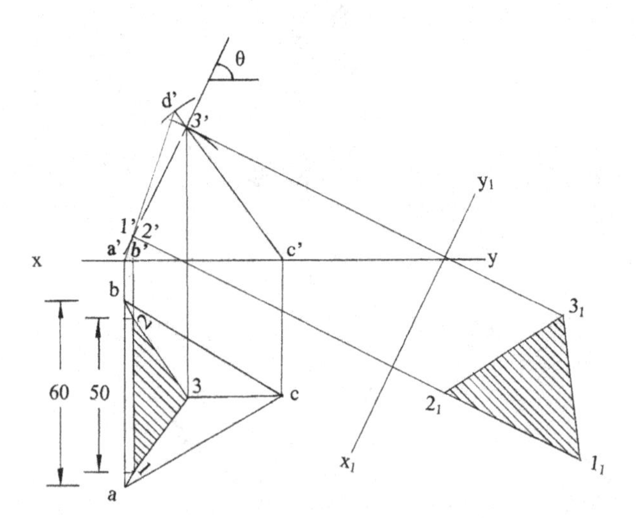

Fig. 11.33 Sectional projections of a tetrahedron

2 are chosen on the edges ad and bd such that the distance between them is 50 mm and the line joining them is perpendicular to xy. These two points are projected to get 1′ and 2′ on the edges a′d′ and b′d′ in the elevation. With 1′ as centre and radius 40 mm, an arc is drawn to cut the edge c′d′ at 3′ as shown in the figure. The section plane passes through the points 1′, 2′ and 3′. The vertical trace of the section plane is extended and the angle θ, inclination of the section plane with the HP, is measured. A projector is drawn perpendicular to xy from 3′ to locate the point 3 on the edge cd in the plan. The portion of the tetrahedron lying to the left of the section plane is removed. The sectioned surface is obtained by connecting the points 1-2-3-1 in the plan. The sectioned surface is hatched by drawing thin lines inclined at 45° to xy. An auxiliary plane parallel to the section plane is chosen at x_1y_1. A projector is drawn perpendicular to x_1y_1 from the point 3′. Point 3_1 is located on this projector such that the distance of 3_1 from x_1y_1 is equal to the distance of 3 from xy. Points 1_1 and 2_1 are located on the respective projectors following the same procedure. The true shape of the section is obtained by connecting the points 1_1-2_1-3_1-1_1. The isosceles triangle is hatched by drawing thin lines inclined at 45° to xy. The inclination of the section plane with the HP, Θ, is 65°.

24. A tetrahedron of edge 60 mm is resting on one of its faces on the HP with an edge of that face perpendicular to the HP. It is cut by a section plane perpendicular to the VP so that the true shape of the section is a square. Draw the elevation, the sectional plan and the true shape of the section.

    The plan and elevation of the tetrahedron resting on the HP on a face with an edge of that face perpendicular to the VP are drawn as shown in Fig. 11.34. The mid-points of the edges a′b′ and d′b′ are located in the elevation. These points are marked as 1′ and 2′ in the elevation. The mid-points of the edges d′c′ and a′c′ coincide with the mid-points of the edges d′b′ and a′b′ already located. These points are marked as 3′ and 4′ in the elevation. If a section plane passes through these four points, the true shape of the section will be a square. The

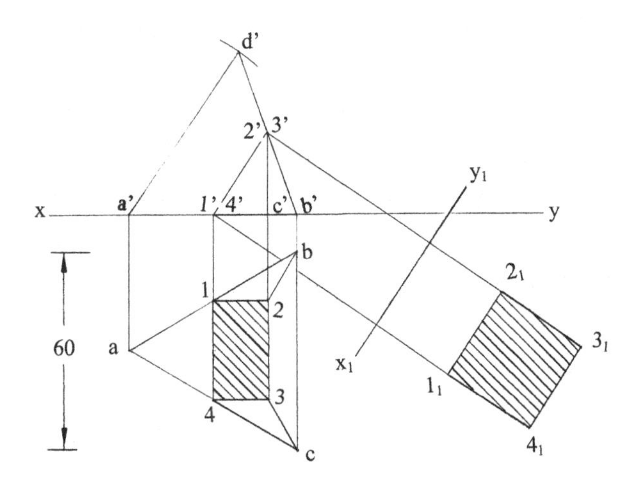

Fig. 11.34 Sectional projections of a tetrahedron

vertical trace of the section plane is drawn passing through the points $1'$, $2'$, $3'$ and $4'$. Projectors are drawn from these four points to locate points 1, 2, 3 and 4 on the respective edges in the plan. The portion of the solid lying to the left of the section plane is removed. The sectioned surface is obtained by connecting the points 1-2-3-4-1 in the plan. The sectioned surface is hatched by drawing thin lines inclined at 45° to xy. An auxiliary plane parallel to the section plane is chosen at x_1y_1. A projector is drawn perpendicular to x_1y_1 from the point $2'$ in the elevation. Point 2_1 is located on this projector such that the distance of 2_1 from x_1y_1 is equal to the distance of 2 from xy. Points 3_1, 4_1 and 1_1 are located on the respective projectors following the same procedure. The true shape of the section is obtained by connecting the points 1_1-2_1-3_1-4_1-1_1. The true shape is a square of side 30 mm (half the edge of the tetrahedron). The sectioned surface is hatched by drawing thin lines inclined at 45° to xy.

25. A cube having an edge of 40 mm is cut by a section plane in such a way that the true shape of the section made is a rhombus having side of maximum length possible. Draw the projections of the sectioned cube showing the true shape of section made.

    The cube is kept on the HP with the vertical faces equally inclined to the VP. The plan of the cube is drawn with sides of square equally inclined to xy as shown in Fig. 11.35. The elevation of the cube is projected from the plan. If the cube is cut by section planes perpendicular to the VP and passing through the centre of cube and cutting the four vertical edges, the true shape of the sections made will always be rhombuses of different sizes. When the section plane passes through the extreme corners c' and e' of the vertical edges $c'g'$ and $a'e'$, a rhombus of maximum size is obtained as the true shape of the section. The vertical trace of the section plane is drawn passing through the corner points c'

Fig. 11.35 Sectional projections of a cube

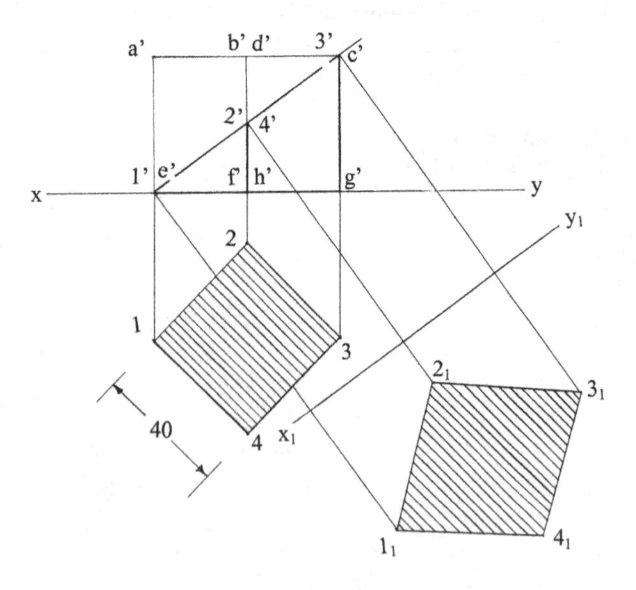

and e′ as shown in the figure. The section plane cuts two vertical edges b′f′ and d′h′ and passes through the points c′ and e′ in the elevation. The intersection points are marked as 1′, 2′, 3′ and 4′. These points are projected to get 1, 2, 3 and 4 in the plan. The sectioned surface is obtained by connecting the points 1-2-3-4-1 in the plan. The sectioned surface is hatched by drawing thin lines inclined at 45° to xy. An auxiliary plane parallel to the section plane is chosen at x_1y_1. A projector is drawn perpendicular to x_1y_1 from point 1′. Point 1_1 is located on this projector such that the distance of 1_1 from x_1y_1 is equal to the distance of 1 from xy. Points 2_1, 3_1 and 4_1 are located on the corresponding projectors following the same procedure. The rhombus of maximum size is obtained by connecting the points 1_1-2_1-3_1-4_1-1_1. The true shape of the section is hatched by drawing thin lines inclined at 45° to xy.

26. A cube of 40 mm edge stands on the HP and is cut by a section plane such that the true shape of the section is a regular hexagon. Draw the projections of the cube and determine the angle of inclination of the section plane with the HP. Also show the true shape of the section.

The cube is kept on the HP with the vertical faces equally inclined to the VP. The plan of the cube is drawn with sides of square equally inclined to xy as shown in Fig. 11.36. The elevation of the cube is projected from the plan. Section planes perpendicular to the VP and passing through the centre of the cube cut the top and bottom ends of the cube and give hexagons as true shapes of sections. If the section plane passes through the mid-points of sides b′c′ and d′c′ at the top end and the mid-points of sides f′e′ and h′e′ in the bottom ends, the true shape of the section will be a regular hexagon. The vertical trace of section plane is drawn as shown in the figure. The inclination of the section plane with the HP, Θ, is found to be 54°. The section plane cuts the edges e′f′, b′f′, b′c′, c′d′, d′h′ and h′e′ at points 1′, 2′, 3′, 4′, 5′ and 6′ in the elevation. These points are projected to get points 1, 2, 3, 4, 5 and 6 on the respective edges in the

Fig. 11.36 Sectional projections of a cube

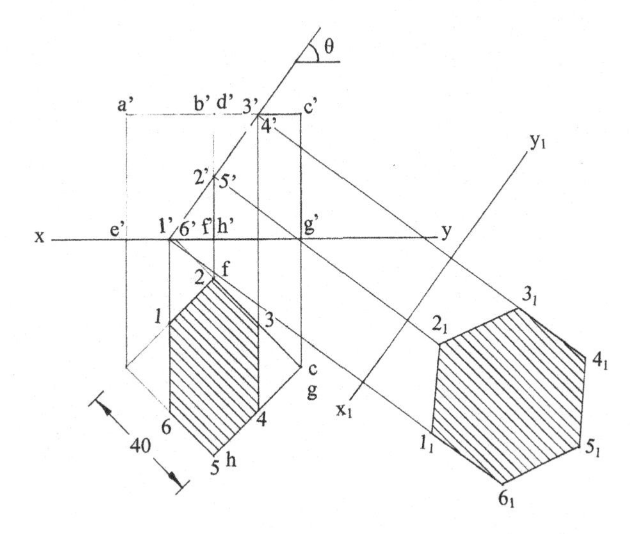

plan. The top portion of the cube is removed. The sectioned surface is obtained by connecting the points 1-2-3-4-5-6-1 in the plan. The sectioned surface is hatched by drawing thin lines inclined at 45° to xy. An auxiliary plane parallel to the section plane is chosen at x_1y_1. A projector is drawn perpendicular to x_1y_1 from point 1′ in the elevation. Point 1_1 is located on this projector such that the distance of 1_1 from x_1y_1 is equal to the distance of 1 from xy. Points 2_1, 3_1, 4_1, 5_1 and 6_1 are located on the respective projectors following the same procedure. The sectioned surface is obtained by connecting the points 1_1-2_1-3_1-4_1-5_1-6_1-1_1. The regular hexagon representing the true shape of section is hatched by drawing thin lines inclined at 45° to xy.

27. A cube of 50 mm edge is cut by a section plane perpendicular to the VP so that the true shape of section is an equilateral triangle of maximum side. Draw the sectional plan and the true shape of the section. Find also the inclination of the section plane with the HP.

    The cube is kept on the HP with the vertical faces equally inclined to the VP. The plan of the cube is drawn with sides of square equally inclined to xy as shown in Fig. 11.37. The elevation is projected from the plan. A section plane is chosen passing through the points b′, d′ and e′. This plane cuts the cube, and the true shape of the section will be an equilateral triangle of maximum side equal to diagonal of face of the cube. The vertical trace of the section plane is drawn passing through the points b′, d′ and e′. The section points are marked as 1′, 2′ and 3′ in the elevation. These points are projected to get 1, 2 and 3 in the plan. The left portion of the cube is removed. The sectioned surface is obtained by connecting the points 1-2-3-1 in the plan. The sectioned surface is hatched by drawing thin lines inclined at 45° to xy. An auxiliary plane parallel to the section plane is chosen at x_1y_1. A projector is drawn perpendicular to x_1y_1 from the point 1′. Point 1_1 is located on this projector such that the distance of 1_1

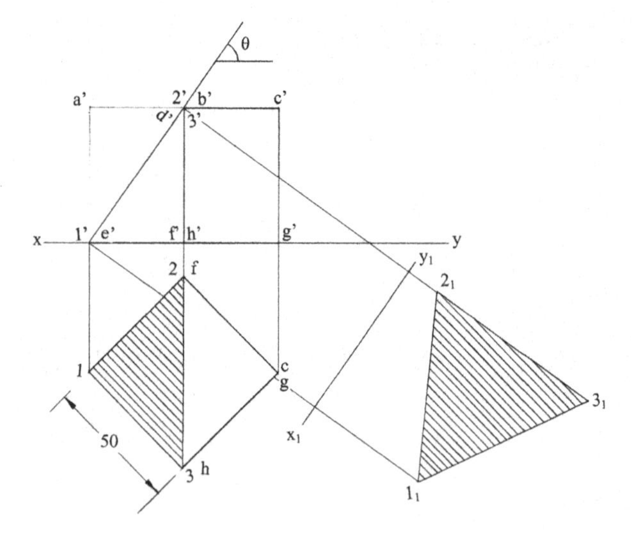

Fig. 11.37 Sectional projections of a cube

from x_1y_1 is equal to the distance of 1 from xy. Points 2_1 and 3_1 are located on the respective projectors following the same procedure. The sectioned surface is obtained by connecting the points 1_1-2_1-3_1-1_1. The equilateral triangle representing the true shape of section of maximum size is hatched by drawing thin lines inclined at 45° to xy. The inclination of the section plane with the HP, Θ, is 55°.

28. A right square prism, side of base 30 mm and altitude 80 mm, is cut into two equal parts by a section plane. The true form of the section is a rhombus of 40 mm side. Draw a sectional plan of one part of the prism on a plane parallel to the section plane.

The square prism is kept on the HP on its base with the vertical faces equally inclined to the VP. The plan of the prism is drawn with the sides of square equally inclined to xy as shown in Fig. 11.38. The elevation is projected from the plan. The centre of the prism is located at the mid-point of vertical edge d′h′. A section plane perpendicular to the VP, inclined to the HP and passing through this point gives rhombus as true shape of the section. Locus lines are drawn parallel to xy from the points a, b and d in the plan. The length of the shorter diagonal of the rhombus is equal to the length of diagonal of the square in the plan. The rhombus of side 40 mm is constructed on the locus lines as shown in the figure. The length of the larger diagonal of the rhombus is measured. An arc of radius equal to half the larger diagonal is drawn from the centre of the prism to cut the vertical edges a′e′ and c′g′ at points 1′ and 3′ in the elevation. The section plane passes through the points 1′ and 3′ and cuts the edges b′f′ and d′h′ at 2′ and 4′. These are projected to get points 1, 2, 3 and 4 in

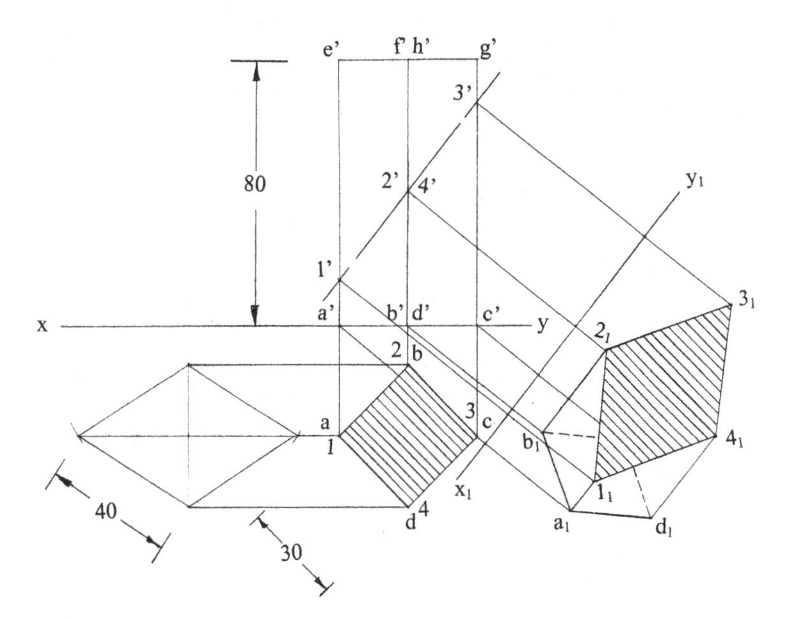

Fig. 11.38 Sectional projections of a square prism

the plan. The top portion of the prism is removed. The sectioned surface is obtained by connecting the points 1-2-3-4-1 in the plan. The sectioned surface is hatched by drawing thin lines inclined at 45° to xy. An auxiliary plane parallel to the section plane is chosen at x_1y_1. A projector is drawn perpendicular to x_1y_1 from point $1'$. Point 1_1 is located on this projector such that the distance of 1_1 from x_1y_1 is equal to the distance of 1 from xy. Section points 2_1, 3_1 and 4_1 are located on the respective projectors following the same procedure. Base points a_1, b_1, c_1 and d_1 are also located on the respective projectors following the same procedure. The sectioned surface is obtained by connecting the points 1_1-2_1-3_1-4_1-1_1. The sectioned surface is hatched by drawing thin lines inclined at 45° to xy. The projections of the base of the prism and the edges connecting the section points with the base points are completed. The invisible edges are shown as dotted avoiding the sectioned surface.

29. A cylinder is cut to the shape of a wedge as shown in Fig. 11.39. Draw the sectional side view and the true shape of the section for one side.

    The cylinder is kept on the HP on its base. The plan, elevation and left side elevation are drawn as shown in Fig. 11.40. Twelve generators are chosen on the lateral surface of the cylinder at equidistant, and these generators are shown in the plan, elevation and side elevation. Two section planes are used in the elevation to cut the cylinder to the shape of a wedge as shown in Fig. 11.40. The section plane cuts seven generators of the cylinder on the left side. The intersection points are marked as $1'$, $2'$, $3'$, $4'$, $5'$, $6'$ and $7'$ in the elevation. These points are projected to get points 1, 2, 3, 4, 5, 6 and 7 in the plan. Locus lines are drawn parallel to xy from points $1'$, $2'$, $3'$, $4'$, $5'$, $6'$ and $7'$ to intersect

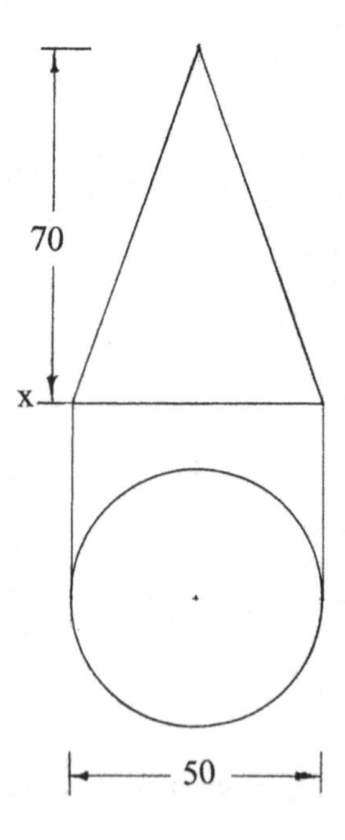

Fig. 11.39 Cut cylinder

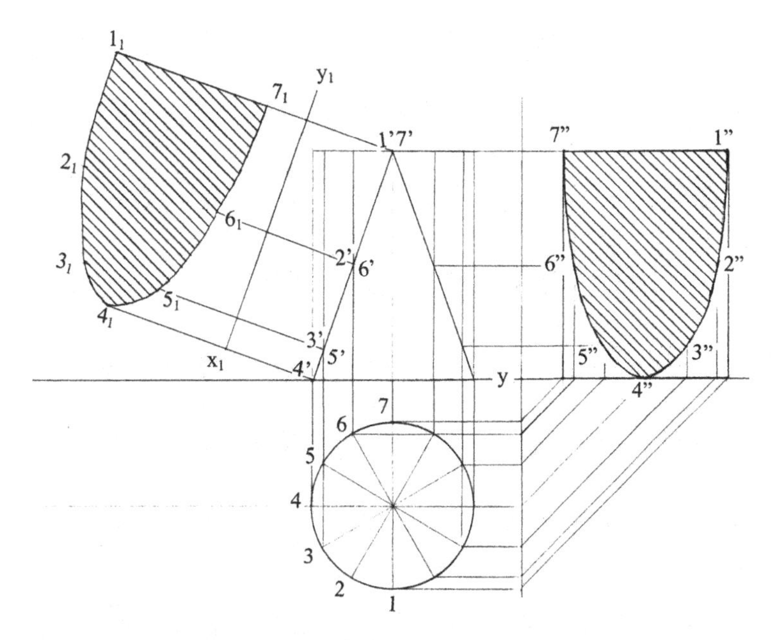

Fig. 11.40 Sectional projections of cut cylinder

the corresponding generators in the side elevation. The intersection points are marked as $1''$, $2''$, $3''$, $4''$, $5''$, $6''$ and $7''$ in the side elevation. The sectioned surface in the side elevation is obtained by drawing a smooth curve passing through the points $1''$-$2''$-$3''$-$4''$-$5''$-$6''$-$7''$ and joining the points $1''$ and $7''$ by a straight line. The sectioned surface is hatched by drawing thin lines inclined at 45° to xy. An auxiliary plane parallel to the section plane on the left side is chosen at x_1y_1. A projector is drawn perpendicular to x_1y_1 from the point $1'$. Point 1_1 is located on this projector such that the distance of 1_1 from x_1y_1 is equal to the distance of 1 from xy. Points 2_1, 3_1, 4_1, 5_1, 6_1 and 7_1 are located on the respective projectors following the same procedure. The true shape of the sectioned surface is obtained by drawing a smooth curve passing through the points 1_1-2_1-3_1-4_1-5_1-6_1-7_1 and joining the points 1_1 and 7_1 by a straight line. The sectioned surface is hatched by drawing thin lines inclined at 45° to xy.

30. A hexagonal pyramid, side of base 30 mm and axis 60 mm, is in the first quadrant. A thread is tied to one corner of its base, and the pyramid is suspended freely so that the vertex of the pyramid is above the HP and the axis is parallel to the VP. A vertical section plane making an angle of 20° with the VP and passing through the vertex of the pyramid cuts it completely. Draw the projections of the larger portion of the pyramid and show the true shape of the section made.

The pyramid base is kept parallel to HP with a side of hexagon parallel to the VP and the vertex below the base. The plan of the pyramid is drawn as shown in Fig. 11.41. The elevation is projected from the plan. The centre of gravity is

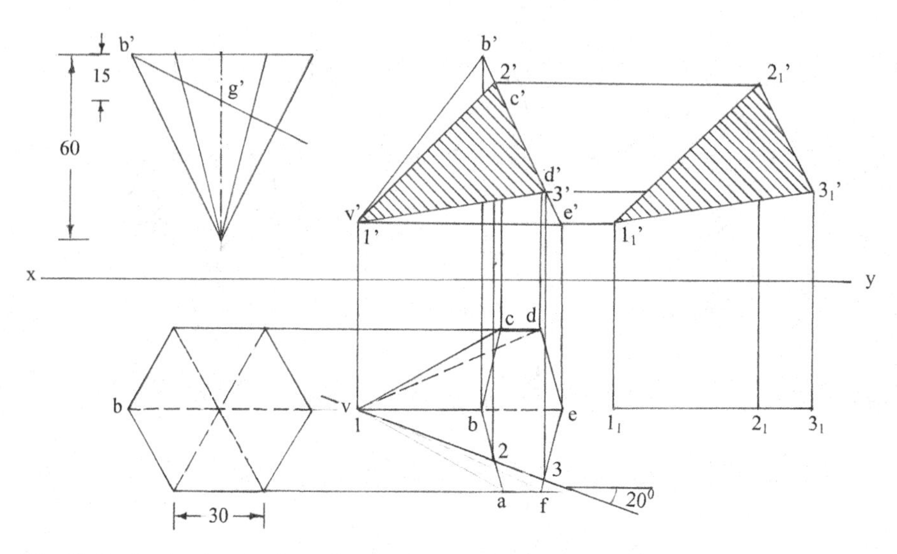

Fig. 11.41 Sectional projections of a hexagonal pyramid

one-fourth the axis from the base of the pyramid and is located at g′ in the elevation. The left extreme base point b′ is connected to the centre of gravity g′. The thread is tied to the point b′ and is freely hung from it. All the points move parallel to the VP. The elevation is redrawn such that the line connecting the points b′ and g′ is perpendicular to xy. Projectors are drawn perpendicular to xy from the angular points and vertex of the pyramid in the elevation. Locus lines are drawn parallel to xy from the angular points and the vertex in the plan. The intersections of the projectors with the locus lines locate the angular points and the vertex in the new plan. The new plan is completed showing the visible and invisible edges of the pyramid. The horizontal trace of the section plane is drawn inclined at 20° to xy and passing through the vertex in the plan. The section plane cuts two base edges ab and ef and passes through the vertex v in the plan. The intersection points are marked as 1, 2 and 3 in the plan. These points are projected to get 1′, 2′ and 3′ on the respective edges in the elevation. The portion of the pyramid in front of the section plane is removed to retain the larger portion of the pyramid. The sectioned surface is obtained by connecting the points 1′-2′-3′-1′ in the elevation. The sectioned surface is hatched by drawing thin lines inclined at 45° to xy. The sectioned surface in the plan is kept parallel to the VP by moving the section points horizontally. The section points move parallel to the HP. The new plan of the sectioned surface is represented as 1_1-2_1-3_1. Projectors are drawn perpendicular to xy from the points 1_1, 2_1 and 3_1. Locus lines are drawn parallel to xy from 1′, 2′ and 3′ in the elevation. The intersection points are marked as $1_1'$, $2_1'$ and $3_1'$. The true shape of section is obtained by connecting the points $1_1'$-$2_1'$-$3_1'$-$1_1'$ and is hatched by drawing thin lines inclined at 45° to xy.

Practice Problems

1. A triangular prism, edge of base 30 mm and axis 50 mm long, is lying on the HP on one of its rectangular faces with its axis inclined at 45° to the VP. It is cut by a section plane parallel to the HP and 10 mm above the HP. Draw its elevation and sectional plan.
2. A pentagonal prism, edge of base 30 mm and axis 80 mm long, lies with one face on the HP and the axis parallel to the VP. The solid is cut by an auxiliary plane inclined at 40° to the HP and passing through the mid-point of the axis. Draw the apparent and true shape of the section.
3. The base of a hollow prism is a regular hexagon of 40 mm side with another concentric hexagon of 20 mm side, the sides of the latter being parallel to those of the former. The prism is 60 mm high and stands on the HP on its base with a face perpendicular to the VP. It is cut by a section plane inclined at 30° to the HP and passing through the mid-point of its axis. Draw the true shape of the section.
4. A cylinder, diameter of base 50 mm and axis 70 mm long, stands on the HP. A section plane perpendicular to the VP and inclined at 60° to the HP bisects the axis. Draw the sectional plan and true shape of the section.
5. A cylinder, diameter of base 60 mm and axis 80 mm long, has a square hole of side 30 mm. The cylinder is resting on its base on the HP. The vertical faces of the square hole are equally inclined to the VP. The axis of the hole coincides with the axis of the cylinder. The cylinder is cut by a section plane perpendicular to the VP. The section plane cuts at a point 10 mm from the top surface of the cylinder and makes an angle of 30° with the HP. Draw the elevation, sectional plan and the true shape of the section.
6. Draw the plan and elevation of a cylinder cut by a section plane in such a manner that the true shape of section is an ellipse of 60 mm and 120 mm as its minor and major axes, respectively. Find the slope of the section plane with the axis of the cylinder. Take the smallest generator length to be 40 mm.
7. A triangular prism, side of base 40 mm and axis 80 mm long, is resting on the HP on its base. It is cut by an auxiliary inclined plane which cuts the vertical edges at heights of 20 mm, 40 mm and 60 mm above the base of the prism. Find the inclination of this plane with the HP. Draw the sectional plan and true shape of the section.
8. A triangular pyramid, edge of base 40 mm and axis 60 mm long, stands on the HP on its base with an edge of base parallel to the VP and nearer to it. A section plane inclined at 60° to the HP and perpendicular to the VP bisects the axis of the pyramid. Draw the elevation, sectional plan and true shape of the section.
9. A square pyramid edge of base 50 mm and altitude 70 mm stands on the HP and is cut by a vertical plane passing through one point half way down one slant edge and another point one quarter way down an adjacent slant edge. Draw its projections and true shape of the section.
10. A regular hexagonal pyramid, base 40 mm side and axis 80 mm long, is resting on its base on the HP with two opposite slant edges parallel to the VP. It is cut

by a section plane perpendicular to the VP and inclined at 45° to the HP. The section plane passes through the axis at a point 30 mm above the base. Draw its elevation, sectional plan and true shape of the section.

11. A right hexagonal pyramid, side of base 30 mm and altitude 60 mm, is cut by a section plane which contains one edge of the base and is perpendicular to the opposite face. Determine the true shape of the section.
12. A cone of 50 mm diameter of base and 70 mm height resting on the HP is cut by a section plane inclined at 40° to the HP and perpendicular to the VP. The section plane passes through a point 30 mm below the apex. Draw the true shape of the section made and name the curve.
13. A frustum of a cone, top diameter 60 mm, base diameter 80 mm and height 80 mm has a coaxial through hole of 40 mm diameter and stands with its base on the HP. It is cut by a section plane whose vertical trace inclined at 40° to reference line xy passes through the mid-point of the axis of the frustum. Draw the sectional plan and true shape of the section.
14. A cone of base 60 mm diameter and height 80 mm resting on its base on the HP is cut by a section plane perpendicular to both the HP and VP at a distance 10 mm from the axis. Draw the sectional side elevation of the cone.
15. A square pyramid, side of base 50 mm and axis 80 mm, is lying on one of its triangular faces on the HP with the axis parallel to the VP. It is cut by a horizontal section plane which bisects the axis. Draw the elevation and the sectional plan of the pyramid.
16. An equilateral triangular prism of 50 mm side of base and height 70 mm is cut by a section plane in such a way that the true shape of the section is a trapezium of parallel sides of 50 mm and 10 mm. Draw the projections of cut prism and the true shape of the section.
17. A hollow square prism, base 50 mm side outside, height 75 mm and thickness 10 mm, is resting on its base on the HP with one of its vertical faces inclined at 30° to the VP. A section plane inclined at 30° to the HP and perpendicular to the VP and passing through the axis 20 mm from its top end cuts the prism. Draw the sectional plan and the true shape of the section.
18. A pentagonal prism, base 35 mm side and axis 90 mm long, is resting on a base edge on the HP with a rectangular face containing that edge being perpendicular to the VP and inclined to the HP at 60°. It is cut by a horizontal section plane whose vertical trace (VT) passes through the mid-point of the axis. Draw the projections of the cut prism.
19. A cylinder of 60 mm diameter of base and axis 90 mm long has its axis inclined at 30° to the VP and parallel to the HP. It is cut by a section plane perpendicular to the HP so that the true shape of the section is an ellipse of major axis 75 mm. Draw the projections of the cut cylinder and the true shape of the section.
20. A triangular prism, side of base 40 mm and height 80 mm, rests with one of its base edges on HP such that the axis is parallel to VP and inclined at 30° to HP. A horizontal section plane passes through the lowest triangular edge of the top end of the prism. Draw the sectional plan showing the true shape of the section.

Chapter 12
Intersection of Surfaces

The need to determine the lines or curves of intersections between two surfaces of similar or different geometric shapes occurs in the fabrication of ducts, boiler mountings, pipe fittings, aircraft and automobile bodies. If two geometric shapes with curved surfaces meet or penetrate each other, the lines of intersection is a curve. The curve of intersection is determined by plotting the projections of points which are common to both surfaces. This is done by choosing section planes which cut the surfaces in lines or circles. The intersections of these lines give points on the required curve.

Classification of intersecting surfaces

(a) Intersection of two curved surfaces

The intersection of curved surfaces is a curve. Cylinder and cylinder, cylinder and cone or cone (penetrating surface) and cylinder belong to this category.

(b) Intersection of two plane surfaces

The intersection of plane surfaces is a straight line. Prism and prism, prism and pyramid or pyramid and pyramid belong to this category.

Methods used to draw the line or curve of intersection

The following two methods are used to draw the line of intersection.

1. Cutting plane method

   In this method, a number of section planes are chosen and cut the intersecting surfaces in lines or circles. The intersections of these lines give points common to both the surfaces. This method is used for cylinder penetrating another cylinder or cone.

2. Line method or piercing point method

   In this method, a number of straight lines are chosen on the penetrating surface, and points of intersection which are common to both surfaces are obtained. This method is used for cone penetrating cylinder and prism penetrating another prism.

K. Rathnam, *A First Course in Engineering Drawing*,
DOI 10.1007/978-981-10-5358-0_12

Solved Problems

1. A horizontal cylinder of 50 mm diameter penetrates the vertical cylinder of 80 mm diameter resting on the HP. The axes of the cylinders are coplanar. The axis of the horizontal cylinder is 50 mm above the HP. Draw the projections of the cylinders showing the curves of intersections of the two cylinders.

   The plan, elevation and left side elevation of the outline of intersecting cylinders with axes coplanar are drawn as shown in Fig. 12.1. It is evident that the horizontal cylinder penetrates the vertical cylinder, and the curves of intersection lie entirely on the surface of the horizontal cylinder. The circumference of the horizontal cylinder in the side elevation is divided into eight equal parts, and the division points are numbered in the three views as shown in the figure. A horizontal section plane is chosen passing through point 1″. This section plane cuts the surface of horizontal cylinder in its topmost generator and the surface of the vertical cylinder in a circle. The plan of the generator passing through point 1 in the plan intersects the circle in points a and p. These points are projected on the generator passing through point 1′ at a′ and p′. These points lie on the curves of intersection. Another horizontal section plane is chosen passing through the points 2″ and 8″. This section plane cuts the surface

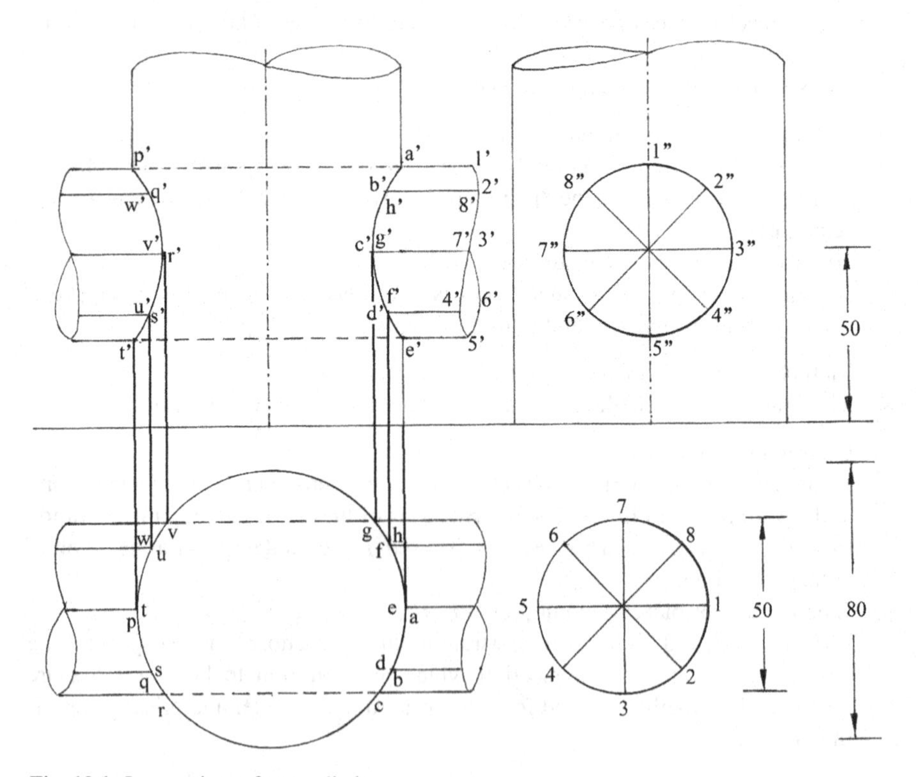

Fig. 12.1 Intersections of two cylinders

of horizontal cylinder along two generators passing through the points 2 and 8 in the plan and the surface of vertical cylinder in a circle. The plans of the two generators intersect the circle in points b, h, w and q. These points are then projected on the generators passing through 2′ and 8′ at b′, h′, w′ and q′. A third horizontal section plane is chosen passing through the points 3″ and 7″. Points c, g, v and r in the plan and points c′, g′, v′ and r′ in the elevation are obtained following the same procedure. The fourth horizontal section plane passing through the points 4″ and 6″ provides four more points, namely d, f, u and s in the plan and d′, f′, u′ and s′ in the elevation. Finally, the fifth horizontal section plane passing through the point 5″ provides two points e and t in the plan and e′ and t′ in the elevation. Draw a smooth curve passing through the points a′, b′ c′, d′, e′, f′, g′ and h′ and another curve passing through the points p′, q′, r′, s′, t′, u′, v′ and w′. The curve of intersection changes its direction at points c′, e′, g′ and a′ on the right side and at r′, t′, v′ and p′ on the left side. These points are called key points or critical points which lie on extreme generators of the penetrating cylinder.

2. A horizontal cylinder of 50 mm diameter penetrates the vertical cylinder of 80 mm diameter resting on the HP. The axis of the horizontal cylinder is parallel to both the HP and VP and is 5 mm in front of the axis of the vertical cylinder. The axis of the horizontal cylinder is 50 mm above the HP. Draw the plan and elevation of the cylinders showing the curves of intersection.

   The plan, elevation and left side elevation of the outline of intersecting cylinders with the axis of the horizontal cylinder 5 mm in front of the vertical cylinder are drawn as shown in Fig. 12.2. The circumference of the horizontal cylinder is divided into eight equal parts in the side elevation, and the division points are numbered in three views as shown in the figure. A horizontal section plane is chosen passing through point 1″. This section plane cuts the surface of horizontal cylinder in its topmost generator and the surface of the vertical cylinder in a circle. The plan of the generator passing through point 1 in the plan intersects the circle in points a and p. These points are projected on the generator passing through point 1′ at a′ and p′. These points lie on the curves of intersection. Another horizontal section plane is chosen passing through the points 2″ and 8″. This section plane cuts the surface of horizontal cylinder along two generators passing through the points 2 and 8 in the plan and the surface of vertical cylinder in a circle. The plans of the two generators intersect the circle in points b, h, w and q. These points are then projected on the generators passing through 2′ and 8′ at b′, h′, w′ and q′. A third horizontal section plane is chosen passing through the points 3″ and 7″. Points c, g, r and v in the plan and points c′, g′ v′ and r′ in the elevation are obtained following the same procedure. The fourth horizontal section plane passing through the points 4″ and 6″ provides four more points namely d, f, u and s in the plan and d′, f′, u′ and s′ in the elevation. Finally, the fifth horizontal section plane passing through the point 5″ provides two points e and t in the plan and e′ and t′ in the elevation. Points 10″ and 11″ are located on the intersections of the middle generator of the vertical cylinder with the circle of the horizontal cylinder in the

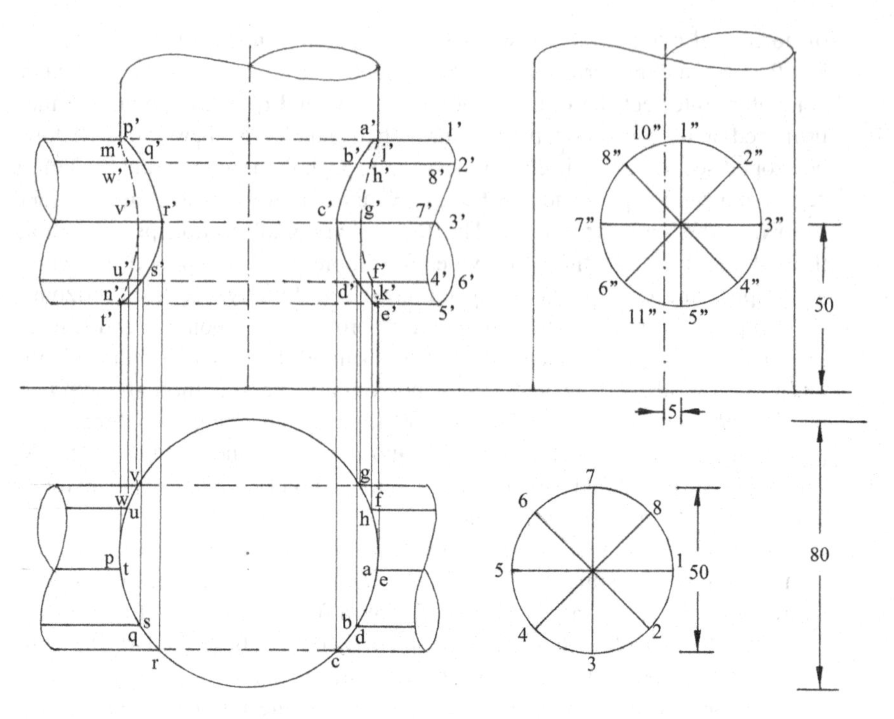

Fig. 12.2 Intersections of two cylinders with offset

side elevation. These points are projected to locate four points j′, k′, m′ and n′ (two on each side) on the extreme generators of the vertical cylinder. These points are also called key or critical points where the curve changes its direction. There are six key points on each side of the vertical cylinder. A smooth curve is drawn passing through points a′, b′, etc. taking into consideration the curve changes its direction at six key points, namely, c′, e′, k′, g′, j′ and a′ on the right side of the vertical cylinder. Another curve is drawn on the left side of the vertical cylinder.

3. A vertical cylinder of 50 mm diameter and 100 mm high is intersected by another cylinder of 40 mm diameter and 100 mm length, and the two axes are bisecting each other. The axis of the penetrating cylinder is inclined at 30° to the HP and parallel to the VP. Draw the intersection curves.

   The plan and elevation of the outline of intersecting cylinders are drawn as shown in Fig. 12.3. The inclined cylinder is the penetrating cylinder in this case. Hence, the section planes are chosen inclined to the HP and parallel to the axis of penetrating cylinder. The circumference of the penetrating cylinder is divided into eight equal parts, and the division points are numbered in two views as shown in the figure. A section plane is chosen parallel to the axis of the penetrating cylinder and passing through point 1′. This inclined section plane cuts the surface of the inclined cylinder in its topmost generator and the surface of the vertical cylinder in an ellipse. This ellipse when projected on the HP becomes a circle of the vertical cylinder. The plan of the topmost generator

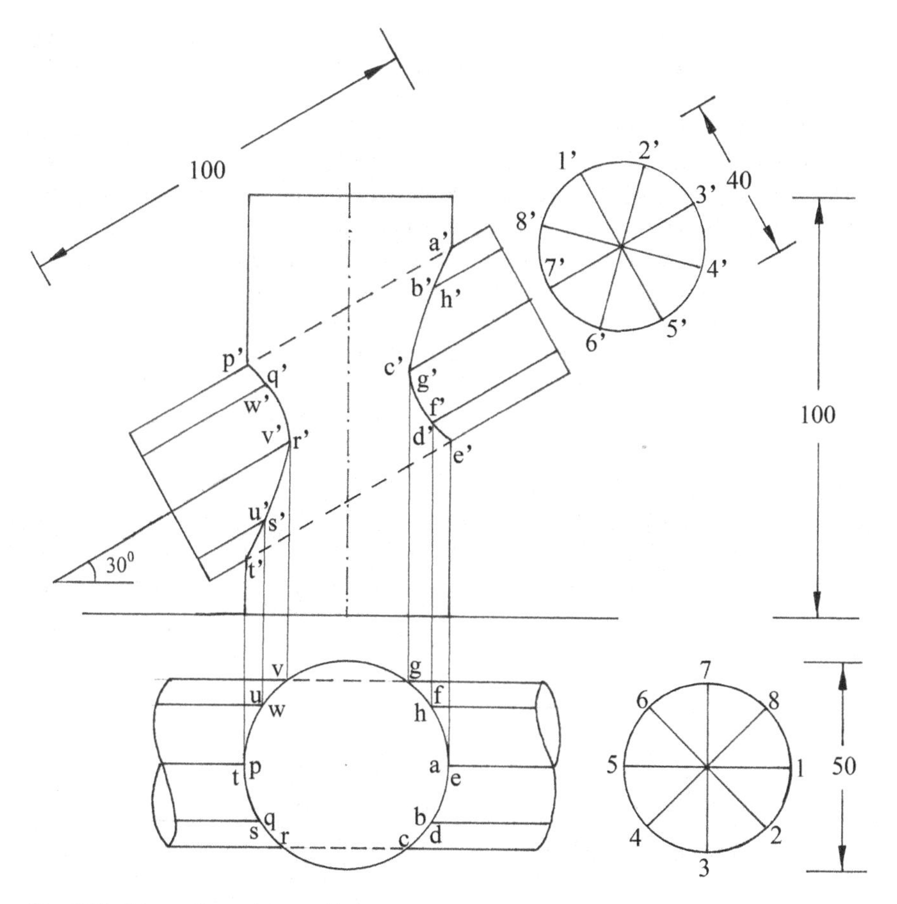

Fig. 12.3 Intersections of two cylinders

intersects the circle in a and p. These points are projected on the generator passing through point 1′ at a′ and p′. Another section plane parallel to the axis of penetrating cylinder is chosen passing through points 2′ and 8′. This section plane cuts the surface of the penetrating cylinder along two generators passing through points 2 and 8 in the plan and the surface of the vertical cylinder in a circle in the plan. The plans of the generators intersect the circle in b, h, w and q. These points are projected on the generators passing through points 2′ and 8′ at b′, h′, w′ and q′. Another section plane parallel to the axis of penetrating cylinder is chosen passing through points 3′ and 7′. Points c, g, v and r in the plan and c′, g′, v′ and r′ in the elevation are obtained following the same procedure. The section plane passing through points 4′ and 6′ provides four more points, namely, d, f, u and s in the plan and d′, f′, u′ and s′ in the elevation. Finally, the section plane passing through point 5′ locates points e and t in the plan and e′ and t′ in the elevation. A smooth curve is drawn passing through the points a′, b′, etc. Another curve is drawn passing through the points p′, q′, etc. These curves represent the curves of intersections.

4. A horizontal cylinder of 50 mm diameter and a vertical cylinder of 40 mm diameter interpenetrate with their axes intersecting at right angles. Draw the projections showing the curves of intersection.

   The plan, elevation and left side elevation of the outline of intersecting cylinders are drawn as shown in Fig. 12.4. The vertical cylinder is penetrating the horizontal cylinder, and the curves of intersections lie entirely on the surface of the vertical cylinder. The circumference of the vertical cylinder is divided into eight equal parts and the division points are numbered as shown in the figure. Section planes are chosen parallel to the axis of the vertical cylinder and perpendicular to both the HP and VP. A section plane is chosen passing through point 1 in the plan. This plane cuts the left extreme generator of the vertical cylinder and the surface of the horizontal cylinder in a circle as shown in the side elevation. The generator line passing through 1″ cuts the circle in a″

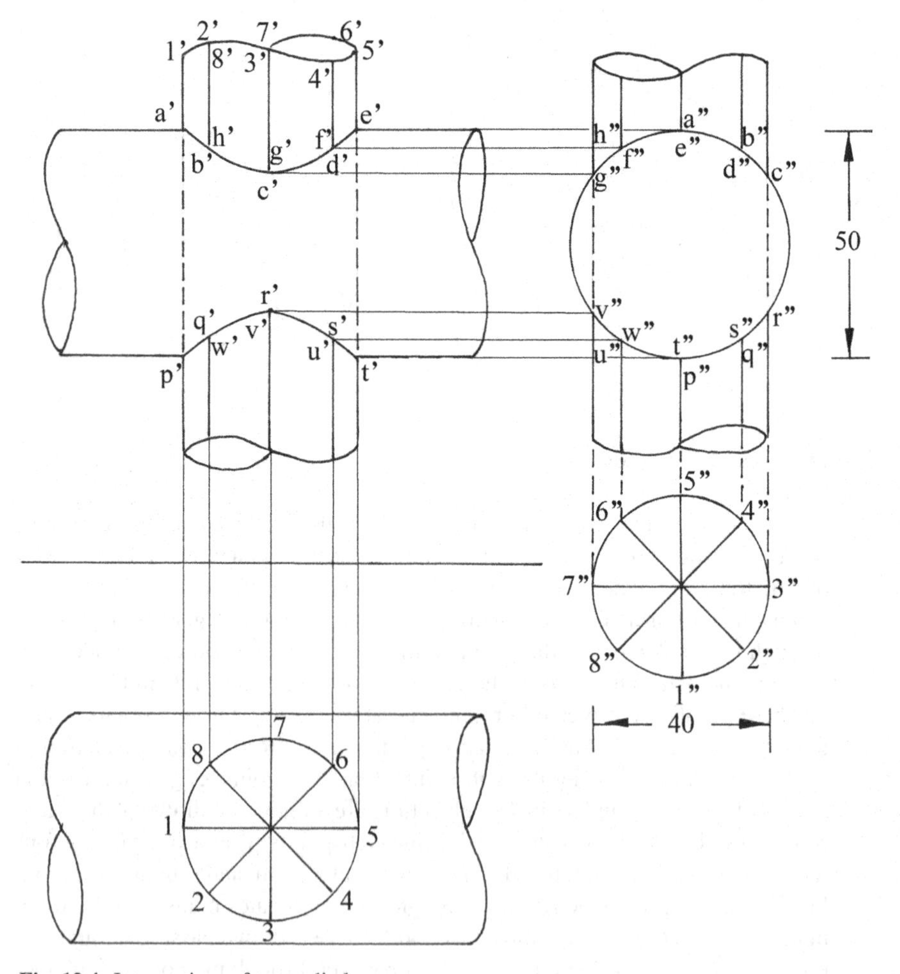

Fig. 12.4 Intersections of two cylinders

and p″. These points are projected on the generator passing through 1′ at a′ and p′ in the elevation. Another section plane is chosen passing through 2 and 8 in the plan. This plane cuts the surface of the penetrating cylinder along two generators passing through points 2″ and 8″ in the side elevation and the surface of the horizontal cylinder in a circle. The generators in the side elevation intersect the circle in b″, h″, w″ and q″. These points are projected on the generators passing through points 2′ and 8′ at b′, h′, w′ and q′. A third section plane is chosen passing through points 3 and 7. Points c″, g″, v″ and r″ in the side elevation and c′, g′ v′ and r′ in the elevation are located following the same procedure. The fourth section plane passing through point 4 and 6 provides four points, namely d″, f″, u″ and s″ in the side elevation and d′, f′, u′ and s′ in the elevation. The fifth section plane passing through point 5 provides two points e″ and t″ in the side elevation and e′ and t′ in the elevation. A smooth curve is drawn passing through points a′, b′, c′, d′, e′, f′, g′ and h′. Another smooth curve is drawn passing through points p′, q′, r′, s′, t′, u′ and v′.

5. A cylindrical pipe of 40 mm diameter has a branch of the same size. The axis of the main pipe is vertical and is intersected by that of the branch at right angles. Draw the projections of the pipe when the two axes lie in a plane parallel to the VP.

   The plan and elevation of the outline of main and branch pipes are drawn as shown in Fig. 12.5. The circumference of the branch pipe is divided into eight equal parts, and the division points are numbered in the two views as shown in

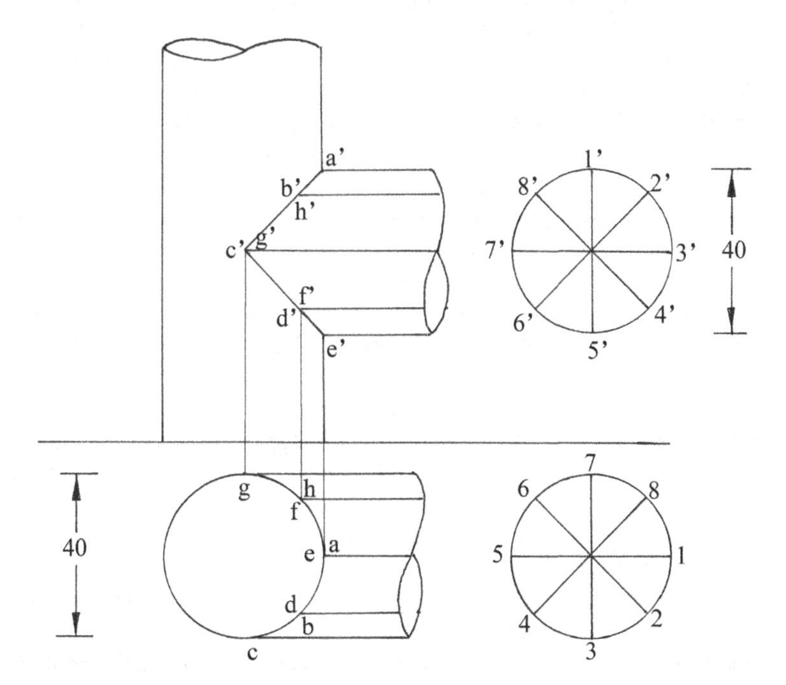

Fig. 12.5 Intersection of pipes

the figure. A horizontal section plane is chosen passing through point 1′. This section plane cuts the surface of the branch pipe in its topmost generator passing through point 1 and the surface of the main pipe in circle in the plan. The plan of the generator intersect the circle in a. This point is projected on the generator passing through point 1′ at a′. The second horizontal section plane is chosen passing through points 2′ and 8′. This section plane cuts the surface of branch pipe along the two generators passing through points 2 and 8 and the surface of the main pipe in a circle. The plans of the generators passing through points 2 and 8 intersect the circle in b and h in the plan. These points are projected on the generators passing through points 2′ and 8′ at b′ and h′. Another horizontal section plane is chosen passing through point 3′ and 7′. Points c and g in the plan and c′ and g′ in the elevation are obtained following the same procedure. The horizontal section plane passing through points 4′ and 6′ provides two more points, namely d and f in the plan and d′ and f′ in the elevation. Finally, the section plane passing through point 5′ locates the point e in the plan and e′ in the elevation. The points a′, b′ and c′ lie on a straight line with the invisible points h′ and g′, respectively, behind b′ and c′. The points c′, d′ and e′ lie on a straight line with the invisible point f′ behind d′. This is a special case of two pipes of same diameter joining on one side with the axes intersecting.

6. A vertical cylinder of 40 mm diameter is penetrated by a horizontal cylinder of same size with their axes intersecting. Draw the curves of intersections if the axis of the horizontal cylinder is inclined at 45° to the VP.

   The plan and elevation of the outline of intersecting cylinders are drawn as shown in Fig. 12.6. The circumference of horizontal cylinder is divided into eight equal parts, and the division points are numbered in two views as shown in the figure. A horizontal section plane is chosen passing through point 1′. This section plane cuts the surface of the horizontal cylinder in its topmost generator passing through point 1 in the plan and the surface of the vertical cylinder in a circle. The plan of the generator intersects the circle in a and p. These points are projected on the generator passing through point 1′ at a′ and p′. Another horizontal section plane is chosen passing through points 2′ and 8′. This section plane cuts the horizontal cylinder along two generators passing through points 2 and 8 and the vertical cylinder in a circle. The plans of the generators passing through 2 and 8 intersect the circle in b, h, w and q. These points are projected on the generators passing through points 2′ and 8′ at b′, h′ w′ and q′. Another horizontal section plane is chosen passing through points 3′ and 7′. Points c, g, v and r in the plan and c′, g′, v′ and r′ in the elevation are obtained following the same procedure. The horizontal section plane passing through points 4′ and 6′ provides four points, namely d, f, u and s in the plan and d′, f′, u′ and s′ in the elevation. Finally, the horizontal section plane passing through point 5′ locates two points e and t in the plan and e′ and t′ in the elevation. These points are joined in proper sequence to show the curves of intersections.

7. A semi-cylinder of 40 mm diameter intersects another semi-cylinder of 60 mm diameter. The cut or flat faces of both the semi-cylinders are on the HP, and the

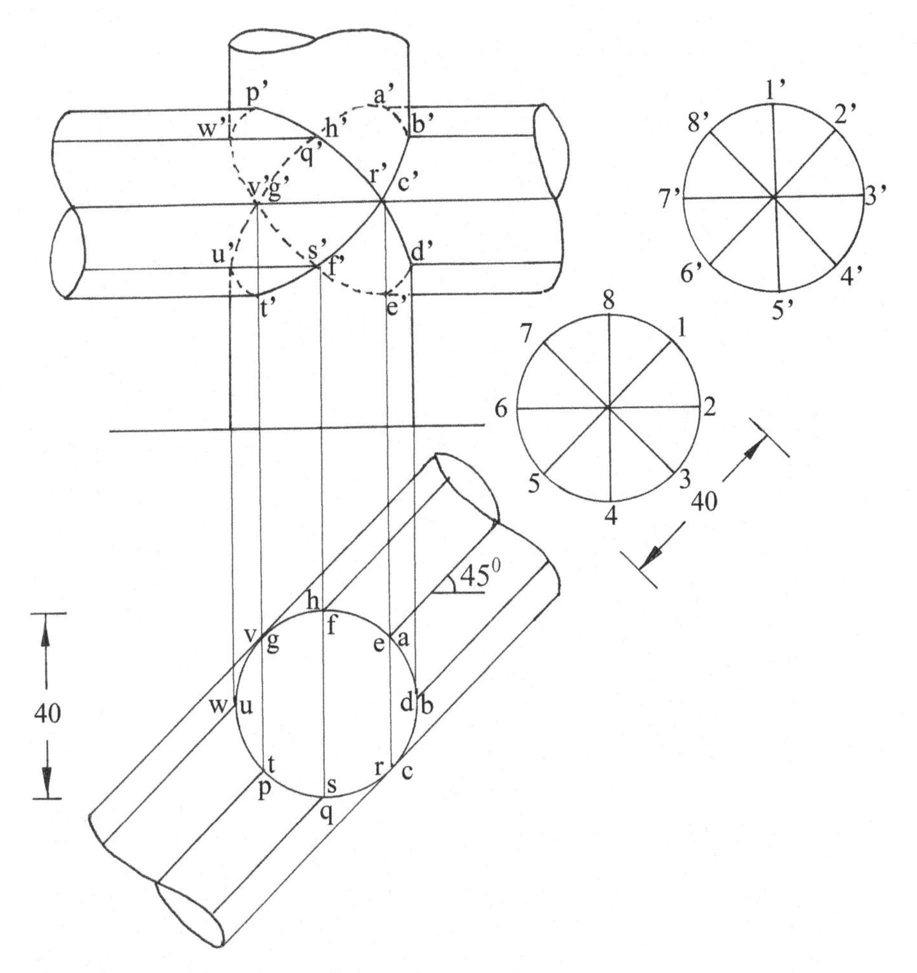

Fig. 12.6 Intersections of two cylinders

axes intersect each other at an angle of 45°. Draw the projections of both the solids clearly indicating the curves of intersection.

The plan and elevation of the outline of intersecting semi-cylinders with the flat faces on the HP are drawn as shown in Fig. 12.7. The semi-cylinder of 40 mm diameter interpenetrates the semi-cylinder of 60 mm diameter. The curves of intersection lie on the surface of the smaller one. The semi-circle of 40 mm diameter is divided into four equal parts, and the division points are numbered as shown in the figure. A horizontal section plane is chosen passing through point 1′. This section plane cuts the surface of smaller cylinder in its topmost generator and the surface of the larger cylinder along the two generators passing through points a′ and p′ in the elevation. These points are projected on the generator passing through point 1 at a and p in the plan. Another horizontal section plane is chosen passing through points 2′ and 5′.

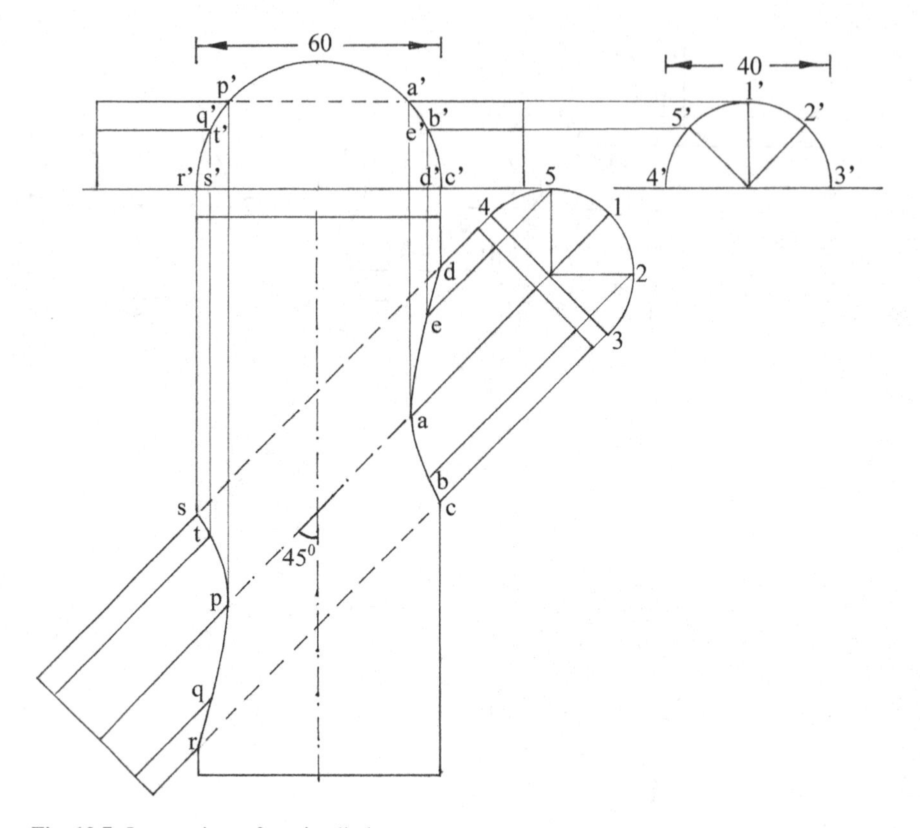

Fig. 12.7 Intersections of semi-cylinders

This section plane cuts the smaller cylinder along the generators passing through points 2′ and 5′ and the larger cylinder along the generators passing through points b′, e′, t′ and q′ in the elevation. These points are projected on the generators passing through points 2 and 5 at b, e, t and q in the plan. Another horizontal section plane is chosen passing through points 3′ and 4′. This plane cuts the smaller cylinder along the generators passing through points 3′ and 4′ and the larger cylinder along the generators passing through points c′, d′, s′ and r′. These points are projected in the plan of the generators passing through 3 and 4 at c, d, s and r. A smooth curve is drawn through points d, e, a, b and c. Another smooth curve is drawn through points s, t, p, q and r.

8. A vertical cone, diameter of base 80 mm and axis 100 mm long, is completely penetrated by a cylinder of 40 mm diameter. The axis of the cylinder is parallel to both the HP and VP and intersects the axis of the cone at a point 30 mm above the base. Draw the projections of the solids showing the curves of intersections.

The plan, elevation and side elevation of the outline of intersecting cone and cylinder are drawn as shown in Fig. 12.8. It is clear that the cylinder penetrates

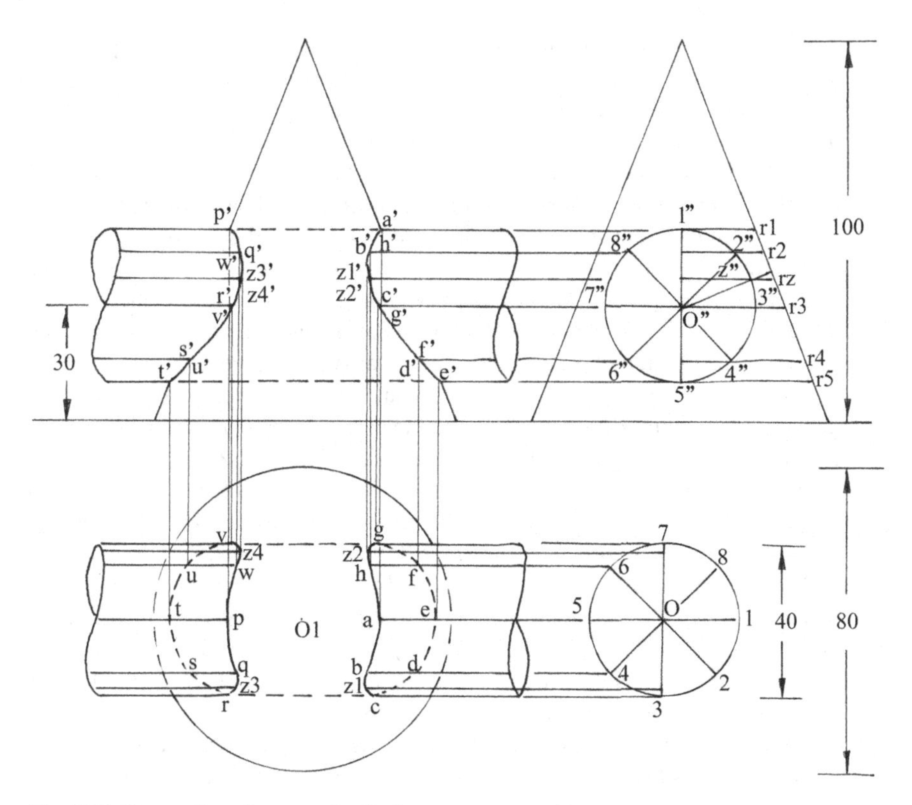

Fig. 12.8 Intersection of cone and cylinder

the cone, and the curves of intersection lie entirely on the surface of the cylinder. The circumference of the cylinder is divided into eight equal parts, and the division points are numbered in the side elevation and the plan as shown in the figure. A horizontal section plane is chosen passing through point 1″. This section plane cuts the surface of the cylinder in the topmost generator and the surface of the cone in a circle of radius r1. The plan of the generator is drawn passing through point 1. With centre O1 and radius r1, draw arcs to intersect the generator in points a and p in the plan. These points are projected on the generator passing through point 1″ at a′ and p′ in the elevation. Another horizontal section plane is chosen passing through the points 2″ and 8″. This section plane cuts the surface of the cylinder along two generators passing through the points 2 and 8 in the plan and the cone in a circle of radius r2. With centre O1 and radius r2, draw arcs to cut the generators in points b, h, w and q in the plan. These points are projected on the generators passing through points 2″ and 8″ at b′, h′, w′ and q′ in the elevation. A third horizontal section plane is chosen passing through points 3″ and 7″. With centre O1 and radius r3, points c, g, v and r are obtained in the plan following the same procedure. These points are projected on the generators passing through points 3″ and 7″ at c′, g′, v′ and r′ in the elevation. The fourth horizontal section plane passing through points 4″ and 6″ provides four more points d, f, u and s in the plan and d′, f′, u′ and s′ in the elevation. The fifth horizontal section plane passing through point 5″ locates two points e and t in the plan and e′ and t′ in the elevation. In the present case, four key points are already obtained at a′, e′, p′ and t′, being the points of intersection between the outline of the cone and the outline of cylinder in the elevation. Four more key points are to be located by choosing another horizontal section plane. A line is drawn perpendicular to the extreme generator of the cone from point O″ in the side elevation. This line intersects the circle of the cylinder at z″. A horizontal section plane is chosen passing through the point z″. This section plane cuts the cylinder along two generators and the cone in a circle of radius rz. With centre O1 and radius rz, draw arcs to cut the generators in points z1, z2, z3 and z4 in the plan. These points are projected on the generators passing through point z″ at z1′, z2′, z3′ and z4′ in the elevation. In the plan, points g, z2, h, a, b, z1 and c are visible and hence are joined to represent the curve of intersection. The points d, e and f are invisible. Hence, the line joining these points is shown as dotted and connected to points c and g. A similar curve is drawn on the left side. In the elevation, points a′, b′, z1′, c′, d′ and e′ are visible and hence are joined to represent the curve of intersection, and invisible points f′, g′, z2′ and h′ are behind the visible points d′, c′, z1′ and b′. A similar curve is drawn on the left side. It may be noted that the hidden portions of base line of the cone are shown as dotted.

9. A vertical cone of base 80 mm diameter and axis 100 mm long is penetrated by a horizontal cylinder of 40 mm diameter in such a way that both the solids envelope an imaginary common sphere and their axes intersect at right angles. Draw the projections of the solids when the axes lie in a plane parallel to the VP.

   The side elevation of cone is drawn first as shown in Fig. 12.9. The axis of the horizontal cylinder is located at O″ by drawing a line parallel to the right extreme generator of the cone and 20 mm from it as shown in the figure. A circle of radius 20 mm is drawn with the centre O″. This circle represents the side elevation of the horizontal cylinder. The elevation and plan of the cylinder are also drawn. The circumference of the cylinder is divided into eight equal parts, and the division points are numbered in the side elevation and the plan as shown in the figure. Horizontal section planes are chosen passing through points 1″, 2″ & 8″, 3″ & 7″, 4″ & 6″ and 5″ to cut the cylinder and the cone. Points a, b, c, d, e, f, g, h, p, q, r, s, t, u, v and w are located in the plan following the procedure detailed in the previous problem (no. 8). These points are then projected to locate points a′, b′, etc. on the respective generators in the elevation. A line perpendicular to the right extreme generator of the cone is drawn and meets the generator at z″. A horizontal section plane is chosen passing through z″. This section plane cuts the surface of the cylinder along two generators and the surface of the cone in a circle of radius rz. Points j and k are located in the plan such that the distances of j and k from the axis of the cone is equal to the radius rz along the projector drawn through the axis of the cone.

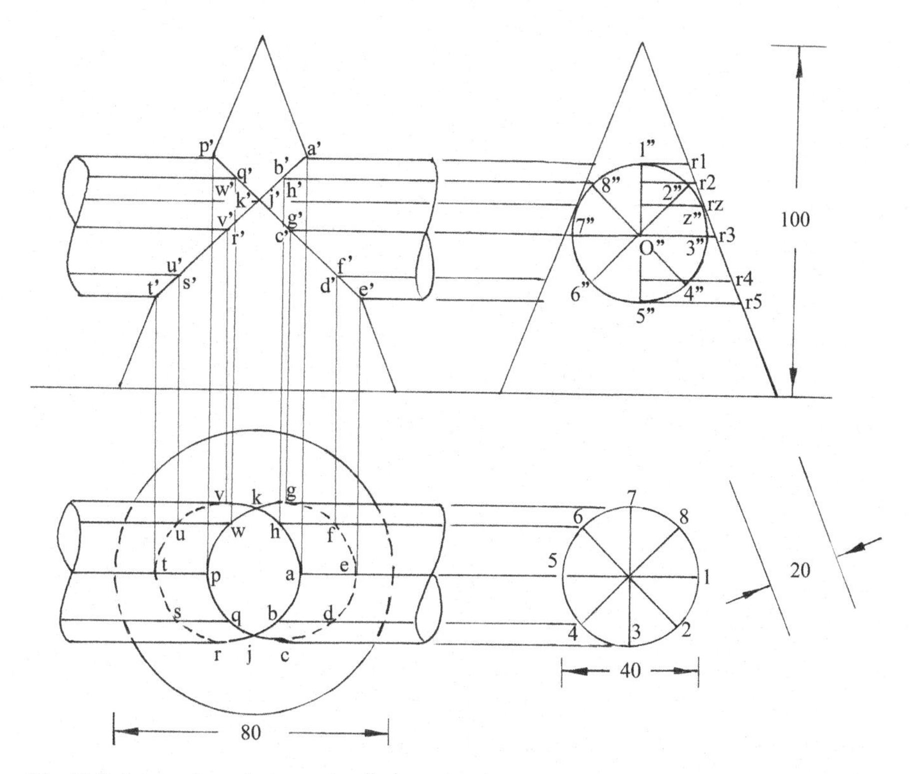

Fig. 12.9 Intersection of cone and cylinder

The generator is drawn passing through z''. Points j' and k' are located at the intersection of the projector through j and k with the generator. It may be noted that points a', b', j', r', s' and t' lie on a straight line and points p', q', j', c', d' and e' lie on another straight line in the elevation. The intersection curves are ellipses in the plan. A part of each curve is invisible and is shown as dotted in the plan. The invisible portions of the base line of cone are shown as dotted.

10. A horizontal cylinder of 60 mm diameter is penetrated by a vertical cone base 100 mm diameter and axis 90 mm long. The axis of cylinder is 40 mm above the base of cone and intersects the axis of cone at right angles. Draw the projections showing the lines of intersections.

The plan, elevation and side elevation of the outline of intersecting cone and cylinder are drawn as shown in Fig. 12.10. It is clear from the side elevation that the cone penetrates the cylinder, and the curves of intersection lie entirely on the lateral surface of the cone. The line method is employed to locate points common to both the solids. Assume that there are 8 generators at equidistant on the lateral surface of the cone. These generators are designated as $O''1''$, $O''2''$, $O''3''$, etc. These generators are also numbered in the elevation and plan of the cone. The generator $O''1''$ intersects the circle at a'' and p'' in the side elevation. These points are projected on the generator $O'1'$ at a' and p' in the elevation.

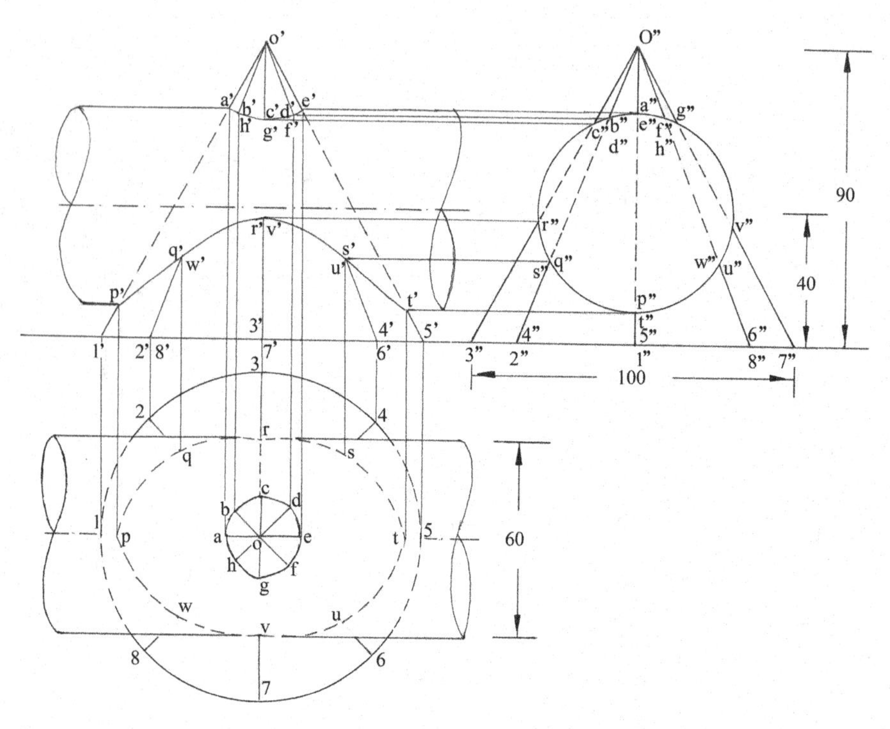

Fig. 12.10 Intersection of cone and cylinder (line method)

Projectors are drawn from a′ and p′ to locate points a and p on the generator O1 in the plan. The generator O″2″ cuts the circle at b″ and q″ in the side elevation. These points are projected to locate points b′ and q′ on the generator O′2′ in the elevation. Projectors are drawn from b′ and q′ to locate points b and q on the generator O2 in the plan. The generator O″3″ intersects the circle at c″ and r″ in the side elevation. These points are projected to locate points c′ and r′ on the generator O′3′ in the elevation. Points c and r are located on the generator O3 following the method of projection detailed in the chapter on the sections of solids. Points d″, e″, f″, g″, h″, s″, t″, u″, v″ and w″ are located in the side elevation. These points are projected in the elevation and then in the plan following the same procedure. Top curve of intersection is obtained joining the points a′, h′, g′, f′ and e′ in the elevation. The bottom curve passes through the points p′, w′, v′, u′ and t′. Two separate curves of intersection are obtained in the plan. The curve passing through p, q, r, s, t, u, v and w is invisible.

11. A vertical square prism base 50 mm side is completely penetrated by a horizontal square prism base 35 mm side so that their axes intersect at right angles. The faces of the two prisms are equally inclined to the VP. Draw the projections of the prisms showing the lines of intersections. Assume suitable lengths for the prisms.

The plan, elevation and side elevation of the outline of intersecting prisms are drawn as shown in Fig. 12.11. The lines of intersection lie entirely on the surface of the horizontal prism. The edges of the horizontal prism are numbered in the three projections as shown in the figure. A horizontal line is chosen passing through the edge 1″ in the side elevation. This line intersects the edges of the vertical prism in points a and p in the plan. These points are projected on the line passing through point 1′ at a′ and p′ in the elevation. Another horizontal line is chosen passing through the edge 2″ in the side elevation. This line intersects the faces of the vertical prism at b and q in the plan. These points are projected on the line passing through point 2′ at b′ and q′ in the elevation. Two more horizontal lines are chosen passing through points 3″ and 4″ in the side elevation. Points of intersections c, r, d and s are obtained on the respective lines in the plan. These points are then projected at c′, r′, d′ and s′ on the respective lines in the elevation. The lines of intersection are obtained connecting the points a′-b′ and b′-c′. The invisible lines of intersection connecting the point d′ with a′ and c′ coincide with the visible lines a′-b′ and b′-c′. The lines of intersection are also obtained connecting the points p′-q′ and q′-r′. The invisible lines of intersection connecting the point s′ with p′ and r′ coincide with the visible lines p′-q′ and q′-r′.

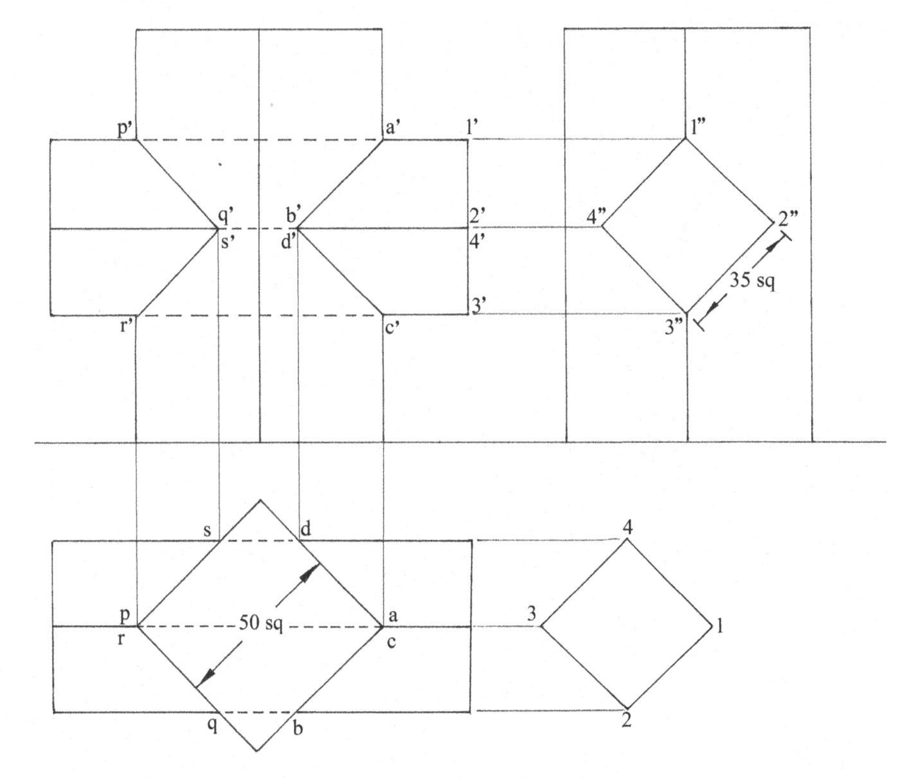

Fig. 12.11 Intersections of square prisms (line method)

12. A vertical square prism base 45 mm side has its lateral faces equally inclined to the VP. It is completely penetrated by a square prism of base 30 mm side, the axis of which is parallel to both the HP and VP and is 5 mm in front of the axis of the vertical prism. The lateral surfaces of the horizontal prism are equally inclined to the VP. Draw the projections of the prisms showing the lines of intersection. Assume suitable lengths for both the prisms.

The plan, elevation and side elevation of the outline of intersecting prisms with offset are drawn as shown in Fig. 12.12. The lines of intersection lie entirely on the surface of the horizontal prism. The edges of the horizontal prism are numbered in the three projections as shown in the figure. A horizontal line is chosen passing through the edge 1″ in the side elevation. This line intersects the faces of the vertical prism in points a and p in the plan. These points are projected on the line passing through point 1′ at a′ and p′ in the elevation. Another horizontal line is chosen passing through the edge 2″ in the side elevation. This line intersects the faces of the vertical prism at b and q in the plan. These points are projected on the line passing through point 2′ at b′ and q′ in the elevation. Two more horizontal lines are chosen passing through points 3″ and 4″ in the side elevation. Points of intersections c, r, d and s are obtained on the respective lines in the plan. These points are then projected at

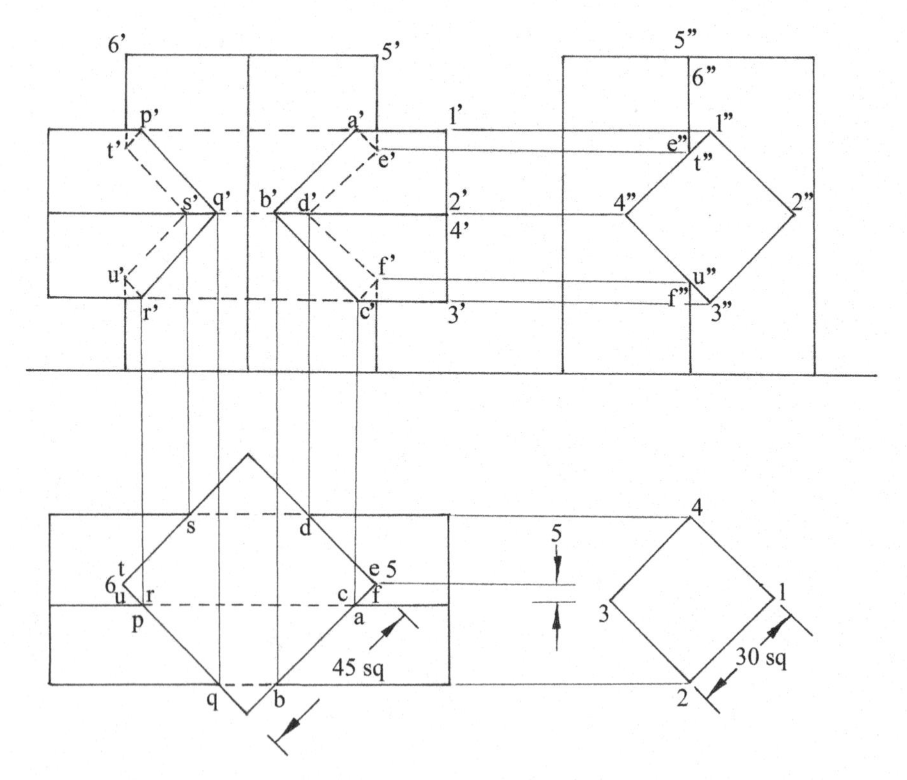

Fig. 12.12 Intersections of square prisms (line method)

c′, r′, d′ and s′ on the respective lines in the elevation. Two vertical lines are chosen passing through points 5 and 6 in the plan. The vertical edges passing through the points 5 and 6 are numbered in the side elevation and elevation. The vertical line passing through point 5″ intersects the surfaces of horizontal prism at e″ and f″ in the side elevation. These points are projected on the vertical line passing through 5′ at e′ and f′ in the elevation. These points are also shown in the plan. The vertical line passing through point 6″ intersects the surfaces of horizontal prism at t″ and u″ in the side elevation. These points are projected on the vertical line passing through point 6′ at t′ and u′ in the elevation. These points are also shown in the plan. There are 12 points that establish the required lines of intersection on both sides of the prisms. The lines of intersection are shown by full lines connecting points a′-b′ and b′-c′, and the invisible lines of intersection are shown as dotted connecting the points a′-e′, e′-d′, d′-f′ and f′-c′ on one side. The lines of intersection are shown by full lines connecting points p′-q′ and q′-r′, and the invisible lines of intersection are shown as dotted connecting the points p′-t′, t′-s′, s′-u′ and u′-r′ on the other side.

13. A square pipe of 40 mm sides has to be connected to another square pipe of sides 30 mm. The axis of the bigger pipe is vertical, and the axis of the smaller pipe intersects the axis of the bigger pipe at 45°. All the faces of both the pipes are equally inclined to the VP. Draw the projections of the arrangement showing clearly the lines of intersection. Assume suitable lengths for the pipes.

    The plan and elevation of the outline of square pipes are drawn as shown in Fig. 12.13 for the given data. The intersection lines lie entirely on the surfaces of the smaller pipe. Four inclined lines are chosen on the edges of the smaller pipe whose axis intersects the axis of the bigger pipe at 45°. The lines are numbered in the plan and elevation as shown in the figure. The line passing through point 1 meets the vertical edge of the bigger pipe at a in the plan. This point is projected on the line passing through point 1′ at a′ in the elevation. The line passing through point 2 meets the surface of the bigger pipe at b in the plan. This point is projected on the line passing through point 2′ at b′ in the elevation. The line passing through point 3 meets the vertical edge of the bigger pipe at c in the plan. This point is projected on the line passing through 3′ at c′ in the elevation. The line passing through point 4 meets the surface of the bigger pipe at d in the plan. This point is projected on the line 4′ at d′ in the elevation. The lines of intersection are shown by full lines connecting the points a′-b′ and b′-c′, and the invisible lines of intersection connecting the points c′-d′ and d′-a′ coincide with the visible lines.

14. A vertical square prism, base 50 mm side and height 100 mm, has one of its rectangular faces inclined at 30° to the VP. It is completely penetrated by a horizontal square prism, base 40 mm side and axis 100 mm long, the rectangular faces of which are equally inclined to the VP. The axes of the two prisms are parallel to the VP and bisect each other at right angles. Draw the projections of the prisms showing the lines of intersection.

    The plan, elevation and side elevation of the outline of intersecting prisms are drawn for the given data as shown in Fig. 12.14. The intersection lines lie

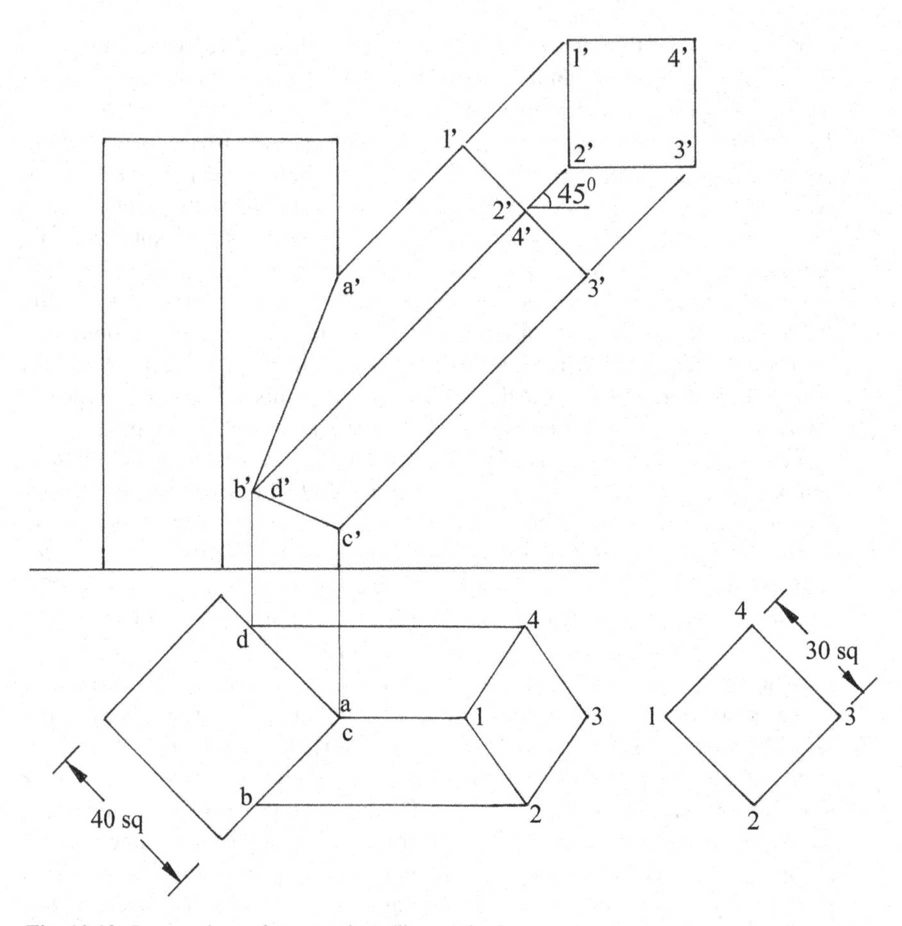

Fig. 12.13 Intersections of square pipes (line method)

entirely on the surface of the horizontal prism. Four lines are chosen on the edges of the horizontal prism. The lines are numbered in the side elevation, plan and elevation as shown in the figure. The line passing through point 1 meets the surfaces of vertical prism at a and p in the plan. These points are projected on the line passing through 1′ at a′ and p′ in the elevation. The line passing through point 2 meets the surfaces of vertical prism at b and q in the plan. These points are projected on the line passing through 2′ at b′ and q′ in the elevation. Lines passing through points 3 and 4 meet the surfaces of vertical prism and provide the points c, r, d and s in the plan. These points are then projected on the respective lines at c′, r′, d′ and s′ in the elevation. Two vertical lines are chosen passing through the point 5 and 6 in the plan. These lines are shown in the side elevation and the elevation. The line passing through 5″ meets the surfaces of the horizontal prism at f″ and e″ in the side elevation. These points are located at f′ and e′ on the line passing through 5′ in the

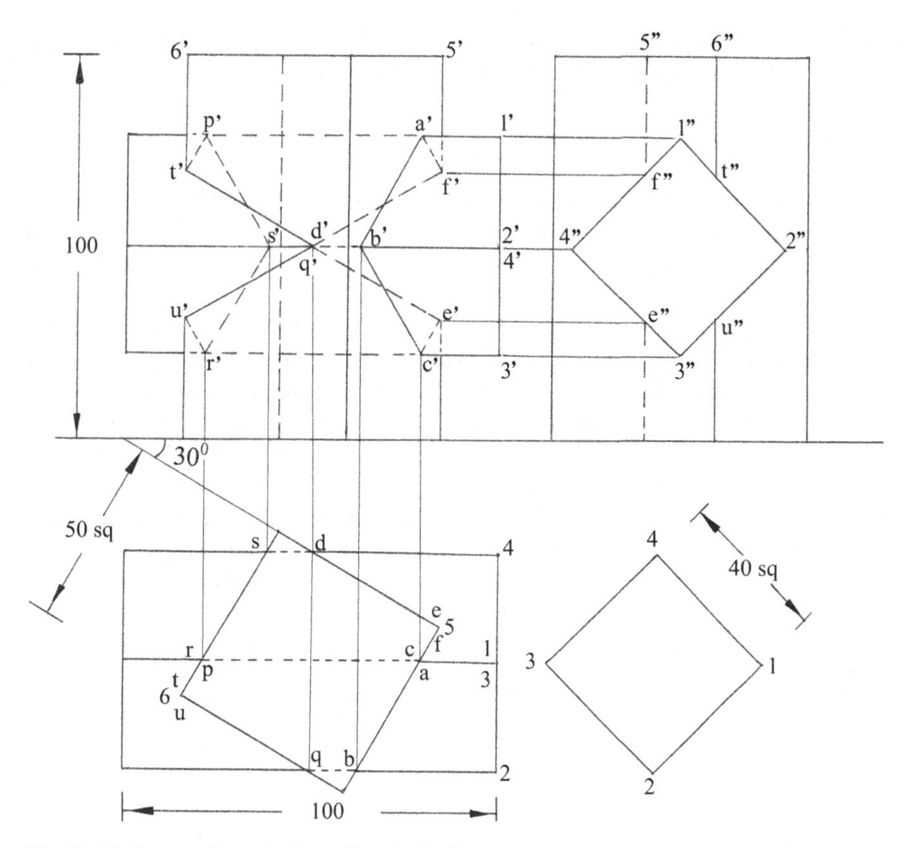

Fig. 12.14 Intersections of prisms (line method)

elevation. These points are also shown in the plan. The line passing through 6″ meets the surfaces of horizontal prism at t″ and u″ in the side elevation. These points are located at t′ and u′ on the line passing through point 6′ in the elevation. These points are shown in the plan. The lines of intersection are shown by full lines connecting the points a′-b′ and b′-c′, and the invisible lines of intersection connecting the points c′-e′, e′-d′, d′-f′ and f′-a′ are shown as dotted on one side. The lines of intersection are shown by full lines connecting the points t′-q′ and q′-u′, and the invisible lines of intersection connecting the points u′-r′, r′-s′, s′-p′ and p′-t′ are shown as dotted on the other side.

15. A conical funnel, top diameter 100 mm, has a horizontal cylindrical branch of 50 mm diameter at its bottom. The cylindrical branch and the conical portion envelope each other such that the axes intersect at right angle. The height of the funnel is 75 mm. Draw the plan and elevation of the funnel showing the curve of intersection.

The side elevation of the outline of funnel is drawn such that the extreme generators of conical portion are tangents to the circle of diameter 50 mm representing the side elevation of cylindrical branch of the funnel as shown in Fig. 12.15. The plan and elevation of the outline of the funnel are drawn. When the cylindrical branch and conical portion of the funnel envelope each other with their axes intersecting at right angle, the curve of intersection is obtained connecting the intersection of extreme elements of cylinder and cone (i.e. connecting the points a′ and e′) as shown in the elevation of the funnel. The circumference of the circle is divided into eight equal parts, and the division points are numbered in the projections of the cylindrical branch of the funnel. Locus lines are drawn through points 1″, 2″, 3″, 4″, 5″, 6″, 7″ and 8″ from the side elevation. Points b′, c′, d′, f′, g′ and h′ are located at the intersections of locus lines with the curve of intersection in the elevation. Locus lines are drawn through points 1, 2, 3, 4, 5, 6, 7 and 8. Projectors are drawn through a′, b′, c′, d′, e′, f′, g′ and h′ to locate points a, b, c, d, e, f, g and h on the respective locus lines in the plan. A smooth curve is drawn passing through the points a-b-c-d-e-f-g-h-a which is the required curve of intersection in the plan.

16. A vertical cone 90 mm diameter and 100 mm long is penetrated by a vertical cylinder of 50 mm diameter and 120 mm long. The axis of the cylinder is 10 mm away from that of the cone, and the plane containing the axes is parallel to the VP. Draw the projections showing the curve of intersection.

    The plan and elevation of the outline of the cone and the cylinder are drawn as shown in Fig. 12.16. The curve of intersection lies on the surface of the cylinder and coincides along the plan of the cylinder. Since the axes of both

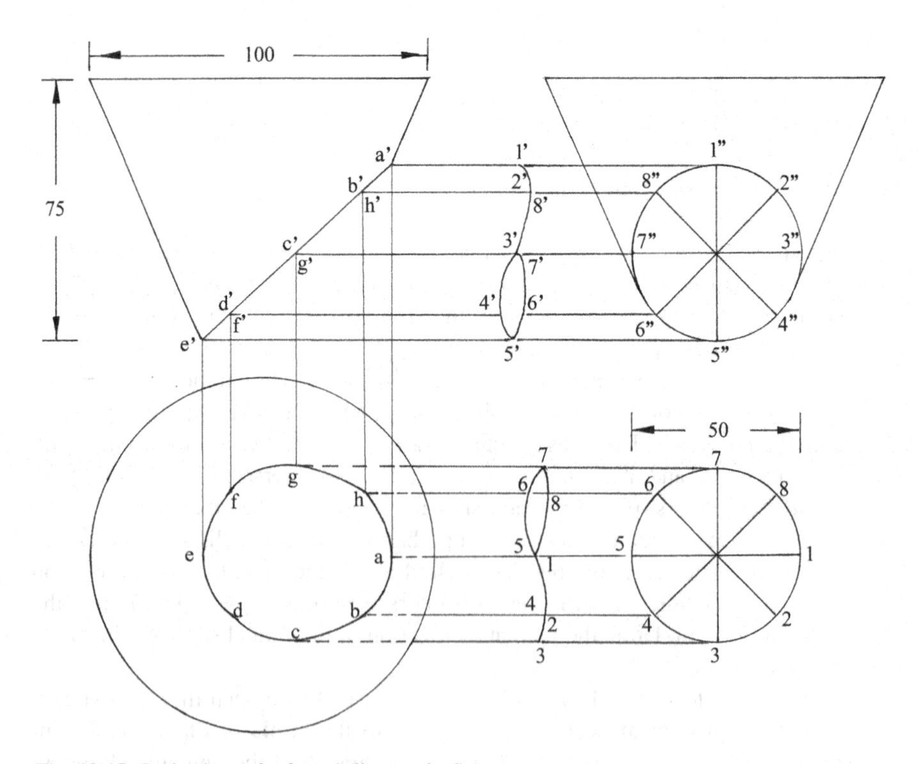

Fig. 12.15 Projections of a funnel (line method)

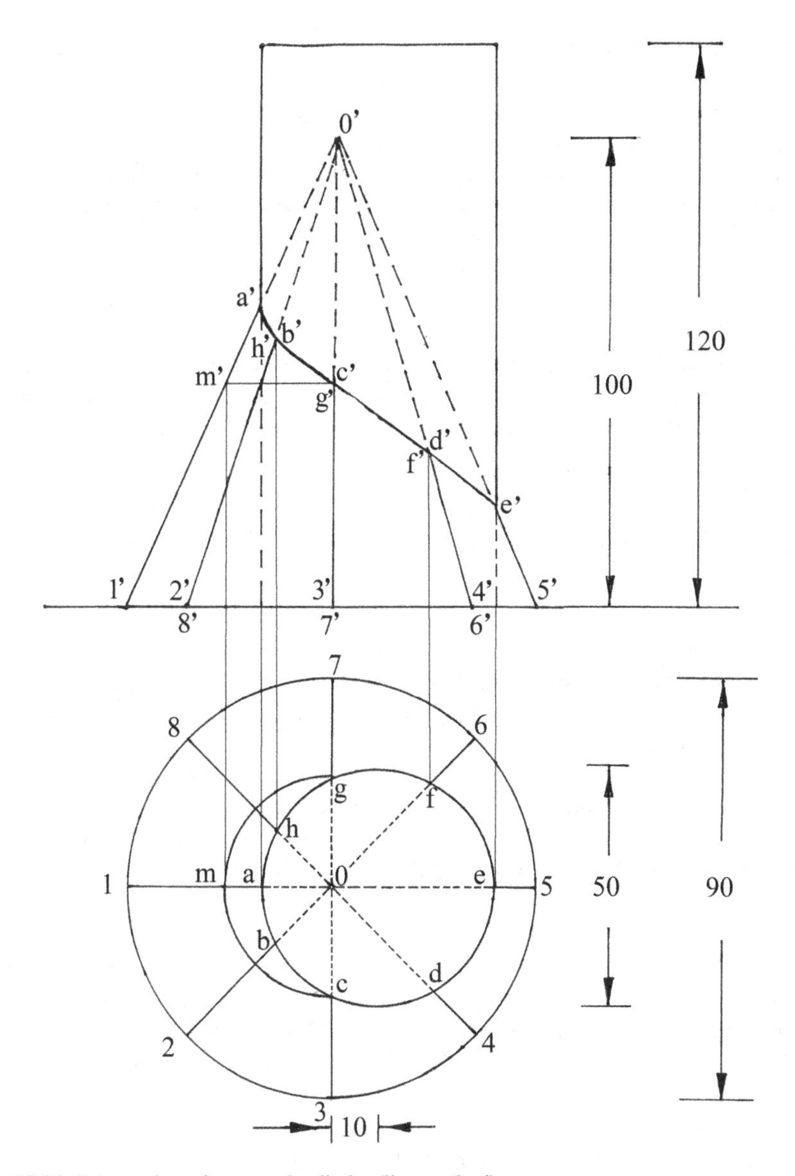

Fig. 12.16 Intersection of cone and cylinder (line method)

solids are parallel, we can choose the line method for obtaining the curve of intersection. Eight generators are chosen on the lateral surface of the cone by dividing the circumference of the base of the cone into eight equal parts. These lines are designated as 01, 02, etc. These generators are projected on the lateral surface of the cone in the elevation. The generator passing through point 1 intersects the surface of the cylinder in point a in the plan. This point is projected on the generator $0'1'$ at a' in the elevation. The generator passing through point 2 intersects the surface of the cylinder in point b in the plan. This point is projected on the generator 0′2′ at b′ in the elevation. The generators passing through points 3 and 7 intersect the surface of the cylinder in points c and g in the plan. These two points lie on the generators whose projections are perpendicular to the reference line. With centre 0 and radius 0c, draw an arc to cut the generator passing through point 1 at m in the plan. A projector is drawn from m to locate m′ on the generator 0′1′. A horizontal line drawn from m′ locates the points c′ and g′ on the projectors 0′3′ and 0′7′ in the elevation. The generators passing through points 4, 5, 6 and 8 intersect the surface of the cylinder in points d, e, f and h in the plan. These points are then projected on the respective generators to locate points d′, e′, f′ and h′ in the elevation. Points a′, b′, c′, d′ and e′ are visible and are joined by a smooth curve. The points f′, g′ and h′ are directly behind the points d′, c′ and b′. The curve connecting these points with a′ and e′ coincide with the visible curve already drawn.

Practice Problems

1. A cylinder of 80 mm diameter is standing vertically on the HP, and it is completely penetrated by another horizontal cylinder of 60 mm diameter. If the axes of the cylinders intersect each other and the axis of the horizontal cylinder is parallel to the HP, draw the projections of the cylinders showing the curves of intersection.
2. A vertical cylinder 80 mm diameter is standing on the HP and is penetrated by another cylinder of 60 mm diameter whose axis is parallel to both the reference planes and 10 mm in front of the axis of the vertical cylinder. Draw the projections of the cylinders showing the curves of intersection.
3. A cone of 100 mm diameter and axis 120 mm long, resting on its base on the HP is penetrated by a horizontal cylinder 50 mm diameter and 120 mm long. The axis of the cylinder is 30 mm above the base of the cone and parallel to the VP. If the axis of the cylinder is 5 mm in front of the axis of the cone, draw the projections of the solids showing the curves of intersection.
4. A cone base 100 mm diameter and axis 120 mm long is resting on its base on the HP. It is completely penetrated by a horizontal cylinder of 50 mm diameter in such a way that the cylinder and cone envelope a common sphere and their axes meet at right angles. Draw the projections of the solids showing the curves of intersection.
5. A vertical cylinder of 80 mm diameter and 100 mm long is penetrated by a horizontal cone of base 90 mm diameter and axis 100 mm long. The axis of cone is 50 mm above the base of the cylinder. The axes of both the solids bisect each other. Draw the projections of the solids showing the curves of intersection.
6. A vertical square prism of sides 50 mm is penetrated by another square prism 40 mm side, the axis of which is inclined at 45° to the HP and parallel to the VP. The axis of vertical prism is 5 mm behind the axis of penetrating prism and the rectangular faces of both the prisms are equally inclined to the VP. Show the lines of intersection. Assume suitable lengths for both the prisms.

Chapter 13
Development of Surfaces

The development of a three-dimensional object made of thick paper or sheet metal is a plane figure obtained by unfolding the surface of the object onto a plane. The development consists of drawing successive surfaces of the object in its true size. As there is no stretching of the surface in the development, every line on the development shows the true length of the corresponding line on the surface. Polyhedrons and single-curved surfaces (cylinder and cone) are developable. Double-curved surfaces cannot be developed to their true sizes, because these surfaces contain no straight lines.

The procedure in development is to select convenient lines on a given surface and then to find the true surface relationship of these lines. Reproduction of the relationship on a plane surface produces the development. For prisms and cylinder, the lines parallel to the axis are convenient. For pyramids and cone, the convenient lines are those passing through the apex or the vertex.

Classification of Developments

1. Parallel line method. This method is used for solid surfaces generated by a line which moves parallel to the axis of solid. Developments of prisms and cylinder can be drawn by drawing stretch out line or girth line. The stretch out line gives the perimeter of the object.
2. Radial line method. This method is used for pyramids and cone.
3. Triangulation development. This method is suitable for transition piece like a hopper that connects two openings of different shapes of cross sections or of the same shaped cross sections with offset. The plane surfaces are divided into triangular areas, and each triangle can be laid down in the development with its three sides.
4. Approximate development. This method is employed for double-curved surfaces such as sphere and anchor ring.

K. Rathnam, *A First Course in Engineering Drawing*,
DOI 10.1007/978-981-10-5358-0_13

The development of a surface is generally drawn with inside measurements of unfolded thick paper or sheet metal. However, it is common practice in aircraft industry and in air conditioning plants to make the layout with outside measurements, so that when the form is completed the folded lines are on the outside. In developments allowances to be given for lap at seams are ignored unless stated otherwise.

Solved Problems

1. Draw the development of the surface of a square prism, side of base 40 mm and axis 60 mm long.

   The plan and elevation of the square prism are drawn as shown in Fig. 13.1 (i). The stretch out line or girth line A-A is drawn as shown in Fig. 13.1(ii). Points B, C and D are marked on this line such that AB = BC = CD = DA = 40 mm to represent the length of sides at the top. The stretch out line E-E is also drawn marking the points F, G and H on it such that EF = FG = GH = HE = 40 mm to represent the length of sides at the base. Join the verticals AE, BF, CG, DH and AE of length 60 mm representing of edges and the fold lines. Two squares ABCD and EFGH are drawn to represent the top and base of the square prism. The complete development of square prism is shown in Fig. 13.1(ii) following the parallel line method.
2. Draw the development of a square pyramid of base 40 mm square and height 60 mm.

   The plan and elevation of the square pyramid are drawn as shown in Fig. 13.2(i). The true length of slant edge is found out by the construction

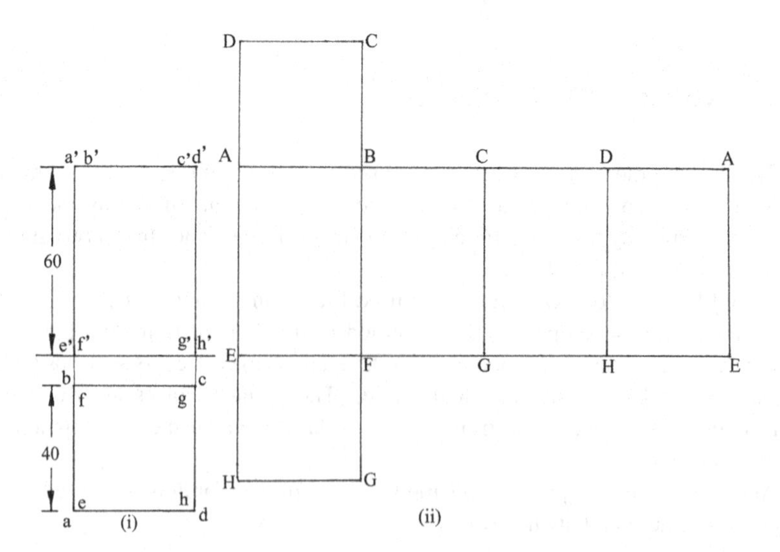

Fig. 13.1 Development of square prism

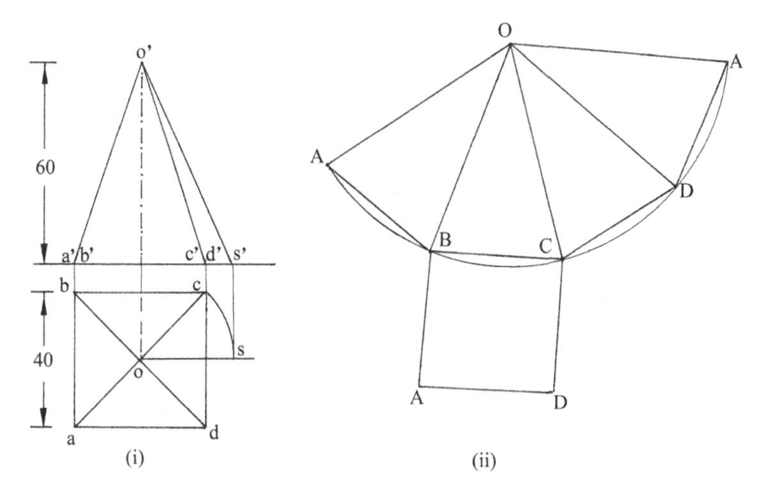

Fig. 13.2 Development of square pyramid

shown in the figure. The true length TL of the slant edge is given by the line o′s′. With centre O and radius o′s′, draw an arc ABCDA. Mark the points B, C and D such that AB = BC = CD = DA = 40 mm, side of base. Join the chords AB, BC, CD and DA. A square ABCD is drawn on the side BC. The square represents the base of the pyramid. The complete development of the pyramid is shown in Fig. 13.2(ii) following the radial line method.

3. Draw the development of a tetrahedron of edge 40 mm.

   The plan and elevation of the tetrahedron are drawn as shown in Fig. 13.3(i). The elevation o′a′ will have true length of 40 mm. The tetrahedron consists of four equilateral triangles of side 40 mm. The development of the tetrahedron is shown in Fig. 13.3(ii). It may be noted that the edges AB, BC, CA, OA, OB and OC are each 40 mm long.

4. Draw the development of an octahedron of edge 40 mm.

   The plan and elevation of the octahedron are drawn as shown in Fig. 13.4(i). The octahedron consists of eight equilateral triangles of side 40 mm. The 12 edges are each 40 mm long. Assume the octahedron is made of thick wrapper. Let us cut the octahedron along the edges AB, EB, AD, CD and CF and open the surfaces unfolding other edges. The surfaces lie on a plane as shown in Fig. 13.4(ii). The development of octahedron is thus obtained. It is to be noted that this is one of the possible solutions.

5. A triangular prism, side of base 30 mm and height 40 mm, stands on the HP on its base with a rectangular face perpendicular to the VP. It is cut by a plane perpendicular to the VP, inclined at 30° to the HP and passing through a point 20 mm above the base along the axis. Draw the development of the lower portion of the prism.

   The plan and elevation of the triangular prism are drawn as shown in Fig. 13.5(i). The stretch out line A-A of length 3×30 mm is drawn, and points B and C are marked as shown in Fig. 13.5(ii). The stretch out line D-D of length

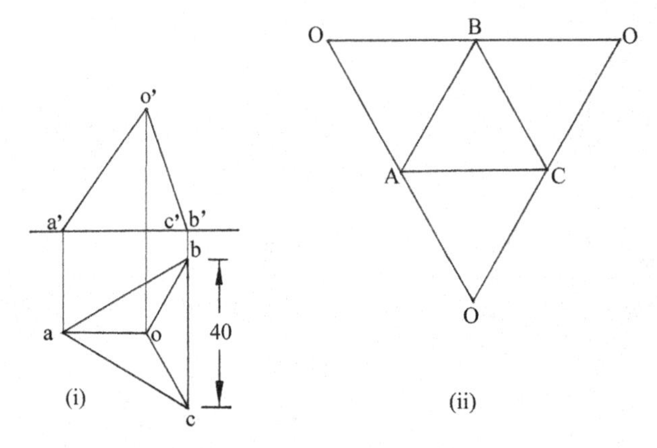

Fig. 13.3 Development of tetrahedron

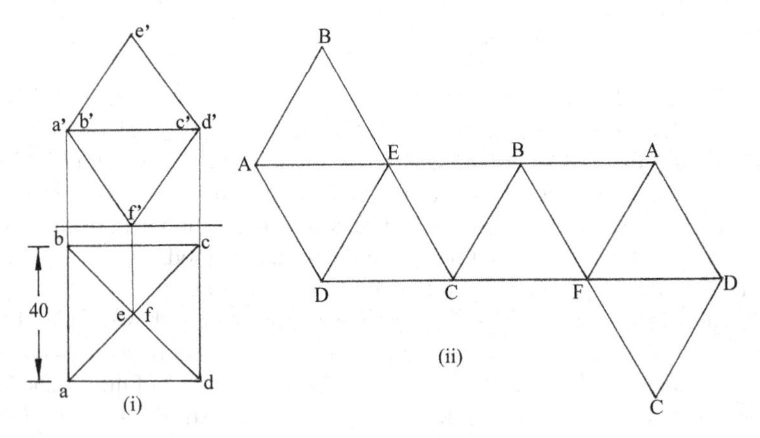

Fig. 13.4 Development of octahedron

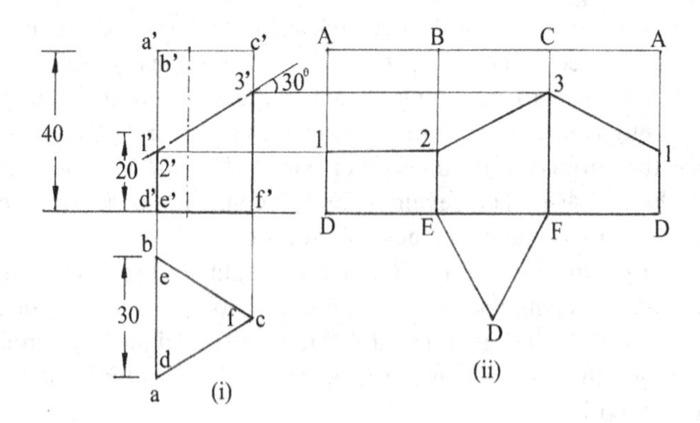

Fig. 13.5 Development of sectioned triangular prism

3×30 mm is also drawn and points E and F are marked. Join the verticals AD, BE, CF and AD of length 40 mm each to represent the vertical edges of the prism. The prism is cut by a section plane perpendicular to the VP, inclined at 30° to the HP and passing through a point 20 mm above the base along the axis as shown in the elevation. The section plane cuts the vertical edges AD, BE and CF of the prism at points $1'$, $2'$ and $3'$. A horizontal line is drawn through point $1'$ to intersect the vertical edge AD at point 1 in the development. The same point is also located at the right extreme edge AD in the development. Points 2 and 3 are located on the respective edges following the same procedure. Join these points to obtain the development of lateral surface of the lower cut prism. A triangle EDF is also drawn to complete the development of the cut prism as shown in Fig. 13.5(ii).

6. A square prism, edge of base 30 mm and axis 60 mm long, has its base on the HP, and its faces are equally inclined to the VP. It is cut by a plane perpendicular to the VP, inclined at 60° to the HP and passing through a point 45 mm above the base along the axis. Draw the development of the lower portion of the prism.

   The plan and elevation of the square prism are drawn as shown in Fig. 13.6 (i). The development of the full prism is drawn by thin lines as shown in Fig. 13.6(ii). The prism is cut by a section plane perpendicular to the VP, inclined at 60° to the HP and passing through a point 45 mm above the base along the axis as shown in the elevation. The section plane cuts three vertical edges and the top end of the prism at two points. The section points are marked as shown in the elevation of the prism. A horizontal line is drawn through point $1'$ to intersect the vertical edge AE at point 1 in the development. The same

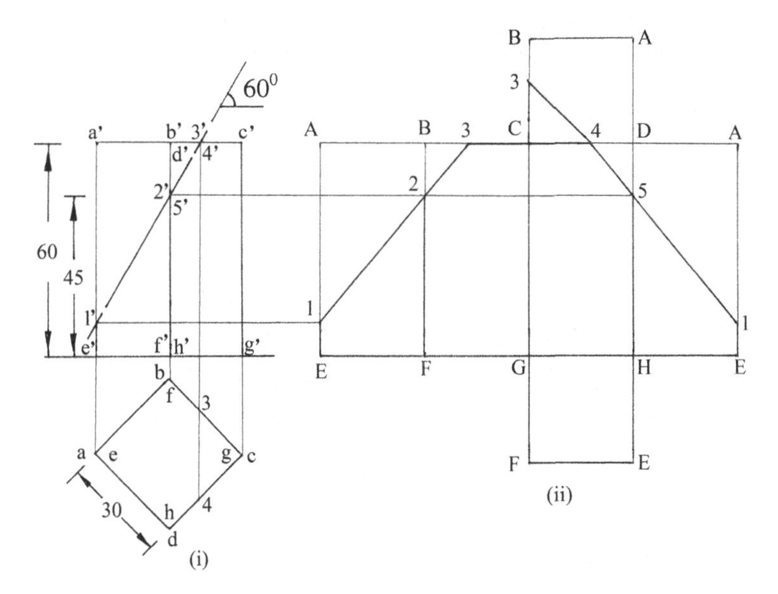

Fig. 13.6 Development of sectioned square prism

point is also located at the right extreme edge AE in the development. Points 2 and 5 are located on the respective edges following the same procedure. Projectors are drawn through points 3′ and 4′ to locate points 3 and 4 in the plan of the prism. The distance of point 3 from B in the development is equal to the distance of 3 from b in the plan. Hence, point 3 can be located in the development. Similarly, point 4 is located on the edge DC in the development. Join the points 1-2-3, 3-4 and 4-5-1 by straight thick lines. The vertical edges connecting the section points with the base corner points are also drawn thick. A square EFGH is drawn to show the base of the prism in the development. The triangle 3-4-C represents the top end of the cut prism in the development.

7. A hexagonal prism of side of base 30 mm and height 60 mm is resting on the HP with a side of base parallel to the VP. Right half of the prism is cut by an upward plane inclined at 60° to the HP and starting from the axis 30 mm below the top end. Left half of the prism is cut by a plane inclined at 30° to the HP downwards from the axis. The two section planes are continuous. Draw the development of the lower portion of the prism.

   The plan and elevation of the hexagonal prism are drawn as shown in Fig. 13.7(i). The development of the prism is drawn by thin lines as shown in Fig. 13.7(ii). The right half of the prism is cut by an upward section plane perpendicular to the VP, inclined at 60° to the HP and passing through a point 30 mm along the axis below the top end. The left half of the prism is cut by a plane inclined at 30° to the HP downwards from the axis. The two continuous section planes cut the vertical edges and the top end of the prism. The section

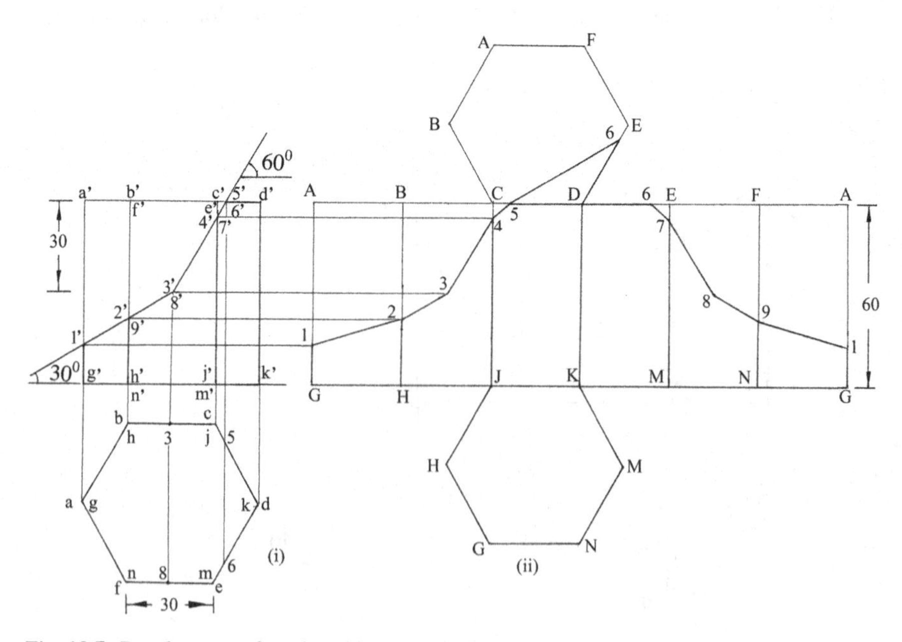

Fig. 13.7 Development of sectioned hexagonal prism

points are marked as shown in the elevation. The points 3 and 8 are common to both section planes. Points 3′, 5′, 6′ and 8′ are projected in the plan. Points 1, 2, 4, 7 and 9 are located on the respective vertical edges of the prism in the development by drawing horizontal lines through 1′, 2′, 4′, 7′ and 9′. Horizontal line drawn through 3′ contains points 3 and 8 in the development. Point 3 lies on the mid-line of the face BCJH and point 8 lies on the mid-line of the face EFNM. These points are located on the intersection of horizontal line through 3′ with mid-line of the faces BCJH and EFNM. Point 5 is located on the edge CD in the development by using the distance of point 5 from c in the plan. Similarly point 6 is located on the edge DE. Join the points 1-2-3-4-5-6-7-8-9-1 by thick lines. The vertical edges connecting the section points with the base corner points are drawn by thick lines. A hexagon GHJKMN on the side JK is also drawn to show the base of the prism in the development. The triangle D-5-6 represents the top portion of the cut prism in the development.

8. A cube of side 40 mm rests on its base on the HP with a vertical face inclined at 30° to the VP. It is cut into two halves by a plane perpendicular to the VP and inclined at 30° to the HP. Draw the development of one portion of the cube.

   The plan and elevation of the cube are drawn as shown in Fig. 13.8(i). The section plane is drawn passing through a point 20 mm below the top end along the axis of cube. Development of the cube is drawn by thin lines without the top end as shown in Fig. 13.8(ii). The section points are marked along the vertical edges in the elevation. A horizontal line is drawn through point 1′ to intersect the edge AE at point 1 in the development. Points 2, 3 and 4 are located on the respective edges following the same procedure. Join the points 1-2-3-4-1 by straight lines, and these points are also connected to the base corner points of

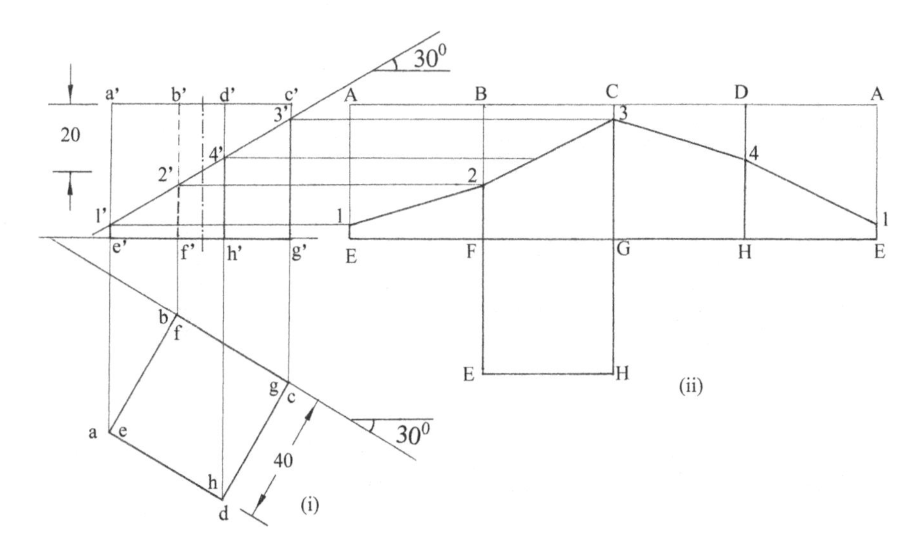

Fig. 13.8 Development of sectioned cube

the cube. A square EFGH is drawn through the edge FG to complete the development of the sectioned cube.

9. Draw the development of the lower portion of a cylinder, diameter of base 50 mm and axis 70 m, when it is cut by a plane perpendicular to the VP, inclined at 40° to the HP and passing through the mid-point of the axis.

   The plan and elevation of the cylinder are drawn as shown in Fig. 13.9(i). The plan is divided into eight equal parts. Through each of these division points, generators are drawn on the lateral surface of the cylinder in the elevation. The stretch out line A-A equal to the circumference of the cylinder is drawn, and this line is divided into eight equal parts with the division points as shown in Fig. 13.9(ii). From the division points, draw parallel lines equal to the length of the axis of the cylinder. These parallel lines are generators of the cylinder in the development. The section plane cuts the cylinder bisecting the axis as shown in the elevation. The intersection points $1'$, $2'$ $3'$, $4'$, $5'$, $6'$, $7'$ and $8'$ are marked in the elevation. Horizontal lines are drawn through these points $1'$, $2'$, $3'$, etc. The horizontal lines intersect and locate points 1, 2, 3, etc. on the corresponding generators in the development. A smooth curve is drawn passing through the points 1-2-3-4-5-6-7-8-1. A circle representing the base of the cylinder is also drawn passing through the point E with the stretch out line as tangent to complete the development of the cut cylinder.

10. A square pyramid, side of base 40 mm and height 60 mm, stands on the HP with a base edge parallel to the VP. It is cut by a plane inclined at 45° to the HP,

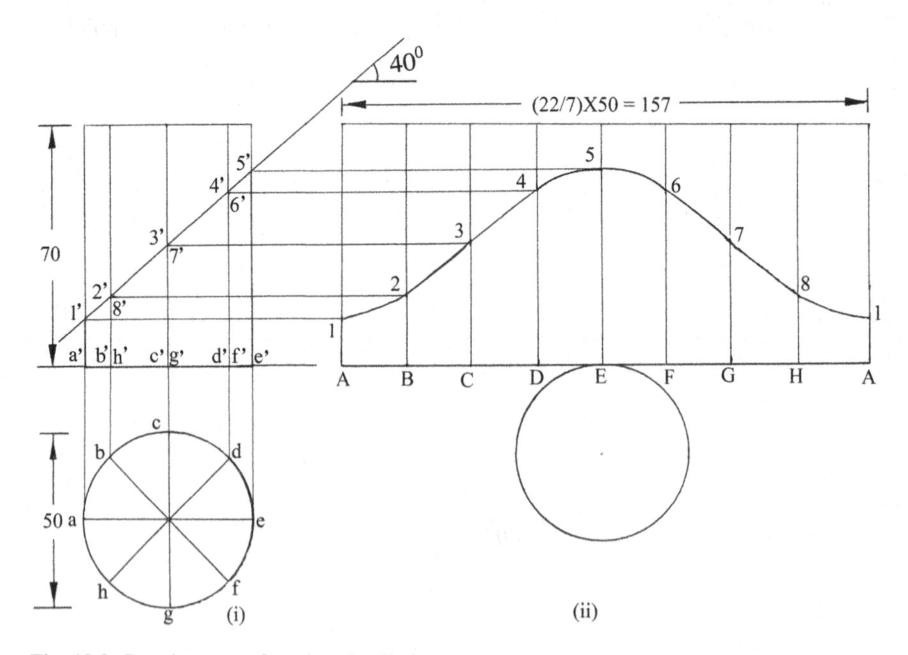

Fig. 13.9 Development of sectioned cylinder

perpendicular to the VP and bisecting the axis. Draw the development of truncated pyramid.

The plan and elevation of the square pyramid are drawn as shown in Fig. 13.10(i). The true length of slant edge is found out by the construction shown in the figure. The true length TL of the slant edge is given by the line o′s′. With centre O and radius o′s′, draw an arc ABCDA as shown in Fig. 13.10 (ii). Mark the points B, C and D such that AB = BC = CD = DA = 40 mm, side of base. Join the chords AB, BC, CD and DA. The section plane cuts the pyramid bisecting the axis as shown in the elevation. The intersection points 1′, 2′, 3′ and 4′ are marked on the slant edges. A horizontal line is drawn through point 1′ to intersect the edge o′s′ at 1 giving the distance of point 1 from O on the edge OA. The point 1 is marked on the edge OA in the development. The same procedure is repeated for the remaining points 2′, 3′ and 4′, and points 2, 3 and 4 are located on the respective edges in the development. These points 1-2-3-4-1 are joined by straight lines. These points are also joined to the base points A, B, C and D in the development. A square ABCD is drawn on the side CD. The square represents the base of the pyramid. The development of the cut pyramid is drawn by thick lines as shown in Fig. 13.10(ii).

11. A right circular cone has base diameter 40 mm and axis 40 mm long. Draw the development of the cone.

The plan and elevation of the cone are drawn as shown in Fig. 13.11(i). The slant length TL is found from the extreme generator o′s′ in the elevation. The development of the cone is a sector of a circle having radius equal to the slant length of the cone and the angle subtended by the circumference of its base. The angle subtended by the circumference of base circle is calculated from the formula $\Theta = (360° \times r)/(TL)$, i.e. $\Theta = (360°) \times (20)/(45) = 160°$, where r is the radius of base circle and TL is the slant length of the cone. The development of the cone is drawn as shown in Fig. 13.11(ii). The line OA is extended and a

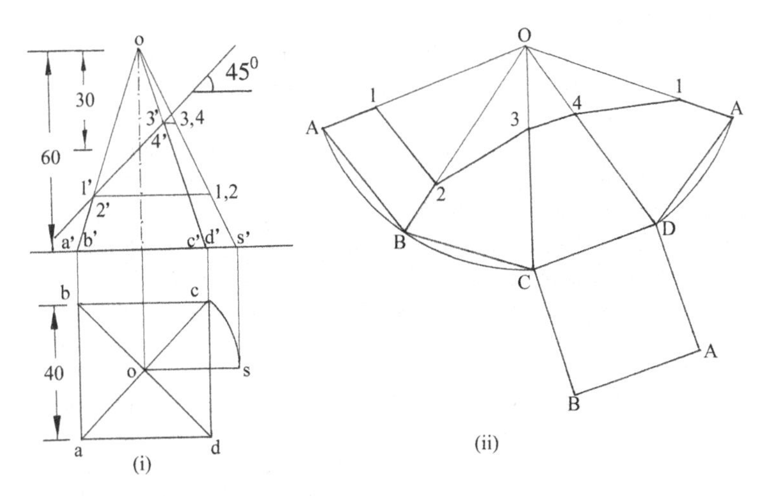

Fig. 13.10 Development of sectioned square pyramid

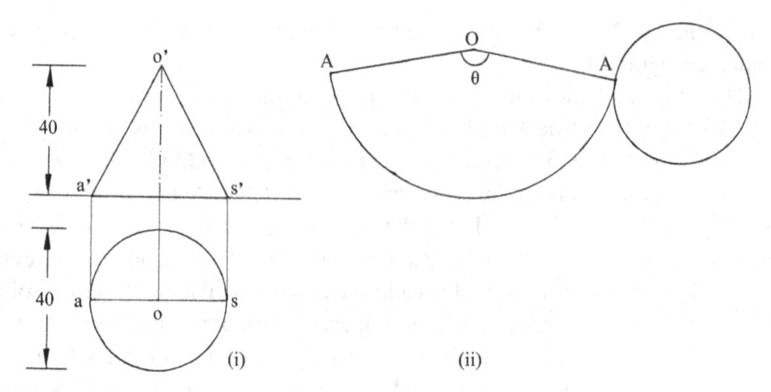

Fig. 13.11 Development of right circular cone

circle of radius of 20 mm is also drawn to complete the full development of the cone.

12. A street lamp shade is formed by cutting a cone of diameter of base 100 mm and axis 200 mm by a HP at 100 mm from the vertex and a plane inclined at 35° to the HP and passing through one extremity of the base. Draw the development of the lamp shade.

    The plan and elevation of the cone are drawn as shown in Fig. 13.12(i) adopting a scale of 1:2. The slant length TL of the right extreme generator, o′s′, is found to be 206 mm. The angle subtended by the circumference of base circle is calculated from the formula $\Theta = (360° \times r)/(TL)$, i.e. $\Theta = (360°) \times (50)/(206) = 87°$, where r is the radius of base circle and TL is the slant length of the cone. The development of the full cone is a sector of a circle having radius equal to the slant length of the cone and the angle subtended by the circumference of its base circle as shown in Fig. 13.12(ii). The cone is cut by a horizontal plane passing through a point 100 mm below the vertex in the elevation. With centre O and radius o′r′, draw an arc to cut the eight generators as shown in the development. The cone is cut by a plane inclined at 35° to the HP and passing through the left extremity of the cone in the elevation. The section plane cuts eight generators at 1′, 2′, 3′, 4′, 5′, 6′, 7′ and 8′. Horizontal lines are drawn through these points to meet the right extreme generator o′s′. The meeting points 1, 2, 3, 4, 5, 6, 7 and 8 furnish the distances of these points from O in the development. These points are marked in the respective generators in the development. Join the points 1-2-3-4-5-6-7-8-1 by a smooth curve to complete the development of the lamp shade.

13. A pentagonal pyramid, edge of base 30 mm and height 50 mm, stands on its base on the HP with an edge of base parallel to the VP. A section plane cuts the pyramid at point 30 mm along the axis and above the base and makes an angle of 50° with the axis. Draw the development of the truncated pyramid.

    The plan and elevation of the pentagonal pyramid are drawn as shown in Fig. 13.13(i). The true length of the slant edge is given by the line o′s′ in the elevation. The development of the pyramid is drawn by thin lines knowing the true length of the slant edge as shown in Fig. 13.13(ii). The pyramid is cut by a

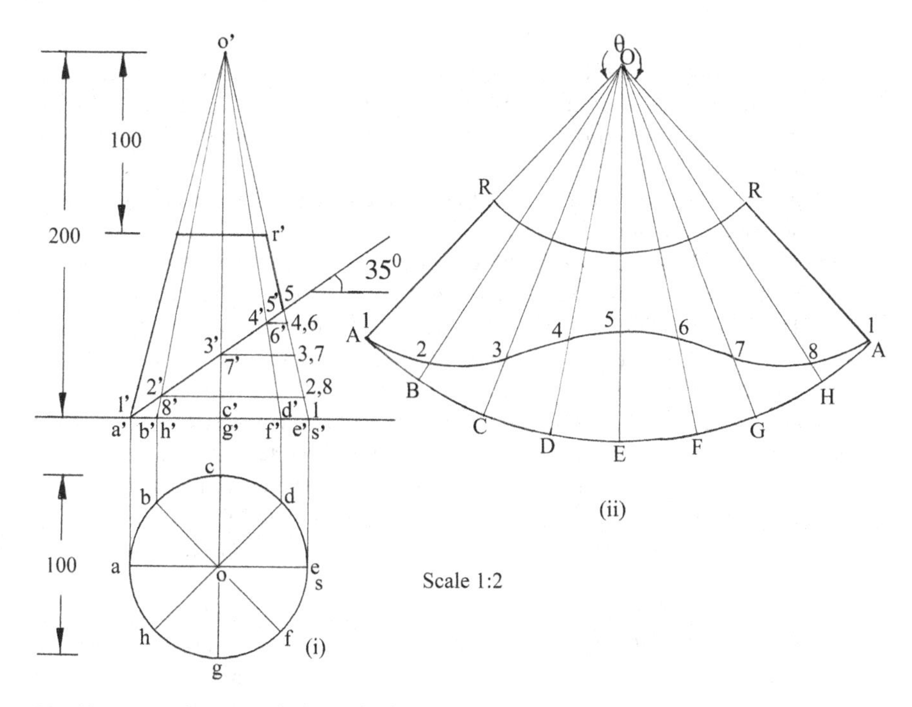

Fig. 13.12 Development of a lamp shade

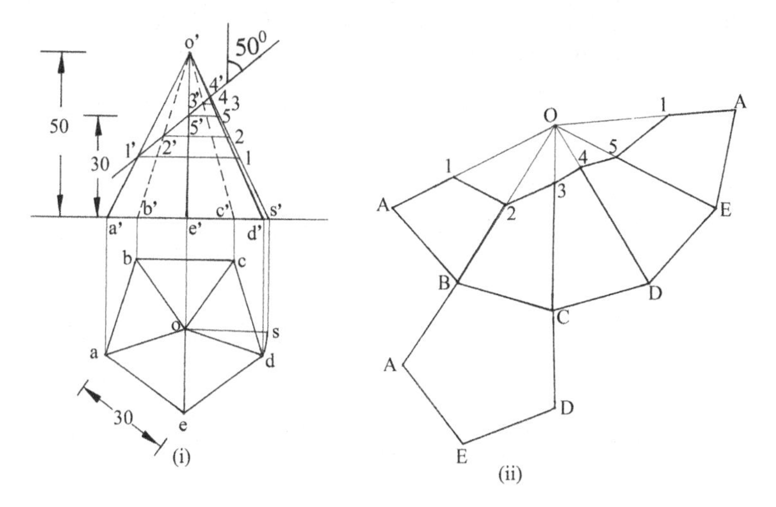

Fig. 13.13 Development of truncated pentagonal pyramid

section plane as shown in the elevation. The section plane cuts five slant edges. The section points 1′, 2′, 3′, 4′ and 5′ are marked on the slant edges. The distances of these points from the vertex/apex of the pyramid are obtained by drawing horizontal lines through these points in the elevation to intersect the edge o′s′ at 1, 2, 3, 4 and 5. These points are marked on the respective slant edges in the development. Join these points 1-2-3-4-5-1 and also the slant edges connecting these points with the base points by thick lines. A pentagon ABCDE on the edge BC is also drawn to complete the development of the truncated pyramid.

14. A hexagonal pyramid, side of base 30 mm and height 70 mm, stands on its base on the HP with a side of base parallel to the VP. It is cut up to the axis by a horizontal plane, 35 mm from the vertex and thereafter by a plane inclined at 60° to the HP. Draw the development of the lower part of the pyramid.

The plan and elevation of the hexagonal pyramid are drawn as shown in Fig. 13.14(i). The extreme edge o′d′ gives the true length of the slant edge of the pyramid. The full development of the pyramid is drawn by thin lines as shown in Fig. 13.14(ii). The left half of the pyramid is cut by a horizontal section plane meeting the axis 35 mm below the vertex and thereafter by a section plane inclined at 60° to the HP as shown in the elevation. The horizontal section plane cuts three slant edges o′a′, o′b′ and o′f′, and the inclined section plane cuts three remaining edges o′c′, o′d′ and o′e′. The two section planes meet at two points 3′ and 7′ on the triangular faces o′b′c′ and o′e′f′. The distances of points 1, 2, 4, 5, 6 and 8 from the vertex are measured from the extreme edge o′s′ which is parallel to the extreme slant edge o′d′. These points are marked on the respective slant edges knowing their distances from the vertex of the pyramid. The points 3 and 7 are to be marked on the medians of isosceles triangles OBC and OEF. It is to be noted that the lines 2-3 and 8-7 are, respectively, parallel to the base edges BC and FE. Join the points 1-2-3-4-5-6-7-8-1 by thick lines. The section points on the slant edges are connected to the

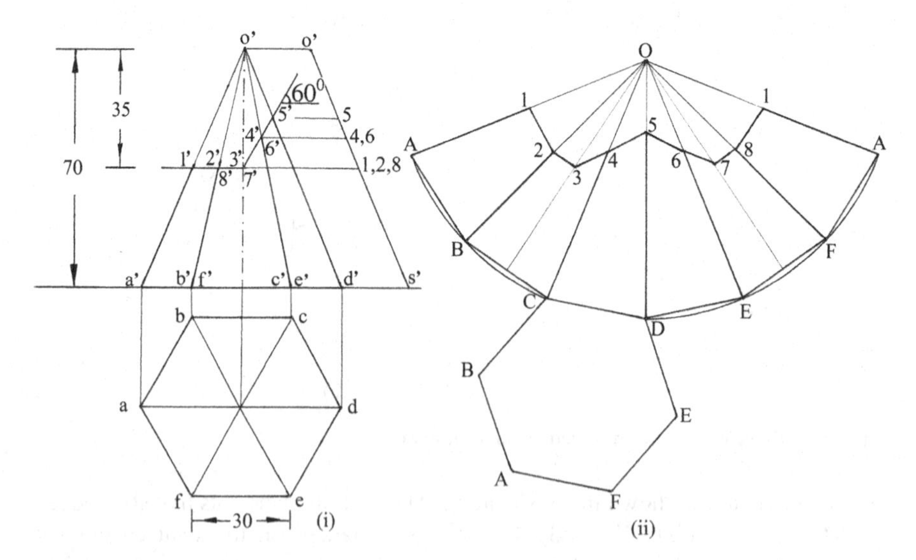

Fig. 13.14 Development of sectioned hexagonal pyramid

respective base points by thick lines. A hexagon ABCDEF on the edge CD is also drawn to complete the development of the lower part of the pyramid.

15. A conical pail has 400 mm diameter at the top and 200 mm diameter at the bottom and its depth is 250 mm. Draw the shape of the sheet metal required to make it.

    The elevation of the conical pail is drawn as shown in Fig. 13.15(i) adopting a scale of 1:5. The extreme generator b′a′ is extended to meet the axis of the pail at o′. The slant length TL of the extreme generator, o′s′, is found to be 550 mm. The angle subtended by the circumference of top circle is calculated from the formula $\Theta = (360^\circ \times r)/(TL)$, i.e. $\Theta = (360^\circ) \times (200)/(550) = 131^\circ$, where r is the radius of top circle and TL is the slant length of the cone. The development of the full conical pail is a sector of a circle having radius equal to the slant length of the cone and the angle subtended by the circumference of its top circle as shown in Fig. 13.15(ii). The distance of a′ from o′ is found to be 225 mm. The point A is located on OB such that the distance OA is equal to 225 mm. With centre O and radius 225, draw an arc to intersect the radial line OB at A. A circle of radius100 mm is also drawn to represent the development of bottom of the pail as shown in Fig. 13.15(ii). The sector A-B-B-A and the circle represent the full development of the conical pail.

16. A flue in the form of a frustum of a square pyramid has the base 6 m, top 2 m and height 12 m. A lightning conductor is taken from the mid-point of one of

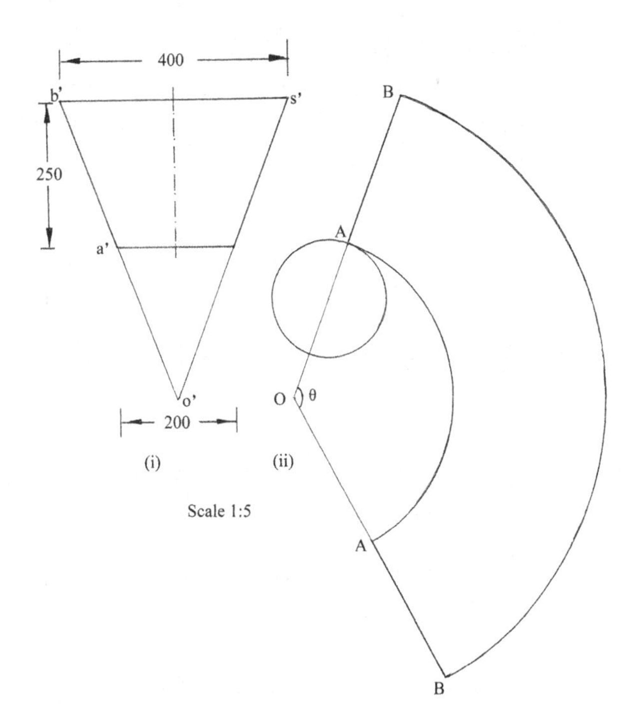

Fig. 13.15 Development of conical pail

the top edges to the mid-point of the opposite edge in the base. Determine the shortest length of the conductor and draw projections of its path.

The plan and elevation of the frustum of pyramid are drawn adopting a scale of 1:200 as shown in Fig. 13.16(i). The extreme edge is extended to cut the axis of the pyramid in the elevation at o′. The true length TL of the slant edge is given by the length o′s′ in the elevation. The development of the frustum of the pyramid is drawn as shown in Fig. 13.16(ii). The mid-point of the top edge EF is marked as point 1. Point 4 is marked on the opposite side CD in the base. Join the two points to determine the shortest length of the lightning conductor. Its length is 13.6 m. The lightning conductor cuts the edges FB and GC at point 2 and 4 in the development. The distances of the point 2 and 3 from O are transferred to the right extreme edge o′s′, and horizontal lines are drawn through these points to intersect the edges f′b′ and g′c′ at 2′ and 3′ in the elevation. Point 1′ is located at the mid-point of the edge e′f′ and point 4′ is located at the mid-point of the edge c′d′ in the elevation. These points are joined in the elevation showing the invisible portion of the line 1′2′ as dotted. These points are projected in the plan and joined. The conductor is visible in the plan.

17. A cone of base 60 mm diameter and height 80 mm is resting on the HP. An insect starts from a point on the circumference of the base, goes round the conical surface and reaches the starting point in the shortest route. Find the distance travelled by the insect and also the projections of the path followed by it.

The plan and elevation of the cone are drawn as shown in Fig. 13.17(i). The true length TL of the extreme generator, o′s′, is found to be 86 mm. Draw an arc A-A with radius equal to the true length of the generator on the cone as shown

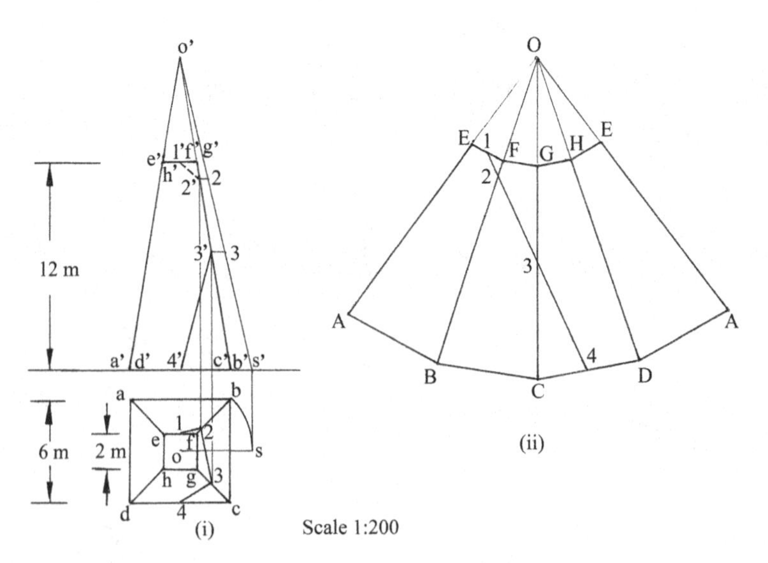

Fig. 13.16 Development of flue

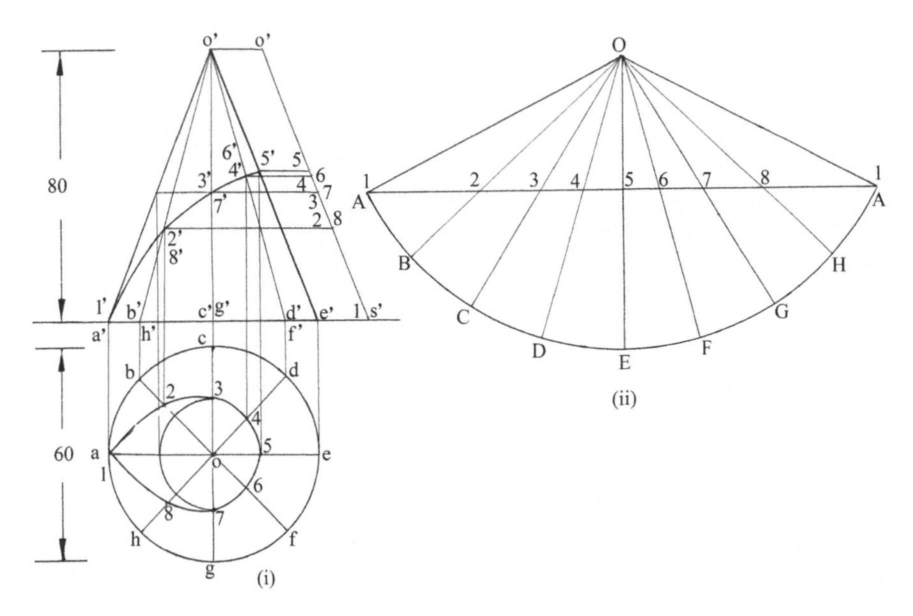

Fig. 13.17 Route map of an insect on a conical surface

in Fig. 13.17(ii). The angle subtended by the arc A-A is calculated from the formula $\Theta = (360° \times 30)/86 = 126°$. Eight generators are chosen at equidistant in the plan. These generators are projected in the elevation and also shown in the development of the cone. The insect starts from point A, goes round the conical surface and reaches the starting point A as shown in the development. Join the points A-A by a straight line that shows the shortest route of the insect. The path intersects the generators OA, OB, OC, etc. at points 1, 2, 3, etc. The distances of these points from O are measured and transferred on the extreme generator o′s′. Horizontal lines are drawn through these points to intersect the corresponding generators at 1′, 2′, 3′, etc. in the elevation. Join the points 1′-8′-7′-6′-5′ to show the visible portion of the path in the elevation. The invisible portion of the path 1′-2′-3′-4′-5′ coincides with the visible portion of the path in the elevation. These points are then projected on the corresponding generators in the plan. Join the points 1-2-3-4-5-6-7-8-1 to show the route map of the insect in the plan.

18. Draw a semicircle of diameter 120 mm and in it inscribe the largest equilateral triangle with its base on the diameter. The semicircle is the development of a cone and the triangle that of a line on its surface. Draw the plan and elevation of the cone and the line when resting with its base on the HP.

    A semicircle of diameter 120 mm is drawn and is divided into eight equal parts with eight generators as shown in Fig. 13.18(i). Taking the line OE as the altitude of the largest equilateral triangle, an equilateral triangle 1-5-1 is drawn with its side 1-1 on the diameter of the semicircle. The semicircle is the development of a cone. The slant length of the cone is 60 mm, and radius of

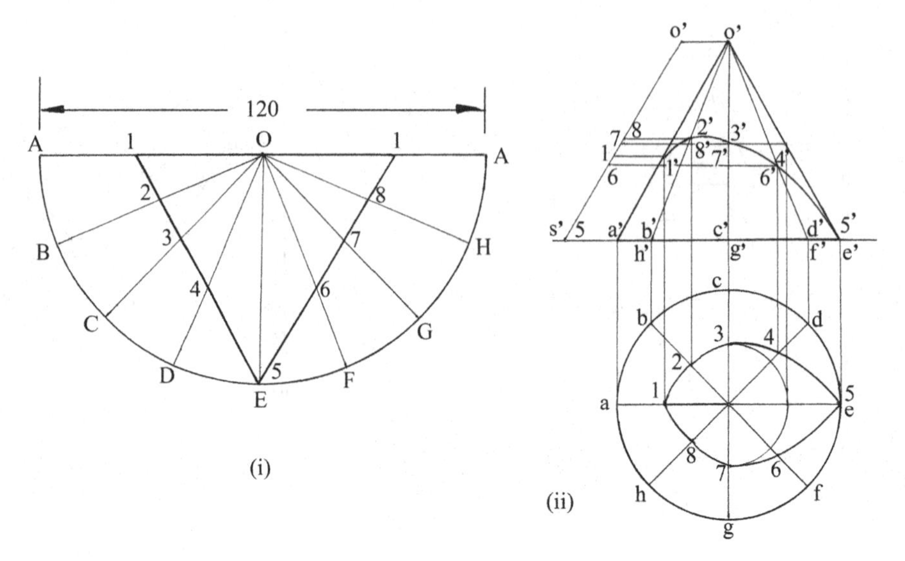

Fig. 13.18 Projections of cone from its development

the cone base is given by the formula, r = (180°/360°) × 60 = 30 mm. The plan and elevation of the cone are drawn as shown in Fig. 13.18(ii). The eight generators are also shown in the plan and elevation of the cone. The equilateral triangle sides 1-5 and 5-1 cut seven generators and its side 1-1 lies on the generator OA. The intersection points are marked on the generators in the development. The points 5, 6, 7, 8 and 1 are located on the slant edge o′s′ taking the distances of these points from O in the development. Horizontal lines are drawn through the points 5, 6, 7, 8 and 1 to intersect the corresponding generators at 5′, 6′, 7′, 8′ and 1′ in the elevation of the cone. A smooth curve is drawn passing through the points 5′-6′-7′-8′-1′. It is to be noted that the points 2′, 3′ and 4′ are directly behind the points 8′, 7′ and 6′. Hence, the line joining the points 1′-2′-3′-4′-5′ will be invisible in the elevation. These points are projected on the respective generators in the plan. A smooth curve is drawn passing through the points 1-2-3-4-5-6-7-8-1 to complete the projection of the triangle.

19. A vertical chimney, 60 cm in diameter, joins the roof of a boiler house, sloping at an angle of 30° to the horizontal. The shortest height over the roof is 30 cm. Determine the shape of sheet metal required to fabricate the chimney. Adopt a suitable scale.

 The plan and elevation of the chimney are drawn from the given data adopting a scale of 1:10 as shown in Fig. 13.19(i). The sloping roof is drawn at 30° to the horizontal and keeping 30 cm as the shortest height of the chimney. Eight generators are shown in the plan and elevation of the chimney. The intersection points of the roof are marked on the generators. A stretch out line A-A of length equal to the circumference of the chimney is drawn with the

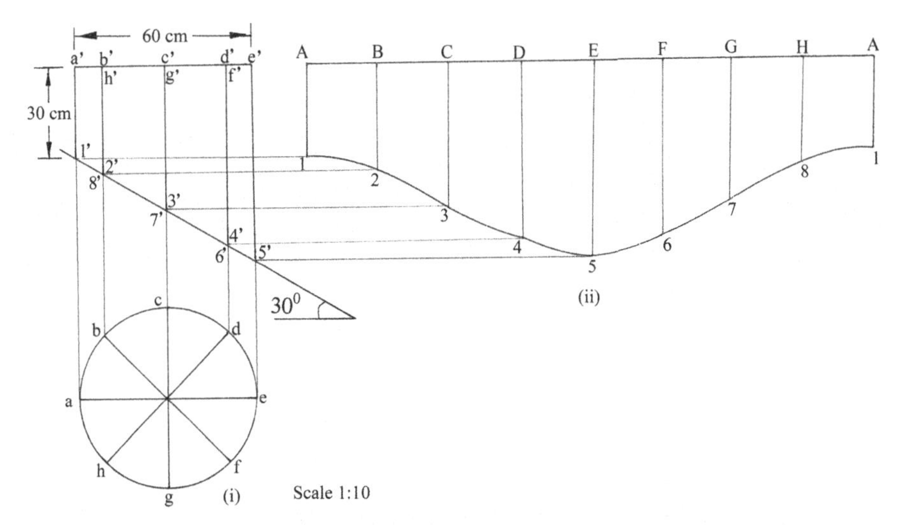

Fig. 13.19 Development of a boiler chimney

generators as shown in Fig. 13.19(ii). Horizontal lines are drawn through points $1'$, $2'$, $3'$, $4'$, $5'$, $6'$, $7'$ and $8'$ to intersect the corresponding generators to locate points 1, 2, 3, 4, 5, 6, 7 and 8 in the development. A smooth curve is drawn passing through the points 1-2-3-4-5-6-7-8-1 to complete the development of the chimney.

20. A horizontal cylindrical boiler shell 1.5 m diameter is surmounted by a cylindrical steam dome 0.75 m diameter. Draw the development of the surface of the steam dome. Assume 1 m as its minimum height.

    The elevation and left side elevation of the boiler shell and the steam dome are drawn adopting a scale of 1:25 as shown in Fig. 13.20(i). Eight generators are chosen on the surface of the steam dome as shown in the figure. The intersection curve is obtained in the side elevation by projecting the points $1'$, $2'$, $3'$ etc. in the elevation. A smooth curve is drawn passing through the points $3''$-$4''$-$5''$-$6''$-$7''$-$8''$-$1''$-$2''$-$3''$. A stretch out line C-C of length equal to the circumference of the steam dome is drawn as shown in Fig. 13.20(ii). The generators are also drawn. Horizontal lines are drawn through points $3''$, $4''$, $5''$, $6''$, $7''$, $8''$, $1''$ and $2''$ to intersect the corresponding generators to locate points 3, 4, 5, 6, 7, 8, 1 and 2 in the development. A smooth curve is drawn passing through the points 3-4-5-6-7-8-1-2-3 to complete the development of the steam dome.

21. A cube, edge 40 mm, stands with a face on the HP with its vertical faces equally inclined to the VP. A horizontal circular hole 30 mm diameter is drilled through the cube. The axis of the hole is perpendicular to the VP and bisects the opposite vertical edges. Draw the development of the lateral surfaces of the cube with the hole.

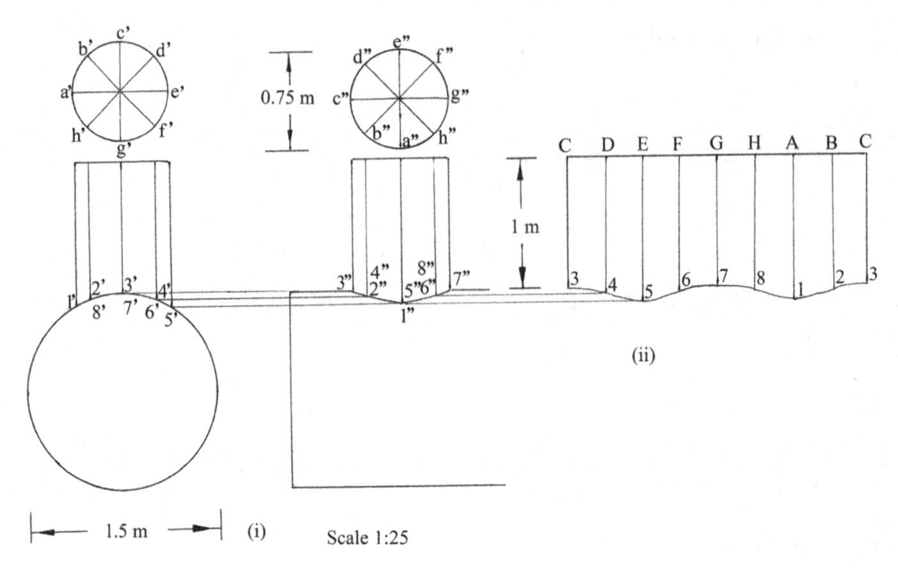

Fig. 13.20 Development of steam boiler dome

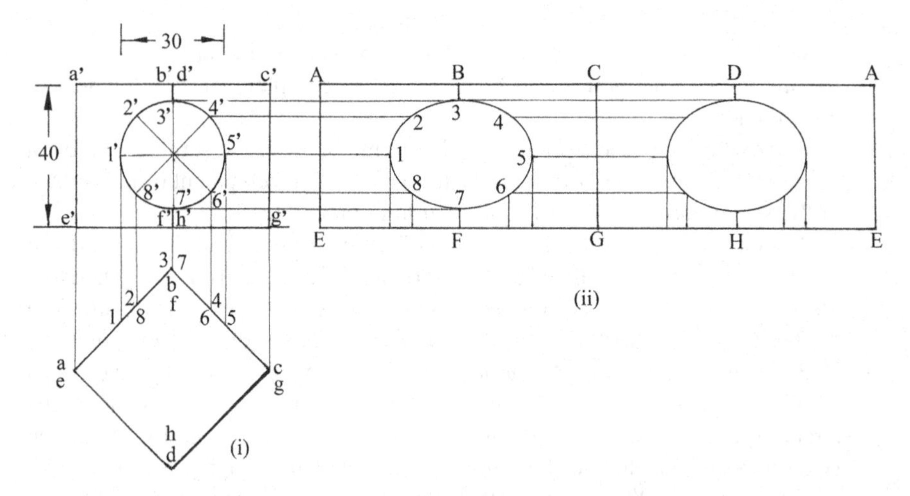

Fig. 13.21 Development of a cube with a circular hole

The plan and elevation of the cube are drawn as shown in Fig. 13.21(i). The lateral surfaces of the cube are developed as shown in Fig. 13.21(ii). A circular hole is drawn in the elevation, and eight points are marked on the circumference of the hole. These points are then projected in the plan on the two faces abfe and cbfg. The distance of the point 1 from e is measured and using this a point is located on the edge EF in the development. A vertical line is drawn through this point. A horizontal line is drawn through $1'$ in the elevation to intersect the vertical line at point 1 in the development. The same procedure is followed to locate points 2 and 8 in the development. Horizontal lines are drawn through points 3′ and 7′ in the elevation to locate points 3 and 7 directly in the development. Points 4, 5 and 6 are located following the procedure detailed for the points 2, 1 and 8. A smooth curve is drawn passing through the points 1-2-3-4-5-6-7-8-1 intercepting the edge BF between the points 3 and 7. Since the circular hole runs through the solid, the curve is redrawn on the other two faces cdhg and adhe to complete the development of the lateral surfaces of the cube with the hole.

22. A square prism of base edge 40 mm is truncated by two section planes equally inclined to its axis at 60° and intersect the axis at a height of 50 mm. Draw the development of its lateral surfaces.

    The plan and elevation of the square prism are drawn taking its height 50 mm as shown in Fig. 13.22(i). The development of lateral surfaces is drawn as shown in Fig. 13.22(ii). The two section planes inclined at 60° to the axis are drawn from the centre of the top of the prism in the elevation. The section planes cut the four vertical edges besides passing through the two top sides of the prism. The section points are marked as shown in the elevation. Horizontal lines are drawn through the points 1′, 2′, 4′ and 5′ to intersect the corresponding vertical edges at 1, 2, 4 and 5 in the development. Points 3 and 6 are located, respectively, at the mid-points of the top sides BC and DA in the development. Points 1, 2, 3, 4, 5, 6, 1 are connected by straight thick lines and also these points with the corresponding base points of the prism to complete its development.

23. A rectangular pyramid of base edges 50 × 40 mm and height 70 mm is sectioned by a plane parallel to one of its lateral faces. Distance between the section plane and the face, which is parallel to the plane, is 10 mm. Draw the development of the truncated pyramid.

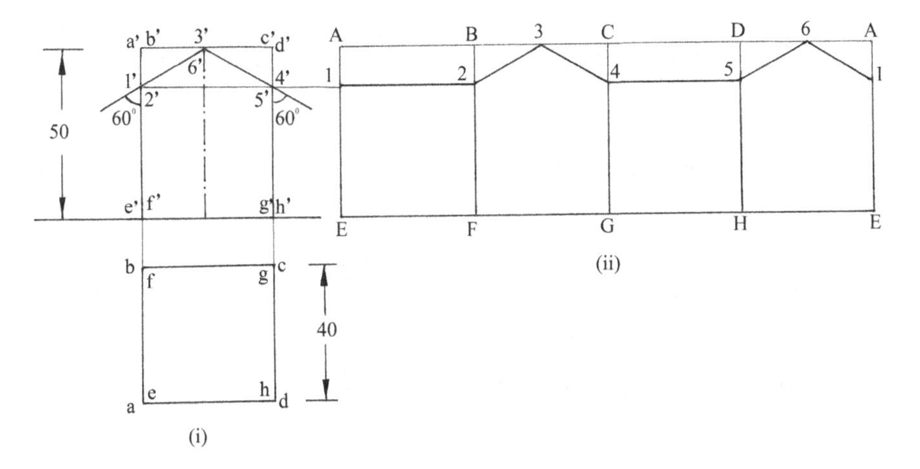

Fig. 13.22 Development of sectioned square prism

The plan and elevation of the rectangular pyramid are drawn as shown in Fig. 13.23(i). Edge oc in the plan is revolved as shown to appear true length in the elevation. The line o′s′ is the true length of the four lateral edges. An arc of radius o′s′ is drawn with O as centre, and the four sides of the base in the plan are transferred directly to the development by marking as successive chords A-B, B-C, C-D and D-A along the arc as shown in Fig. 13.23(ii). Radial lines are drawn from O to the base points A, B, C and D. A rectangle ABCD is also drawn on the side CD to show the base of the pyramid in the development. The section plane is drawn parallel to the face a′o′b′ and 10 mm from it in the elevation. The section plane cuts the two slant edges o′c′ and o′d′ and the two base edges bc and da. The section points 1′, 2′, 3′ and 4′ are marked in the elevation. Horizontal lines are drawn through points 4′ and 3′ to intersect the line o′s′ at points 4 and 3 to give the true length of the points from the vertex of the pyramid. Points 4 and 3 are then marked on the slant edges OD and OC in the development. The distance of point 1 from point a in the plan is transferred to locate point 1 on the edge AD in the development. Point 2 is located on the edge BC in the development following the same procedure. Points 1 and 2 are also located in the rectangle ABCD in the development. The points 2-3-4-1 are connected by straight lines. The development of the truncated pyramid is completed by thick lines in Fig. 13.23(ii).

24. The development of a cylinder is a rectangle of 110 mm length and 55 mm width. A semicircle of diameter 110 mm is inscribed in the rectangle. Draw the projections of the cylinder showing the profile of the semicircle on the elevation.

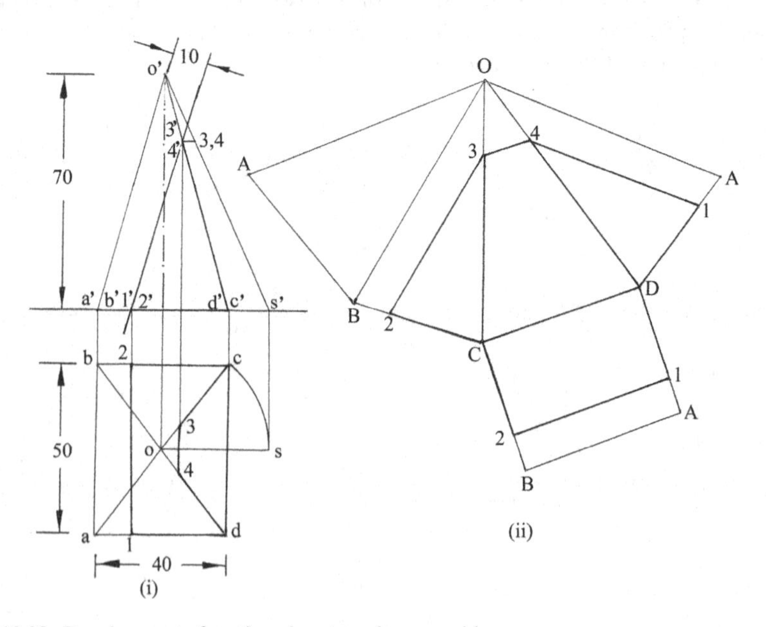

Fig. 13.23 Development of sectioned rectangular pyramid

A rectangle of 110 mm length and 55 mm width is drawn as shown in Fig. 13.24(i). The rectangle represents the development of a cylinder, and the rectangle is divided into eight equal parts along its length with the generators marked therein. A semicircle of diameter is inscribed in the rectangle, and the intersections of the semicircle with the generators are marked as shown in the figure. The diameter of the cylinder is found to 35 mm and the height of the cylinder is 55 mm. The plan and elevation of the cylinder are drawn as shown in Fig. 13.24(ii). The generators are also drawn by dividing the plan into eight equal parts. Point 1′ is located on the generator A by drawing a horizontal line through point 1. The same procedure is followed to locate the remaining points 2′, 3′ etc. The visible profile of the semicircle 1′-8′-7′-6′-5′ is shown by a smooth curve, and the invisible profile is directly behind the visible profile.

25. The plan and part elevation of a dome, horizontal sections of which are regular octagons, are shown in Fig. 13.25(i). Complete the elevation. Draw a development of one of the eight-curved faces of the dome.

The plan and part elevation of the dome is shown in Fig. 13.25(i). The given projections of the dome are redrawn as shown in Fig. 13.25(ii). The dome oab is chosen to complete the elevation of the dome. A projector is drawn through a to intersect the base of the dome at a′. Since the horizontal sections are regular octagons, five lines are drawn parallel to the side ab in the plan. A projector is drawn through d to intersect the dome edge o′b′ at d′. A horizontal line is drawn from d′ to intersect the projector drawn through c at c′. Point c′ lies on the dome edge o′a′. Points are located on the dome edges o′b′ and o′a′ following the same procedure. The points on the edge o′a′ are transferred on the other side of the dome to complete the elevation of the dome. It is evident that the arc length o′b′ is equal to one-fourth the circumference of the circle passing through the base points of the dome. The arc length OB is found to be 5.5 m. The development of one of the eight-curved faces is an isosceles triangle of base AB (equal to ab in

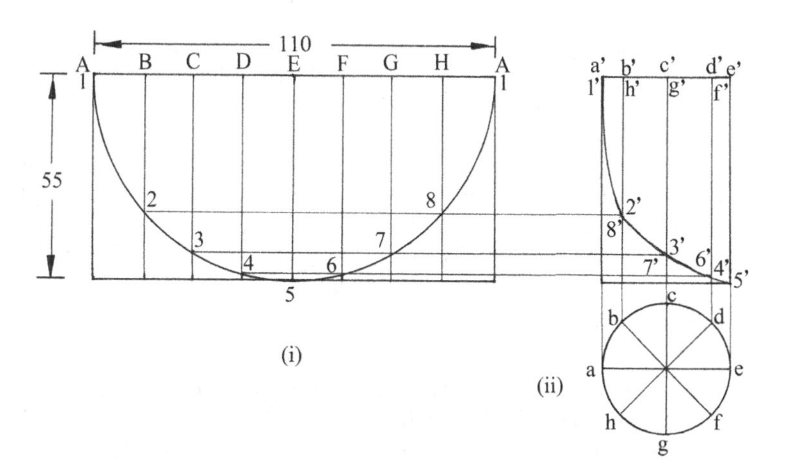

Fig. 13.24 Projections of a cylinder from its development

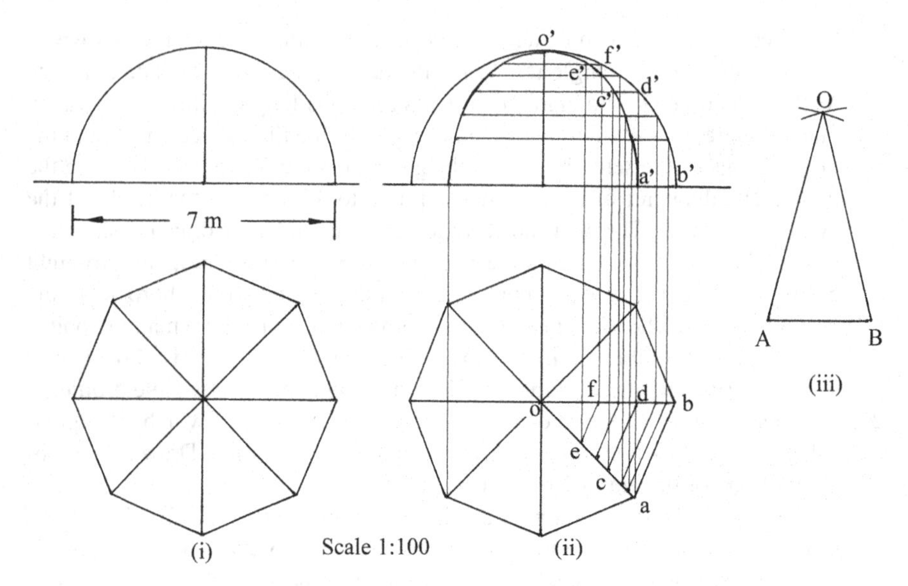

Fig. 13.25 Development of a dome

this case) and sides OA and OB. The isosceles triangle OAB drawn in Fig. 13.25(iii) is the development of one face of the dome.

26. A cube of 40 mm side rests on one of its faces on the HP and the lateral faces equally inclined to the VP. It is sectioned by two section planes parallel to the VP and perpendicular to the HP such that they bisect edges of its bases. Draw the development of the surface of the cube.

    The plan and elevation of the cube are drawn as shown in Fig. 13.26(i). The development of the cube is drawn by thin lines with the top and bottom faces as shown in Fig. 13.26(ii). The section planes are chosen parallel to the VP, and the two planes cut the eight horizontal edges. The section points are marked in the elevation. The section points are located in the development on the respective edges. The middle portion of the cube is retained, and the development of the retained portion of the cube is drawn by thick straight lines including its top and bottom faces.

27. Draw the development of post office envelope of size 220 mm × 110 mm with overlaps of 10 mm for glueing/gumming when the development is folded up to the form of envelope.

    The front and back of the envelope are drawn as shown in Fig. 13.27(i) and (ii). The development of the envelope with overlaps of 10 mm on the two sides of the triangles ADE and BCE is drawn as shown in Fig. 13.27(iii).

28. Figure 13.28(i) shows the elevation of a sheet metal pipe S with two sheet metal branches R and T. Draw the development of the surface of pipe S.

    A circle of diameter 50 mm is drawn below the elevation of metal pipes to represent the plan view of pipe S. The circle is divided into eight equal parts, and the eight generators are marked in the elevation. These generators are cut at

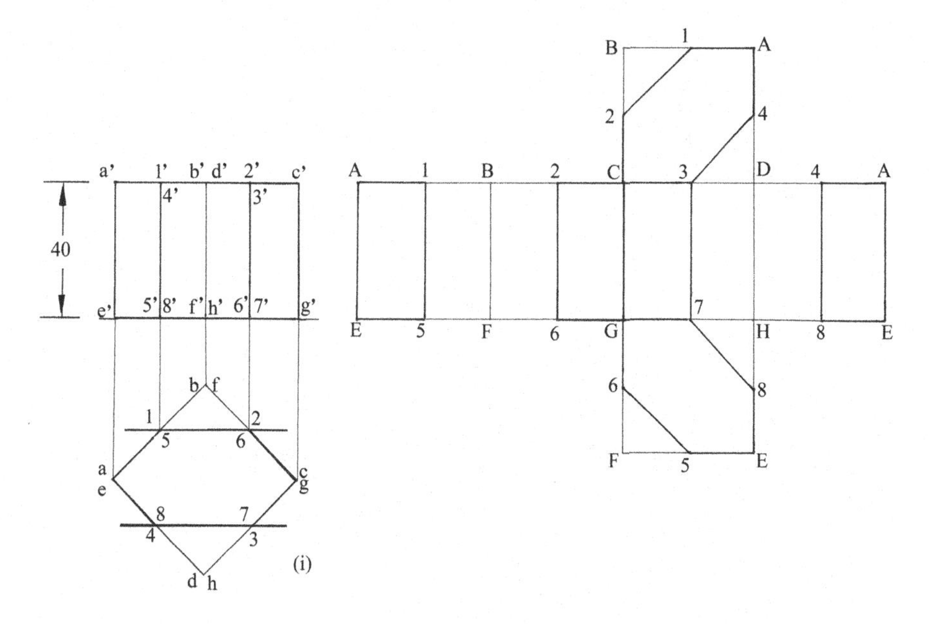

Fig. 13.26 Development of sectioned cube

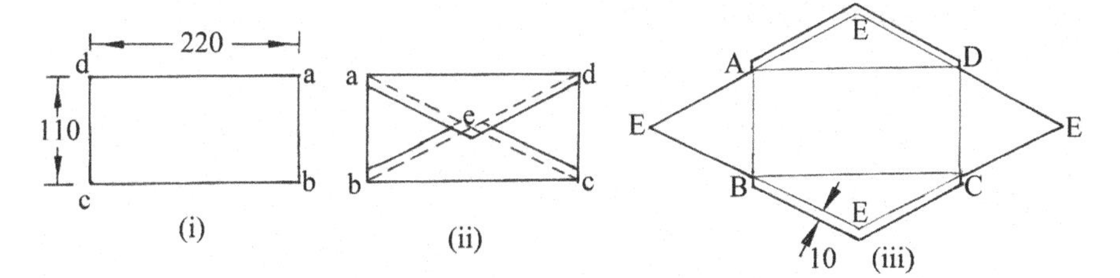

Fig. 13.27 Development of post office envelope

the two ends of the pipe, and the intersection points are numbered as shown in Fig. 13.28(i). The stretch-out line A-A equal to the circumference of the circle of diameter 50 mm is drawn, and the line is divided into eight equal parts to represent the generators of the pipe S as shown in Fig. 13.28(ii). The intersection points are marked on the respective generators in the development by drawing horizontal lines through the points in the elevation. Two smooth curves are drawn passing through the points 1-2-3-4-5-6-7-8-1 and 9-10-11-12-13-14-15-16-9. The extreme lines 1-9 on the generator A are drawn thick to complete the development of the metal pipe S.

29. Figure 13.29(i) shows the plan of compound solid comprising of a half of cylinder and a half of hexagonal prism, the common axis being 60 mm long. It is cut by a section plane perpendicular to the VP, inclined at 45° to the HP and bisecting the axis. Draw the development of the sectioned compound solid.

    The plan and elevation of the compound solid are drawn as shown in Fig. 13.29(ii). The full development of the lateral surfaces of the compound

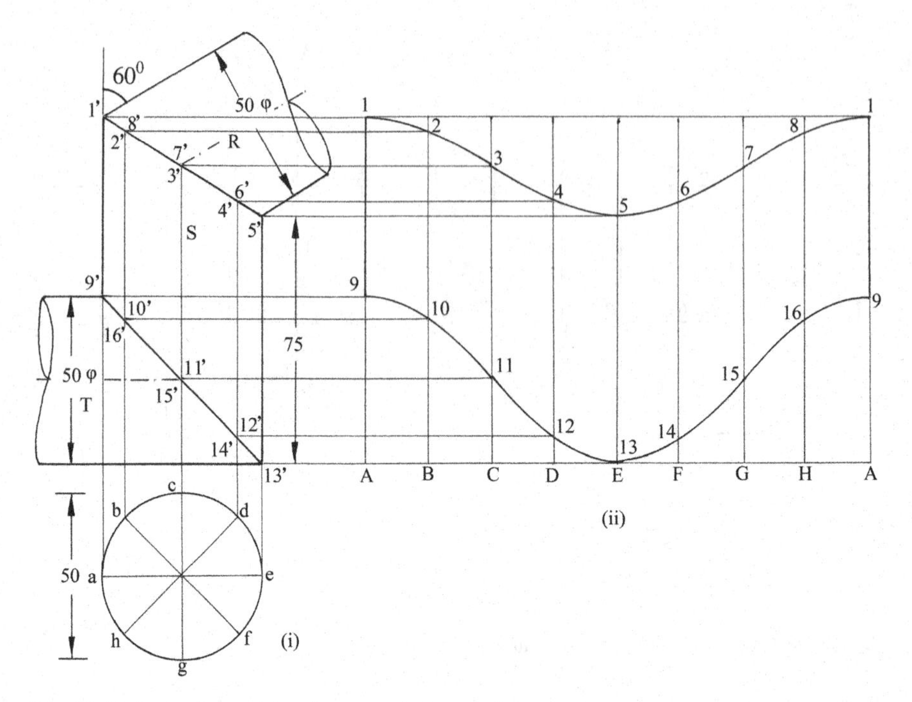

Fig. 13.28 Projection and development of a metal pipe

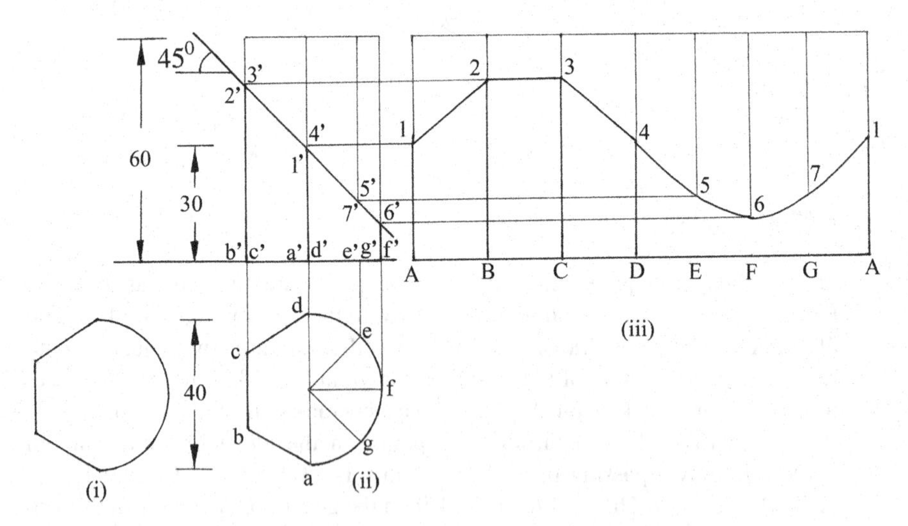

Fig. 13.29 Projections and development of compound solid

solid is drawn by thin lines as shown in Fig. 13.29(iii). The section plane is drawn in the elevation of the solid. The plane cuts two edges of the prism, two common edges or generators of the compound solid and three generators of the cylinder. The seven points are numbered in the elevation. Horizontal lines are drawn through these points to intersect the respective edges or the generators in the development. The points 1-2-3-4 are connected by straight lines, and a smooth curve is drawn passing through the points 4-5-6-7-1. The vertical edges A1, B2, C3 and D4 are drawn by thick lines. The base line of the compound solid is also drawn thick to complete the development of the cut compound solid.

30. A vertical conical pipe in the form of a frustum of a cone with base 100 mm in diameter, top 40 mm in diameter and height 50 mm has a side opening made by a section plane perpendicular to the VP and inclined to the HP at 60° and passing through the centre of the base rim. Draw the development of its lateral surface. Draw its plan showing the points lying on its base and the section plane.

    The plan and elevation of the frustum of the cone are drawn as shown in Fig. 13.30(i). The extreme generators of the cone are extended to meet at o′ in the elevation. The slant length TL is found from the extreme generator o′d′ (i.e. o′s′) of the cone, and the development of the frustum of the cone is drawn by thin lines as shown in Fig. 13.30(ii). The section plane, which is used to make side opening, is drawn inclined at 60° to the HP and passing through the centre of the base rim in the elevation. Since the section plane cuts one-half of the cone, the right half of the cone is divided into four equal parts in the plan,

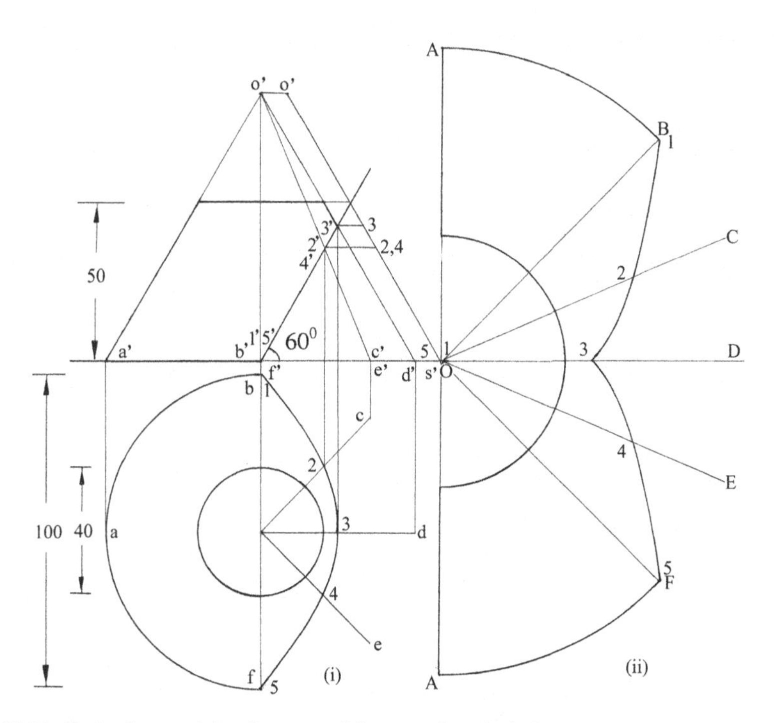

Fig. 13.30 Projections and development of frustum of conical pipe

and the generators are marked in both the projections. The intersection points are numbered in the elevation. Horizontal lines are drawn through these points in the elevation to cut the extreme generator o′s′ giving the true distances of these points on the respective generators from the point O in the development. Points are located in the development. A smooth curve is drawn passing through the points 1-2-3-4-5 to complete the development of the conical pipe. Projectors are drawn through the points 1′, 2′, 3′, 4′ and 5′ in the elevation to locate the points 1, 2, 3, 4 and 5 on the respective generators in the plan. A smooth curve is drawn passing through these points. The semicircle fab and the curve 1-2-3-4-5 show the points lying on the base of the cone and the section plane.

31. Draw the development of the lateral surfaces of an oil can whose elevation is given in Fig. 13.31(i).

    The extreme generators of the top and bottom cones are extended to meet at o′ and p′ to get the true length of the generators o′s′ and p′r′. The top circle of conical part A is drawn and is divided into eight equal parts to show the eight generators in the elevation. The intersection points are marked and are projected on the slant length o′s′ to get the true length of these points from O in the development. The development of cone A is shown in Fig. 13.31(ii). The development of frustum of cone B is shown in Fig. 13.31(iii). The development of the lateral surface of the cylindrical part C is shown in Fig. 13.31(iv). The

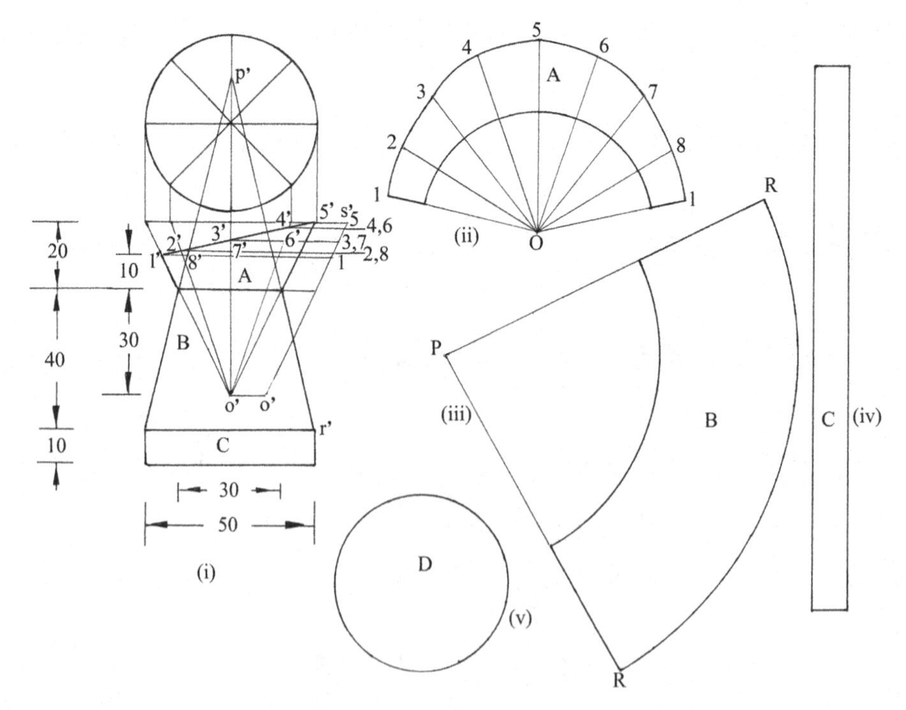

Fig. 13.31 Development of oil can

bottom circular plate D is shown in Fig. 13.31(v) to complete the development of the lateral surfaces of the oil can.

32. Draw the plan and elevation of a cylinder whose development with the line PQ is shown in Fig. 13.32(i). Show the line PQ in the elevation. Does the line PQ when transferred on the elevation represent any curve? If so, name the curve.

    The diameter of the cylinder whose development is given in Fig. 13.32(i) is 35 mm and the height of the cylinder is 60 mm. The plan and elevation of the cylinder are drawn as shown in Fig. 13.32(ii). The rectangle is divided into eight equal parts along its length, and the plan of the cylinder is also divided into eight equal parts. The generators are drawn in the elevation and also in the development. The line PQ is drawn in the development. The line intersects the generators at points 0, 1, 2, 3, etc. Horizontal lines are drawn through the points 0, 1, 2, 3, 4, etc. to intersect the respective generators at $0'$, $1'$, $2'$, $3'$, $4'$ etc. in the elevation. A smooth curve is drawn passing through these points. The part of the curve from $0'$ to $4'$ is on the front half of the cylinder and is therefore visible; from $4'$ to $8'$ the curve is hidden. The curve shown in the elevation is a helix.

    The curve is a double-curved line generated by a point that moves at uniform rate around and also parallel to the axis. The distance the point moves parallel to the axis during one revolution is called the pitch (p) or lead of the helix. The curve shown is right hand helix. For a right hand helix beginning at point $0'$, the curve moves upward around the cylinder in a counter-clockwise direction as seen in the plan. The inclination of the helix or the pitch angle of the helix is the angle Θ on the development in Fig. 13.32(i). It can be shown that $\tan(\Theta) = (p)/(\Pi d)$, where p is the pitch and d is the diameter of the cylinder.

    The most common application of the helix is screw threads of V and square types. The helical grooves of square shape is used on cylindrical and drum

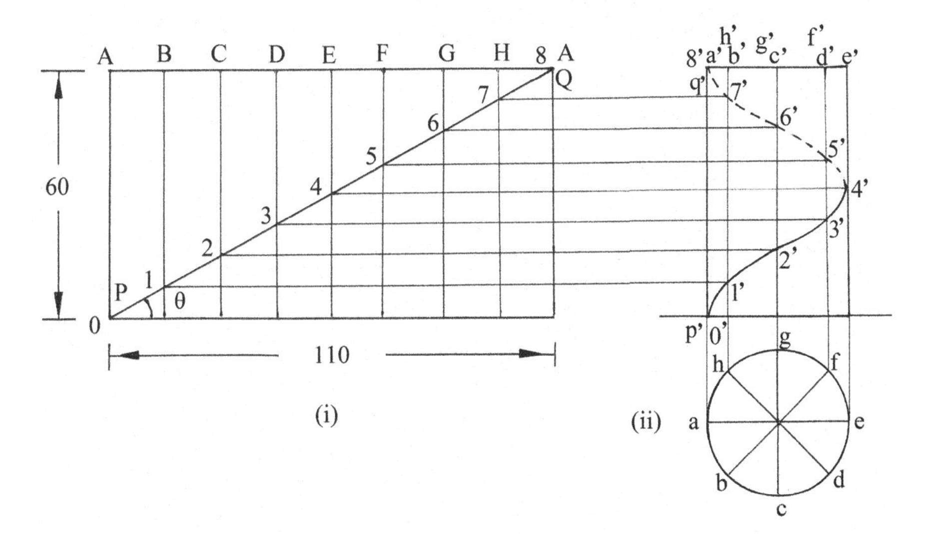

Fig. 13.32 Development of a cylinder

cams and screw conveyors. The springs of square or round shapes are produced by bending wires of respective cross sections into helical shapes.

33. A cylinder, diameter 50 mm and axis 72 mm, has three right hand helices on its surface, starting from three equiangular points on its base. Draw the projections of the helices and also their developments.

    The plan and elevation of the cylinder, diameter of base 50 mm and axis 72 mm, are drawn as shown in Fig. 13.33(i). The plan is divided into 12 equal parts, and the 12 generators are shown in the elevation. The axis is also divided into 12 equal parts in the elevation. Three angular points a, c and e are chosen in the plan. The point a moves upward around the cylinder in a counter clockwise direction as seen in the plan and reaches the point b to complete the helix a′-b′ in the elevation. The point c moves around the cylinder and reaches the point d to complete the helix c′-d′ in the elevation. The point e moves around the cylinder and reaches the point f to complete the helix e′-f′ in the elevation. A rectangle of length equal to the circumference of the base circle of the cylinder and width equal to the axis of the cylinder is drawn as shown in Fig. 13.33(ii). Twelve generators are drawn in the development of the cylinder. Horizontal lines are drawn from the points on the helix in the elevation to intersect the respective generators in the development. Straight lines A-B, C-D and E-F are drawn connecting the respective points to show the helices in the development.

34. Draw one complete coil of a helical spring of square section of 10 mm side. The outside diameter of the spring is 120 mm and the pitch is 60 mm.

    The helical curves starting from four corners of the square section are to be drawn. The helical curves which start from the outside corners of the square move around the curved surface of an imaginary cylinder of diameter equal to

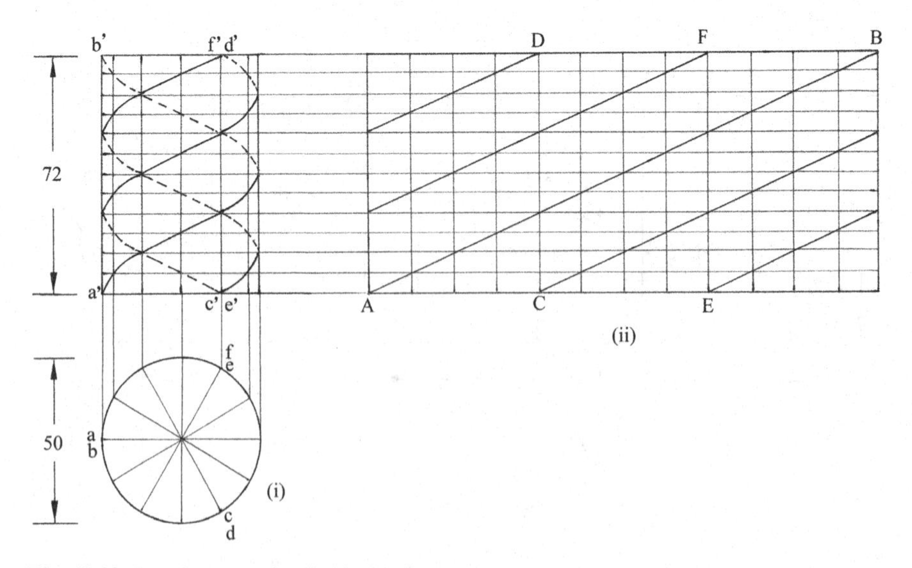

Fig. 13.33 Development of cylindrical helices

the outside diameter of the spring. The helical curves which start from the inside corners of the square move around the curved surface of an imaginary cylinder of diameter equal to the inside diameter of the spring. The inside diameter is given by the formula: d = D − 2 × side of the square. The inside diameter is 100 mm. Two concentric circles of diameter 120 mm and 100 mm are drawn in the plan as shown in Fig. 13.34. The circles are divided into 12 equal parts. Draw two vertical lines through the extremities of the outside circle. A length of pitch of the helix is placed, and the line is divided into 12 equal parts. The outer helices a′-b′ and c′-d′ are drawn for the imaginary cylinder of 120 mm diameter. The inner helices e′-f′ and g′-h′ are drawn for the imaginary cylinder of 100 mm diameter. The pitch for both the helices is 60 mm. The visible helical portions of the spring alone are shown in the elevation.

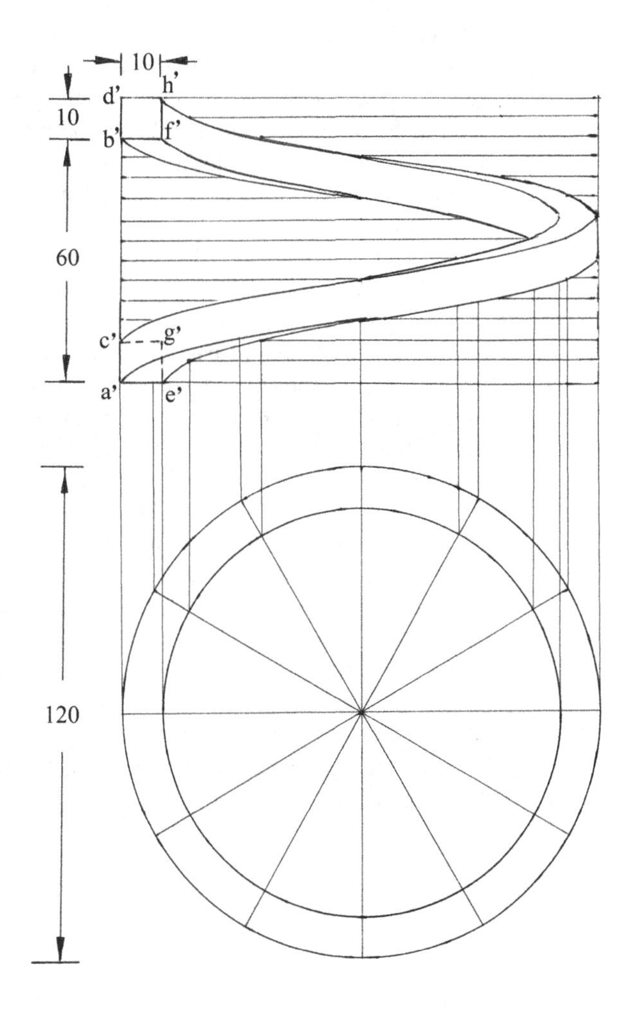

Fig. 13.34 Helical spring (square in cross section)

Practice Problems

1. Develop the entire surface of a pentagonal prism, side of base 30 mm and axis 50 mm.
2. Draw the development of the lateral surface of a square pyramid, side of base 30 mm and height 50 mm.
3. A hexagonal prism, side of base 30 mm and axis 50 mm, rests with its base on the HP such that a rectangular face is parallel to the VP. It is cut by a section plane inclined at 45° to the HP, perpendicular to the VP and passing through the right extreme corner of the top end of the prism. Draw the development of the truncated prism.
4. A vertical cylinder, 50 mm diameter and 60 mm height, rests with its base on the HP. A square hole of 20 mm side is made so that the axis of the hole is parallel to the HP and perpendicular to the VP. The faces of square hole are equally inclined to the HP, and its axis bisects the axis of the cylinder. Draw the development of the lateral surfaces of the cylinder with the square hole.
5. A cone, diameter of base 50 mm and height 65 mm, rests with its base on the HP. It is cut by a section plane perpendicular to the VP, inclined at 30° to the HP and intersecting the axis at a point 30 mm above the base. Draw the development of truncated cone.
6. Draw a semi-circle of diameter 100 mm and in it inscribe the largest possible circle. The semi-circle is the development of a cone and the circle that of a line on its surface. Draw the projections of the cone showing those of circle also.
7. A rectangular pyramid, side of base 30 mm $\times$ 40 mm and axis 50 mm long, stands with its base on the HP and a diagonal of its base parallel to the VP. It is cut by a section plane perpendicular to the VP, inclined at 45° to the HP and intersecting the axis at a point 20 mm distant from the vertex. Draw the development of the lateral surface of the truncated pyramid.
8. A funnel is to be made of sheet metal. The funnel tapers from 100 mm diameter to 40 mm diameter to a height of 30 mm and from 40 mm diameter to 20 mm diameter to a height of 50 mm. The bottom of the funnel is levelled off to a plane inclined at 45° to the axis. Draw the development of the funnel.
9. A cylinder of base diameter 50 mm is so truncated that its elevation is a semi-circle of diameter 50 mm and its plan is a circle of diameter 50 mm. Draw the development of its surface.
10. An equilateral triangle of 80 mm sides having a 20 mm diameter at its centre represents the elevation of a cone with a circular hole. Draw the development of its lateral surface.

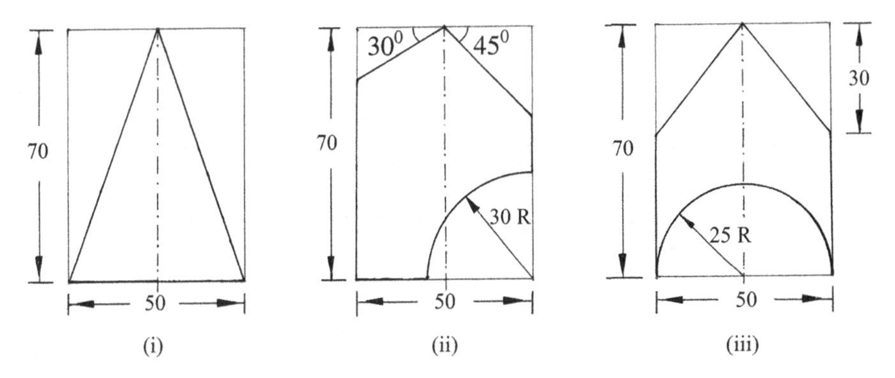

Fig. 13.35 Elevations of cut cylinders

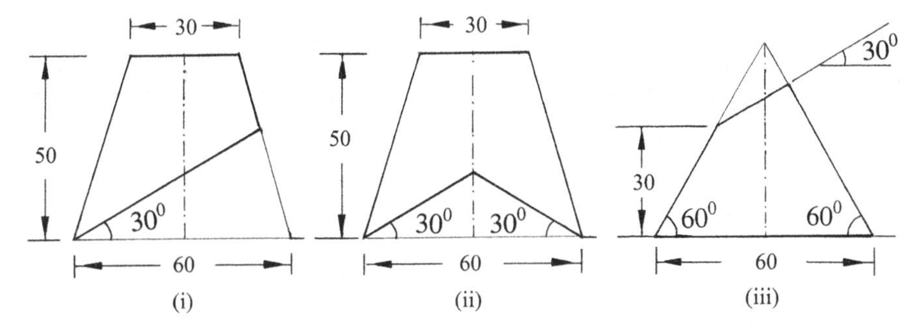

Fig. 13.36 Elevations of cut cones

11. Draw the development of the lateral surfaces of the cut cylinder of 50 mm diameter and height 70 mm cut in different ways shown in Fig. 13.35(i)–(iii).
12. Draw the development of the lateral surfaces of the cut cones of given dimensions shown in Fig. 13.36(i)–(iii).

Chapter 14
Isometric Projections

It is easy to draw the two principal views (plan and elevation) of an object from its pictorial representation. The reverse is not always easy to visualize the object from two of its principal views. A third view on the profile plane aids to visualize the exact shape of the object. Isometric projection is an orthographic projection of an object on the vertical plane showing its three dimensions in one view (elevation). To understand the principles of isometric projection, let us consider the auxiliary projection of a cube. The problem is one of drawing the projections of a cube when one of its solid diagonals is perpendicular to the vertical plane. The solution for this problem, which is self-explanatory, is given in Fig. 14.1.

The special features on the final elevation are summarized as follows:

1. The 12 edges of the cube are distinct.
2. Three edges d′h′, e′h′ and g′h′ are invisible.
3. The 12 edges are equal in length and hence they are equally inclined to the VP.
4. The edges a′b′, c′d′, e′f′ and g′h′, which are parallel, are also parallel in the final elevation.
5. The edges a′d′, b′c′, f′g′ and e′h′, which are parallel, are also parallel in the final elevation.
6. The vertical edges a′e′, b′f′, c′g′ and d′h′ are also vertical in the final elevation.
7. The faces of the cube are equally inclined to the vertical plane.
8. The diagonal a′c′ on the face a′b′c′d′ is not foreshortened and is perpendicular to the edge b′f′.
9. The diagonal a′f′ on the face a′b′f′e′ is not foreshortened and is perpendicular to the edge b′c′.
10. The diagonal c′f′ on the face b′c′g′f′ is not foreshortened and is perpendicular to the edge b′a′.
11. The edges b′f′, b′c′ and b′a′ make equal inclination of each 120°.

The final elevation is the ***isometric projection*** of the cube.

Another special feature of the projection is given below:

K. Rathnam, *A First Course in Engineering Drawing*,
DOI 10.1007/978-981-10-5358-0_14

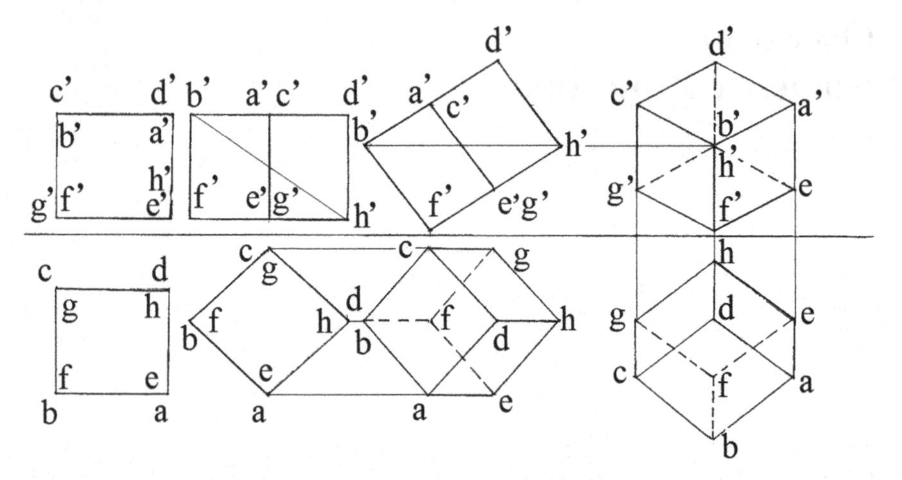

Fig. 14.1 Auxiliary projections of a cube

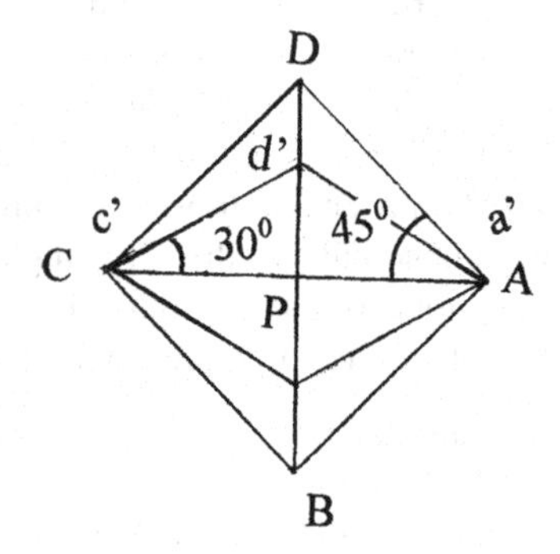

Fig. 14.2 Isometric projection of a square

The foreshortened length c′-d′ is called the ***isometric length*** of the edge CD (refer to Fig. 14.2). Consider the triangles CDP and c′-d′-P which are the right angled triangles and have the common adjacent edge c′P,

$$\frac{c'P}{c'd'} = \cos\left(30^\circ\right) = \frac{\sqrt{3}}{2}$$

$$\frac{c'P}{CD} = \cos\left(45^\circ\right) = \frac{1}{\sqrt{2}}$$

Hence, $\frac{c'd'}{CD} = \frac{\sqrt{2}}{\sqrt{3}} = 0.816$

$$\frac{\text{isometric length}}{\text{actual length}} = 0.816$$

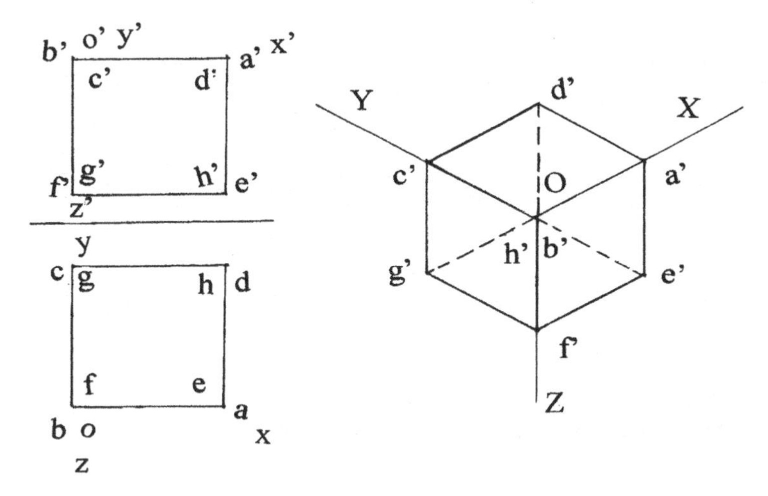

Fig. 14.3 Projections of a cube and its isometric projection

The final elevation of the cube is redrawn together with the projections of the cube in its simple position as shown in Fig. 14.3.

The three mutually perpendicular co-ordinate axes OX, OY and OZ are introduced in the plan and elevation of the cube as shown in the figure. These axes which appear inclined to each other at 120° in the isometric projection are called as the ***isometric axes***. Any plane parallel to two of the isometric axes is called an ***isometric plane***. The isometric axis OZ is always oriented in the vertical direction in the isometric projection. The three horizontal edges parallel to the edge a′-b′ are also parallel to the isometric axis OX, and the three horizontal edges parallel to b′-c′ are parallel to the isometric axis OY. The four vertical edges remain vertical and parallel to the isometric axis OZ. The vertical faces a′-b′-f′-e′ and d′-c′-g′-h′ are parallel to the isometric plane XOZ. The vertical faces b′-c′-g′-f′ and a′-d′-h′-e′ are parallel to the isometric plane YOZ. The horizontal faces a′-b′-c′-d′ and e′-f′-g′-h′ are parallel to the isometric plane XOY.

The final isometric projection of the cube and the ***isometric scale*** are shown in Fig. 14.4.

Since the isometric projection is a single view representation of an object, capital letters are used to represent the angular points of the object. The invisible edges are shown by dotted lines. Lengths of edges are marked along the isometric axes using the isometric scale which is also as shown in Fig. 14.4. A horizontal line OA is drawn, and two more lines OP and OQ inclined, respectively, at 30° and 45° to the horizontal are also drawn. The actual scale in mm is marked on OQ. Vertical lines are drawn from each of the division points on the actual scale to cut OP at the corresponding division. The division point on OP measures the distance reduced to isometric length. Lines which are not parallel to the co-ordinate axes cannot be represented directly using the isometric scale as there will be distortion. It is to be noted that the difference in length between the projections of equal diagonals BD and AC is considerable.

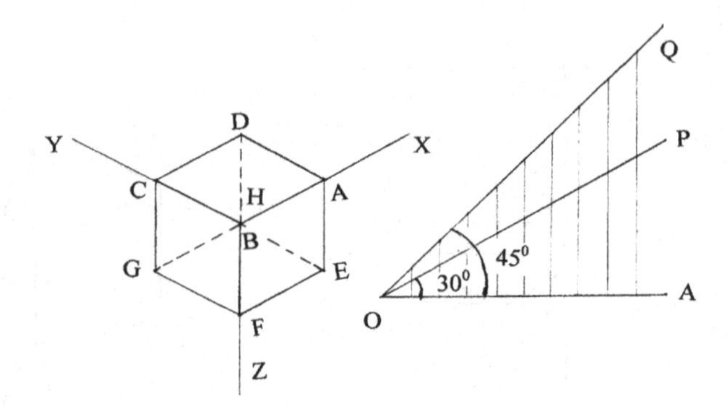

Fig. 14.4 Isometric projections of a cube and isometric scale

Isometric Projection and Isometric View

Isometric projection of a solid is drawn by orienting the edges of the solid parallel to the co-ordinate axes to the extent possible. The dimensions of the edges parallel to the axes are reduced using the isometric scale and laid along the isometric axes. The view obtained is the isometric projection of the solid. The solid can also be represented on the isometric axes using the actual/true lengths instead of the isometric lengths. The view drawn on the isometric axes using the actual length is called ***Isometric View/Drawing***, and the view drawn using the isometric scale is called ***Isometric Projection***. The projection obtained in isometric view will be larger in size.

Solved Problems

1. Draw the isometric projection of an equilateral triangle of side 40 mm when it is placed with its surface vertical and a side horizontal.

   The plan and elevation of the vertical lamina are drawn keeping one of its sides horizontal as shown in Fig. 14.5. The lamina is enclosed in a rectangle a′o′d′c′. The co-ordinate axes o′x′ and o′z′ are chosen passing through the edges of the rectangle as shown in the figure. The isometric projection can be drawn using the two isometric axes OX and OZ with the included angle of 120° and the corresponding isometric plane XOZ. The isometric length of side o′d′ is obtained from the isometric scale, and using this length the point D is located along OX. Similarly, the isometric length of side o′a′ is obtained, and using this length the point A is located along the axis OZ. Draw from D and A lines parallel, respectively, to the isometric axes OZ and OX to locate the point of intersection at C. Now AODC is the required parallelogram representing the isometric projection of the rectangle a′o′d′c′. The mid-point of OD locates the point B of the triangle. Join the points BA and BC. The triangle ABC is the

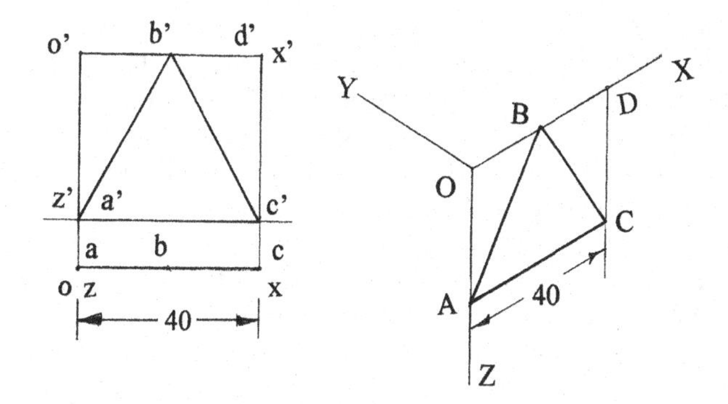

Fig. 14.5 Projections of a lamina and its isometric projection

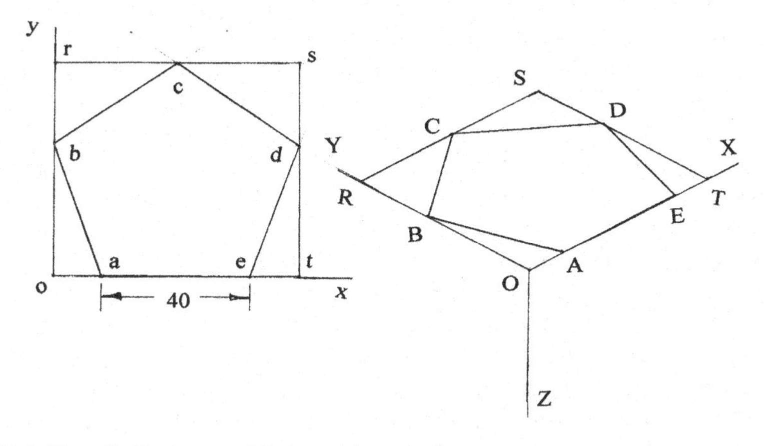

Fig. 14.6 Plan of a Pentagon and its isometric projection

isometric projection of the given equilateral triangle. The projection can also be obtained on the isometric plane YOZ.

2. Draw the isometric projection of a regular pentagon of 40 mm side when it is placed with its surface horizontal and a side parallel to the vertical plane.

   The plan of the pentagon of side 40 mm is drawn with a side parallel to VP, and this plan is enclosed in a rectangle orst as shown in Fig. 14.6. The co-ordinate axes ox and oy are drawn passing through the sides of rectangle. The isometric projection can be drawn using the isometric plane XOY. The isometric length of side ot is obtained, and using this length the point T is located along OX. Similarly, the isometric length of side or is obtained, and using this length the point R is located along OY. From points T and R, draw lines parallel, respectively, to the isometric axes OY and OX to locate the intersecting point S. The angular points of the pentagon lie on the circumference of the parallelogram ORST. The mid-point of RS locates the point C. The isometric length of side ob is obtained and the point B is located on OY. Since the distance of d from t is the same as the distance of b from o, the point D is located on TS by drawing a line parallel to OX from B. The isometric length of at is obtained and the point A is located on TO measuring its distance from T. The point E is also located from this isometric length on line OT measuring its distance from O. (Note: the error involved in obtaining the isometric length at is small compared to the length oa). Join the points A, B, C, D, E and A to give the isometric projection of the pentagon.

3. Draw the isometric projection of a semi-circular lamina of 60 mm in diameter standing on its diameter with its surface vertical.

   The elevation of the semi-circular lamina is drawn with its centre at d′ and the diameter on HP as shown in Fig. 14.7. The lamina is enclosed in a rectangle a′o′b′c′, and the co-ordinates o′x′ and o′z′ are chosen as indicated in the figure. Join the point o′ and d′ and also b′ and d′. From d′, draw a vertical line to meet o′x′ at point 2′. The intersection of the line o′d′ on the circle is indicated as point 1′. The intersection of the line b′d′ on the circle is indicated as point 3′. Lines are drawn perpendicular to a′c′ from 1′ and 3′ to meet the line a′c′ at 1″ and 3″.

   The isometric projection can be drawn using the isometric plane XOZ. Since the sides o′b′ and o′a′ of the rectangle are aligned along the o′x′ and o′z′ axes, their isometric lengths are obtained and points B and A are located along the isometric axes as indicated. From points A and B, draw lines, respectively, parallel to the isometric axes OX and OZ to locate the intersecting point C. The mid-point of OB locates point 2 on the isometric projection of the semi-circular lamina and points A and C are also points on the lamina. To locate point 1 on the isometric projection, the distance between 1″ and c′ is measured and the isometric length of this distance is kept from C towards A to fix A1 on the isometric line CA. From this point, a line is drawn parallel to the isometric axis OZ and point 1 is located knowing the isometric length of the distance between the points 1″ and 1′ in the elevation. The same procedure is repeated to locate point 3 on the isometric projection. A smooth curve is drawn passing through the points A, 1, 2, 3 and C. This curve represents the isometric projection of the vertical semi-circular lamina standing on its diameter AC.

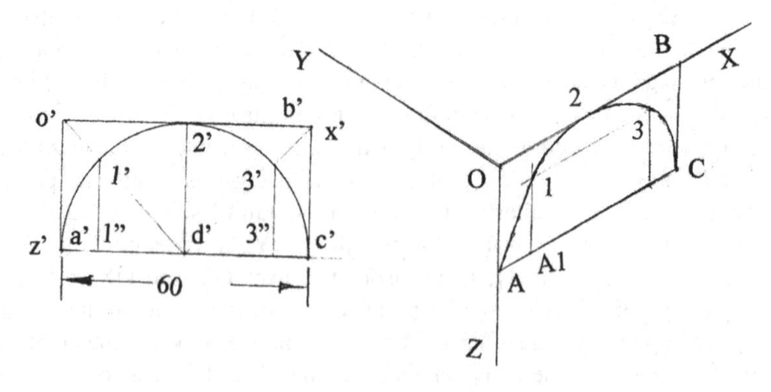

Fig. 14.7 Elevation of a semi-circular lamina and its isometric projection

4. Draw the isometric projection of a circular lamina of 80 mm in diameter when its surface is vertical.

   The circular lamina is enclosed in a square r′s′t′u′ as shown in Fig. 14.8. Connect the diagonals r′t′ and s′u′ to intersect the circle at four points. The mid-points of sides r′s′, s′t′, t′u′ and u′r′ are also marked as shown in the figure. Name all the points as 1, 2, 3, 4, etc. The isometric projection can be drawn using the isometric plane XOZ. As the sides t′u′ and t′s′ are aligned along the axes o′x′ and o′z′, the isometric length of either t′u′ or t′s′ is obtained and points U and S are located along the isometric axes OX and OZ as indicated. From points U and S, draw lines, respectively, parallel to the axes OZ and OX to locate the intersecting point R. The rhombus RSTU represents the isometric projection of the square. The diagonals RT and SU are connected to locate the centre of circle. As the diagonal SU is perpendicular to the isometric axis OY, the distance between S and U will be equal to the distance between s′ and u′. The distance of the point 2 and 6 from the circle centre will be equal to actual radius of the circle. The points 2 and 6 are located on the diagonal SU. From points 2 and 6, lines are drawn parallel to OX and OZ axes. These lines intersect and locate points 4 and 8. The points 1, 3, 5 and 7 are marked as mid-points of the sides. A smooth curve is drawn through these points to show the isometric projection of the vertical circle.

5. A rectangular prism of size 30 × 20 × 80 mm lies on HP on one of its largest faces with its axis parallel to both HP and VP. Draw its isometric projection.

   The projections of the prism are drawn, and the co-ordinate axes are oriented as shown in Fig. 14.9. The three edges b′a′, b′c′ and b′f′ of the prism are to be oriented along the three isometric axes OX, OY and OZ. The isometric axes OX, OY and OZ are drawn inclined to each other at 120° with the axis OZ oriented in the vertical direction. The isometric lengths s_1, s_2 and s_3 of the three

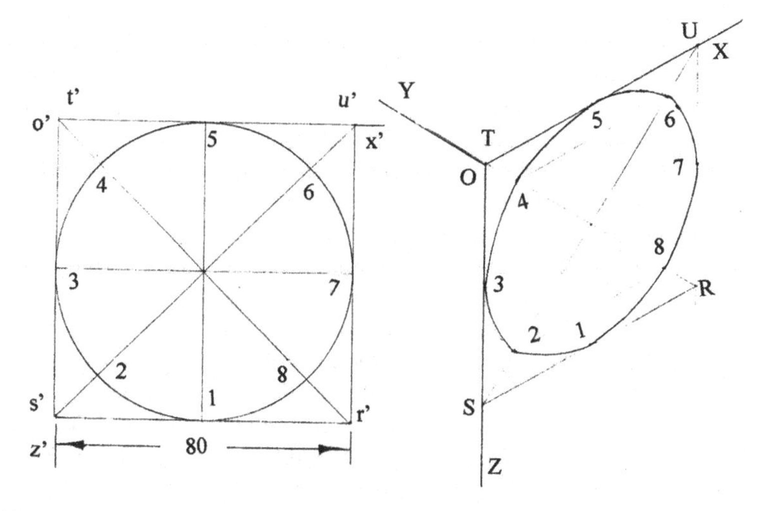

Fig. 14.8 Elevation of a circular lamina and its isometric projection

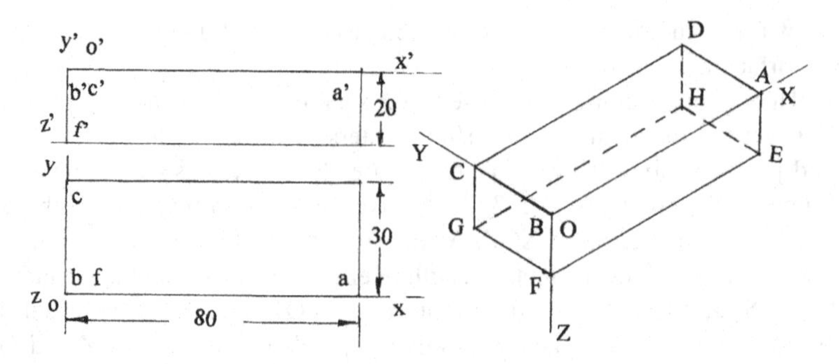

Fig. 14.9 Projections of a rectangular prism and its isometric projection

edges 80 mm, 30 mm and 20 mm are obtained from the isometric scale. Point B is located at O and point A is located on the isometric axis OX at a distance s_1 mm from B. Point C is located on the isometric axis OY at a distance s_2 mm from B. Point F is located on the isometric axis OZ at a distance s_3 mm from B. Lines are drawn from A and F, respectively, parallel to OZ and OX axes to get the intersecting point E which represents isometric point of an angular point of the prism. Lines are drawn from A and C, respectively, parallel to OY and OX axes to locate other angular point D. Lines are drawn from C and F, respectively, parallel to OZ and OY to locate yet another point G. From D, draw a line parallel to OZ. From G, draw a line parallel to OX. From E, draw a line parallel to OY. The intersection of these lines locates the point H. All the eight points are located, and the isometric projection is obtained by joining these points. The hidden edges DH, EH and GH are shown by dotted lines.

6. Draw the isometric projection of a pentagonal prism of edge of base 30 mm and axis 70 mm long when the axis is horizontal.

   The elevation of the pentagonal prism is drawn and is enclosed in a rectangle o′r′s′t′ as shown in Fig. 14.10. The co-ordinate axes are also introduced in the elevation. The isometric projection of the rectangle o′r′s′t′ is to be drawn on the isometric plane XOZ, and the axis is to be oriented parallel to the isometric axis OY. The isometric length of o′r′ is obtained and point R is located on OX. Similarly, the isometric length of o′t′ is obtained and point T is located on OZ. From R and T lines are drawn, respectively, parallel to OZ and OX axes to locate point S. The parallelogram ORST represents the isometric projection of the rectangle. The isometric length of the axis of prism is obtained, and point U is located on the isometric axis OY as shown in the figure. From U, a line is drawn parallel to OX and another line is drawn parallel to OZ. From R, a line is drawn parallel to OY to get the intersecting point P. From T, a line is drawn parallel to OY to get the intersecting point Q as shown in the figure. From points P and Q, lines are drawn, respectively, parallel to OZ and OX to locate the intersecting point V. Join the points V and S. On the parallelogram ORST, points A, B, C, D and E are located following the procedure detailed in Problem 2. From these points, lines are drawn parallel to OY to get the intersecting

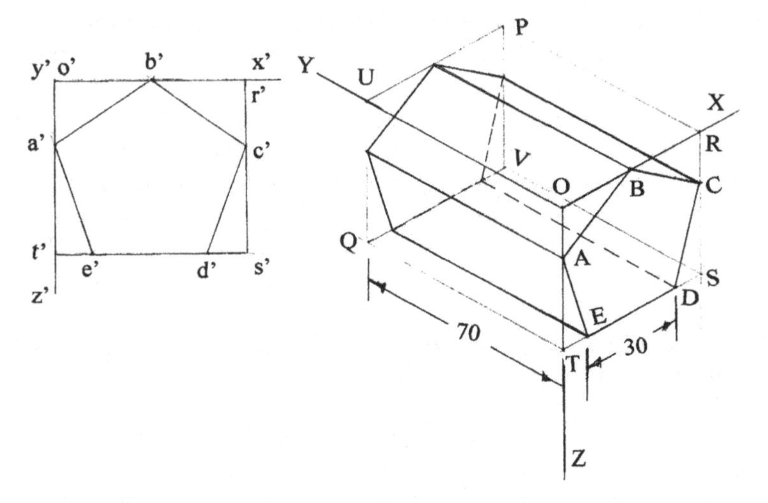

Fig. 14.10 Elevation of a pentagon and Isometric projection of pentagonal prism

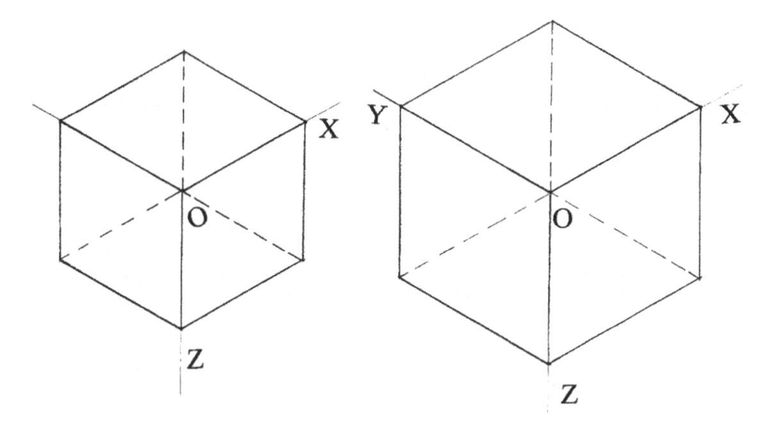

Fig. 14.11 Isometric projection and isometric view of a cube

points on the parallelogram UPVQ. These points represent the angular points on the rear end of the prism. These points are joined to complete the isometric projection of the pentagonal prism with the hidden edges shown by dotted lines.

7. Draw the isometric projection and isometric view of a cube of edge 50 mm.

   The solution can be obtained directly without the orthographic projection of the cube. The isometric projection of cube is drawn using isometric length of edge as shown in Fig. 14.11. The isometric view of cube is also drawn using the edge (true length) dimension as shown in Fig. 14.11.

8. Draw the isometric projection and isometric view of a sphere of diameter 50 mm and mark the lowest and highest points on its surface.

The isometric projection of a sphere will be a circle of true radius r and it can be drawn directly as shown in Fig. 14.12. Shown also in the figure is the isometric view of the sphere with a circle of apparent enlarged radius r_e obtained on the isometric scale. The isometric length of the radius r (i.e. r_i) is placed above and below the centre of circle in the isometric projection to indicate the highest and lowest points on the surface of the sphere. The true radius is placed above and below the centre of the circle in the isometric view to show the highest and lowest points on the surface of the sphere. The lowest point will be the point of contact of the sphere on another solid in the isometric projection/view of the solids in combination.

9. Draw the isometric projection of a circle of diameter 60 mm whose surface is (i) horizontal and (ii) vertical. Use the four centre method.

   The isometric length of the diameter of the circle is obtained from the isometric scale, and the isometric projection of the square enclosing the circle is drawn as rhombus RSTU on the isometric plane XOY as shown in Fig. 14.13. The mid-points 1, 3, 5 and 7 on the four sides of the rhombus are marked. Draw

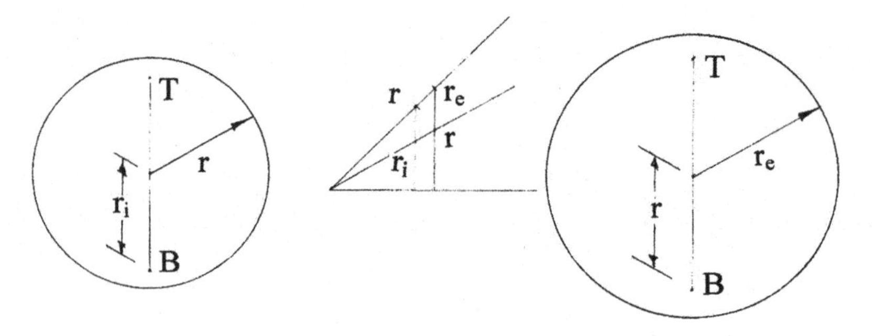

Fig. 14.12 Isometric projection and isometric view of a sphere

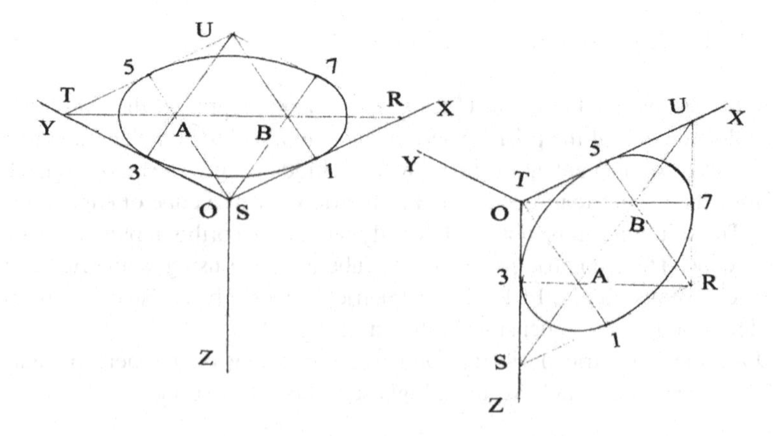

Fig. 14.13 Isometric projections of a circle (four centre method)

the diagonal RT. Join the points U and 3. This line intersects the diagonal at A. Similarly obtain the point B. With centre A and radius A3, draw an arc 35. Also draw an arc 17 with centre B. With centre U and radius U3, draw an arc 13. Also draw an arc 57 with centre S. The ellipse passing through the four points represents the isometric projection of horizontal circle. The isometric projection of vertical circle is drawn on the isometric plane XOZ as shown in the figure. The same procedure is repeated in locating the points A and B on the longer diagonal of the rhombus, and four circular arcs are drawn from the four centres A, B, R and T to complete the isometric projection of the vertical circle.

10. Draw the isometric projection of a cylinder of diameter 50 mm and axis 70 mm long when the axis is vertical.

    The plan of cylinder is drawn and is enclosed in a square with the co-ordinates ox and oy as shown in Fig. 14.14. The two diagonals ac and bd are connected. Points 1, 2, 3, 4, 5, 6, 7 and 8 are indicated on the circumference of the circle as shown in the figure.

    The isometric projection of the square is drawn on the isometric plane XOY using the isometric length of side oa or oc. Lines are drawn from A and C, respectively, parallel to OY and OX to locate the point of intersection D. Join the diagonals AC and BD to locate the circle centre. As the rhombus diagonal AC which is perpendicular to the isometric axis OZ does not undergo foreshortening, the points 4 and 8 are located on either side of the circle centre using the true radius on the diagonal AC. From point 4, lines are parallel to OX and OY. From 8 also lines are drawn parallel to OX and OY. Points 2 and 6 are located on their intersections as shown in the figure. The mid-points of the sides AB, BC, CD and DA are marked as 1, 3, 5 and 7. A smooth curve is drawn passing through the points 1, 2, etc. The top end of the cylinder is now shown as ellipse in the isometric plane XOY. Lines are drawn parallel to OZ from the eight points as shown in the figure. Points are located on these lines using the isometric length of the axis of cylinder. A smooth curve is drawn to represent

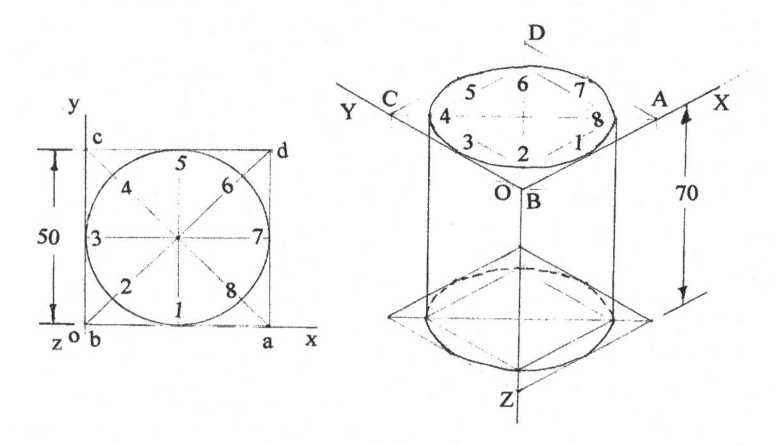

Fig. 14.14 Plan of cylinder and Its isometric projection

the isometric projection of the bottom end of the cylinder. Two common tangents representing the extreme generators of the cylinder are drawn to the ellipses, and the hidden portion of the bottom ellipse is shown as dotted.

11. Draw the isometric projection of a hexagonal pyramid of edge of base 40 mm and axis 70 mm long when its axis is vertical.

    The orthographic projections of the hexagonal pyramid are drawn as shown in Fig. 14.15. The solid is enclosed in a rectangular box, and the co-ordinate axes are also oriented as shown in the figure. The base of the pyramid is enclosed in a rectangle rstu. The isometric length of axis of pyramid is obtained to locate the point S from O. From S, lines are drawn parallel to OX and OY axes. The isometric length of rs is marked from S to locate point R on the line drawn parallel to OX. Similarly, point T is located on the line drawn parallel to OY using the isometric length of st. From R, a line is drawn parallel to OY. From T, a line is drawn parallel to OX to intersect the line drawn from R at point U. The parallelogram RSTU represents the isometric projection of rectangle rstu. The mid-points of sides ST and RU locate the points B and E. The distances between the points ar, fs, dt and cu are all equal. The isometric length of ar is obtained and is marked from R towards S to locate point A. This isometric length can be used to locate points F, C and D. The distance x_1 of the base point g from oy axis in the plan and the distance y_1 of the same point from ox axis are measured, and the isometric lengths are obtained. The isometric length of x_1 is marked along the line SR, and a line is drawn parallel to OY from this point. The base point of the apex of the pyramid is located along this line from the isometric length of y_1. Draw a vertical line above this base point, and using the isometric length of the axis of the pyramid the apex G is located on

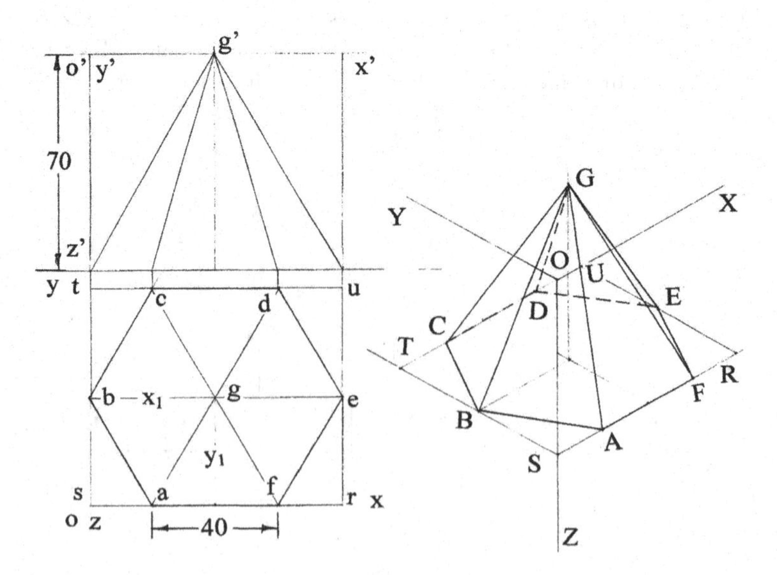

Fig. 14.15 Projections of a hexagonal pyramid and its isometric projection

the vertical line. The apex is joined to the base points to complete the isometric projection of the given pyramid. The hidden edges are shown by dotted lines.

12. Draw the isometric projection of cone of base diameter 50 mm and altitude 70 mm when the base is on HP.

    This problem can be solved without its orthographic projections. The cone is enclosed in a square prism of side 50 mm and axis 70 mm long. The isometric length of the axis of cone is obtained and placed on the isometric axis OZ from O to locate S the base point of the prism as shown in Fig. 14.16. From S, lines are drawn parallel to OX and OY. The isometric length of the diameter of cone is placed on both the lines from S to locate points R and T. From R and T, lines are drawn, respectively, parallel to OY and OX to intersect at point U. The rhombus RSTU represents the isometric projection of base of the prism. The mid-points of the sides locate four points 1, 3, 5 and 7 on the base of the cone. The diagonals RT and SU are connected to locate circle centre. Point 4 is located at a distance equal to the radius of circle from the circle centre on the line RT. Similarly, point 8 is located on the other side. From points 4 and 8, lines are drawn parallel to OX and OY to locate the intersecting points 2 and 6 on the diagonal SU. A line is drawn parallel to OZ from the circle centre, and the apex A is located using the isometric length of the axis of the pyramid. A smooth curve is drawn passing through the base points of the cone. Two tangents are drawn from the apex A to the ellipse representing the extreme generators of the cone. The invisible portion of the curve is drawn dotted.

13. The hexagonal prism edge of base 30 mm and axis 70 mm long rests with one of its rectangular faces on HP and axis perpendicular to VP. It is cut by a plane perpendicular to HP and inclined at 30° to VP bisecting the axis. Draw the isometric projection of one part of the prism showing the section.

    The plan and elevation of the hexagonal prism are drawn as shown in Fig. 14.17. The prism is enclosed in a rectangular prism with m, n, p, q, r, s, t and u as angular points, and the co-ordinates are also drawn as shown in the

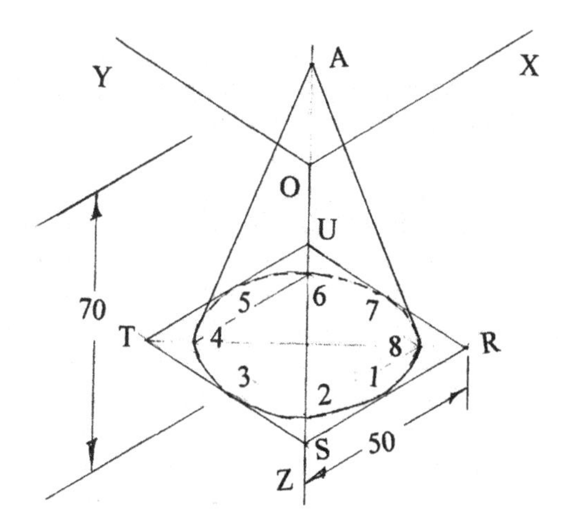

Fig. 14.16 Isometric projection of a cone

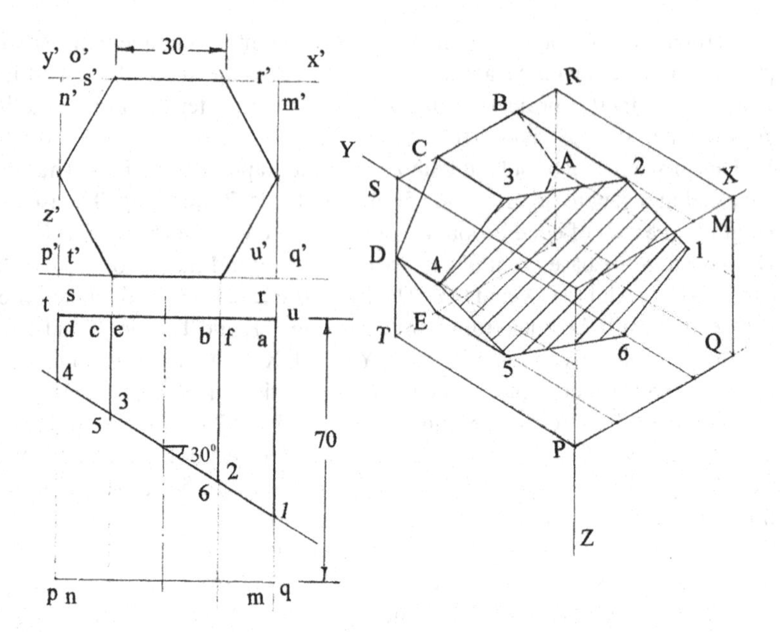

Fig. 14.17 Projections of a sectioned hexagonal prism and its isometric projection

figure. The section plane is drawn and the cutting points 1, 2, 3, 4, 5 and 6 are marked on the longer edges. As the front portion is to be removed for obtaining the isometric projection of the other part, the angular points a, b, c, d, e and f on the rear end of the prism are marked as shown in the figure. The isometric projection of the rectangle mnpq is drawn on the isometric plane XOZ using the isometric lengths of its sides. The point S is located on the isometric axis OY using the isometric length of the axis of the prism. From S and M, lines are drawn, respectively, parallel to OX and OY to locate the intersecting point R. From S and P, lines are drawn, respectively, parallel to OZ and OY to locate point T. Lines are drawn from R and T parallel to OZ and OX to locate point U (point U is not shown in the figure). The isometric projection of the rectangular prism is obtained, and the angular points A, B, C, D, E and F on the rear end of the prism are located following the procedure detailed for the base of the hexagonal pyramid in problem 11. Draw a line parallel to OY from A and mark point 1 using the isometric length of edge a1 in the plan. Lines are drawn from other points B, C, D, E and F, and points 2, 3, 4, 5 and 6 are located using the isometric lengths of the latter points from the former in the plan. The section points are joined and the sectioned portion is hatched. The angular points on the rear end of the prism are also connected to the respective section points. The required isometric projection of one part of the prism is shown with the invisible edges shown by dotted lines avoiding the sectioned portion.

14. A square pyramid is placed over a cube of edge 50 mm such that the corners of the base of the pyramid touch the mid-points of the top face of the cube. Draw the isometric projection of the assembly. The axis of the pyramid is 60 mm.

    The isometric projection of the cube is drawn on the isometric axes using the isometric length of the cube edge as shown in Fig. 14.18. The mid-points of the top face of the cube are located. The centre of the top face is obtained by drawing the diagonals of the rhombus RSTU. A line is drawn parallel to OZ above the top face, and the apex G of the pyramid is located on this line using the isometric length of the axis of the pyramid. The slant edges of the pyramid are joined and also the base edges. The invisible edges are shown by dotted lines. This completes the isometric projection of the assembly of solids.
15. The outside dimensions of a box made of 5 mm thick wooden planks are 80 × 60 × 50 mm. The depth of the lid on outside is 10 mm. Draw the isometric view of the box with the lid open at right angle.

    The elevation of the box is drawn with the lid open at right angle as shown in Fig. 14.19. The co-ordinate axes are indicated as shown in the figure. The three isometric axes OX, OY and OZ are drawn. The isometric view of the bottom of the box is drawn taking its true outside dimensions. The lid is also shown taking its outside dimensions as shown in the figure. Since the thickness of the wooden planks is 5 mm, the inside actual dimensions are reduced, and the isometric view is drawn. The hidden edges are NOT shown in this case. The problem can also be solved without the elevation.
16. Draw the isometric projection of a cylinder of 80 mm diameter and 100 mm long with a 20 mm coaxial square hole. The cylinder is lying on HP with its axis parallel to VP and the sides of hole making an angle of 45° with HP.

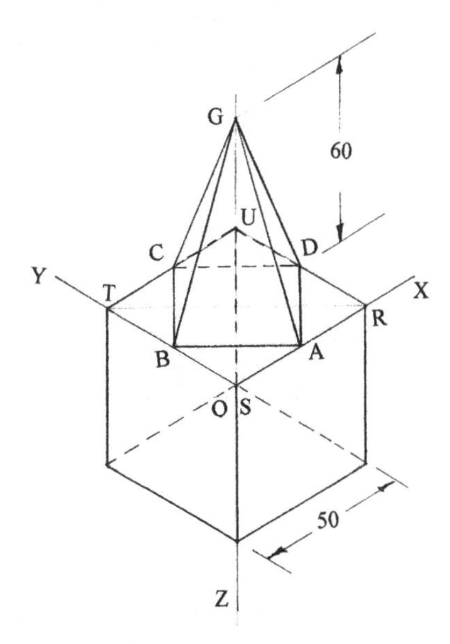

Fig. 14.18 Isometric projections of composite solids (cube and square pyramid)

Fig. 14.19 Projections of a box and its isometric view

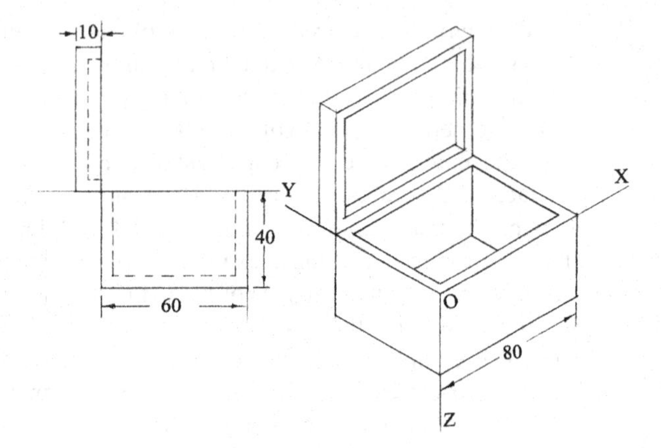

Fig. 14.20 Side elevation of cylinder with square hole and its isometric projection

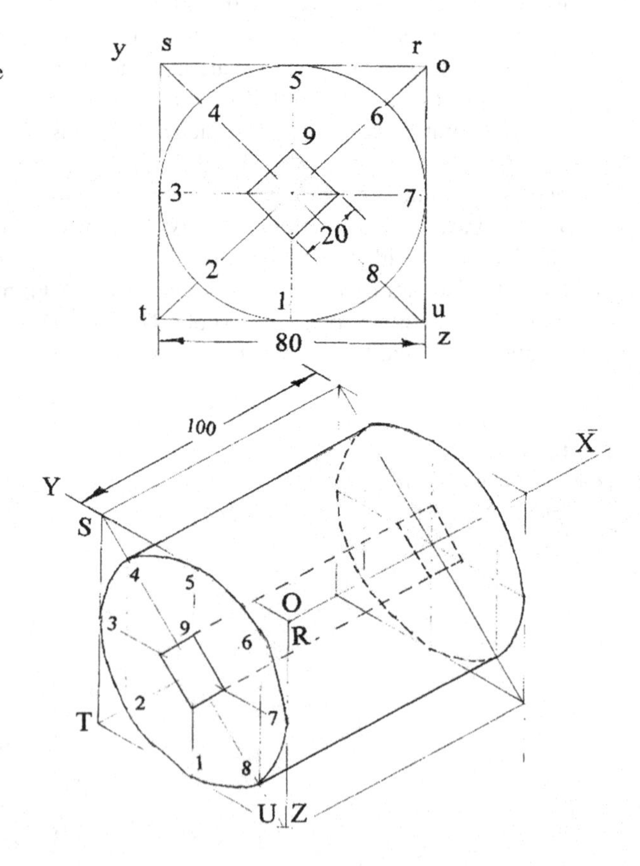

As the axis of the cylinder is parallel to VP, the side elevation of the cylinder with hole is drawn as shown in Fig. 14.20. The circle is enclosed in a square, and the co-ordinate axes are oriented along the sides of the square. The isometric projection of the square rstu is drawn on YOZ plane with isometric length of the side rs. The diagonals of the rhombus RT and SU are connected to locate the circle centre. The mid-points of sides UT, TS, SR and RU are marked to locate points 1, 3, 5 and 7. Since the diagonal SU is perpendicular to the isometric axis OX, the true diameter is placed on this diagonal to locate points 4 and 8 from the circle centre as indicated. From 4 and 8, lines are drawn parallel to OY and OZ axes to locate points 2 and 6 on the diagonal RT. Lines are drawn parallel to OX from each of the points 1, 2, 3 etc. The isometric length of the axis of cylinder is obtained and is used to mark the respective points on the respective lines to locate points on the other end of the cylinder as shown in the figure. Connect points 15 and 37. The isometric length of 59 is placed from points 1, 3, 5 and 7 as indicated to locate the angular points of the square hole on one side of the cylinder. The respective points on the other side of the cylinder are also marked by drawing lines parallel to OX from the angular points of the hole obtained. Two common tangents are drawn to the two ellipses representing the extreme generators, and the edges of the hole are also connected to complete the isometric projection of the cylinder with square hole. The invisible portion of the ellipse and edges of the hole are shown by dotted lines.

17. A square prism, side of base 50 mm and height 60 mm, surmounts coaxially a circular disc of 80 mm diameter and 40 mm thick. A sphere of 40 mm radius is placed centrally over the square prism. Draw the isometric view of the assemblage of solids.

    The isometric view of top of the circular disc is drawn following the four centre method (see problem no. 9) on the isometric plane XOY as shown in Fig. 14.21. The isometric view of bottom of the circular disc is also drawn below the top end with the extreme generators joining the top and bottom ends as shown in the figure. The centre of the top of the disc is located at O1. A rhombus of sides 50 mm is drawn keeping its sides parallel to the isometric axes OX and OY. A line is drawn parallel to OZ through O1 and point O2 is located on this line such that the distance of O2 from O1 is equal to the height of the square prism. Another rhombus of sides 50 mm is drawn keeping the sides parallel to the isometric axes OX and OY. This rhombus represents the isometric view of the top of the prism. The vertical edges of the prism are shown connecting the top corner points with the respective bottom corner points of the prism. A line is drawn through O2 and point O3, the center of the sphere, is located on it such that the distance of O3 from O2 is equal to the radius of the sphere. With centre O3 and the apparent radius, ra (to be obtained from the isometric scale, see problem no. 8), a circle is drawn to show the isometric view of the sphere. This completes the isometric view of the assemblage of solids.

18. Draw the isometric projection of the object shown in Fig. 14.22.

    The object is enclosed in a square prism as shown in Fig. 14.22. The isometric projection of the square prism is drawn on the isometric axes OX, OY and OZ as shown in Fig. 14.23. The isometric projection of the base of the cylinder is completed with eight points as shown in the figure. The points a, b,

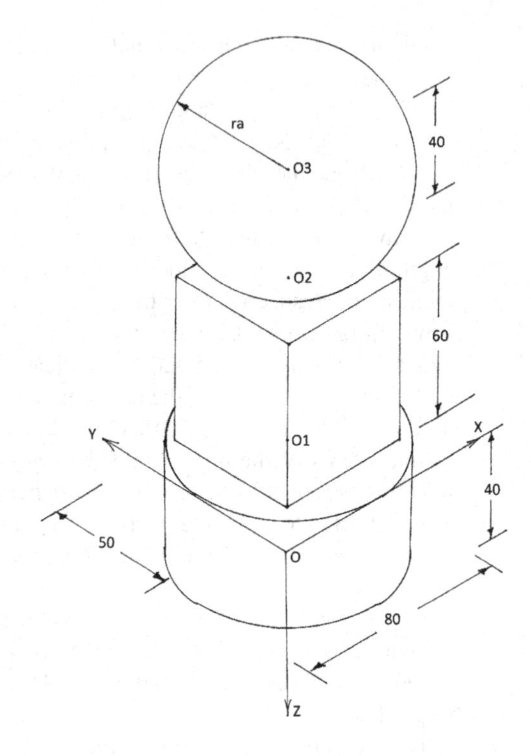

Fig. 14.21 Isometric view of solids

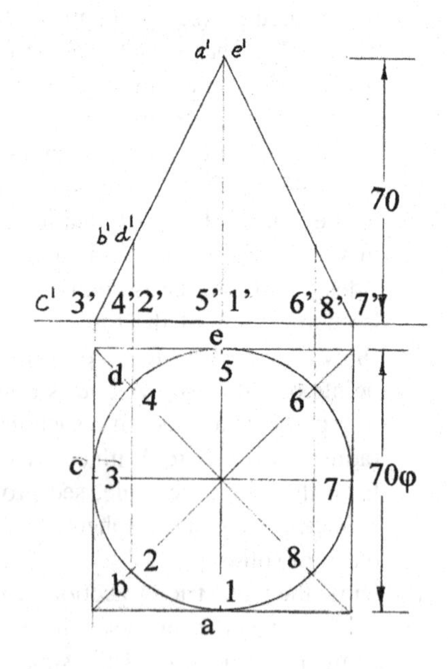

Fig. 14.22 Projections of an object

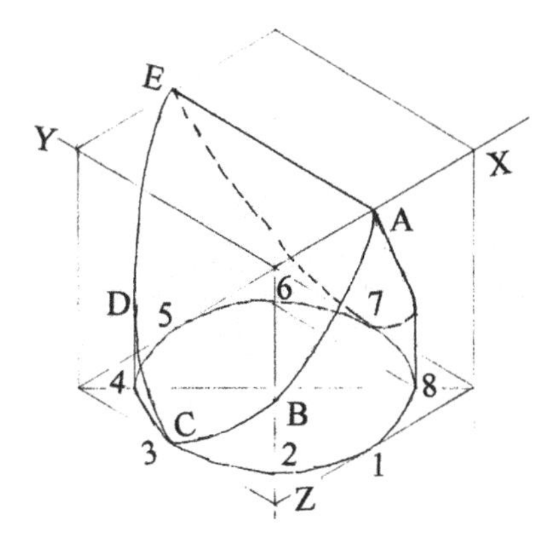

Fig. 14.23 Isometric projection of the object

c, d and e on one side of cylinder are located in the plan and elevation of the object. The points A and E are marked in the isometric projection as shown in the figure. The point C is also marked as indicated. The isometric length of the height of b′ from 2′ is obtained and placed on the line drawn parallel to OZ from 2 to locate point B in the isometric projection. The same isometric length is placed on another line drawn parallel to OZ from 4 to locate point D. A smooth curve is drawn passing through A, B, C, D and E. The line AE is joined. The same procedure is followed on the other side of the cylinder. The wedge-shaped object representing the isometric projection of the given object is completed with the invisible curves shown as dotted.

19. A pedestal consists of a square slab, side of base 60 mm and thickness 20 mm, surmounted by a frustum of a square pyramid, side of base 40 mm, that at top 20 mm and height 60 mm, which is surmounted by a square pyramid having side of base 20 mm and height 40 mm. All three solids are coaxial and are similarly situated. Draw the isometric view of the pedestal.

A rhombus of sides 60 mm is drawn keeping its sides parallel to the isometric axes OX and OY as shown in Fig. 14.24. This rhombus represents the isometric view of top surface of the square slab. The thickness of the slab is placed below the corners of the rhombus to complete the isometric view of the square slab. On the top of the slab, draw a centralized rhombus of side 40 mm keeping its sides also parallel to the axes OX and OY. The inner rhombus represents the isometric view of the base of the frustum of the square pyramid. A line is drawn parallel to OZ from O_1, the centre of the rhombus, and point O_2 is marked on this line such that the distance of O_2 from O_1 is equal to the height of the frustum of the pyramid. A rhombus of sides 20 mm is drawn with centre O_2 and sides parallel to OX and OY. This rhombus represents the top of the frustum of the pyramid and also the base of the square pyramid. A line is drawn parallel to OZ from O_2. The apex of the pyramid is located 40 mm above O_2. The base points are joined to the apex of the pyramid. The slanting edges of the frustum of the pyramid are also joined to complete the isometric view of the pedestal.

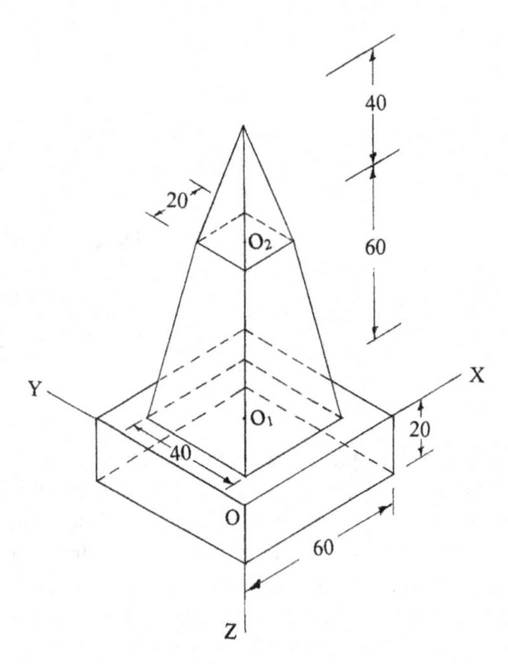

Fig. 14.24 Isometric view of a pedestal

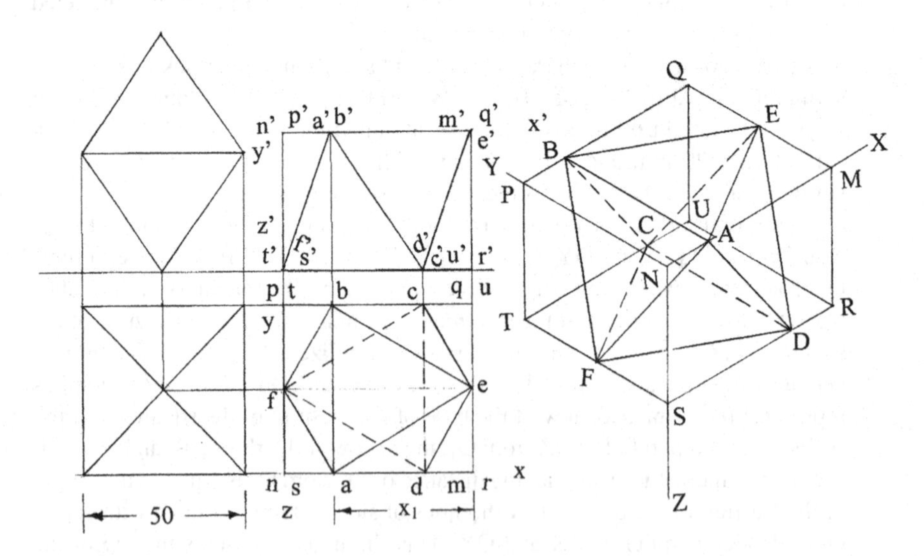

Fig. 14.25 Isometric view of an octahedron

20. Draw the isometric view of an octahedron of edge 50 mm with a face on the HP and a solid diagonal parallel to the VP.

The plan and elevation of the octahedron with a face on the HP and a solid diagonal parallel to the VP are drawn, and the projections are enclosed in a transparent rectangular box as shown in Fig. 14.25. The coordinate axes are also shown. The isometric view of the box is drawn measuring its dimensions from the projections. The six angular points of the octahedron lie on the top and bottom larger faces of the box. Point E is marked at the mid-point of the edge MQ. Point F is marked at the mid-point of the edge ST. The horizontal x_1 between m and a in the plan is measured. Point A is located on MN such that the distance AM is equal to x_1 in the plan of orthographic projection. Point B is located on QP using the same distance. Points C and D are also located on the lines TU and SR using the same distance as shown in the figure. The angular points A, B, C, D, E and F are joined to complete the isometric view of the octahedron. The invisible edges BC, CD, CE and CF are shown by dotted lines.

21. Draw the isometric view of a hexagonal pyramid, side of base 25 mm and axis 70 mm long, when a face is on the HP with the axis parallel to the VP.

The plan and elevation of the hexagonal pyramid lying with a face on the HP are drawn, and the projections are enclosed in a transparent rectangular box as shown in Fig. 14.26. The coordinates are also shown in the projections. The isometric view of the rectangular box is drawn measuring its dimensions from its projections and orienting its edges parallel to the isometric axes OX, OY and OZ. Point A is marked at the mid-point of the edge RU. Points D and E lie on the edge ST and are located easily measuring the distances of d and e from s and t in the plan. Point R_1 is located on RS such that the distance of R_1 from R is equal to the horizontal distance x_1 in the plan. Point U_1 is located on UT using the same horizontal distance x_1. Lines are drawn parallel to OZ from R_1 and U_1 to meet the top edges MN and QP at M_1 and Q_1. Point C is located at the mid-point of R_1M_1. Point F is located at the mid-point of U_1Q_1. Point R_2 is located on RS such that the distance of R_2 from R is equal to the horizontal distance x_2 in the plan. Point U_2 is located on UT using the same distance x_2.

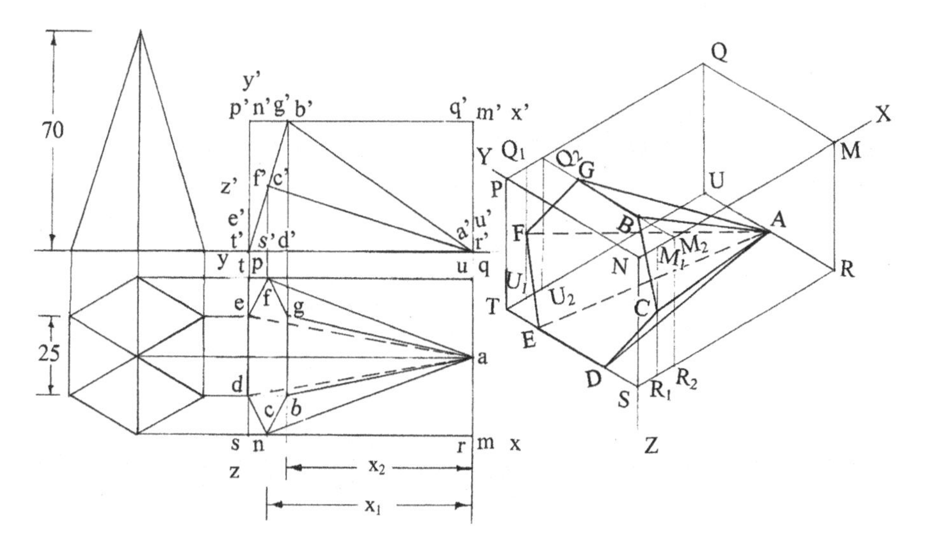

Fig. 14.26 Isometric view of a hexagonal pyramid

Lines are drawn parallel to OZ from R_2 and U_2 to meet the edges MN and QP at M_2 and Q_2. Join the line M_2Q_2. Point B is located on M_2Q_2 such that the distance of B from M_2 is equal to the distance of D from S. Point G is located on M_2Q_2 such that the distance of G from Q_2 is equal to the distance of E from T. Join these points to complete the isometric view of the hexagonal pyramid. The invisible edges AE and AF are shown by dotted lines.

22. The plan and elevation of a wedge block are given in Fig. 14.27. Draw an isomeric view of the block.

    The isometric view of the rectangular prism is drawn taking the overall dimensions from the projections as shown in Fig. 14.28. Points are marked on the respective isometric axes to show the wedge portion from the dimensions given in the projections. The isometric view is completed for the wedge block as shown in the figure.

23. The plan and elevation of a recessed block are given in Fig. 14.29. Draw an isometric view of the block.

    The isometric view of the square prism is drawn taking the overall dimensions from the projections as shown in Fig. 14.30. Points are marked on the respective isometric axes to show the recessed portion from the dimensions given in the projections. The isometric view is completed for the recessed block as shown in the figure.

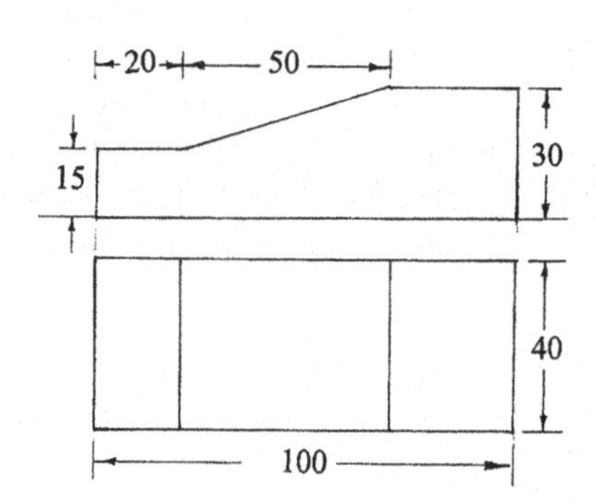

Fig. 14.27 Projections of a wedge block

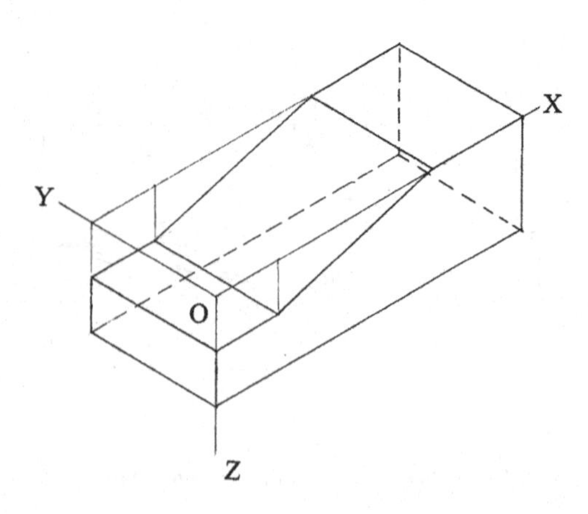

Fig. 14.28 Isometric view of the block

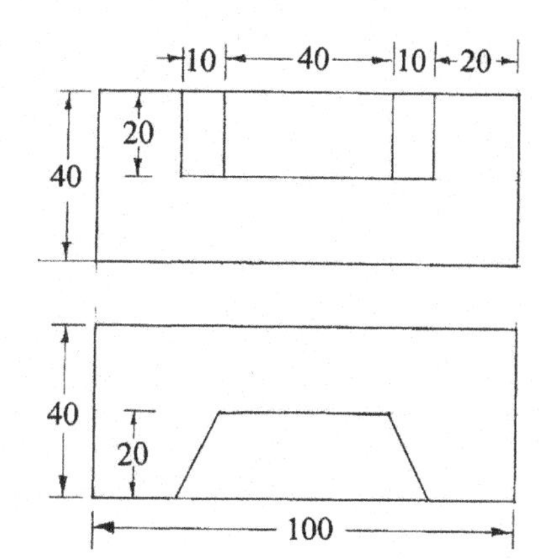

Fig. 14.29 Projections of a recessed block

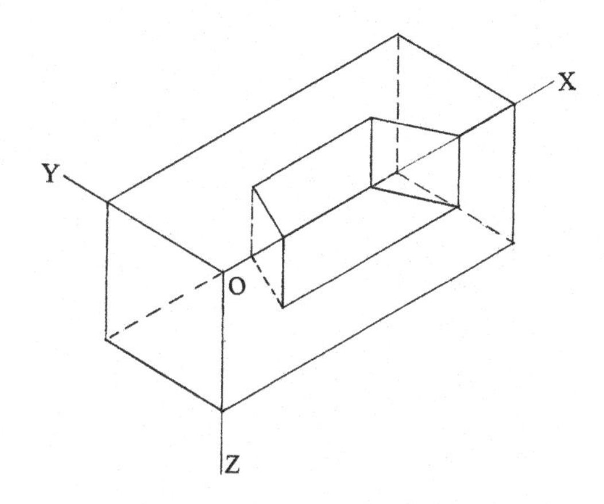

Fig. 14.30 Isometric view of the block

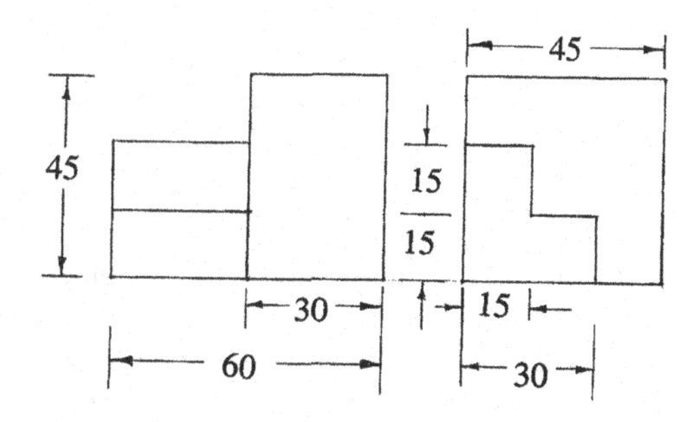

Fig. 14.31 Projections of an end stop

24. The elevation and side elevation of an end stop are given in Fig. 14.31. Draw an isometric view of the end stop.

    The isometric view of the square prism is drawn taking the overall dimensions from the projections as shown in Fig. 14.32. Points are marked on the respective isometric axes and the isometric lines to show the steps from the dimensions given in the projections. The isometric view is completed for the end stop as shown in the figure.

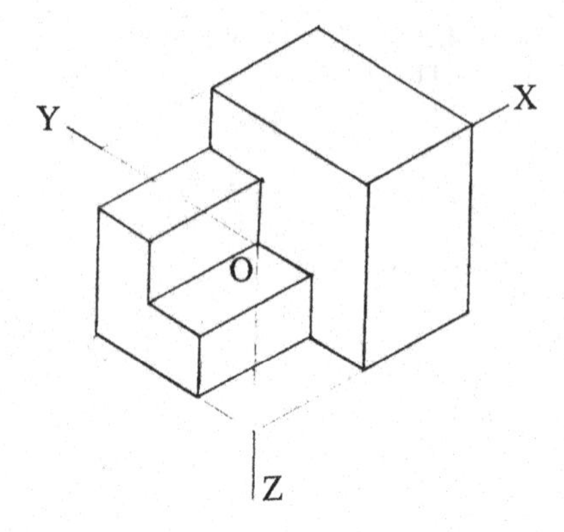

Fig. 14.32 Isometric view of the stop

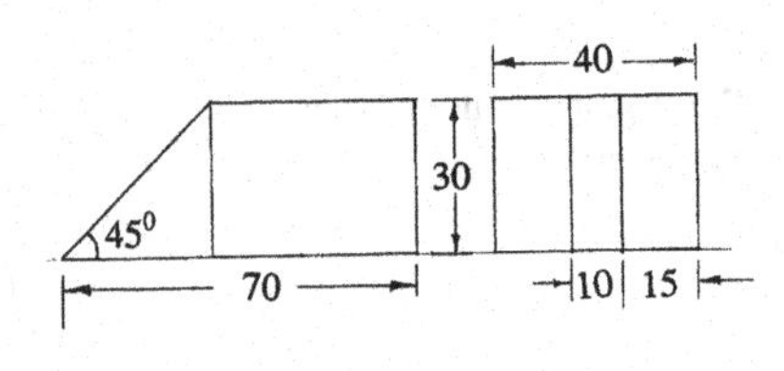

Fig. 14.33 Projections of an angle stop

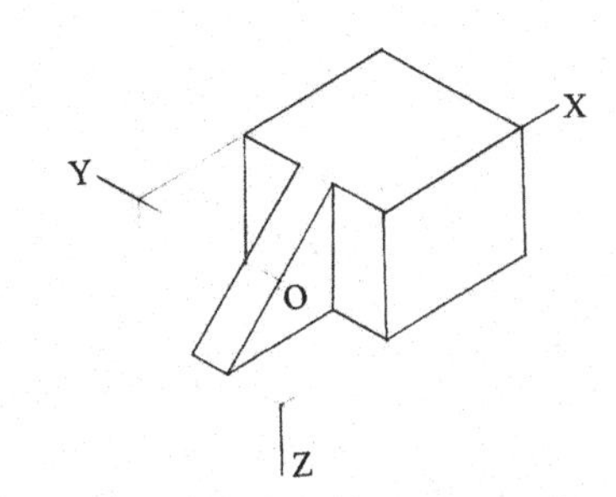

Fig. 14.34 Isometric view of the angle stop

25. The elevation and side elevation of an angle stop are given in Fig. 14.33. Draw an isometric view of the angle stop.

    The isometric view of the rectangular prism is drawn taking the overall dimensions from the projections shown in Fig. 14.34. Points are marked on the respective isometric axes and the isometric lines to show the end points of angle from the dimensions given in the projections. The isometric view is completed for the angle stop as shown in the figure.

26. Six equal cubes of 40 mm side are arranged one on each face of a cube of 40 mm side. Draw an isometric view of the solid formed by these cubes.

    Let the six cubes of 40 mm side be designated as A, B, C, D, E and F. These cubes are arranged one on each face of the seventh cube G. Cube A is on the left face of cube G. Cube B is on the top face of cube G. Cube C is on the right face of cube G. Cube D is on the bottom face of cube G. Cube E is on the front face of cube G. Cube F is on the back face of cube G. The isometric view of the combination of cubes is shown in Fig. 14.35.

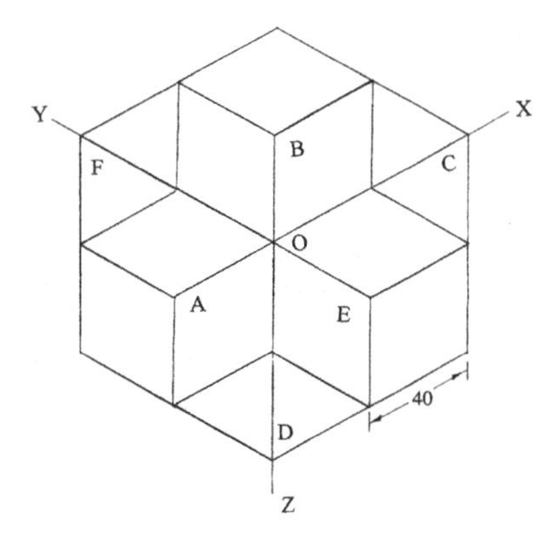

Fig. 14.35 Isometric view of cubes combination

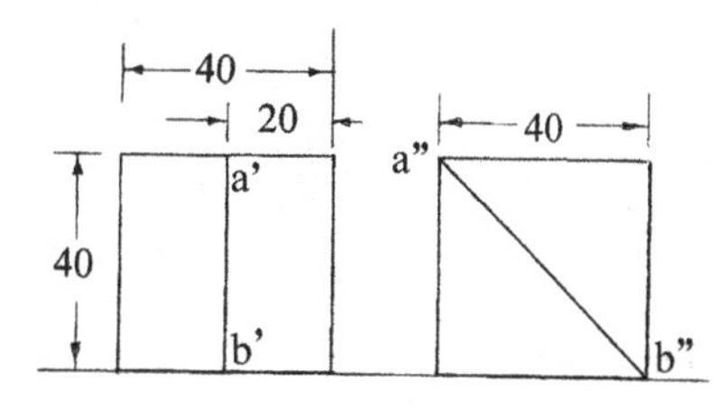

Fig. 14.36 Projections of cubical object

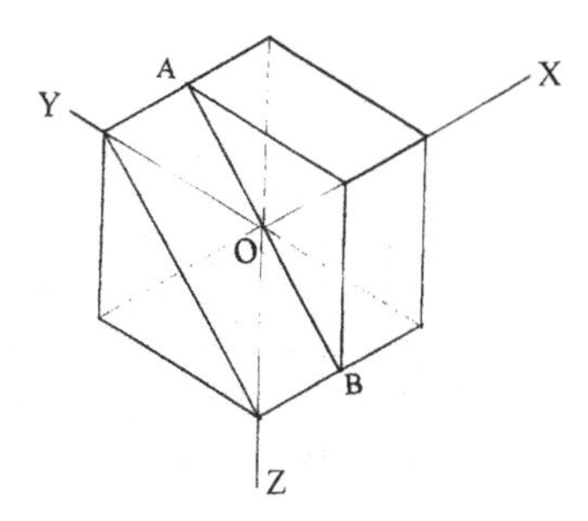

Fig. 14.37 Isometric view of the object

27. The elevation and side elevation of a cubical object are given in Fig. 14.36. Draw its isometric view.

    The isometric view of the cube is drawn in thin lines as shown in Fig. 14.37. Point A is marked at the mid-point of top edge on the back face of the cube. Point B is marked at the mid-point of the bottom edge on the front face of the cube. These two points are joined by thick line. The top corner and the bottom corner of the left face of the cube are also joined by thick line. The visible edges of the cube are shown by thick lines to complete the isometric view of the cubical object.
28. The elevation and side elevation of a cubical object are given in Fig. 14.38. Draw its isometric view.

    The isometric view of the cube is drawn in thin lines as shown in Fig. 14.39. Point A is marked at the centre of the left face of the cube. Point B is marked at

Fig. 14.38 Projections of cubical object

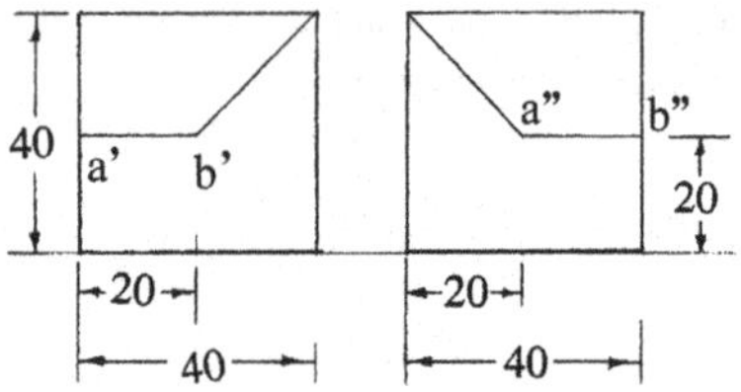

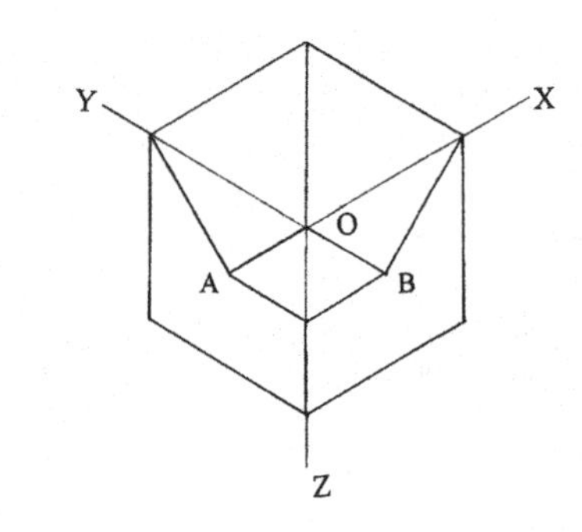

Fig. 14.39 Isometric view of the object

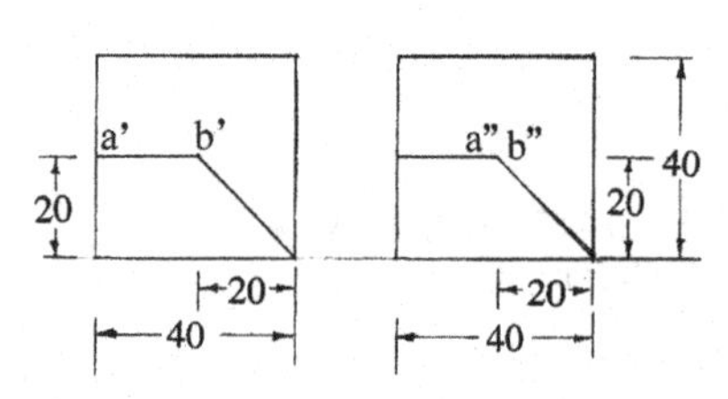

Fig. 14.40 Projections of cubical object

the centre of the front face of the cube. A rhombus is constructed by drawing thick lines parallel to OX and OY axes from A and B as shown in the figure. Point A is joined by thick line to the top corner on the left face of the cube. Point B is joined by thick line to the top corner on the front face of the cube. The centre of the cube is joined by thick line to the top corner on the top face of the cube. The visible edges of the cube are drawn by thick lines to complete the isometric view of the cubical object.

29. The elevation and side elevation of a cubical object are given in Fig. 14.40. Draw its isometric view.

    The isometric view of the cube is drawn in thin lines as shown in Fig. 14.41. Point A is marked at centre of the left face of the cube. Point B is marked at the centre of the cube. Join the points A and B by thick line. The mid-point of the rear vertical edge on the left face of the cube is joined to points A and B by thick lines as shown in the figure. The corners of the bottom edge on the front face are joined to points A and B by thick lines. The visible edges of the cube are shown by thick lines to complete the isometric view of the cubical object.

30. The elevation and side elevation of a cubical object are given in Fig. 14.42. Draw its isometric view.

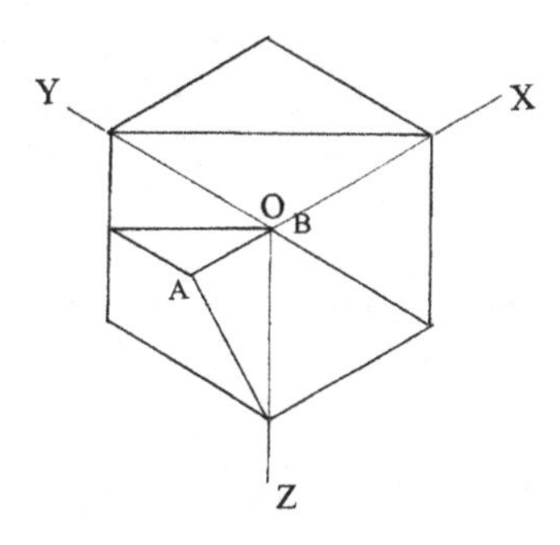

Fig. 14.41 Isometric view of the object

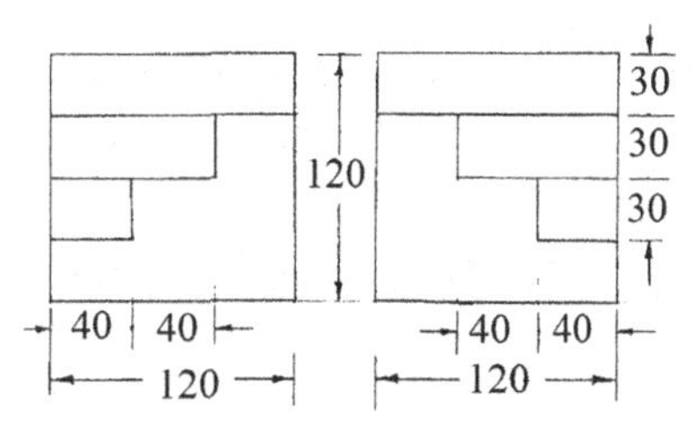

Fig. 14.42 Projections of cubical object

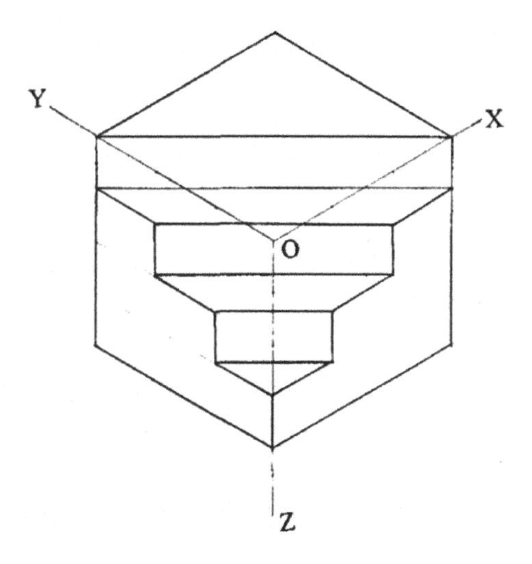

Fig. 14.43 Isometric view of the object

The isometric view of the cube is drawn in thin lines as shown in Fig. 14.43. Points are marked at intervals of 30 mm on the edge common to the left and front faces of the cube. Thin lines are drawn parallel to OX on the front face from the points on the common edge. Thin lines are drawn parallel to OY on the left face from the points on the common edge. Points are marked at intervals of 40 mm on the top edges of the left and front faces of the cube. Thin lines are drawn parallel to OZ on both the faces from points on the top edges. The intersections of the lines on both the faces locate the end points of the steps. Thick lines are drawn to show the steps as shown in the figure. The visible edges of the cube are shown by thick lines to complete the isometric view of the cubical object.

31. A waste paper basket is in the form of an inverted frustum of a hexagonal pyramid with base 40 mm sides hexagon and top 70 mm sides hexagon. Draw the isometric projection if its height is 120 mm.

    Draw a hexagon abcdef of side 70 mm and enclose it in a rectangle rstu as shown in Fig. 14.44 with the co-ordinate axes ox and oy as indicated. Draw another hexagon of side 40 mm concentric with the former. The isometric projection of the rectangular prism is drawn using the isometric lengths of sides rs and st and the isometric length of the height of the pyramid as shown in Fig. 14.45. The points A, B, C, D, E and F are marked on the circumference of

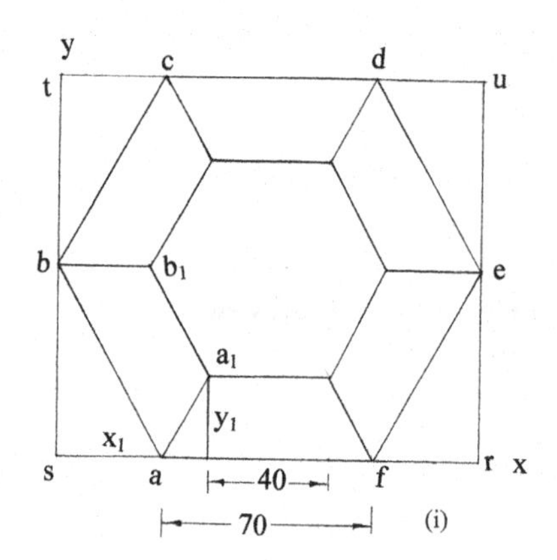

Fig. 14.44 Plan of basket

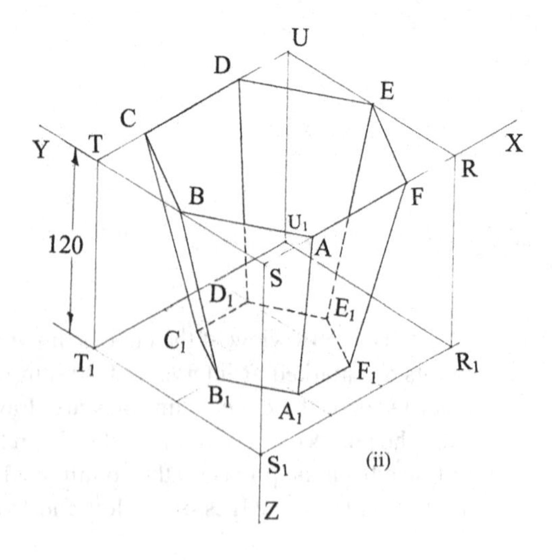

Fig. 14.45 Isometric projection

the parallelogram RSTU. The mid-points of sides S_1T_1 and R_1U_1 on the lower parallelogram are marked and connected. The isometric length of bb_1 in the plan is obtained and placed from the mid-point of S_1T_1 to locate point B_1. The same length is used to locate point E_1 from the mid-point of R_1U_1. The point A_1 is located from the isometric lengths of the distances of point a_1 from ox and oy axes. The same procedure is followed to locate other points on the bottom of the basket. The isometric projection is completed by joining the two sets of the points on the top and bottom of the hexagon. The invisible base edges and part of invisible slant edges are shown by dotted lines.

Note: Refer to problems 28, 29 and 30. The isometric views of the cubical objects show their forms more clearly than the corresponding orthographic projections. This is the chief advantage of the isometric projections/views. The drawbacks of the isometric projections/views are that the drawings take more time to prepare than orthographic views and that the drawings are not easily dimensioned.

Practice Problems

1. Draw the isometric projection of a triangular prism, edge of base 50 mm and axis 70 mm long, when the axis is perpendicular to VP.
2. Draw the isometric projection of a square prism, edge of base 50 mm and axis 70 mm long, when the axis is vertical.
3. Draw the isometric projection of a tetrahedron of edge 50 mm when one of the faces is on HP with an edge parallel to VP.
4. Draw the isometric projection of an octahedron of edge 50 mm when its axis/solid diagonal is vertical.
5. A square pyramid, edge of base 40 mm and axis 60 mm long, is lying on one of its triangular faces on HP and its axis parallel to VP. Draw the isomeric projection of the pyramid.
6. A cylinder of diameter 50 mm and height 60 mm is resting on one of its ends on HP. It is cut by a plane perpendicular to VP and inclined at 45° to HP. The cutting plane passes through a point on its axis located at 20 mm from the top end. Draw the isometric projection of the cut cylinder showing the cut surface.
7. A pentagonal pyramid, edge of base 30 mm and height 70 mm, stands on HP such that an edge of base is parallel to VP and nearer to it. A section plane perpendicular to VP and inclined at 30° to HP cuts the pyramid passing through a point on the axis at a height of 35 mm from the base. Draw the isometric projection of the truncated pyramid showing the cut surface.
8. A sphere of radius 30 mm is placed centrally on a square slab of size 60 mm × 10 mm. Draw the isometric projection and isometric view of the solids in the given position.
9. A hemisphere of 40 mm diameter is nailed on the top surface of a frustum of a square pyramid. The sides of the top and bottom faces of frustum are,

respectively, 20 mm and 40 mm and its height is 50 mm. The axes of both the solids coincide. Draw the isometric projection of the solids.

10. The basement of a pillar consists of three square blocks of size 3 m, 2.5 m and 2 m and height each equal to 0.5 m placed one over the other centrally with largest block at the bottom. The pillar consists of frustum of a cone of bottom diameter 2 m, top diameter 1.5 m and height 3 m, and on the top is placed centrally a sphere of diameter 2 m. Draw the isometric projection of the pillar and the basement. Adopt a suitable scale.
11. A hemisphere of radius 30 mm rests centrally on a cube of 60 mm side such that the circular face of the hemisphere is at the top. Draw the isometric projection of the solids in the given position.
12. Draw the isometric projection of a hexagonal pyramid, side of base 30 mm and axis 60 mm, when a face on X-Y plane and a side of base contained by that face recede to the left.
13. A frustum of a cone, diameter of base 60 mm, diameter at the top end 40 mm and axis 50 mm, lies on the HP on a generator. Draw its isometric projection when its top end is nearer to the observer than its base.
14. A cube of 60 mm edge is surmounted by a cylinder, 40 mm diameter and 80 mm axis coaxially. Draw the isometric view of the solids in the given position.
15. A waste paper basket is in the form of a hollow inverted frustum of a square pyramid. The upper end is a square of 100 mm side, the lower end is a square of 70 mm and the depth is 120 mm. Draw its isometric projection.

Chapter 15
Perspective Projections

Perspective projection is a method of pictorial representation of an object on a transparent picture plane as seen by an observer stationed at a particular position relative to the object. The object is placed behind the picture plane, and the observer is stationed in front of the picture plane. The visual rays emanating from the eye of the observer to the object pierce the picture plane and locate the position of the object on the picture plane. The visual rays emanate from a common point known as the station point or the eye of the observer. This type of projection is called perspective projection. This is also known as scenographic projection or convergent/conical projection. The method of preparing a perspective view differs from the various other methods of projections discussed earlier.

Nomenclature of Perspective

The elements of perspective projection are shown pictorially in Fig. 15.1. The important terms used in the perspective projections are defined below:

1. Ground plane (GP). This is the plane on which the object is assumed to be placed.
2. Auxiliary plane (AGP). This is any plane parallel to the ground plane (not shown in the figure).
3. Station point (SP). This is the position of the observer's eye from where the object is viewed.
4. Picture plane (PP). This is the transparent vertical plane positioned in between the station point and the object to be viewed. Perspective is formed on this vertical plane.
5. Horizon plane (HP). This is the imaginary horizontal plane perpendicular to the picture plane and passing through the station point. This plane lies at the level of the observer.

K. Rathnam, *A First Course in Engineering Drawing*,
DOI 10.1007/978-981-10-5358-0_15

Fig. 15.1 Nomenclature of perspective projection

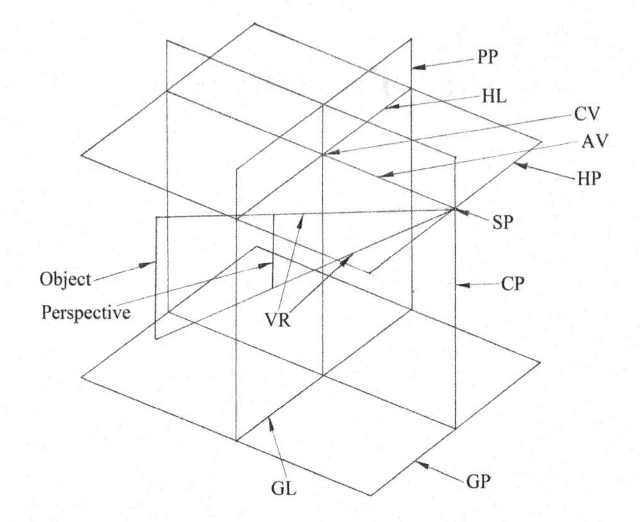

6. Ground line (GL). This is the line of intersection of the picture plane with the ground plane.
7. Auxiliary ground line (AGL). This is the line of intersection of the picture plane with the auxiliary plane (not shown in the figure).
8. Horizon line (HL). This is the line of intersection of the horizon plane with the picture plane. This line is parallel to the ground line.
9. Axis of vision (AV). This is the line drawn perpendicular to the picture plane and passing through the station point. The axis of vision is also called the line of sight or perpendicular axis.
10. Centre of vision (CV). This is the point through which the axis of vision pierces the picture plane. This is also the point of intersection of horizon line with the axis of vision.
11. Central plane (CP). This is the imaginary plane perpendicular to both the ground plane and the picture plane. It passes through the centre of vision and the station point while containing the axis of vision.
12. Visual rays (VR). These are imaginary lines or rays joining the station point to the various points on the object. These rays converge to a point.

The orthographic projections of the perspective elements in third angle projection are shown in Fig. 15.2. The top portion represents the top view/plan of the perspective elements. The bottom portion represents the front view/elevation of the perspective elements. The distance of the eye or the station point from the picture plane will be the distance of SP from PP. The height of the eye or the station point is the distance of HL from GL or the distance of CV from GL. Though the distance between PP and HL is arbitrary, it should preferably be longer than the distance of the station point/eye from the picture plane. The top view or plan of the object is drawn considering its position with respect to the picture plane. Let us assume a vertical line AB as our object and proceed to show its top view a-b and front view

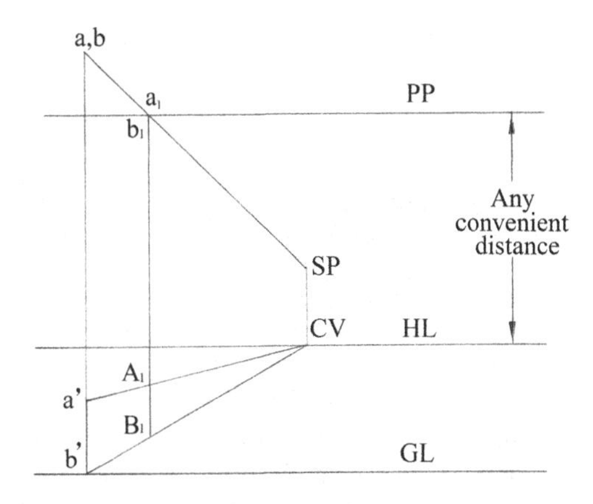

Fig. 15.2 Orthographic projections of perspective elements

a′-b′. The top views of the visual rays are then drawn by joining the points on the top view of the object with SP. In this case, both the rays coincide and pierce PP at one point in the top view. The front views of the visual rays, which connect CV with the front view of the points, are then drawn. From the point of intersection of the top view of the visual ray with PP at a_1, a projector is drawn to cut the front view of the corresponding visual ray to show the perspective of the point A_1 of the object. In this case, we can extend this projector to cut the elevation of the other visual ray also to get the perspective the other point B_1. The line joining these two points is the perspective or perspective view A_1 B_1 of the object.

We have shown in the Fig. 15.2 the orthographic projections of the perspective elements which are needed to obtain the perspective of the line AB. The aforementioned procedure is the centre of vision or visual ray method. In this method, the top view and the front view of the visual rays are drawn. The piercing point of the top view of the visual ray is located on PP, and a projector is drawn from this point to intersect the front view of the corresponding visual ray to fix the perspective of the point on the object. Though this method can be used for solving many problems in perspective, we may have to use many front view lines to get the required points in the perspective. Also, one has to work in a limited space to get the final perspective view, and it will be very difficult to locate the points accurately in the perspective. In the vanishing point method, the top view of the visual rays will stay as such. Instead of the front view of the visual rays, we locate the vanishing point on HL by identifying the parallel lines on the top view of the object and drawing a line parallel to them from SP to cut PP to locate the top view of the vanishing point. A projector is drawn from this point to cut HL, and this will give the front view of the vanishing point. The parallel lines on the top view are extended to meet PP, and projectors are drawn from the meeting points to locate the front views of the corresponding points having the same height as the points on the object. Then these points on the front views are connected to the vanishing point. The perspective of the points will lie on these lines, and their exact location can be obtained by drawing projectors from the piercing points of the top views of the visual rays with the PP.

The centre of vision or visual ray method is used to solve the first five problems, and the vanishing point method is employed for the remaining problems.

Solved Problems

1. A point A lies 15 mm within the picture plane, 40 mm to the left of the eye and on the ground. Another point B lies 20 mm within the picture plane, 15 mm to the left of the eye and 10 mm above the ground. A third point C lies 30 mm within the picture plane, 30 mm to the right of eye and at a height of 30 mm from the ground. A fourth point D lies 15 mm without the picture plane, 40 mm to the right of the eye and 50 mm above the ground. The eye is 75 mm from the picture plane and 40 mm from the ground. Draw the perspective of the four points.

   We shall draw the top view of picture plane (PP) and also the station point/eye which is 75 mm from the picture plane as shown in Fig. 15.3. The top view of the point A, which is 15 mm from PP and 40 mm to the left of SP, is also located as shown in the figure. Similarly, the top views of B and C are located. The point D, which lies 15 mm in front of the picture plane, is also located as shown in the figure. The top views of the visual rays are drawn, and the piercing point a_1, b_1, c_1, and d_1 are located on PP. Since the point D lies in front of PP, the top view of this visual ray should be extended to meet PP.

   The GL and HL, which are, respectively, intersections of horizon plane (HP) and ground plane (GP) with the picture plane (PP), are drawn keeping a

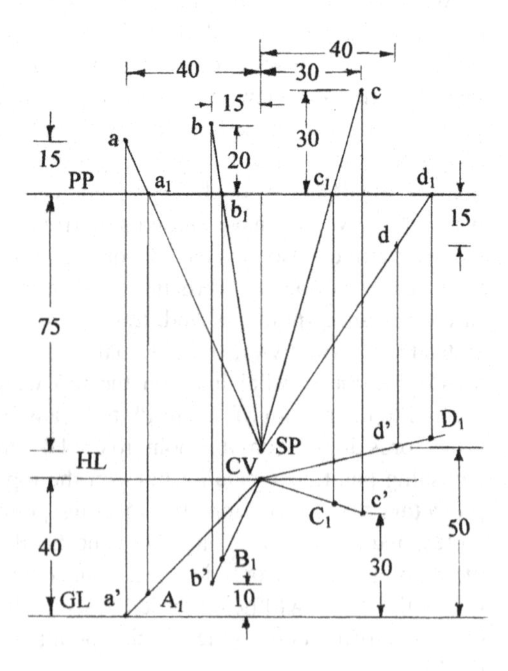

Fig. 15.3 Perspective of points in space

distance 40 mm being the height of the eye from the ground. The distance of HL from PP should be longer than the distance of SP from PP. The centre of vision CV, which is the front view of the station point, is located on HL. The front views of the points are then located on the projectors drawn from the top views of the points and taking the heights of the points from ground into account. The front views of visual rays are drawn from the front views of the points to the centre of vision (CV). Projectors are drawn from a_1, b_1, c_1 and d_1 to cut the respective front view of the visual ray to get perspective of the points A_1, B_1, C_1 and D_1.

2. Five parallel lines, at intervals of 20 mm, lie on a horizontal plane 30 mm above the ground. The lines are 80 mm long and parallel to the picture plane. They have their left ends 20 mm to the left of the eye, which is 60 mm from the picture plane and 50 mm above the ground. The first line is 10 mm within the picture plane. Draw the perspective of these lines.

   The top view and front view of the perspective elements which are needed for the problem (i.e. PP, SP, HL and GL) are drawn as shown in Fig. 15.4. As the

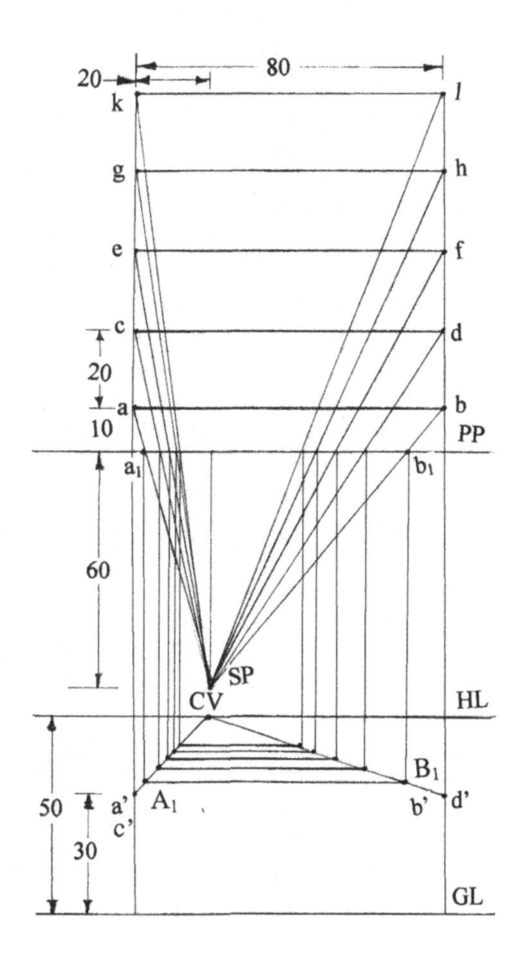

Fig. 15.4 Perspective of horizontal lines parallel to picture plane

lines are horizontal and parallel to the picture plane, the top views of the lines are 80 mm long and parallel to one another at intervals of 20 mm. The top views of the lines are drawn parallel to PP with their left ends 20 mm to the left of SP and the first line being 10 mm from PP. The front view of the eye (CV) is located on HL by drawing a projector from SP. Since the horizontal lines are 30 mm above ground and parallel to the picture plane, their front views will coincide and lie 30 mm above GL. The front views of the visual rays are drawn by joining the ends of the front views with CV. Piercing points of top views of visual rays on PP are obtained in the top view, and these points are designated as a_1, b_1, c_1, d_1, etc. Projectors are drawn from these points to get the perspective of the points on the respective visual rays in the front view. The end points of the lines are joined to show the perspective of the lines. A_1 B_1 represents the perspective of the line AB. One can observe that the perspective of horizontal line parallel to the picture plane will also be horizontal.

3. Six parallel lines, all perpendicular to the picture plane and each 80 mm long, lie on a horizontal plane 12 mm above the ground. The nearer ends of these lines lie 10 mm within the picture plane. The interval between each pair of adjacent lines is 15 mm. The eye is at a distance of 70 mm from the picture plane, 50 mm from the ground and directly in front of the third line from the left. Draw the perspective of these lines.

   The top view and front view of the perspective elements PP, HL and GL are drawn as shown in Fig. 15.5. The top views of the six lines, which are perpendicular to the picture plane with their nearest ends 10 mm from the picture plane, are drawn keeping a distance of 15 mm between them as shown in the figure. Since the eye is directly in front of the third line from the left, the top view of the eye SP is located from PP as shown in the figure. The front view of the eye is located on HL by drawing a projector from SP. The front view of each line will be a point 12 mm above GL, and front views of the six lines lie on a horizontal line 12 mm above GL as shown in the figure. The front views of visual rays are drawn from the front views of these points to CV. As the front view of a line coincides, the visual ray drawn from the front view of the line contains the perspective of the end points and hence the line. The piercing points a_1 and b_1 of the visual rays are located on PP in the top view. Projectors are drawn from these points to intersect the corresponding visual ray in the front view. The intersecting points are perspective points A_1 and B_1. As the points are on the same visual ray, the perspective is distinguished by darkening it. The same procedure is followed for other lines noting that end points of the lines lie on horizontal lines as shown in the figure.

4. Draw the perspective view of a tetrahedron of 50 mm edge when resting on ground plane on one of its faces with one of resting edges parallel to the picture plane. The vertical axis of the tetrahedron is in between this parallel edge and the picture plane, and 35 mm behind it. The station point is 40 mm in front of the picture plane, 50 mm to the left of the axis and 30 mm above the ground.

   The top view of tetrahedron is drawn with the edge a-b parallel to the picture plane and the other two edges a-c and b-c pointing towards the picture plane as

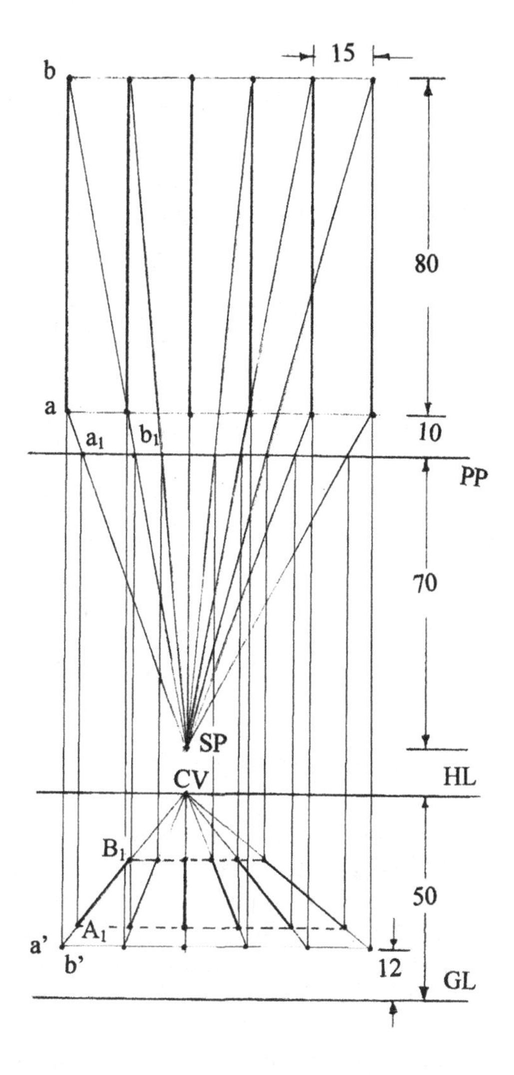

Fig. 15.5 Perspective of horizontal lines perpendicular to picture plane

shown in Fig. 15.6. The vertical axis of the tetrahedron passes through the point d. The height of d′ (i.e. h) from the face/base A-B-C of tetrahedron is found out as shown in the figure. The top view of the picture plane PP is drawn 35 mm from the axis of tetrahedron. The top view of station point SP is located at a distance of 40 mm from PP and 50 mm to the left of d (the axis of tetrahedron) as shown in the figure.

The front views of perspective elements HL and GL are drawn keeping a distance of 30 mm between them. The front view of station point is located on HL at CV. The front view points a′, b′ and c′ are located on GL, and the front view d′ is located above GL knowing the height of d′ from the face a-b-c. Four lines are drawn from CV connecting front view points a′, b′, c′ and d′. The top

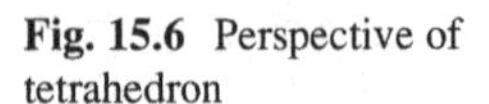

Fig. 15.6 Perspective of tetrahedron

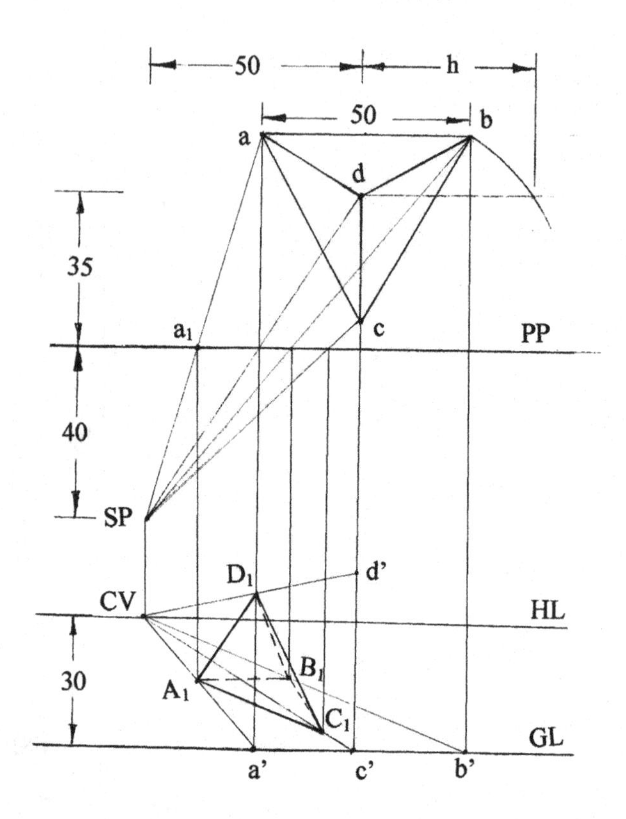

views of visual rays are drawn from SP to the top view points a, b, c and d. Piercing point of the visual ray SP-a is located on PP at a_1, and a projector is drawn from this point to intersect the line CV-a′ giving the perspective point A_1. The perspectives of other points are obtained following the same procedure. These points are joined to show the perspective view of the tetrahedron. The invisible edges are shown by dotted lines.

5. A hexagonal prism 25 mm side and 50 mm long is lying on one of its rectangular faces on the ground plane. Draw the perspective view of the prism if one end of the hexagonal face is on the picture plane and the station point is 80 mm in front of the picture plane, 65 mm above the ground plane and lies in a central plane which is 70 mm to the right of the axis of the prism.

   The front views of perspective elements HL and GL are drawn as shown in Fig. 15.7. The top view of the picture plane is also drawn. The front view of the prism is drawn with a side of hexagon on GL, and the top view is projected from the front view with one end of hexagonal face on PP. The top view of the station point SP is located at a distance of 80 mm from PP and 70 mm to the right of the axis of prism as shown in the figure.

   The front view of the station point is located on HL at CV. Six lines are drawn from CV joining the front views of the angular points of the prism. The front view points which are on the picture plane represent the perspective of one end

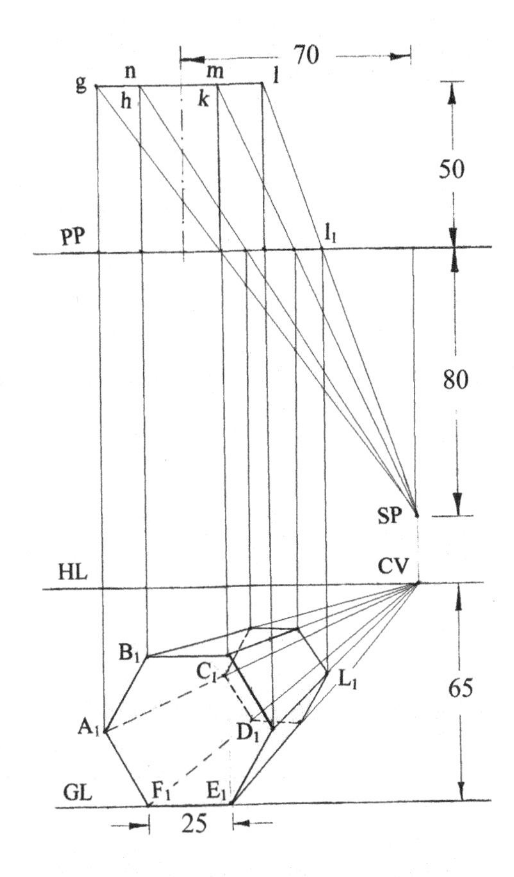

Fig. 15.7 Perspective of hexagonal prism

of the hexagonal prism. The perspectives of other six points which are on the other end of the prism are to be located on the six lines already drawn from CV. The top views of visual rays are drawn from SP to meet the top views g, h, k, l, m and n. The piercing point of visual ray 'SP-l' is located on PP at l_1, and a projector is drawn from this point to intersect the visual ray CV-D_1 at L_1. The point L_1 represents the perspective of the point L. The perspectives of the remaining points are obtained following the same procedure. The perspective points are joined to show the perspective view of the prism. The invisible edges are shown by dotted lines.

6. A straight line AB, 40 mm long, is parallel to the ground and is 15 mm above it. The line is inclined at 30° to the picture plane, and its end A is 20 mm behind the picture plane. The station point is 40 mm above the ground, 50 mm in front of the picture plane and is in front of the mid-point of AB. Draw the perspective of AB. Another line CD lies directly below AB and on the ground. CD is also 40 mm long and is parallel to AB. Draw the perspective of CD also.

   The top views of the straight lines AB and CD, which are inclined at 30° to the picture plane, are drawn on a line inclined at 30° to horizontal as shown in Fig. 15.8. The top view of picture plane is now drawn keeping a distance of

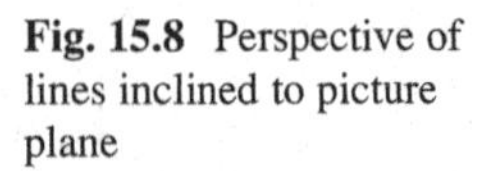

Fig. 15.8 Perspective of lines inclined to picture plane

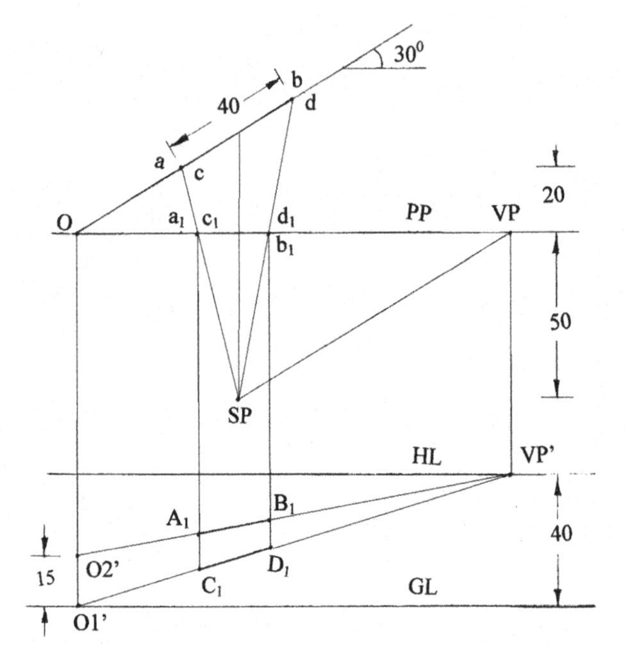

20 mm from the nearest end of the lines. The top view SP of the station point, which lies in front of the mid-point of the lines, is located at a distance of 50 mm from PP. The front view of the perspective elements HL and GL are drawn keeping a distance of 40 mm between them.

The top view of vanishing point is located on PP by drawing a line from SP parallel to the top views of the lines. The intersection of this line locates VP on PP. The front view of vanishing point VP′ is located on HL by drawing a projector from VP.

The line on which the top views of lines are shown is extended to cut PP at O. A projector is drawn from O and on this projector are shown two points O1′ and O2′. The point O1′ lies on GL indicating the locus of front views of the end points of the line CD and the point O2′ lies 15 mm above GL indicating the height of the end points of the line AB. Lines are drawn joining VP′ to points O1′ and O2′. Locate piercing points b_1 and a_1 on PP by drawing visual rays from SP to the top view points b and a. Projectors are drawn from a_1 and b_1 to intersect the line VP′-O2′ to give perspective A_1 and B_1 of the line AB. The line A_1 B_1 represents the perspective of the line AB. The projectors are extended to cut the line VP′-O1′ to give the perspective of the line CD (i.e. C_1 D_1).

7. Five vertical lines, each 50 mm long, stand in a row inclined at 30° to the picture plane, receding to the right, with their lower ends on the ground, and at intervals of 20 mm. The station point is 30 mm from the ground and 60 mm from the picture plane. The station point is in front of the second line from the left which

is 20 mm within the picture plane. Draw the perspective projections of these lines.

The top views of five lines standing in a row are drawn on a line inclined at 30° to the horizontal as shown in Fig. 15.9. The top view of the picture plane PP is now drawn keeping its distance 20 mm from the second line on the left. The top view SP of station point, which is directly in front of the second line, is located at a distance of 60 mm from PP. The front view of perspective elements HL and GL are drawn keeping a distance of 30 mm between them.

The top view of vanishing point is located on PP by drawing a line from SP and parallel to the orientation of lines in the top view (i.e. inclined at 30° to PP). The point of intersection of this parallel line locates the top view of vanishing point VP on PP. The front view of vanishing point VP′ is located on HL.

The orientation line is extended to cut PP at O. A projector is drawn from O and on this projector are shown two points O1′ and O2′. The point O1′ lies on GL indicating the locus of front view of the points lying on ground and the point O2′ is 50 mm above GL indicating height of the lines. Lines are drawn joining VP′ to the points O1′ and O2′. Locate piercing points on PP by drawing visual rays from SP for the line AB. From the piercing point, a projector is drawn to cut VP′-O2′ to give perspective B_1. The same projector is extended to cut the line VP′-O1′ to give the perspective A_1. The line A_1B_1 represents the perspective of the line AB. The perspectives of other lines EF, GH and KL are obtained following the above procedure. The perspective of the ends of the lines lies on VP′-01′ and VP′-02′. Hence, a projector is drawn from the top view of line CD, which is

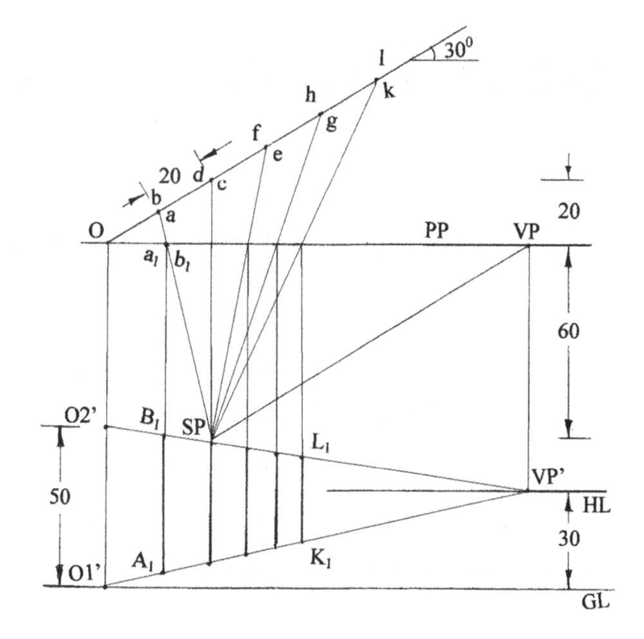

Fig. 15.9 Perspective of vertical lines aligned to picture plane

directly in front of the station point. The intersections of this projector with VP′-01′ and VP′-02′ locate the ends C_1 and D_1 of the line.

8. A square ABCD of 40 mm side lies on the ground with the side AB inclined at 30° to the picture plane. The corner A lies 10 mm within the picture plane and 35 mm to the right of the station point. The corners B, C and D lie within the picture plane. The station point is 70 mm from the picture plane and 45 mm from the ground. Draw the perspective of the square ABCD.

   The top view of the lamina is drawn with one side inclined at 30° to the horizontal as shown in the Fig. 15.10. The top view PP of picture plane is drawn keeping a distance of 10 mm from the nearest corner of the lamina. The top view SP of station point is located 35 mm to the left of nearest corner and 70 mm from PP. The front view of perspective elements HL and GL are drawn keeping a distance of 45 mm between them. Since the top views of side a-b and d-c are parallel, the vanishing point is located by drawing a line parallel to a-b from SP to intersect PP at VP. The front view of vanishing point VP′ is located on HL. The top views of the lines a-b and c-d are extended to cut PP at O and M. Since the points lie on the ground, projectors are drawn from O and M to locate O1′ and M1′ on GL. Draw lines joining VP′ to O1′ and M1′. Piercing point a_1 of the visual ray 'SP-a' is located on PP. A projector is drawn from a_1 to intersect the line VP′-O1′ to fix the perspective A_1 of the point A. The perspectives of other points are obtained on the respective lines VP′-O1′ and VP′-M1′ following the same procedure. The four points are joined to give the perspective of the square ABCD.

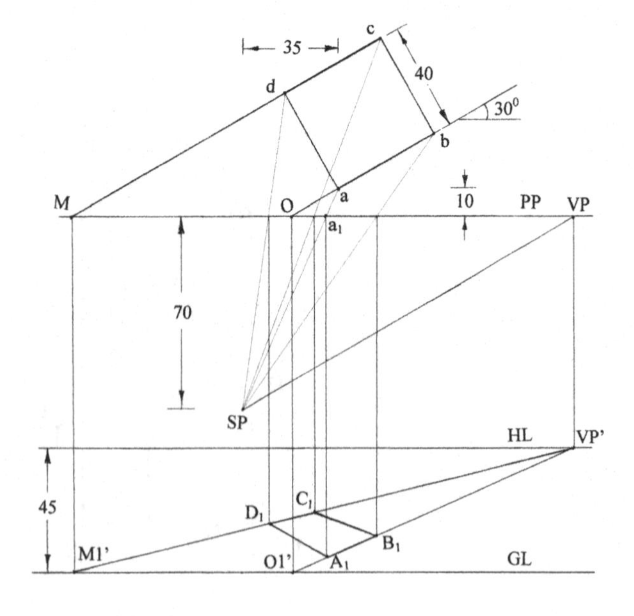

Fig. 15.10 Perspective of square lamina

9. A regular hexagon of 30 mm side lies on a vertical plane inclined at 30° to the picture plane, with a side on the ground, its nearest angular point being 25 mm within the picture plane and 20 mm to the right of the eye. The eye is 70 mm from the picture plane and 40 mm from the ground. Draw the perspective of the hexagon.

   The top view PP of the picture plane is drawn as shown in Fig. 15.11. The top view SP of the station point is located at distance of 70 mm from PP. The front view of perspective elements HL and GL is also drawn keeping a distance of 40 mm between them.

   Since the hexagon lies on a vertical plane with a side on ground, its front view is drawn with one side on GL. The top view of the hexagon is obtained on PP. This top view is redrawn on a line inclined at 30° to PP with the nearest angular point a being 25 mm from PP and 20 mm to the right of SP.

   The vanishing point is located on a line drawn parallel to the top view of the hexagon from SP to meet PP at VP. The front view VP′ of vanishing point is located on HL. The top view of hexagon is extended to meet PP at O. A projector is drawn from O to meet GL. Horizontal lines are drawn from the front view of the hexagon already drawn on GL to intersect this projector at three point O1′, O2′ and O3′. The point O1′ represents the locus of angular points e′ and f′ which lie on the ground. The point O2′ represents the locus of points a′ and d′, and the point O3′ represents the locus of b′ and c′. Lines are drawn from VP′ to meet these points O1′, O2′ and O3′. Piercing point of visual ray SP-a is located on PP, and a projector is drawn from this point a_1 to intersect the line VP′-O2′ to give the perspective A_1 of angular point A. The perspectives of other angular points are obtained following the same procedure. These points are joined to show the perspective of the hexagon.

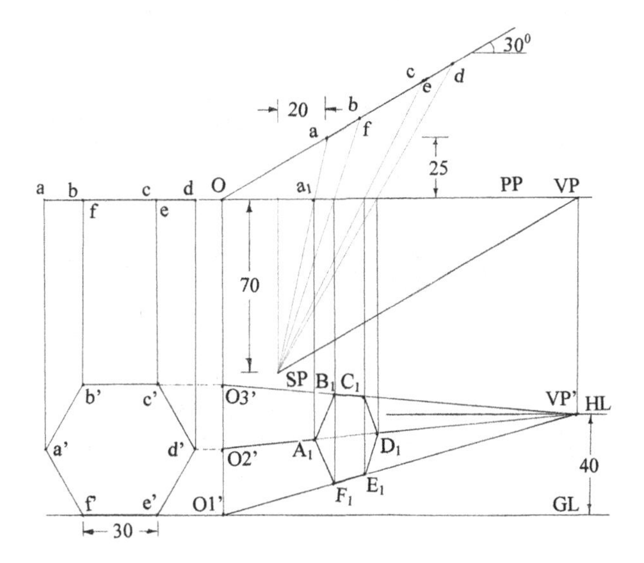

Fig. 15.11 Perspective of hexagonal lamina

10. A cube, edge 40 mm, lies with a face on the ground and an edge on the picture plane. All the vertical faces are equally inclined to the picture plane. The station point is 80 mm from the picture plane and 55 mm above the ground. The edge of the cube on the picture plane is 10 mm to the right of the station point. Draw the perspective of the cube.

    The top view PP of the picture plane is drawn as shown in Fig. 15.12. The top view SP of station point is located at a distance of 80 mm from PP. The front view of perspective elements HL and GL are drawn keeping a distance of 55 mm between them.

    The front view of the cube is drawn on GL as the solid lies with a face on the ground. The top view obtained from the front view is redrawn with a vertical edge on PP, 10 mm to the right of SP and the vertical faces equally inclined to the picture plane.

    The vanishing point is located by drawing a line parallel to the line a-d from SP to cut PP at VP. The front view of vanishing point VP′ is located on HL. Since the top view of edges b-c, f-g and e-h are also parallel to a-d, the perspective of all eight points can be located from the vanishing point VP′. The top view of edge b-c is extended to meet PP at M. Since the top view a of edge a-b lies on PP itself, there is no need to extend this line. A projector is drawn from a to meet GL, and the front views of points a′ and e′ are located as shown in the figure. Draw a projector from M to meet GL at M1′ representing the locus

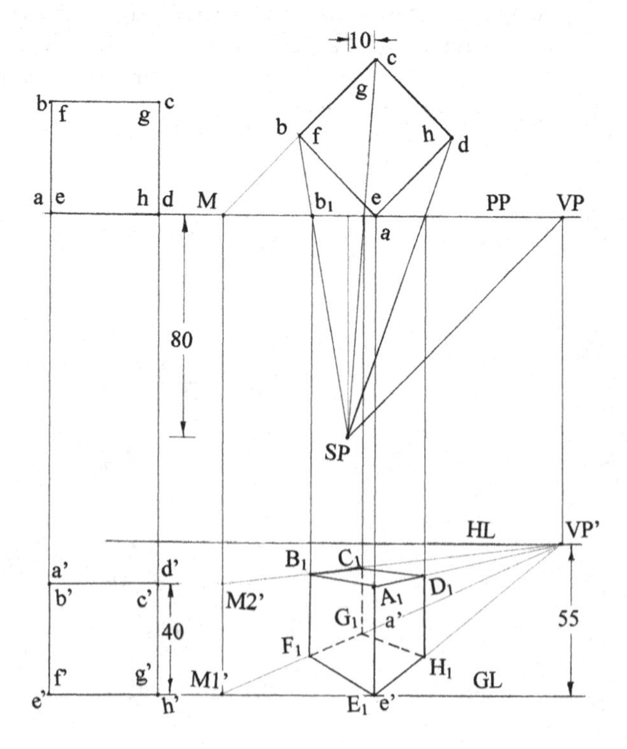

Fig. 15.12 Perspective of cube figure

of front views of points f′ and g′, and another point M2′ is located representing the locus of b′ and c′. Four lines are drawn from VP′ to meet the points a′, e′, M1′ and M2′. As the points a′ and e′ lie on the picture plane, these points represent the perspective A_1 and E_1 of the edge AE of the cube. Piercing point of the visual ray 'SP-b' is located on PP at b_1. A projector is drawn from b_1 to intersect the line VP′-M2′ to give perspective B_1. The perspectives of other points are obtained following the same procedure. These points are joined to show the perspective of the cube. The invisible edges are shown by dotted lines.

11. A square prism, side of base 30 mm and axis 60 mm, lies with a face on the ground, the nearest side of the base being 20 mm within the picture plane, and 30 mm to the left of the eye. The eye is 60 mm from the picture plane and 40 mm from the ground. Draw the perspective of the prism when its edges recede to the right at an angle of 40°.

    The top view PP of the picture plane is drawn as shown in Fig. 15.13. The top view SP of station point is located at a distance of 60 mm from PP. The front views of perspective elements HL and GL are drawn keeping a distance of 40 mm between them. The front view of the square prism is drawn on GL, and the top view is also drawn on PP as shown in the figure. The top view is redrawn on lines receding to the right at 40° with its nearest side 20 mm from PP and 30 mm to the left of SP.

    The vanishing point is located by drawing a line parallel to a-e and b-f from SP to meet PP at VP. The front view of vanishing point VP′ is located on HL. The top view lines a-e and b-f are extended to meet PP at O and M. Projectors are drawn from O and M to meet GL, and points O1′ and M1′ are located on GL indicating the locus of points d′, h′, c′ and g′. On the respective projectors, points O2′ and M2′ are also located at a distance 30 mm above GL representing the locus of points a′, e′, b′ and f′. Four lines

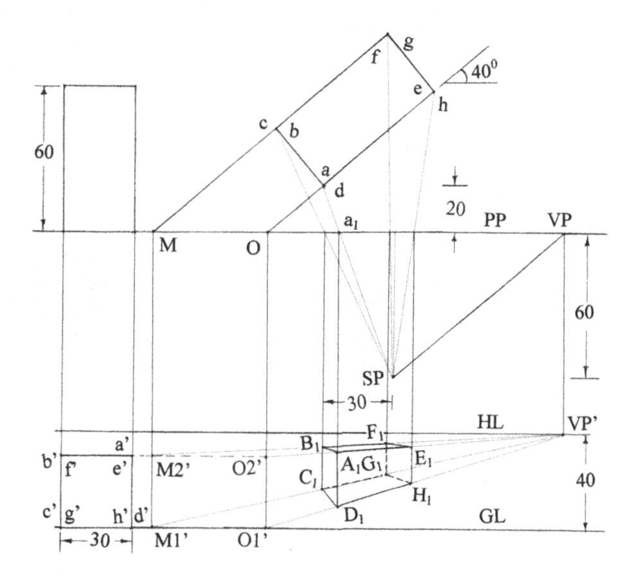

Fig. 15.13 Perspective of square prism

are drawn from VP′ to meet O1′, M1′, O2′ and M2′. Piercing point of visual ray 'SP-a' is located on PP at a_1, and a projector is drawn from a_1 to intersect the line VP′-O2′ to give the perspective A_1. The perspectives of other points are obtained following the same procedure. The perspective points are joined to show the perspective view of the square prism. The invisible edges are shown by dotted lines.

12. Draw the perspective of a hexagonal pyramid, side of base 25 mm and axis 65 mm, when its base is on the ground and a side of its base parallel to the picture plane. The axis of the pyramid is 30 mm within the picture plane and 10 mm to the right of the eye. The eye is 80 mm from the picture plane and 60 mm from the ground.

    The top view of the hexagonal pyramid is drawn with a side parallel to the picture plane (i.e. on a horizontal line) as shown in Fig. 15.14. The top view PP of the picture plane is drawn at a distance of 30 mm from the apex of pyramid. The top view SP of station point is located at a distance of 80 mm from PP and 10 mm to the left of apex. The front views of perspective elements HL and GL are drawn keeping a distance of 60 mm between them.

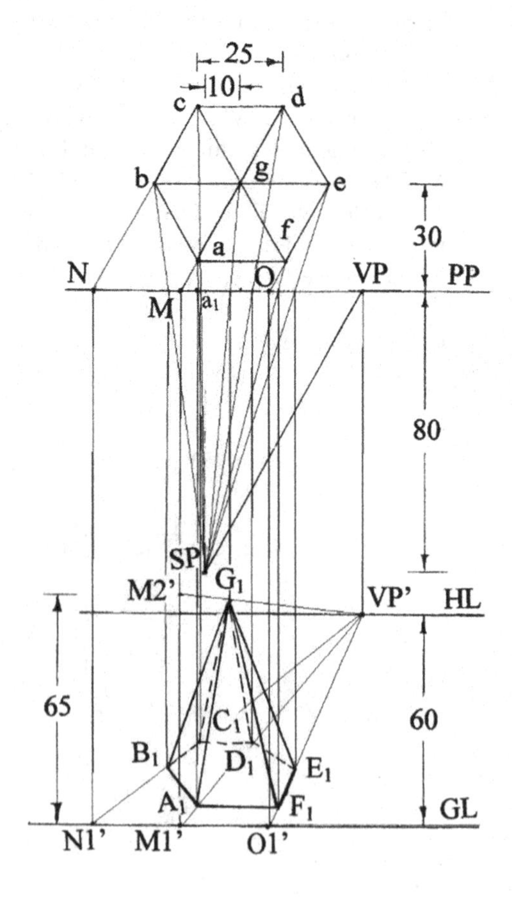

Fig. 15.14 Perspective of hexagonal pyramid

A close examination of the top view of pyramid shows that lines f-e, a-d and b-c are parallel to one another. The vanishing point is located by drawing a line parallel to a-d from SP to meet PP at VP as shown in the figure. The front view of vanishing point VP′ is located on HL. The top view lines f-e, a-d and b-c are extended to meet PP at O, M and N. Projectors are drawn from O, M and N to meet GL at points O1′, M1′ and N1′. These points represent the locus of base points of pyramid which lie on the ground. The point M2′ is located on the projector drawn from M at a distance of 65 mm from GL representing the height of the pyramid. Four lines are drawn from VP′ to meet O1′, M1′, M2′ and N1′. Piercing point of the visual ray 'SP-a' is located on PP at a_1 and a projector is drawn from this point to intersect the line VP′-M1′ to represent the perspective A_1 of the base point A of the pyramid. Perspectives of other base points are located following the same procedure. The perspective of apex is located on VP′-M2′ following the same procedure. The points are joined to show the perspective view of the pyramid. The invisible edges are shown by dotted lines.

13. A square pyramid of side of base 30 mm and height 50 mm rests on the ground with an edge of base inclined at 30° to the picture plane and the nearest corner 20 mm within the picture plane. The station point is on the central plane passing through the apex of pyramid and is located 50 mm in front of the picture plane and 60 mm above the ground. Draw the perspective view of the pyramid.

The top view PP of the picture plane is drawn as shown in Fig. 15.15. The top view of square pyramid is drawn with a side of base inclined at 30° to PP and the nearest corner of the base 20 mm from PP. The top view of station point SP is located at a distance of 50 mm from PP on a projector drawn from the apex of the pyramid in the top view. The front views of perspective elements HL and GL are drawn keeping a distance of 60 mm between them. The vanishing point is located by drawing a line parallel to a-d from SP to meet

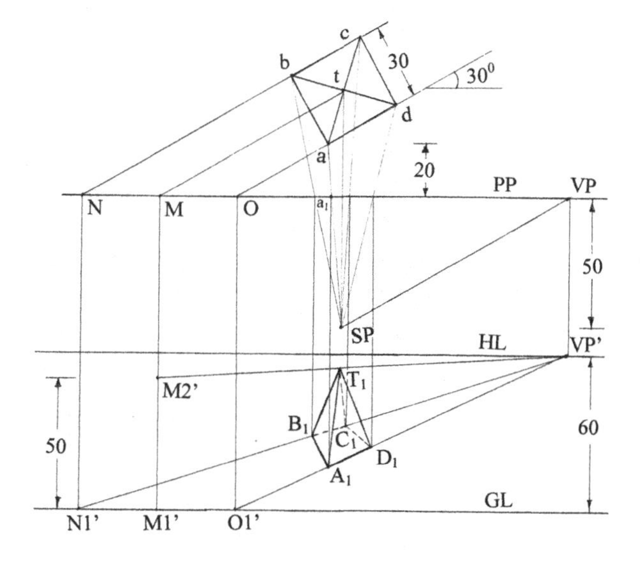

Fig. 15.15 Perspective of square pyramid

PP at VP as shown in the figure. The front view of vanishing point VP′ is located on HL. The top view lines a-d and b-c are extended to meet PP at O and N. Projectors are drawn from O and N to meet GL at O1′ and N1′. One more line is drawn parallel to a-d from t to meet PP at M. A projector is drawn from M to meet GL, and a point M2′ is located 50 mm above GL (i.e. the height of the pyramid). Lines are drawn from VP′ to meet O1′, N1′ and M2′. The piercing point of visual ray 'SP-a' is located on PP at a_1, and a projector is drawn from this point to intersect the line VP′-O1′ at A_1 to give the perspective of the base point A of the pyramid. The perspectives of other points are obtained following the same procedure. The perspective points are joined to show the perspective view of the square pyramid. The invisible edges are shown by dotted lines.

14. A circle having a diameter of 50 mm and lying on the ground has its centre 30 mm within the picture plane, and 10 mm to the right of the station point, which is 80 mm from the picture plane and 60 mm from the ground. Draw the perspective of the circle.

    The top view PP of the picture plane is drawn as shown in Fig. 15.16. The top view of the circle which lies on the ground is drawn keeping its centre 30 mm within PP. The top view SP of the station point is located 80 mm from

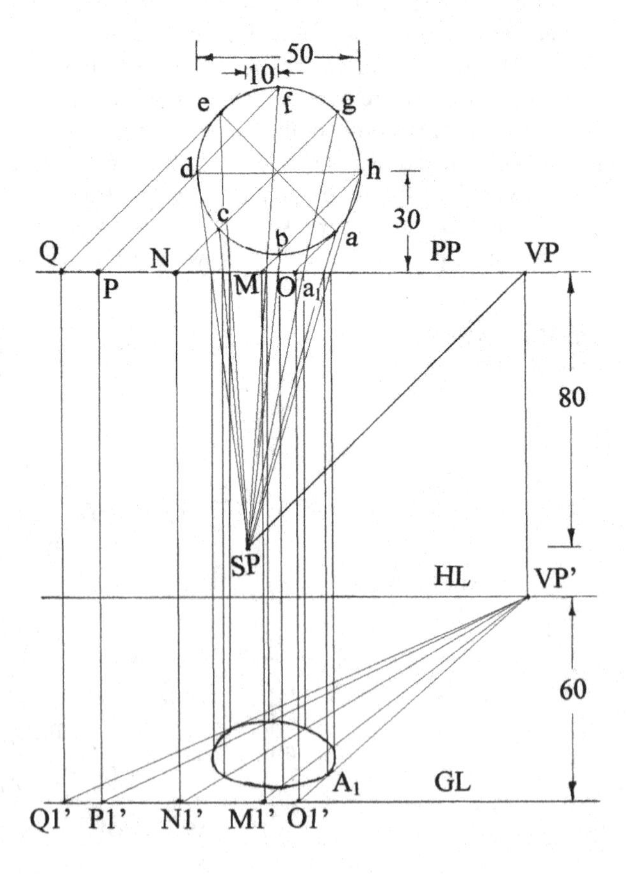

Fig. 15.16 Perspective of circle

PP and 10 mm to the left of the centre of circle. The front views of perspective elements HL and GL are drawn keeping a distance of 60 mm between them. The circle is divided into eight parts, and the chords are designated as b-h, c-g and d-f as shown in the figure. This problem is essentially one of finding perspective of eight points lying on the circumference of the circle. The vanishing point is located by drawing a line parallel to top views of the chords from SP to meet PP at VP. The front view of vanishing point VP′ is located on HL. The chords in the top view are extended to meet PP at M, N and P. Two more lines are drawn parallel to the chords from e and a to meet PP at Q and O. Projectors are drawn from these points to meet GL at O1′, M1′, N1′, P1′ and Q1′. Lines are drawn from VP′ to connect O1′, M1′, N1′, P1′ and Q1′. The piercing point of visual ray ‘SP-a’ is located on PP at a_1, and a projector is drawn from a_1 to intersect the line VP′-O1′ to give the perspective A_1 of the point A. The perspectives of other points are obtained following the same procedure. These points are joined by a smooth cure representing the perspective of circle.

15. A cylinder 50 mm in diameter and 70 mm long rests on the ground plane with its axis horizontal. The front end of the axis is 20 mm from the picture plane and 20 mm to the left of the eye. The axis recedes to the right at an angle of 45°. The eye is 60 mm from the picture plane and 40 mm from the ground plane. Draw the perspective of the cylinder.

The top view PP of the picture plane is drawn as shown in Fig. 15.17. The front views of perspective elements HL and GL are drawn keeping a distance of 40 mm between them. The front view and the top view of the cylinder are drawn at the left extreme with eight generators at equal distance along its lateral surface. The top view is redrawn with its axis front end 20 mm from PP and the generators inclined at 45° to PP. The top view of station point SP is located 60 mm from PP and 20 mm to the right of front end of cylinder axis as shown in the figure. The vanishing point is located by drawing a line parallel to the generators from SP to meet PP at VP. The front view VP′ is located on HL. The generators are extended to meet PP at five points, and projectors are drawn from these points to meet GL. This problem is one of obtaining perspective of eight generator ends (16 points) of the cylinder and joining them by smooth curves. The generator extended from ‘a’ meets PP at O. A projector is drawn from O and on this projector are shown two points O1′ representing the locus of generator lying on the ground and another point O2′ representing the locus of top-most generator. Lines are drawn from VP′ to connect O1′ and O2′. The piercing point of visual ray ‘SP-a’ is located at a_1, and a projector is drawn from this point to intersect the line VP′-O2′ giving the perspective A_1. The same projector on extension intersects the line VP-O1′ to give the perspective of the point lying on the ground. The same procedure is repeated to locate perspective of the points at the front end and back end of each generator. The perspective points are joined by smooth curves to represent the perspective of both ends of the cylinder. Common tangents are drawn to the curves to show the enveloping generators. The invisible portion of the back end of the cylinder is shown as dotted.

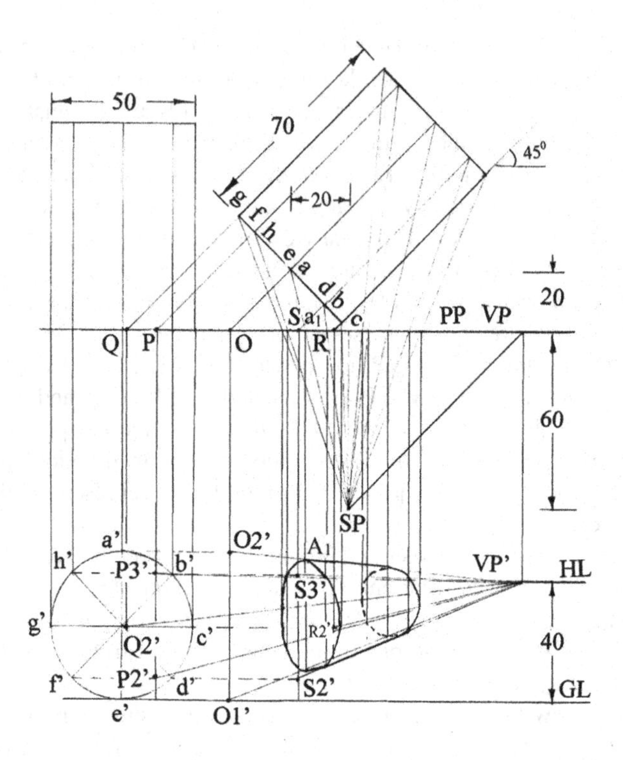

Fig. 15.17 Perspective of cylinder

Practice Problems

1. Represent in perspective a point A on the ground plane, 20 mm to the right of the eye and 10 mm from the picture plane; also a point B 20 mm to the left of the eye, 10 mm from the picture plane and 30 mm above the ground plane. Join AB. AB is the nearest side of a rectangle of which two other sides are horizontal and 60 mm long. Complete the perspective representation of the rectangle. Position of eye is 100 mm from the picture plane and 40 mm above the ground plane.
2. A rectangular prism 90 mm × 50 mm wide and 40 mm high rests with one of its largest rectangular faces on the ground. Nearest vertical edge of the prism is 10 mm behind the picture plane and is 30 mm to the right of the station point. Longer edges are inclined towards the right at 30° to the picture plane. Draw the perspective of the prism, if the station point is 45 mm above the ground, 100 mm from the picture plane.
3. A pentagonal prism, side of base 20 mm and axis 60 mm, rests with a rectangular face on the ground and a corner of one of its ends touching the picture plane. Its axis recedes to the right at an angle of 30°. The eye lies in front of the midpoint

of the axis and 65 mm from the picture plane. If the eye is in level with the top edge of the prism, draw the perspective of the prism.

4. A pentagonal pyramid, side of base 20 mm and axis 50 mm, stands on the ground with the nearest corner of its base 10 mm within the picture plane and 30 mm to the right of the eye. The base edges containing the nearest corner are equally inclined to the picture plane. The eye is 65 mm from the picture plane and 50 mm above the ground. Draw the perspective of the pyramid.
5. A square prism of 20 mm side of base and height 40 mm rests with its base on ground such that one of the rectangular faces is parallel to the picture plane and 10 mm behind it. The station point lies opposite to the axis of the prism, 60 mm above the ground and 50 mm in front of the picture plane. Draw the perspective view of the square prism.
6. A square pyramid of side of base 30 mm and height 50 mm rests centrally on the top of a square prism, side of base 50 mm and height 20 mm with an edge of base of either of the solids inclined at 30° to the picture plane and the nearest corner 20 mm from it. The station point is on the central plane passing through the apex of the pyramid and is located 80 mm in front of the picture plane and 60 mm above the ground. Draw the perspective view of the solids.
7. A rectangular pyramid, base measuring 60 mm × 30 mm and the axis 50 mm, rests with its base on the ground plane such that the longer base edges recede to the right at angle of 30°. The nearest base point is 20 mm within the picture plane. The station point is 65 mm in front of the picture plane, 30 mm to the left of the axis of the pyramid and 60 mm above the ground plane. Draw the perspective view of the pyramid.
8. A hexagonal prism, side of base 30 mm and axis 60 mm, lies with a face on the ground, the nearest corner of the base being 20 mm within the picture plane, and 30 mm to the right of the eye. The eye is 60 mm from the picture plane and 40 mm from the ground. Draw the perspective of the prism when its edges recede to the right at an angle of 40°.
9. A hexagonal pyramid, side of base 30 mm and axis 50 mm, stands on the ground with the nearest corner of its base 10 mm within the picture plane and 30 mm to the right of the eye. The base edges containing the nearest corner are equally inclined to the picture plane. The eye is 65 mm from the picture plane and 50 mm above the ground. Draw the perspective of the pyramid.
10. A combination solid consists of a cylinder of 30 mm diameter and 50 mm height placed centrally on a square prism of 60 mm side and 40 mm height. The axes of the two solids are collinear. The combination is placed such that one of the vertical faces of the square prism is inclined at 30° to the PP and a vertical edge touches the PP. The station point is 80 mm in front of the PP, 100 mm above the ground and is exactly in front of the vertical edge touching the PP. Draw the perspective of the combination solid.

Chapter 16
Objective Type Questions

1. The least number of straight lines required to form a rectilinear/recti-lineal figure is
   (i) 3 (ii) 4 (iii) 5 (iv) 6
2. Which of the figures are always similar?
   (i) Polygons (ii) Rectangles (iii) Squares (iv) Triangles
3. If sum of the exterior angles of a polygon is equal the sum of the interior angles, then the polygon has
   (i) four sides (ii) six sides (iii) eight sides (iv) ten sides
4. If each of the interior angle of a regular polygon is 160°, then the polygon has
   (i) 12 sides (ii) 14 sides (iii) 16 sides (iv) 18 sides
5. Through what angle does the hour hand of a clock rotate between Twelve O'clock and ten minutes to One?
   (i) 6° (ii) 12° (iii) 25° (iv) 30°
6. Which triangle with the given sides, in cm, is a right angled triangle?
   (i) 1, 2 and 3 (ii) 3, 4 and 5 (iii) 4, 5 and 6 (iv) 5, 7 and 9
7. If the hypotenuse and one side of a right angled triangle are 6.5 cm and 2.5 cm, then the other side, in cm, is
   (i) 4 (ii) 5 (iii) 6 (iv) 7
8. If the area of an equilateral triangle is $25\sqrt{3}$ cm^2, then the length of its side, in cm, is
   (i) 5 (ii) 6 (iii) 8 (iv) 10
9. Which of the statements is incorrect for the diagonals of a rhombus?
   (i) Bisect each other (ii) Are equal (iii) Are at right angles (iv) Bisect opposite angles
10. Which of the geometric figures shown in Fig. 16.1 is a rhombus?

K. Rathnam, *A First Course in Engineering Drawing*,
DOI 10.1007/978-981-10-5358-0_16

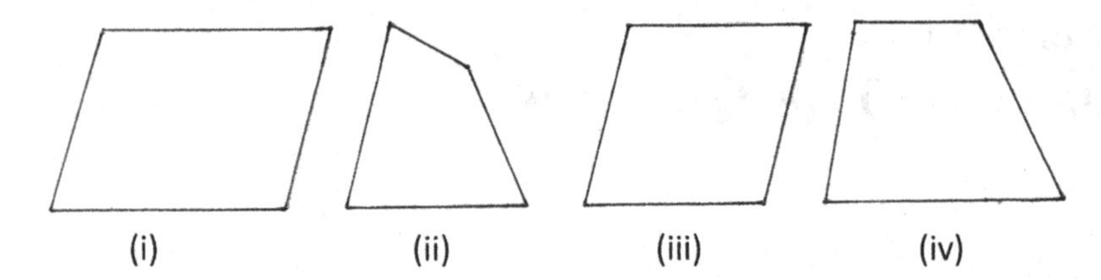

Fig. 16.1 Geometric figures

11. Which one of the angles is a straight angle?
    (i) 30° (ii) 90° (iii) 120° (iv) 180°
12. Identify the triangle whose orthocentre, centroid, centre of circumscribing circle and centre of inscribing circle all coincide.
    (i) Acute angled triangle (ii) Equilateral triangle
    (iii) Isosceles triangle (iv) Right angled triangle
13. Points which lie on the circumference of a circle are said to be
    (i) concurrent (ii) concyclic (iii) collinear (iv) congruent
14. How many escribed circles can be drawn to a triangle?
    (i) One (ii) Two (iii) Three (iv) Four
15. The centroid of a triangle lies
    (i) at its orthocentre (ii) at the centre of the circumscribing circle
    (iii) at the intersection of its medians (iv) at the centre of the inscribing circle
16. How is I section of connecting rod shown in section?
    (i) Full section (ii) Broken section (iii) Revolved section (iv) Half section
17. Which of the following is represented by thin continuous line in a view?
    (i) Visible line (ii) Hidden line (iii) Centre line (iv) Locus line
18. Which of the following is represented by thin chain lines in a view?
    (i) Extension lines (ii) Dimension lines (iii) Centre lines (iv) Hidden lines
19. Which instrument is used to draw parallel lines?
    (i) Mini-drafter (ii) French curve (iii) Divider (iv) Protractor
20. Which of the section symbols shown in Fig. 16.2 indicates the material steel?

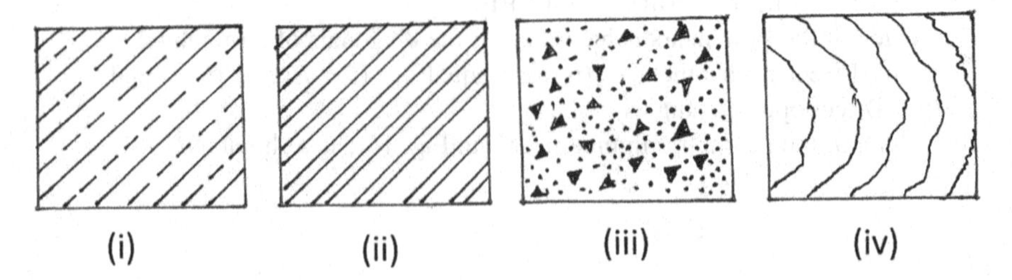

Fig. 16.2 Material section symbols

21. The shape of rice kernel is approximated as
   (i) oblate ellipsoid (ii) prolate ellipsoid (iii) hyperboloid (iv) paraboloid
22. Any material point moving in space in the gravity field will describe
   (i) a cycloid (ii) an ellipse (iii) a hyperbola (iv) a parabola
23. Name the path of planet earth around the sun in our solar system.
   (i) Circle (ii) Involute (iii) Ellipse (iv) Trochoid
24. The path of water jet issuing from a vertical orifice is
   (i) elliptic (ii) parabolic (iii) hyperbolic (iv) cycloid
25. The shape of natural draft cooling tower in a thermal power plant is
   (i) cycloid of revolution (ii) prolate ellipsoid
   (iii) elliptical hyperboloid (iv) elliptical paraboloid
26. The shape of head lamp reflector of motor car is
   (i) cycloid of revolution (ii) prolate ellipsoid (iii) Oblate ellipsoid (iv) paraboloid
27. XY is a fixed straight line of indeterminate length. A part of it, BC, is the base of an equilateral triangle ABC. The triangle rolls over, without slipping, on XY, until AC lies on XY. The locus of point B describes
   (i) a cycloid (ii) an arc of circle (iii) a straight line inclined to XY (iv) a cissoid
28. The terminology latus rectum is associated with the curve
   (i) cycloid (ii) helix (iii) involute (iv) parabola
29. Which curve finds its application in ballistics?
   (i) Cycloid (ii) Ellipse (iii) Hyperbola (iv) Parabola
30. The terminology asymptote is associated with the curve
   (i) circle (ii) ellipse (iii) hyperbola (iv) parabola
31. Which curve is associated in the representation of Boyle's gas law?
   (i) Cycloid (ii) Ellipse (iii) Hyperbola (iv) Parabola
32. Identify the central conic curves.
   (i) Circle and parabola (ii) Parabola and ellipse
   (iii) Ellipse and hyperbola (iv) Hyperbola and circle
33. Which of the conditions is true for similar ellipses?
   (i) Same directrices (ii) Same eccentricity (iii) Same foci (iv) Same latus rectum
34. The liquid surface in a vessel rotating about its vertical axis with constant velocity takes the shape of
   (i) an ellipsoid (ii) a paraboloid (iii) a hyperboloid (iv) a cycloid
35. The locus of the middle pints of a system of parallel chords of an ellipse is
   (i) a straight line (ii) an ellipse (iii) an evolute (iv) an involute

36. If the area of the figure AQNL shown in Fig. 16.3 is one-third of the area of the rectangle APNL, then the curve AQN is

Fig. 16.3 Geometric figure

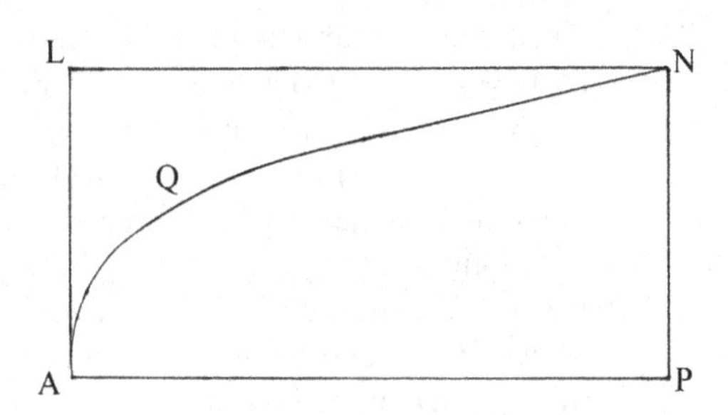

(i) a cycloid (ii) a parabola (iii) a cardioid (iv) an ellipse

37. The epicycloid will also be called as cardioid when the ratio of radius of rolling circle to directing circle is
(i) 1/4 (ii) 1/3 (iii) 1/2 (iv) 1
38. The path followed by insects while approaching a burning candle is
(i) an Archimedean spiral (ii) a hyperbola (iii) a logarithmic spiral (iv) a parabola
39. Of all curves, connecting two points A and B, the curve through which a material point, accelerated by gravity, will move in the shortest time is
(i) parabola (ii) hyperbola (iii) ellipse (iv) cycloid
40. Which of the curves is used in the design of volute casing of centrifugal pump?
(i) Cycloid (ii) Helix (iii) Hyperbola (iv) Spiral
41. Which curve finds an application in screw thread formation?
(i) Cycloid (ii) Helix (iii) Involute (iv) Spiral
42. The path followed by a moving insect at constant speed along an oscillating/swinging lever of a pendulum, when shown on Cartesian co-ordinates, is
(i) an Archimedean spiral (ii) a hyperbola (iii) a logarithmic spiral (iv) a parabola
43. Consider the following conic curves obtained from a cone by different section planes inclined to the axis of the cone:
1. Circles 2. Ellipses 3. Parabolas
Which of the curves are similar?
(i)1, 2 and 3 (ii) 1 and 2 (iii) 2 and 3 (iv) 1 and 3
44. Which of the surfaces is the most common form of a double-curved type?
(i) Right cylinder (ii) Oblique cylinder (iii) Cone (iv) Sphere
45. Which of the curves is used in the development of toothed quadrant gearing found in measuring instruments?
(i) Archimedean spiral (ii) Logarithmic spiral (iii) Cylindrical helix (iv) Cycloidal curve
46. Which curve is used in the development of tooth profiles in gearing?
(i) Archimedean spiral (ii) Logarithmic spiral (iii) Cylindrical helix (iv) Involute of circle

47. The ends of an epicycloid meet at a point when the ratio of the rolling circle radius to the base circle radius is equal to
    (i) 1/4 (ii) 1/2 (iii) 3/4 (iv) 1
48. The epicycloid developed by a rolling circle of radius 40 mm on a base circle of radius 100 mm will be similar to another epicycloid developed by a rolling circle of radius 30 mm on the base circle of radius
    (i) 90 mm (ii) 85 mm (ii) 80 mm (iv) 75 mm
49. Which of the statements is true in orthographic projection of an object?
    (i) First angle projection is obsolete everywhere
    (ii) Third angle projection pertains to the fact that three views of an object are shown
    (iii) A reference/folding line is the line of intersection of two adjacent projection planes
    (iv) The elevation/front view shows the height and depth of the object
50. Which of the statements is true in the first angle method of orthographic projection?
    (i) The plan and elevation are designated as related views
    (ii) The plan and right side elevation are designated as adjacent views
    (iii) The elevation and the right side elevation are designated as adjacent views
    (iv) The elevation and the right side elevation are designated as related views
51. Assertion (A): The object is kept in the first/third quadrant only in the orthographic projection to get the projections of the object.
    Reason (R): The projections of the object will lie on one side of the reference line if object is kept in the second/fourth quadrant.
    (i) Both A and R are true and R is the correct explanation of A
    (ii) Both A and R are true but R is not the correct explanation of A
    (iii) A is true but R is false
    (iv) A is false but R is true
52. The maximum number of basic views employed to describe completely an object in the first/third angle method of projections is
    (i) 2 (ii) 3 (iii) 4 (iv) 6
53. Which of the following methods belongs to the category 'Orthographic projection'?
    (i) Aerial projection (ii) Isometric projection
    (iii) Oblique projection (iv) Perspective projection
54. In which projection a minimum of two views of an object are shown?
    (i) Isometric projection (ii) Oblique projection
    (iii) Orthographic projection (iv) Dimetric projection
55. Which of the statements is true in orthographic projection of an object with three principal views obtained on the principal planes?
    (i) Horizontal plane appears edgewise in plan
    (ii) Elevation shows the height and depth of the object
    (iii) Profile plane appears edgewise in elevation

(iv) Visible line in plan of an object must also be visible in side view

56. In orthographic multi-view projections, the vertical and horizontal dimensions are assigned as height and width with depth as third dimension in each view. Which statement is correct in the projection of an object having three dimensions?
    (i) The depth of plan is same as the width of side elevation
    (ii) The depth of elevation is same as the height of side elevation
    (iii) The height of elevation is same as the depth of side elevation
    (iv) The width of plan is same as the depth of side elevation
57. Which statement is true in orthographic projection of a straight line?
    (i) The straight line can project longer than the line itself
    (ii) The inclination of the straight line with HP is always observed from its elevation
    (iii) The straight line may lie in only two principal planes at the same time
    (iv) The inclination of the straight line with VP is always observed from its plan
58. The plan and elevation of an object with two inclined surfaces in the first angle projections are shown in Fig. 16.4

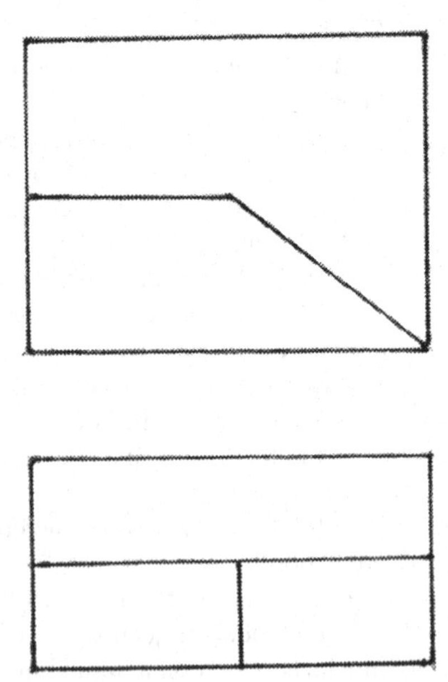

Fig. 16.4 Plan and elevation of an object

Which of the views shown in Fig. 16.5 represents correctly the left side elevation/view?

Fig. 16.5 Side elevations/ views of the object

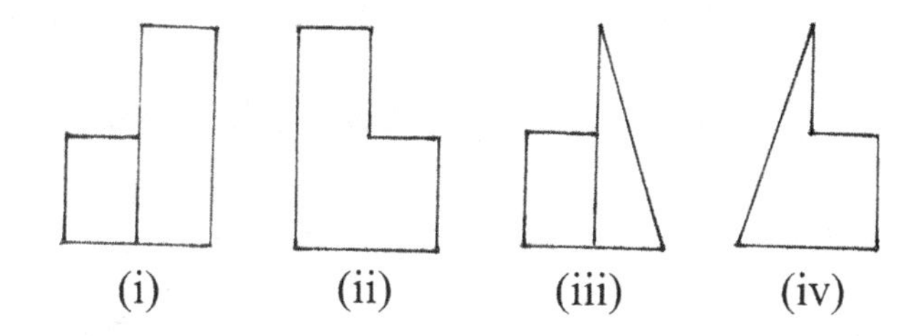

59. The point whose elevation and plan are, respectively, 30 mm and 20 mm above the reference line xy lies in
    (i) first quadrant (ii) second quadrant (iii) third quadrant (iv) fourth quadrant
60. The point whose elevation and plan are, respectively, 30 mm and 20 mm below the reference line xy lies in
    (i) first quadrant (ii) second quadrant (iii) third quadrant (iv) fourth quadrant
61. A point L is 30 cm above HP and 40 mm in front of VP. The shortest distance of the point from reference line xy, in mm, is
    (i) 45 (ii) 50 (iii) 55 (iv) 60
62. The projections of a point coincide and both lie below the reference line xy. In which quadrant, the point is situated?
    (i) First quadrant (ii) Second quadrant (iii) Third quadrant (iv) Fourth quadrant
63. Which statement is true in the orthographic projections of straight lines?
    (i) The plans of all horizontal lines are parallel
    (ii) Three parallel lines must also be equidistant
    (iii) For a line to be parallel to a plane, the line must be parallel to a line on that plane
    (iv) For a horizontal line to be parallel to VP, the projection on VP is a point
64. Which statement is true in the orthographic projections of straight lines?
    (i) Perpendicular lines do not necessarily intersect on their projections
    (ii) The locus of points from the ends of a line is a circle
    (iii) Only one line can be drawn perpendicular to a given line through a specified point on the given line
    (iv) Two lines perpendicular in space will appear perpendicular in any orthographic view
65. Which statement is true for a straight line parallel to VP with its ends are respectively in first and fourth quadrants?
    (i) The line has no horizontal trace
    (ii) The horizontal trace lies above the reference line xy
    (iii) The horizontal trace lies on the reference line xy
    (iv) The horizontal trace lies below the reference line xy

66. The pictorial representation of four straight lines are shown in Fig. 16.6
Which line has horizontal trace only?

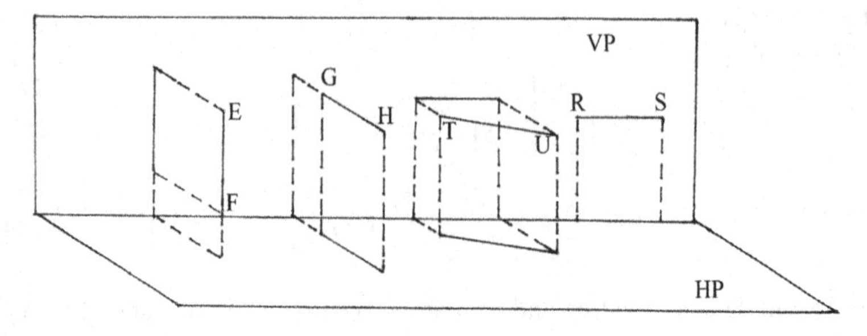

Fig. 16.6 Pictorial representation of four lines

(i) Line EF (ii) Line GH (iii) Line TU (iv) Line RS

67. The pictorial representation of a straight line AB is shown in Fig. 16.7 with the distances marked in mm. What should be the distance of B from VP so that the two traces of the line coincide and lie on the reference line xy?
(i) 6 mm (ii) 8 mm (iii) 10 mm (iv) 15 mm

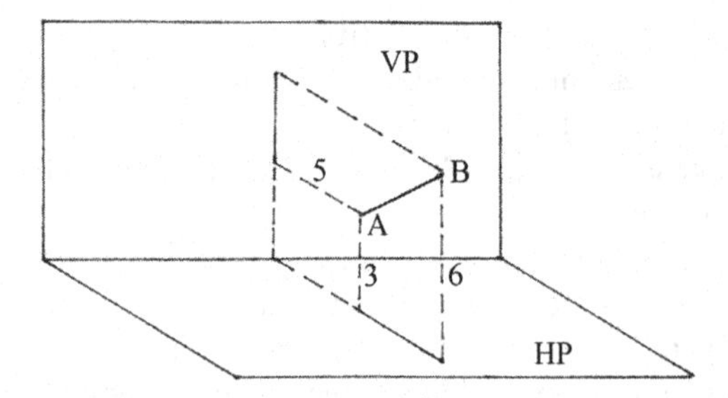

Fig. 16.7 Pictorial representation of a straight line AB

68. When will traces of a line with its ends on two different quadrants lie on its projections?
(i) One end in the first quadrant and the other end in the second quadrant
(ii) One end in the first quadrant and the other end in the third quadrant
(iii) One end in the first quadrant and the other end in the fourth quadrant
(iv) One end in the second quadrant and the other end in the third quadrant

69. Which one of the principal projections of a straight line AB shown in Fig. 16.8 needs an auxiliary projection to get its true length?

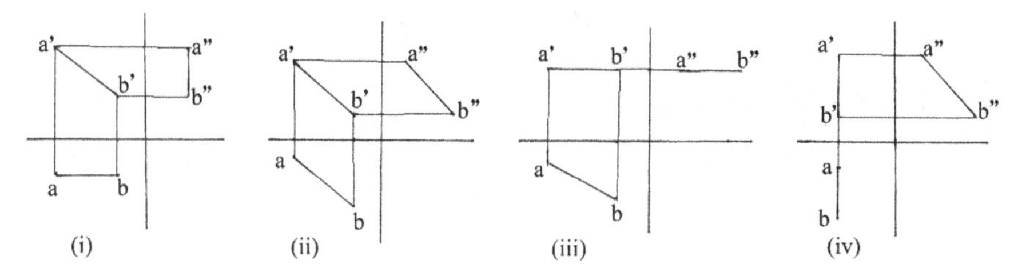

Fig. 16.8 Principal projections of a straight line AB

70. Which plane, whose projections are shown by its traces in Fig. 16.9, corresponds to oblique group?

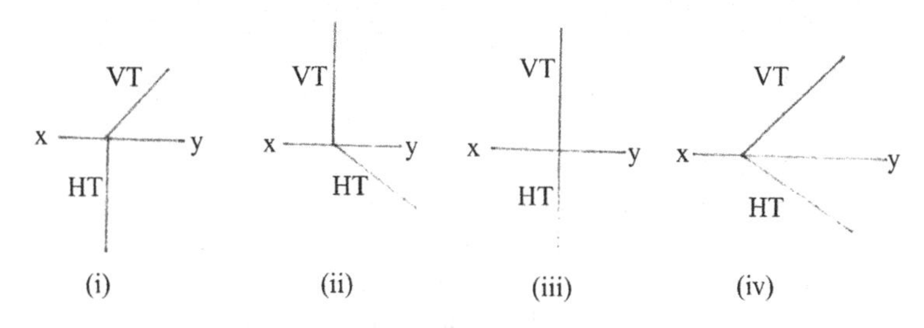

Fig. 16.9 Projections of planes

71. The sides of a 30°–60° set square, arranged in increasing order, are
    (i) S, $\sqrt{2}$ S, $\sqrt{3}$ S (ii) S, $\sqrt{2}$ S, 2 S (iii) S, $\sqrt{3}$ S, 2 S (iv) S, 2 S, 3 S
72. The projections of a plate are squares as shown in Fig. 16.10.

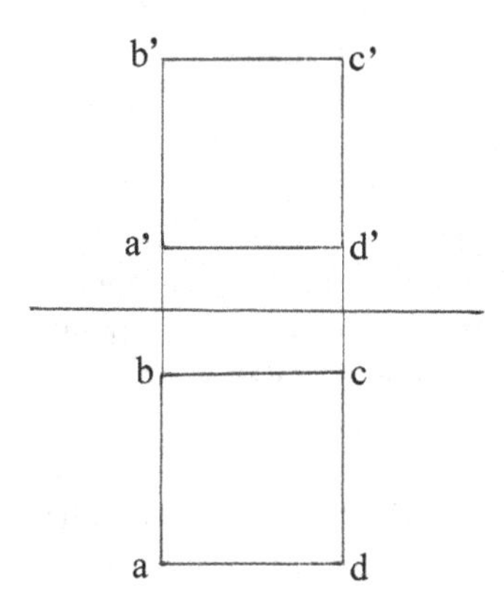

Fig. 16.10 Projections of a plate

The exact/true shape of the plate is

(i) square (ii) rectangle (iii) parallelogram (iv) rhombus

73. The vertical trace and the horizontal trace of the plane figure ABC whose projections are shown in Fig. 16.11 lie respectively

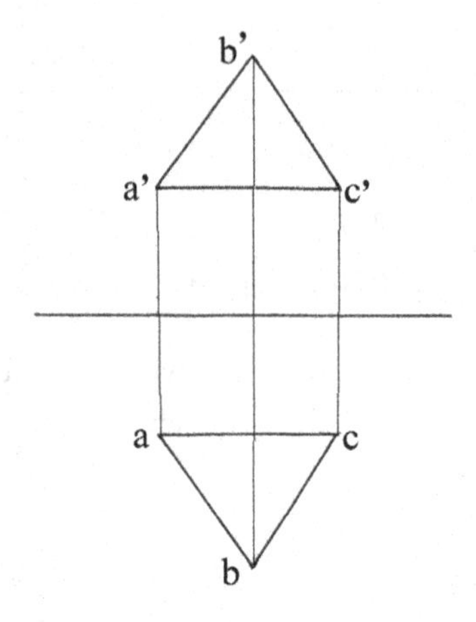

Fig. 16.11 Projections of a plane figure

(i) above the elevation and above the plan
(ii) below the elevation and above the plan
(iii) below the elevation and below the plan
(iv) above the elevation and below the plan

74. Identify the polyhedron which has 6 angular points and 12 edges.
(i) Dodecahedron (ii) Hexahedron (iii) Octahedron (iv) Tetrahedron

75. Identify the polyhedron which has 8 angular points and 12 edges.
(i) Dodecahedron (ii) Hexahedron (iii) Octahedron (iv) Tetrahedron

76. Which solid is referred to as solid of revolution?
(i) Cone (ii) Polyhedron (iii) Prism (iv) Pyramid

77. Which solid does not possess straight edges?
(i) Tetrahedron (ii) Prism (iii) Pyramid (iv) Cone

78. An octahedron rests on HP on one of its faces. The number of angular points visible in the plan is
(i) 3 (ii) 4 (iii) 5 (iv) 6

79. Which projection of solid is considered as simple?
(i) A cube is hung freely from an angular point
(ii) An octahedron is hung freely from an angular point
(iii) A square pyramid is hung freely from a base corner point
(iv) A cone is hung freely from a point on its base

80. The centre of gravity of a right circular cone of height, h, from its apex is
(i) h/4 (ii) h/3 (iii) 2h/3 (iv) 3h/4

81. An inverted conical pail which contains water initially is tilted so that water oozes out. The outline of water surface inside the pail is
    (i) circle (ii) cycloid (iii) ellipse (iv) concoid
82. The plan of a regular octahedron is halved by a section plane A-A parallel to VP as shown in Fig. 16.12.

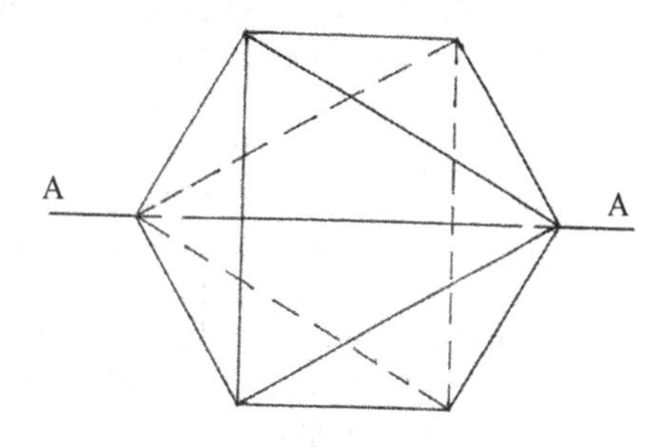

Fig. 16.12 Sectional plan of the octahedron

The true shape of the section obtained is
(i) an equilateral triangle (ii) a square (iii) a rhombus (iv) a hexagon
83. The plan of a regular octahedron is halved by a section plane A-A parallel to VP as shown in Fig. 16.13.

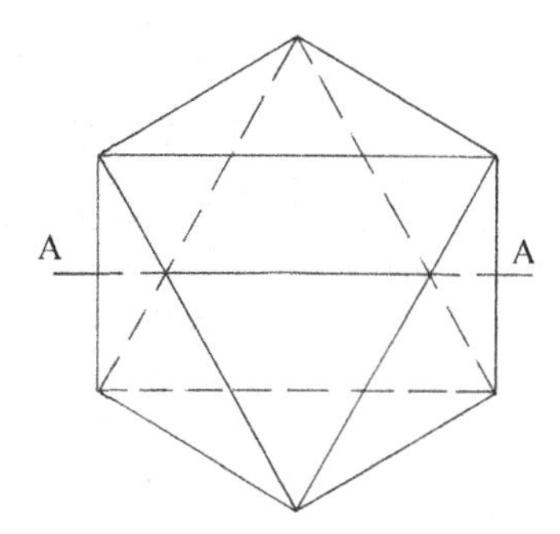

Fig. 16.13 Sectional plan of the octahedron

The true shape of the section obtained is
(i) an equilateral triangle (ii) a square (iii) a rhombus (iv) a hexagon
84. Four identical cubes are cut by four section planes to obtain figures of different sizes. Which figure has the maximum surface area?
    (i) Equilateral triangle (ii) Hexagon (iii) Rectangle (iv) Rhombus

85. The true shape of section obtained on a tetrahedron when the cutting plane passes through mid-points of any two adjacent edges and the centre of gravity is
    (i) an equilateral triangle (ii) an isosceles triangle (iii) a square
    (iv) a trapezium
86. The individual line method is the easiest one to obtain the curve of intersection when
    (i) vertical cylinder penetrates a horizontal cylinder
    (ii) horizontal cylinder penetrates a vertical cylinder
    (iii) vertical cone penetrates a horizontal cylinder
    (iv) horizontal cylinder penetrates a vertical cone
87. The location of critical/key points on the interpenetrating curve/curve of intersection needs extra construction lines in the case of
    (i) vertical cylinder penetrating and bisecting a horizontal cylinder
    (ii) horizontal cylinder penetrating and bisecting a vertical cylinder
    (iii) vertical cone penetrating and bisecting a horizontal cylinder
    (iv) horizontal cylinder penetrating and bisecting a vertical cone
88. Assertion (A): In obtaining the curves of intersection of two solids with curved surfaces section planes are chosen to cut the penetrating solid.
    Reason (R): The curves of intersection lie entirely on the surface of the penetrating solid.
    (i) Both A and R are true and R is the correct explanation of A
    (ii) Both A and R are true but R is not the correct explanation of A
    (iii) A is true but R is false
    (iv) A is false but R is true
89. Which solid surface cannot be developed accurately?
    (i) Cone (ii) Cylinder (iii) Sphere (iv) Tetrahedron
90. Which method is used to develop a hopper?
    (i) Parallel line method (ii) Radial line method
    (iii) Triangulation method (iv) Zone/polyconic method
91. The isometric projection of a horizontal square lamina with its sides parallel to the horizontal reference axes will be
    (i) a parallelogram (ii) a rhombus (iii) a rectangle (iv) a square
92. The isometric projection of a horizontal square lamina with its sides equally inclined to VP will be
    (i) a parallelogram (ii) a rhombus (iii) a rectangle (iv) a square

93. The plan of a circular lamina in an enveloping square with its sides parallel to horizontal reference axes is shown in Fig. 16.14 for obtaining isometric projection of the circle.

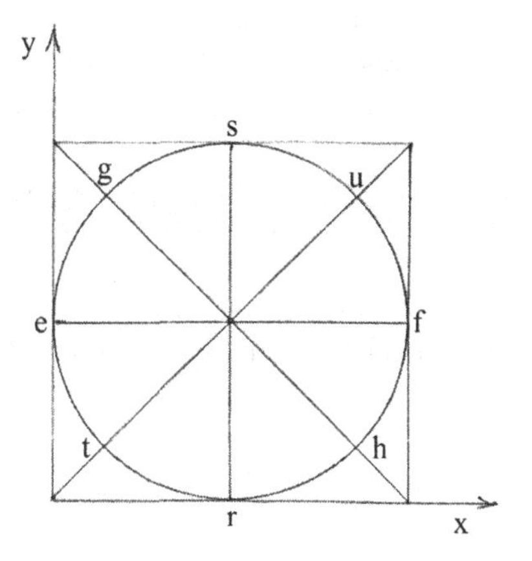

Fig. 16.14 Plan of a square lamina

Which of the diameters does not undergo foreshortening in isometric projection?

(i) RS (ii) TU (iii) EF (iv) GH

94. Assume that there are seven identical cubes and six cubes that are attached to each face of the seventh cube. The outline of the isometric projections of the cubes will be

(i) a parallelogram (ii) a rhombus (iii) a square (iv) a hexagon

95. The isometric projection of a cube is shown in Fig. 16.15.

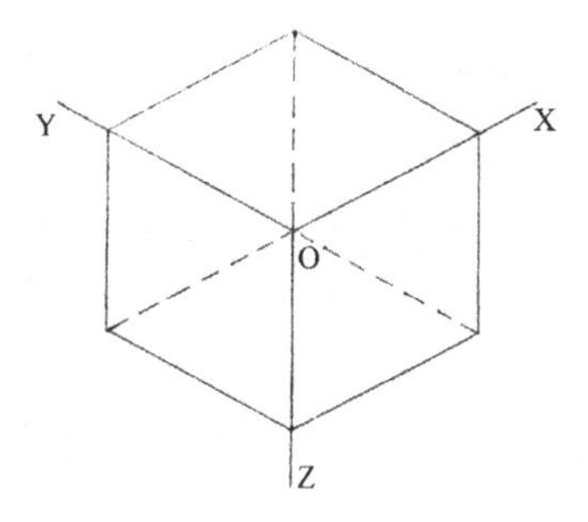

Fig. 16.15 Isometric projection of a cube

How many diagonals on its faces do not undergo foreshortening?

(i) Three (ii) Four (iii) Six (iv) Eight

96. The nomenclature 'centre of vision' is associated with

(i) orthographic projection (ii) oblique projection
(iii) isometric projection (iv) perspective projection

97. Perspective projection belongs to the category

(i) parallel projection (ii) orthographic projection
(iii) isometric projection (iv) conical projection

98. Which projection/view is needed in getting the perspective view of an object following the centre of vision method or the vanishing point method?
(i) Elevation (ii) Plan (iii) Right side elevation (iv) Left side elevation

99. Which types of lines converge towards the centre of vision in perspective projection?
(i) Vertical lines in a row inclined to the picture plane
(ii) Horizontal lines parallel to the picture plane
(iii) Horizontal lines perpendicular to the picture plane
(iv) Horizontal lines inclined at 45° to the picture plane

100. Identify the perspective projection/view of a cube from the views shown in Fig. 16.16.

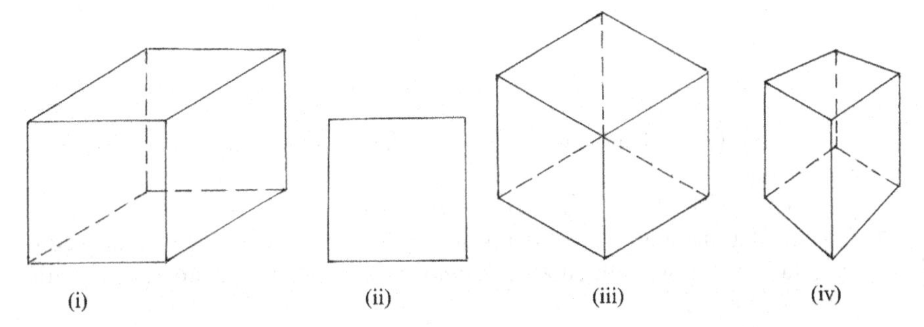

Fig. 16.16 Views of a cube

Answers

1. i	2. iii	3. i	4. iv	5. iii	6. ii	7. iii	8. iv	9. ii	10. iii
11. iv	12. ii	13. ii	14. iii	15. iii	16. iii	17. iv	18. iii	19. i	20. ii
21. ii	22. iv	23. iii	24. ii	25. iii	26. iv	27. ii	28. iv	29. iv	30. iii
31. iii	32. iii	33. ii	34. ii	35. i	36. ii	37. iv	38. iii	39. iv	40. iv
41. ii	42. i	43. iv	44. iv	45. iv	46. iv	47. iv	48. iv	49. iii	50. iii
51. i	52. iv	53. ii	54. iii	55. iii	56. iv	57. iii	58. iv	59. ii	60. iv
61. ii	62. iv	63. iii	64. iii	65. iii	66. i	67. iii	68. ii	69. ii	70. iv
71. iii	72. ii	73. ii	74. iii	75. ii	76. i	77. iv	78. iv	79. ii	80. iv
81. iii	82. iii	83. iv	84. iii	85. iii	86. iii	87. iv	88. i	89. iii	90. iii
91. ii	92. iii	93. iv	94. iv	95. iii	96. iv	97. iv	98. ii	99. iii	100. iv

Bibliography

1. P. Abbott, *Teach Yourself Geometry* (The English Universities Press, London, 1959)
2. W. Abbott, *Practical Geometry and Engineering Graphics* (Blackie & Sons (India), Bombay, 1961)
3. W. Abbott, *Technical Drawing* (Blackie & Sons, London, 1976)
4. N.D. Bhatt, *Engineering Drawing* (Charotar Publishing Co., Anand, 1964)
5. K.R. Gopalakrishna, *Engineering Drawing* (Subhas Stores, Bangalore, 2001)
6. K.C. Jhon, P.I. Varghese, *Engineering Graphics* (Jovast Publishers, Trichur, India, 1989)
7. D.A. Low, *Practical Geometry and Graphics* (Longmans, Green and Co., London, 1948)
8. W.J. Luzadder, *Fundamentals of Engineering Drawing* (Prentice Hall of India, New Delhi, 1988)
9. E.G. Pare, R.V. Loving, I.L. Hill, R.C. Pare, *Descriptive Geometry: Metric* (Macmillan, New York, 1982)
10. N.S. Parthasarathy, V. Murali, *Engineering Drawing* (Oxford University Press, New Delhi, 2015)
11. M.B. Shah, B.C. Rana, *Engineering Drawing* (Pearson, Noida, 2014)
12. A.N. Siddique, Z.A. Khan, M. Ahmad, *Engineering Drawing* (Prentice Hall of India, New Delhi, 2004)
13. B.L. Wellman, *Technical Descriptive Drawing* (McGraw Hill, New York, 1948)
14. C. Zwikker, *Advanced Plane Geometry* (North-Holland, Amsterdam, 1950)

K. Rathnam, *A First Course in Engineering Drawing*,
DOI 10.1007/978-981-10-5358-0

Printed in the United States
By Bookmasters